Models, Mysteries and Magic of Molecules

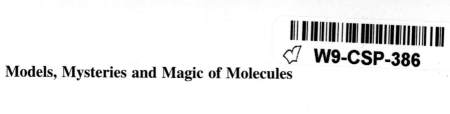

Models, Mysteries and Magic of Molecules

Edited by

Jan C.A. Boeyens
University of Pretoria, South Africa

and

J.F. Ogilvie
Universidad de Costa Rica, Costa Rica

Springer

A C.I.P. Catalogue record for this book is available from the Library of Congress.

ISBN 978-1-4020-5940-7 (HB)
ISBN 978-1-4020-5941-4 (e-book)

Published by Springer,
P.O. Box 17, 3300 AA Dordrecht, The Netherlands.

www.springer.com

Printed on acid-free paper

TABLE OF CONTENTS

PREFACE

A century ago Chemistry was the queen of the sciences. Today Molecule is king.

Many of the exciting scientific theories of a hundred years ago originated with chemists such as Stoney, who postulated the existence of electrons, and those who formulated the periodic law of the elements. Van't Hoff, the father of stereochemistry, was also the first Nobel laureate. A chemist named van den Broek introduced the concept of atomic number, and another chemist named Harkins postulated the neutron and first mooted the idea of a mass defect as the origin of nuclear binding energy. Not only the chemists of the era but also the subject area of chemistry produced spectacular theoretical advances. The physicist Lord Rutherford, who is purported to have stated

"in science there is physics and stamp collecting"

must have been mortified to have his research recognized with a Nobel prize for chemistry.

A hundred years on, significant progress in chemistry arises only at the bench, through the synthesis of high T_c superconductors, nanomaterials, 'bucky balls' and other magical molecules, with little appreciation of the underlying theories. Chemistry is no longer a priority as a career option, and the emphasis is shifting constantly towards molecular sciences. Students, who vote with their feet, sense that, through lack of theoretical innovation, chemistry is becoming a sterile science. By comparison, models, mysteries and magic abound in molecular sciences.

It was demonstrated more than sixty years ago [1] that a major intramolecular migration of chiral groups, during Beckmann rearrangements, occurs with retention of chirality. A reasonable mechanism to account for this amazing effect has never been formulated; like the optical activity of chiral molecules, it is simply accepted as a fact of life, not deserving of further theoretical analysis. The reason for this indifference is not difficult to find. The models of chemical bonding, enthusiastically embraced during the twentieth century, might seem plausible, but they lack the authority to address fundamental issues. The soporific that stupefied chemistry is the lore of orbitals. By now it has been so thoroughly entrenched in the world's textbooks for undergraduate chemistry and in computational models, widely accepted as the ultimate in chemical theory, that it borders on the futile to resist its use. A German-language text on Elementary Quantum Chemistry [2] laments (my translation),

1. "Modern quantum chemistry allows us to calculate bond energies (more correctly: dissociation energies) and bond lengths (more correctly: internuclear separations), but does not tell us what the chemical bond 'actually' is. We are not even sure whether it is stupid or intelligent to enquire into the nature of chemical bonding.

2. The definition or characterization of orbitals found in many textbooks for chemistry in secondary school is a scandal. It is an insult to our youth to confront them with such trash".

Another expert [3] remarks

In reality, chemical bonding is a molecular property, not a property of atomic pairs.

The chemistry community finds itself burdened with an operational theory, based on outdated and discredited classical concepts, and without predictive power in the quantum world of the twenty-first century. Few fascinating observations documented in this volume can be explained convincingly in terms of orbital jargon. Topics such as polymorphism, nanomaterials and biopolymers are of immense commercial importance, but scientifically poorly understood. These phenomena are as mysterious as crystal growth or the structure of snow flakes, quasi crystals or viruses. To state that interactions responsible for biological modifications of life, the circadian clock or the colour of cooked lobster are of chemical origin is tantamount to an admission of complete ignorance. Models that feature in the description of chemically more familiar systems such as transition-metal complexes, host-guest interactions and solvation are largely empirical and rely on experimental results obtained with diffraction of X-rays and neutrons.

Aspects of all these various topics were discussed at the fifth international *Indaba* workshop of the International Union of Crystallography at Berg-en-Dal, Kruger National Park, South Africa, 20–25 August 2006. In most instances the analysis terminated before a fundamental interpretation of the results, because there is no fundamental theory of chemistry. As all chemical interactions are mediated by electrons it is clear that such a theory must be quantum-mechanical. The only alternative is a phenomenological simulation of chemical processes with empirical classical models. There is no middle ground.

The classical approach appropriately known as molecular mechanics has been used with conspicuous success to predict molecular geometries, chemical reactivities and even magnetic, electronic and spectral properties of molecular systems. Molecular mechanics functions with no intention or pretence to elucidate the essential nature of molecules; it applies concepts that pertain to the nineteenth-century classical model of the molecule, *i.e*, bond length, bond order, force constant, torsional rigidity and steric congestion. Transferable numerical values are empirically assigned to the corresponding parameters to ensure that minimization of steric energy converges to observed molecular structures. Any remaining discrepancies are assumed to arise from quantum-electronic effects, incorrectly simulated with a classical model. Additional parameters to compensate for particular effects such as planarity of aromatic rings or Jahn-Teller distortions are then introduced.

There is no evidence that any classical attribute of a molecule has quantum-mechanical meaning. The quantum molecule is a partially holistic unit, fully characterized by means of a molecular wave function, that allows a projection of derived properties such as electron density, quantum potential and quantum torque. There is no operator to define those properties that feature in molecular mechanics. Manual introduction of these classical variables into a quantum system is an unwarranted abstraction that distorts the non-classical picture irretrievably. Operations such as orbital hybridization, LCAO and Born-Oppenheimer separation of electrons and nuclei break the quantum symmetry to yield a purely classical picture. No amount of computation can repair the damage.

The previous statement does not imply that the computational chemistry of the previous half century is all wasted effort; it simply means that what became known as *quantum chemistry* is based on classical mechanics, without necessarily invalidating the useful results that flowed from this pursuit. This conclusion should not be read as an insult, but as an encouragement to renew the quest in a true spirit of quantum theory, which is bound to produce unthought-of new insights into the mysteries and magic of molecules. A modest start has been made [4] in recognizing the power of quantum potential to explain the stability and reactivity of atoms and molecules and of quantum torque as the agent responsible for molecular shape, torsional rigidity and optical activity. In principle, the same approach might produce a winning strategy to solve the mysteries of intramolecular rearrangement, photochemistry, protein folding and the unwinding of DNA.

The book is aimed at students of chemistry and general readers interested in molecular sciences, structural chemistry, the fundamental basis of the life sciences, and computational chemistry. Not only is it intended to highlight some of the many challenges in molecular science, but also it serves as an introductory text for students who might subsequently wish to specialize in any of these fascinating fields. Written at a level for advanced undergraduates and graduate students in chemistry, most concepts are expected to be familiar to readers with a first degree in chemistry and modest expertise in mathematics and physics.

The material is organized under four main headings. The first section includes specialized methods used in the study of molecules, followed by partially solved molecular mysteries and poorly understood phenomena with a magical quality. The final section treats various models that might contribute to an improved understanding of molecular behaviour. The confusion in chemistry is largely due to an indiscriminate mixing of classical and quantum concepts and the distortion of the latter by the science writers of the twentieth century, more interested in sensation than science. These Proceedings might assist to eliminate some confusion and show a new direction for theoretical chemistry.

I wish to express my sincerest appreciation to all authors for producing their contributions under serious temporal constraints, to J.F. Ogilvie for editing all texts under even greater constraint, and to Thereza Botha for organizing more than a workshop and the production of these Proceedings.

November 15, 2006
Jan C.A. Boeyens
Unit for Advanced Study, University of Pretoria
e-mail: jan.boeyens@up.ac.za

REFERENCES

1. J. Kenyon and D.P. Young, Retention of asymmetry during the Curtius and the Beckmann change, J. Chem. Soc., 1941, 263–267.
2. H. Primas and U. Müller-Herold, Elementare Quantenchemie, 1984, Teubner, Stuttgart.
3. P.G. Mezey, Shape in Chemistry, 1993, VCH, New York.
4. J.C.A. Boeyens, New Theories for Chemistry, 2005, Elsevier, Amsterdam.

LIST OF CONTRIBUTORS

Howell G.M. Edwards and Michael D. Hargreaves University Analytical Centre / Chemical & Forensic Science, School of Life Sciences, University of Bradford, Bradford, BD7 1DP, UK, e-mail: h.g.m.edwards@bradford.ac.uk

Jacqueline M. Cole Department of Physics, Cavendish Laboratory, University of Cambridge, J. J. Thomson Avenue, Cambridge, CB3 0HE, UK, e-mail: jmc61@cam.ac.uk

Ashwini Nangia School of Chemistry, University of Hyderabad, Hyderabad 500 046, India, e-mail: ashwini_nangia@rediffmail.com

Alessia Bacchi Dipartimento di Chimica Generale ed Inorganica, Chimica Analitica, Chimica Fisica, Università di Parma, Viale G.P. Usberti 17A – Parma, Italy, e-mail: alessia.bacchi@unipr.it

Yuji Ohashi Industrial Application Division, Japan Synchrotron Radiation Research Institute (SPring-8), Okayama, Meguro-ku1-1-1 Kouto, Sayo, HYOGO, 679-5198, Japan

Ivan Bernal Chemistry Department, University of Houston, Houston, TX 77204-5003 USA

Elena Boldyreva Institute of Solid-State Chemistry and Mechanochemistry, SB RAS, REC-008 Novosibirsk State University, Russia, e-mail: boldyrev@nsu.ru

John R. Helliwell and Madeleine Helliwell School of Chemistry, The University of Manchester, Manchester M13 9PL and CCLRC Daresbury Laboratory, Warrington WA4 4AD, UK

Andrzej Katrusiak Faculty of Chemistry, Adam Mickiewicz University, Grunwaldzka 6, 60-780 Poznań, e-mail: hpc.amu.edu.pl

Gert J. Kruger, Dave G. Billing and Melanie Rademeyer University of Johannesburg, Johannesburg, South Africa; University of the Witwatersrand, Johannesburg, South Africa; University of KwaZulu-Natal, Pietermaritzburg

Aloysio Janner Theoretical Physics, Radboud University, Toernooiveld, NL-6525 ED Nijmegen, The Netherlands, e-mail: A.Janner@science.ru.nl

Zorka Papadopolos, Oliver Gröning and Roland Widmer   Institut für Theoretische Physik, Univ. Tübingen, Auf der Morgenstelle 14/D8, D-72076 Tübingen, Germany, e-mail: zorka.papadopolos@uni - tuebingen.de; EMPA, Federal Laboratories for Materials Testing and Research, Feuerwerkerstrasse 39, CH-3602 Thun, Switzerland, e-mail: Oliver.Groening@empa.ch

Martin Egli, Rekha Pattanayek and Sabuj Pattanayek   Department of Biochemistry, Vanderbilt University, School of Medicine, Nashville, Tennessee 37232, USA

Andy Becue, Nathalie Meurice, Laurence Leherte and Daniel P. Vercauteren Laboratoire de Physico-Chimie Informatique (PCI), Facultés Universitaires Notre-Dame de la Paix (FUNDP), Rue de Bruxelles 61, B-5000 Namur, Belgium

Gideon Steyl and Andreas Roodt   Department of Chemistry, University of the Free State, P.O. Box 339, Bloemfontein, 9300, South Africa, e-mail: geds12@yahoo.com

J.F. Ogilvie and Feng Wang   Centre for Molecular Simulation, Faculty of Information and Communication Technologies, Swinburne University of Technology, P. O. Box 218, Hawthorn, Melbourne, Victoria 3122, Australia; permanent address – Escuela de Quimica, Universidad de Costa Rica, Ciudad Universitaria Rodrigo Facio, San Pedro de Montes de Oca, San Jose 2060, Costa Rica

Giovanni Ferraris and Marcella Cadoni   Dipartimento di Scienze Mineralogiche e Petrologiche, Università di Torino, and Istituto di Geoscienze e Georisorse, CNR, Via Valperga Caluso, 35, I-10125, Torino, Italy, e-mail: giovanni.ferraris@unito.it

Peter Comba and Marion Kerscher   Universität Heidelberg, Anorganisch-Chemisches Institut, Im Neuenheimer Feld 270, D-69120 Heidelberg, Germany, e-mail: peter.comba@aci.uni-heidelberg.de

Mihail Atanasov, Peter Comba, Claude A. Daul and Frank Neese   Institute of General and Inorganic Chemistry, Bulgarian Academy of Sciences, Acad.Georgi Bontchev Str. Bl.11, 1113 Sofia, Bulgaria; Anorganisch Chemisches Institut, Universität Heidelberg, Im Neuenheimer Feld 270, D-69120 Heidelberg, Germany; Département de Chimie, Departement für Chemie, Ch.du Musée 9, CH-1700 Fribourg, Switzerland; Institut für Physikalische und Theoretische Chemie, Universität Bonn, D-53115 Bonn, Germany

Jan C.A. Boeyens   Centre for Advanced Studies, University of Pretoria, South Africa

CHAPTER 1

RAMAN SPECTROSCOPY

The biomolecular detection of life in extreme environments

HOWELL G.M. EDWARDS AND MICHAEL D. HARGREAVES

Abstract: The strategic adaptation of extremophiles (organisms which can survive where humans cannot) to survival in hostile terrestrial environments depends critically upon their synthesis of protectant biomolecules in geological niches to combat low wavelength radiation insolation, desiccation, and extremes of temperature and pH.

Each year sees the discovery of novel geological scenarios in which organisms have successfully created a tenacious colonisation in "limits of life" habitats. Terrestrial analogues such as hot and cold deserts, volcanoes and geothermal springs provide ***models*** for the extraterrestrial study of the evolution of life in our Solar System – astrobiology. In particular, the current robotic exploration of the surface and subsurface of Mars, our neighbouring planet which has held ancient ***magical*** significance for our ancestors and is still shrouded in ***mysteries***, is now indicative of the importance of a range of terrestrial scenarios which can be considered as Mars analogues. The next phase of Martian exploration must address the robotic search for extinct or extant life in the geological record, which is essential for the proposed human missions in the next two decades. A key factor in the armoury of remote analytical instrumentation that is now envisaged for inclusion on extended-range Mars landers and rovers is the identification of the chemical and biochemical protectants that might have been produced by extremophiles for survival in the Martian deserts. The ***molecular*** signatures evidenced from the Raman spectra of these key protectants will be fundamental for the detection of biological activity on Mars and the adoption of miniaturised Raman spectrometers for Martian exploration is now being seriously considered by NASA and ESA. In this context, the evaluation of prototype Raman spectroscopic instrumentation for the detection of molecules of relevance to a wide range of terrestrial extremophilic activity is now being addressed and forms the subject of this article. Exemplars from various geological scenarios in the Arctic and Antarctic cold deserts and from relevant hot desert habitats, such as volcanic geothermal springs and salt pan evaporites, will reinforce the tenet of this book – that the ***molecular*** studies of extremophilic ***models*** will be pivotal in the understanding of the ***magic*** and ***mysteries*** of evolution of life on Earth and the search for life on Mars

1

J.C.A. Boeyens and J.F. Ogilvie (eds.), Models, Mysteries and Magic of Molecules, pp. 1–28.
© 2008 *Springer.*

INTRODUCTION

Prologue

Mars (Figure 1-1) epitomises the theme of this Indaba Five meeting: from the birth of recorded history, the observation of our nearest neighbour planet was steeped in astrological *magic* which the ancient Greeks associated with evil. The earliest scientific astronomical recordings of the Martian planetary and orbital motions in the 16th and 17th Centuries enhanced the *mystery* of the planet, which intensified with the first telescopic observations of *"this ochre coloured orb"* by Galileo and Huygens in the mid-17th Century. It is intriguing that Mars today still remains the only planet in our Solar System whose surface can be observed using terrestrial-based telescopes [1].

The mysterious nature of Mars received impetus in the late 18th Century when in 1783 Herschel, the discoverer of the infrared region in the electromagnetic spectrum, observed clouds on Mars and correlated Martian planetary "waxing and waning" effects with surface vegetation and seasonal growth. The dawn of the 19th Century witnessed the age of geological discovery of relic, fossilised life in ancient terrestrial strata which with the biological revolution echoed by Darwin and Russell Wallace

Figure 1-1. Mars: the epitome of the Indaba Five theme – *mystery, models, magic and molecules* (see color plate section)

indicated that life had not always been constant and that evolutionary adaptation was the key to a successful survival strategy for biological species. In this scientific atmosphere, it seemed perfectly logical that Schiaparelli's observation of some forty "canali" on Mars and Lowell's plotting of up to 160 "irrigation ditches" designed to bring water from the Martian polar regions to the equatorial vegetation brought to a close the 19th Century *mystery* phase of Mars that had transcended nearly two centuries. Hence by the end of the 19th Century, informed opinion was that there was indeed life on Mars – as exemplified by the following statement [2] from Lord Kelvin in his August 1871 Presidential Address to the British Association for the Advancement of Science.

"Should the time come when this earth comes into collision with another body comparable in dimensions to itself... many great and small fragments carrying seeds of living plants and animals would undoubtedly be scattered through space. Hence, and because we all confidently believe that there are at present and have been from time immemorial, many worlds of life besides our own, we must regard it as probable in the highest degree that there are countless seed-bearing meteoritic stones moving about through space. If at the present instance no life existed on the earth, one such stone falling upon it might, by what we call natural causes, lead to it becoming covered with vegetation"

This can be considered an early vision of *panspermia*, a vigorously debated topic in current astrobiology. However, not everyone was convinced of the presence of life elsewhere in the Universe; Lord Kelvin was lampooned in *"Punch"* just three weeks later:

"A Conjurer's Conjecture:
Could a Meteoritic Stone
  Pray, Sir William Thomson,
Fall with Lichen Overgrown,
  Say, Sir William Thomson,
From its Orbit having Shot
Would it, Coming Down Red-Hot
Have All Life Burnt off it not?
    Eh, Sir William Thomson?
Not ? Then Showers of Fish and Frogs
  Too, Sir William Thomson,
Fall, it might Rain Cats and Dogs.
  Pooh, Sir William Thomson,
That they do cone down we are told
As for Aerolite with Mould,
That's at Least too Hot to Hold.
    True, Sir William Thomson!"

Contrast then this reaction in the 19th Century to the thoughts of Giordano Bruno in 1600, which resulted in his incurring the wrath of Pope Clement VIII and his being burned alive at the stake for his heretical beliefs and suggestion "that God's realm could include other worlds (Mars) comparable with Earth", an action which incidentally was believed to have resulted in the timely recantation of Galileo; yet, in 1900, the Guzman Prize was offered for the first evidence of extraterrestrial life outside Earth – with the specific exclusion of Mars, because it was obviously populated! This prize still awaits a claimant. Even then, Alfred Russell Wallace, the co-author of evolutionary biology with Charles Darwin, in 1907 declared that Mars could not possibly sustain life.

Early 20th Century thinking resulted in the quest for terrestrial *models* for the life forms that might exist on Mars with the eventual realisation that the planet, being significantly smaller than Earth, could have cooled sooner and harboured developing life from a "Primæval soup" at a much earlier phase than could have occurred terrestrially—and that Mars therefore could be the home for an even more advanced form of life than ours [3]! The challenge to find Mars analogues on Earth was launched in an age where there was a genuine public scientific perception that it was inevitable that we would soon attract the rather unwelcome attention of extraterrestrial intelligence in the form of Martian inhabitants (e.g. H.G. Wells' *"War of the Worlds"* and the famous Orson Welles broadcast in 1938 that sent the public into a wild panic, fearing that a Martian attack on Earth was imminent!). The repetition of the 19th Century experiments initiated by respected scientists such as Gauss, Cros and von Littrow to signal the Martian inhabitants by lighting fires in the Sahara, drawing geometric patterns in the Siberian snow and heliosignalling would hence have been strongly discouraged publicly in the 1930s.

The mid-to-late 20th Century, 1965–1976, heralded the death-knell of the "intelligent life on Mars" idea with the discovery by remote observation and planetary lander platforms such as Mariner and Viking that Mars was a "dead planet", debunking the surface vegetation and irrigation ditches philosophy, whilst confirming the earlier but hotly reviled conclusions that Mars was a heavily cratered, rather inhospitable place with solid carbon dioxide at the poles and devoid of a life supportive atmosphere.

The dawn of the 21st Century has seen the birth of *astrobiology:* the study of astronomical and planetary context within which life on earth has evolved and the implications for the nature and prevalence of life elsewhere in the Universe – a multi-disciplinary science whose name was highlighted in 2000 as "A *new science for the new Millennium*" (Lord Sainsbury, Minister of Science, HM Government, UK). The most recent, advanced explorations of Mars from orbit and from surface landers and rovers has revitalised the possibility of finding the signs of life on Mars – extinct or extant. We now have evidence of images of a planetary surface [4, 5] that clearly has suffered water erosion, subsurface aquifers, the possibility of subglacial polar lakes (similar to Ellsworth and Vostok in Antarctica) and a tantalising glimpse of geological niche scenarios that match terrestrial *models*, for example Mars Oasis, Antarctica [6] (Figure 1-2), where biological adaptation and survival to these harsh Martian environments may have occurred. The new phase for Martian exploration now firmly rests on the *molecules* that have been synthesised by extremophiles for their protection against low wavelength radiation and desiccation and the associated auxiliary molecules that are required to harness the limited energetics and nutrients that are available in their protected niche environments [7], [8], [9]; finally, in a "limits of life" scenario, where life has become extinct, the remote analytical observation of residual molecular signatures in the

Figure 1-2. Mars Oasis, Antarctica; a terrestrial niche for the survival of endolithic cyanobacteria in a *"limits of life"* situation; a translucent Beacon sandstone outcrop containing endolithic microbial communities on top of a dolerite sill. Dr David Wynn-Williams, Head of Antarctic Astrobiology at the British Antarctic Survey, Cambridge, is prospecting for endoliths. Reproduced from the book Astrobiology: The Quest for the Conditions of Life, Eds: Gerda Horneck, Christa Baumstark-Khan, 2003, Springer. With kind permission of Springer Science and Business Media (see color plate section)

geological record may perhaps indicate that the evolutionary experiment on Mars commenced but was not successful in its survival (Figure 1-3).

This journey has taken 5000 years of human observation of Mars and scientific endeavour to make the transit from:

Magic – Mystery – Models – Molecules

In this quest it is clear that the themes ***Models*** and ***Molecules*** are pivotal to the current and future instrumental advances necessary for the detection and

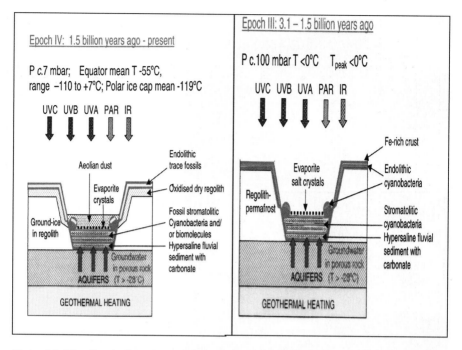

Figure 1-3. Mars Epochs III and IV; highlighting the geological survival possibility of endoliths on the planet. Reproduced from ICARUS, 144, David Wynn-Williams, Howell G.M. Edwards, Proximal Analysis of Regolith Habitats, pp. 486–503, 2000, with permission from Elsevier

remote interpretation of Mars life signature data and eventually should provide an answer in the next two to three decades to that most haunting and mysterious of questions:

"Is there life on Mars?"

Extremophiles

Extremophiles are organisms that can survive where humans cannot; their survival in hostile terrestrial environments depends critically upon their ability to synthesise sophisticated suites of protectant biomolecules (Table 1-1) and the adaptation of geological and hydrological niches for colonisation to combat low wavelengths of insolar radiation, desiccation, and extremes of temperature, pH, salinity, pressure and heavy metal toxins [9, 10]. The search for life on Mars is linked closely to the discovery of novel strategies adopted by extremophiles in terrestrial scenarios where previously the existence of life was believed impossible. A selection of extremophiles and their terrestrial habitats will be considered here, which has provided astrobiologists with the term *"terrestrial Mars analogues"* on which the experimental remote analytical projects envisaged for robotic Mars

Table 1-1. Raman spectral signature bands and UV-visible absorption maxima for functional microbial pigments in extreme terrestrial Antarctic habitats

Physiological Function	Pigment	Pigment type	Raman vibrational bands (wavenumber cm^{-1})					Absorption peaks (nm)			
								UVC <280	UVB 280–320	UVA 320–400	Visible >400
Ultraviolet Screening pigments	Usnic acid	**Cortical acid**	2930	1607	1322	1289		>220	290	325	<400
	Pulvinic dilactone	**Pulvinic acid derivative**	1672	1405				>246	290	367	
	Paretin	**Anthraquinone**	1675	1099	551			>257	288		431
	Calycin	**Pulvinic acid derivative**	1611	1379				269			422
	Atranorin	**para-depside**	2942 1666	1303	1294	1266		<274			>400
	Gyrophoric acid	**Tri-depside**	1661	1290				275	304		
	Fumarprotocetraric acid	**Depsidone**	1642 1630	1290	1280			273	315		
	Emodin	**Quinone**	1659						291		440
	MAA (*Nostoc*) 7437	**Mycosporine Amino Acid**	2920 1400	820					(>310)	330	
Anti-oxidants	Scytonemin	**8-ring dimer**	1590 1549	1323	1172			252	300	370	>400
	β-carotene	**Carotenoid**	1524	1155				<246	283	384	429
	Rhizocarpic acid	**Isoprenoid**	1665 1620 1596	1453				>200			
Photo-synthesis	Porphyrin	**Tetrapyrrole ring**									
	Chlorophyll, (Cyano bacteria)	**Tetrapyrrole ring**	1360	1320							680 700
	BCh1$_a$ (*Rhodopseudomonas*)	**Tetrapyrrole ring**	na								850 870
Accessory Light-harvesting	*Chlorobium* Ch1										
	Phycocyanin	**Tetrapyrrole ring**	na	1638	1369						650 660
	Phycoerythrin	**Phycobilin**	na								560 620
		Phycobilin	na								544

explorations are now being evaluated. The driving force behind the detection of life signatures on Mars is really two-fold, in that the search for a biological evolutionary occurrence other than our own will undoubtedly enhance the definition of our own early prebiotic chemistry but it will also engage with the efforts of space exploration in the placement of a human being on the surface of Mars, which will require a fuller knowledge of the planetary composition and its potential biological dangers than we have currently [11, 12].

In this respect, the study of extremophile survival strategies, which operate in some of the most inhospitable places on Earth, will further our understanding of our own precarious existence in a changing natural world.

Extremophiles that have been studied in our laboratories and which address several scenarios that could be expected on Mars, i.e. *terrestrial Mars analogues*, include the following:

Life in rocks: Antarctic endolithic microbes;
Meteorite impact craters: halotrophic cyanobacteria;
Lacustrine sediments: microbial colonies;
Volcanic lava deposits: vacuole colonies;
Shocked rocks: pore colonisation;
Stromatolites: cyanobacterial residues;
Subglacial lakes: polar colonisation.

Raman Spectroscopy

Raman spectroscopy has several advantageous characteristics which make it potentially valuable for the molecular characterisation of extremophiles, their protectant biomolecules and extraterrestrial exploration [13–17]. The Raman spectrum will contain molecular information from both inorganic molecular ions and organic molecules of some complexity, which renders it suitable for the derivation of data relevant for the chemical interaction between the biological and geological entities which comprise extremophilic colonisation of rock substrates. Little or no chemical or mechanical sample preparation is required and Raman spectra can be obtained from macroscopic and microscopic samples effectively down to about 1 cubic micron specimen volume confocally, representing picogram quantities of material. Coupling of the Raman spectrometer with a flexible optical remote sensing probe of several metres in length permits the access of the irradiating laser beam and the collection of the scattered Raman radiation from inaccessible sample locations, especially on planetary surfaces and subsurface holes. In this context it is hardly surprising that the development of miniaturised Raman spectrometers with mass less than about 1 kg is already far advanced and space mission trials are envisaged for the adoption of Raman spectroscopy as a novel technology for molecular characterisation, either alone or in conjunction with laser ablation techniques, on future planned Mars robotic missions as part of lander and rover life-detection instrumentation [18–20].

The Raman effect is a molecular scattering process which is manifest from the interaction between a laser beam and a chemical system from which the shift in wavenumber of the exciting radiation scattered by vibrating molecules and the incident electromagnetic radiation can be related to the structure, composition and identification of the scattering molecules [21]. The Raman scattering is significantly weaker in intensity than Rayleigh scattering, where the incident electromagnetic radiation is scattered without change in wavenumber. Although the wavenumber shifts observed in Raman scattering are independent of the excitation wavelength, the scattering intensity is inversely proportional to the wavelength of excitation; hence, with all other instrumental effects remaining the same, a Raman spectrum obtained in the ultraviolet at 250 nm is inherently nearly 300 times stronger than that excited with near infrared radiation at 1064 nm. Hence, many Raman spectra which have been obtained hitherto have been recorded in the visible region of the electromagnetic spectrum. However, the onset of fluorescence emission, which is several orders of magnitude larger than Raman scattering, occurring at lower wavelengths and higher laser excitation energies, can completely swamp the observation of the weaker Raman bands especially of organic molecules which have low energy electronic states. For this reason, the recording of Raman spectra using longer wavelength laser excitation to obviate the occurrence of potentially troublesome fluorescence has been finding much favour despite the problems caused by detection of weaker spectral features. Improvements in experimentation, especially in sample illumination and in detection of the long-wavelength shifted Raman bands, have now created several possibilities for the recording of good quality Raman spectra from difficult specimens with effective suppression of troublesome fluorescence backgrounds [13], [22].

Raman spectroscopic equipment

FT-Raman spectra were recorded using a Bruker IFS66 spectrometer with FRA 106 Raman module attachment and dedicated microscope. The wavelength excitation was at 1064 nm, using a Nd^{3+}/YAG laser. The spectral resolution was $4 cm^{-1}$ and from 2000 to 4000 scans were accumulated over about 30–60 minutes to improve the spectral signal-to-noise ratios. For analyses with 785, 633, 514.5 and 488 nm laser excitation a Renishaw InVia *Reflex* Raman Microscope coupled to a Leica DMLM microscope with 5X, 20X, and 50X objective lenses were utilized. 30–70 accumulations at 10 s exposure time for each scan with a laser power between 0.5 to 50 mW were typically used to collect spectra.

RESULTS AND DISCUSSION

Chasmolithic Community

A chasmolithic community living in a fracture inside a marble rock was collected during the AMASE expedition (Arctic Mars Analogue Svalbard Expedition) in August 2004 to the Vest Spitzbergen Island in the Svalbard Archipelago, sited

80° N, inside the Arctic Circle. Although the temperatures are not as cold as those in Antarctica (at Spitzbergen Island, the winter temperature reaches minus 30–35° C), the site and the samples are considered extremophiles because of the low temperatures and the absence of sunlight for almost six months each year.

The sample was found in a glacier moraine in Bockfjorden, at an altitude of 30 metres above sea level, and shows an epilithic community on the crust which gives a greyish tonality to the white stone substrate, and a chasmolith inside a crack where several coloured strata, namely pinkish, black, green and brown areas, are visible.

The Raman spectrum of the substratum presents bands at $1097 \, cm^{-1}$, which is seen as a shoulder on the stronger signature at $1086 \, cm^{-1}$, $713 \, cm^{-1}$ with a shoulder at $725 \, cm^{-1}$ and $157 \, cm^{-1}$. This provides an excellent example of the use of Raman spectroscopy to distinguish between isomorphic minerals. The bands at 1097, 725, 299 and $177 \, cm^{-1}$ are characteristic of dolomite, a calcium magnesium carbonate ($CaMg(CO_3)_2$), whereas calcite, the calcium carbonate isomorph, shows signatures at 1086, 713, 281 and $157 \, cm^{-1}$, very close in wavenumber, but readily distinguishable from the spectral signatures of the dolomite. Figure 1-4 gives a Raman spectral stackplot of magnesite, $MgCO_3$, with Raman bands at 1094, 738, 326, 242 and $119 \, cm^{-1}$, dolomite, $CaMg(CO_3)_2$, with bands at 1098, 725, 299 and $177 \, cm^{-1}$, aragonite, $CaCO_3$, with bands at 1086, 704, 208 and $154 \, cm^{-1}$, and calcite, $CaCO_3$, with bands at 1086, 713, 283 and $156 \, cm^{-1}$.

On the surface crust, the grey tonality is the result of cyanobacterial colonisation, which gives the characteristic Raman spectral bands of scytonemin. This UV-protective pigment is not found in the chasmolithic organisms inside the crack and has probably been produced as a strategic response to the increased radiation levels experienced at the rock surface.

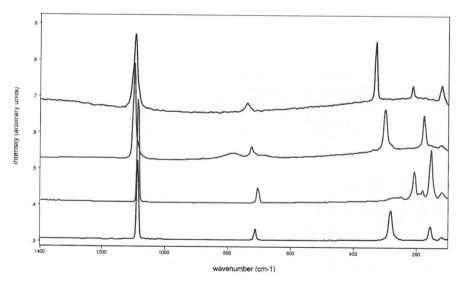

Figure 1-4. Raman spectra of, from the top: magnesite, dolomite, aragonite and calcite-indicating the spectral discrimination between these carbonates

Figure 1-5 shows a section of chasmolithic feldspar from Lake Hoare, Antarctica, in which two zones of bacterial colonisation can be clearly seen, from which the Raman spectrum of calcite can be clearly identified closely associated with the colonisation zones signatures of quartz and feldspar can also be seen in the spectrum.

In the chasmolithic pink coloured zone area, the Raman spectra show bands of chlorophyll and a carotenoid, but the fluorescence emission is very strong here, which reduces the quality of the Raman signatures. The green layer gives a strong fluorescence emission with 785 nm laser excitation and no spectral signatures are visible, but selection of 514.5 nm laser excitation provides bands from a carotene which are clearly recognizable at 1524, 1157 and $1000\,\text{cm}^{-1}$ (Figure 1-6); these

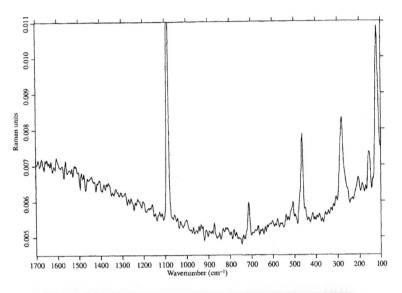

Figure 1-5. Chasmolith in feldspar, Lake Hoare, Antarctica: with the Raman spectrum of calcite, quartz and feldspar from the cyanobacterial colonisation zone (see color plate section)

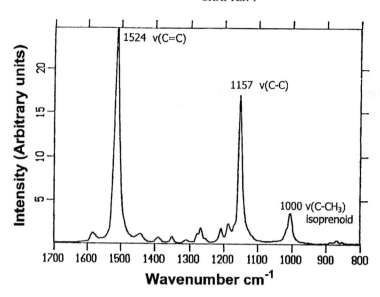

Figure 1-6. FT-Raman spectrum of a carotene found in an Antarctic Beacon sandstone endolith, Battleship Promontory, Alatna Valley

signatures are consistent with an assignment to astaxanthin. The same applies to the black zone in the chasmolith, but now the carotene gives bands at 1516, 1154 and $1004 \, \text{cm}^{-1}$ and so it can be identified as beta-carotene. Although no spectral signatures could be recorded for this system with the near-infrared laser at 1064 nm, three different micro-organisms are differentiated with 514 nm excitation because of the different carotenes produced, two discrete organisms occurring in the green band the third in the black band.

Endolithic Community

This specimen was collected at Mars Oasis in Antarctica, the coldest desert on Earth, during the summer season of 2002. The sandstone shows a normal orange-red colouration but a depletion of the red mineral (haematite) is clearly visible in the vicinity of the endolithic organism (Figure 1-7).

Quartz is the main component of the rock, with signatures at 128, 206, 355, 542, 696, 795, 807, 1064, 1081, 1161 and $1227 \, \text{cm}^{-1}$; the red colour is recognizable as haematite because of its Raman bands at 223, 291, 404, 495 and $609 \, \text{cm}^{-1}$. This is abundant below the organism stratum but has been removed from the immediate vicinity, probably to permit photosynthetic-active radiation to reach endolithic community. The specificity of Raman spectroscopy permits one to distinguish the very weak signature of haematite in the crustal spectrum (Figure 1-8); haematite has been proved to be a UV-screening compound and it is interesting that the organisms can maintain this mineral on the crust for the purpose of protecting against radiation that reaches the terrestrial surface (exacerbated by the depletion

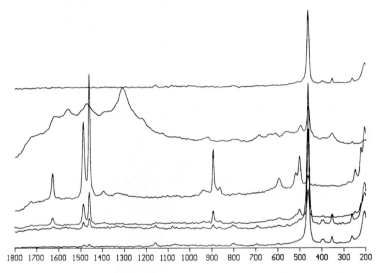

1800 1700 1600 1500 1400 1300 1200 1100 1000 900 800 700 600 500 400 300 200

Figure 1-7. The Raman spectral stackplot shows the spectra obtained from a vertical transect through the surface crust, the biological hyaline layer, algal layer, depletion zone and basal rock, from the top. Endolithic colonisation of Beacon Sandstone, Mars Oasis, Antarctica, showing zonation. The lower interface of the algal zone dominated by cyanobacteria occurs 8 mm below the upper surface of the rock

of Antarctic ozone). This particular crust colouration, modified by the underlying biological colony, was used by the prospecting geologist as a visual clue to the presence of endolithic communities inside the rock.

The stackplot of spectra collected from the endolithic specimen shows bands of calcium oxalate monohydrate at 1490, 1463, 896, 502 and 206 cm^{-1} in the depleted pigment area. Two different calcium oxalates are identified by Raman spectroscopy because of their signatures in extremophile exemplars; whereas in the calcium oxalate monohydrate (whewellite) Raman spectrum, apart from some weaker signatures, two bands are shown at 1490 and 1463 cm^{-1}, for calcium oxalate dihydrate (weddellite) only one characteristic strong signature, centred at 1475 cm^{-1}, is visible. Even when they appear in admixture whewellite and weddellite are clearly distinguishable using the Raman spectroscopic analytical technique. In addition, beta-carotene with bands at 1516, 1154 and 1004 cm^{-1} and chlorophyll,

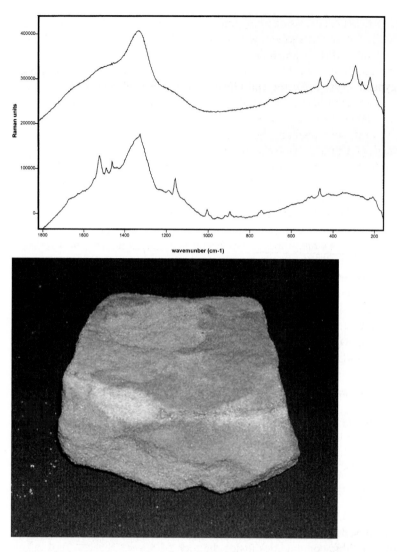

Figure 1-8. Raman spectrum of haematite crust on sandstone endolith. With Raman spectra of the crust (top) and cyanobacteria layer (bottom)

with a strong and broad signature at $1327\,cm^{-1}$ and weaker signals at 915, 744 and $517\,cm^{-1}$, have also been identified, as well as quartz.

Halotrophs

The presence of salts in SNC (Shergotty-Nakhla-Chassigny) Martian meteorites [23–26] and the acknowledgement of the presence of sulphates in the Martian regolith [27–30], together with the recognition of the existence of lacustrine water sediments at or below the surface of Mars [31–36] makes terrestrial halotrophic organisms potential Martian analogues.

Apart from the Raman spectroscopic identification of two different types of bacteria in a gypsum crystal (*Nostoc* and *Gloeocapsa*), it was also possible to detect organic signatures from bacterial colonies sited several millimetres below the surface in a transparent crystal of selenite from a 26 Mya meteoritic impact crater [37] at Devon Island in the Canadian High Arctic (Figure 1-9).

The Raman spectrum of the selenite crystal surface shows the strongest gypsum band at $1006\,\text{cm}^{-1}$ and other minor gypsum signatures; Raman spectra taken from a vertical transect through the crystal from the surface, using a confocal laser beam to obtain spatial information from within the crystal, provides evidence of additional features in which the Raman bands, can be identified with the cyanobacterial signatures. Confocal laser probing of the dark-coloured inclusion within the crystal reveals that it is a halotrophic cyanobacterial colony containing the radiation protectant scytonemin (Figure 1-10).

A parallel study of an Antarctic gypsum endolith (Figure 1-11) shows evidence of the same radiation protectant, scytonemin, in the lower spectrum with that of the surface gypsum for comparison in the upper spectrum; the different strategies adopted for the chemical protection of the colonies against low wavelength radiation for the two Antarctic endolithic systems, involving in one case the presence of haematite at the surface and in the other gypsum, is noteworthy.

Volcanic Rock Colonisation

A primary colonization in fumarole minerals from a recently active and still warm (two week-old) lava flow was collected by the author during a field

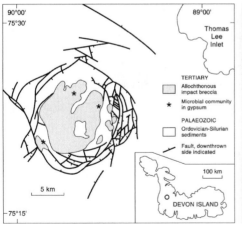

Figure 1-9. (Selenite var. gypsum) deposits, Haughton meteorite impact crater, Devon Island, Canadian High Arctic, 26 Mya. Also shown are details of selenite crystals within surface melt breccia deposits

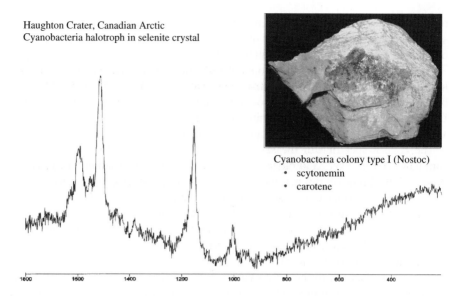

Haughton Crater, Canadian Arctic
Cyanobacteria halotroph in selenite crystal

Cyanobacteria colony type I (Nostoc)
• scytonemin
• carotene

Figure 1-10. Raman spectra of halotrophic cyanobacterial colony inside selenite crystal, located approximately 5–8 mm below the upper surface. The Raman spectrum of the cyanobacterial inclusion obtained using confocal laser imaging shows the presence of scytonemin and a carotenoid

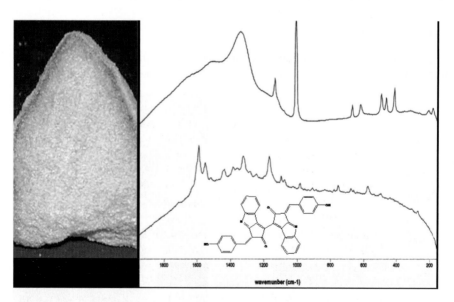

Figure 1-11. Antarctic endolith in gypsum, with Raman spectra of, upper spectrum, the surface (gypsum) and lower spectrum, the underlying cyanobacterial zone (scytonemin). The chemical structure of scytonemin is also given here

trip to the Kilauea volcano, Hawaii, in 2004 and was analyzed using Raman spectroscopy about seven days after collection. Besides identifying the mineral components, a carotenoid was recognized because of the Raman signatures at 1530 and $1143\,cm^{-1}$, and a chlorophyll-like compound could be related to the signatures at 1449, 1342, 952, 834, 748, 680, 595, 438, 257 and $173\,cm^{-1}$. This result indicates that the organisms needed only a very short time indeed to colonize an inhospitable niche such as a fumarole in a still very warm and active lava flow.

A second sample of lava (Figure 1-12) was collected during the AMASE expedition of 2004 in Scott Keltiefiellet, in the Vest Spitzbergen island of Svalbard [38] at 80° N at about 1000 m altitude. Organisms were found inside a vacuole of about 2 cm diameter in a black lava basalt matrix; the vacuole possessed a simple orifice of less than 1 mm diameter through which the organisms, sun, water and gases could penetrate. Technically, endoliths are organisms living inside the pores of a sedimentary rock, but this new kind of "endolithic" community affords a novel opportunity to look for life inside volcanic rocks that would not normally have been considered to be a geological niche for endolithic colonisation.

Three different carotenes were observed in the vacuole when 514.5 nm laser excitation was used, but there were no chlorophyll signals. With 785 nm excitation only two carotenes were detected but chlorophyll signals appeared alongside the carotenes and characteristic bands from c-phycocyanin, an accessory light-harvesting pigment, were also detected.

Stromatolites

Shade adaptation is required by bottom-dwelling cyanobacteria in ice-covered lakes, whilst enabling them to avoid UV-stress. During seasonal changes in the flow of melt-water, the thick cyanobacterial mats at the bottom of ice-covered Antarctic Dry Valley lakes becomes buried in silt. However, they permeate through the mineral layer to form a new mat each season, resulting in stratified stromatolites. The translucency of the 3 m thick ice cover on Lake Vanda permits penetration of solar radiation through 70 m of water column to the bottom mats where the receipt of photosynthetically active radiation (PAR) is remarkably up to $\sim 120\,\mu mol\,m^{-2}\,s^{-1}$. Even the PAR level of $\sim 15\,\mu mol\,m^{-2}\,s^{-1}$ measured at the bottom of Lakes Hoare and Fryxell, which have thicker ice covers (about 4.5 m), is more than enough to sustain active photosynthesis by oxygenic mats. This has resulted in a wide diversity of cyanobacterial mat communities in many Antarctic Dry Valley lakes. Analogous habitats on Mars, such as Gusev and Noachis Craters, may provide suitable sources of fossil microbes and biomolecules such as porphyrins and cyanobacterial scytonemin. Modern stromatolites are considered to be analogues of *Conophyton*, which is a columnar stromatolite abundant in Precambrian rock when UV stress would have been much greater.

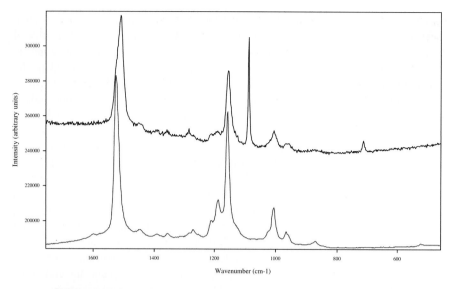

Figure 1-12. Novel endolith in a vacuole in a volcanic basalt lava matrix, Svalbard, Spitsbergen, Norwegian Arctic, showing two different carotenoids in admixture. Accessory photopigments are also identified in this system (not shown here) (see color plate section)

We have studied columnar endolithic colonisation of carbonates from Salda Golu Lake in Turkey, where stromatolitic outcrops of magnesite and hydromagnesite have formed (Figure 1-13). Here, the identification of the biological component in the Raman spectrum of the inorganic carbonate matrix is a good terrestrial model for the survey of potential stromatolite sites on Mars – such as that identified at Juventus Chasmae in Sabaea Terra (Figure 1-14) near the Schiaparelli Crater; in the figure, the light-coloured area in the meteorite impact crater is believed to be a magnesium carbonate deposit.

With climatic changes on a geological scale, these lakes may dry up and their cyanobacteria become desiccated fossils. This process may have helped

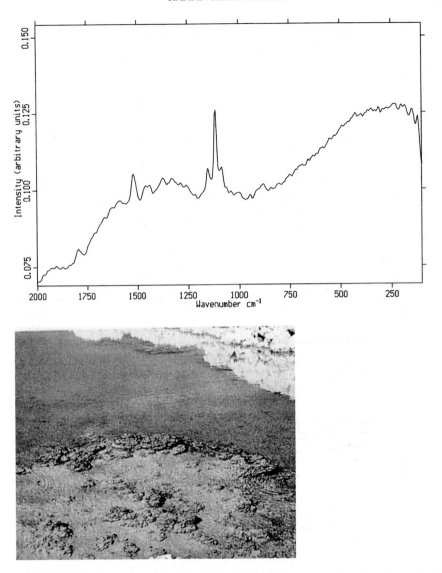

Figure 1-13. Salda Golu Lake, Turkey: hydromagnesite colonised stromatolites in a saltern. The FT-Raman spectrum of the cyanobacterial colonised zone in a stromatolite is shown (see color plate section)

to preserve cyanobacteria-like organisms in chert (flint) from the Australian Apex formation in Queensland for 3.5 Gya. Their pigments, especially early porphyrins which are recalcitrant components of chlorophylls, are recognizable in oil-bearing shales. A primitive photosynthetic system capable of delivering an electric charge across a membrane bi-layer could have consisted of a porphyrin

Figure 1-14. Juventus Chasmae, potential stromatolite region, on Mars. Reproduced from the Journal of the Geological Society, 156, Michael J. Russell et al., Search for signs of ancient life on Mars: expectations from hydromagnesite microbialites, Salda Lake, Turkey, 1999, with permission from the Geological Society (see color plate section)

(pigment), a quinone (electron donor) and a carotene (electron acceptor) to provide a chain of conjugated bonds. Such molecules are valuable biomarkers on Earth and may fulfil a similar function for their recognition by remote life-detection systems on Mars. A recently recorded microRaman spectrum of the Trendall Pilbara stromatolite has identified a carotene and a porphyrin together in a fissure inside the rock (Figure 1-15). This stromatolite is one of the oldest rocks found on Earth, with an age of 3800 M years, dating from a very early period in the geological evolutionary history of our planet. However, whilst it would be facile to assume that the bioorganic signatures that have been detected here are also of this age, this example does demonstrate very nicely the viability of Raman spectroscopy for the detection of organic signatures in ancient rocks. In another sampling region of the same specimen, there is clear spectroscopic evidence for the presence of beta-carotene, scytonemin, dolomite and goethite (Figure 1-16).

Salt Pan Evaporites

The dried terrestrial salt pans with their extremophilic halotrophic cyanobacterial colonisation are good models for the study of Martian sites such as Gusev Crater.

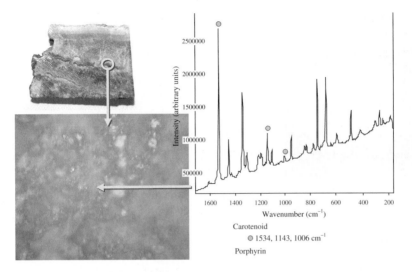

Figure 1-15. Trendall Pilbara stromatolite, Western Australia, 3800 Mya; Raman spectroscopic evidence for protective biochemicals, carotene and porphyrin, produced in a protected niche

Here, an example from the Rhub-al-Khalil in the Arabian Desert shows a cyanobacterial colony in a dolomitic zone, several cm below the surface of a large gypsum crystal embedded in a halite matrix (Figure 1-17). The Raman spectra of the biological component clearly demonstrate the presence of photoprotective pigments such as scytonemin and carotenoids.

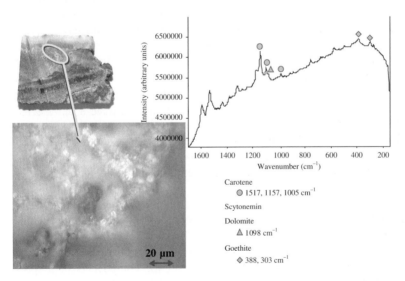

Figure 1-16. Specimen of Trendall Pilbara stromatolite shown in Figure 1-15, but different niche; different chemical suite identified – scytonemin, carotene, dolomite and goethite

Sabkha Evaporite Saltpan
United Arab Emirates

- Surface Crust of Gypsum and Halite
- Colonies at Interface with Subsurface
  Dolomitized Calcite
- Organic Signatures at the Subsurface

Figure 1-17. Sabkha surface crust with gypsum and halite crystals, Rhub-al-Khalil, Arabian desert; cyanobacterial colonisation at interface with subsurface dolomitized calcite. The Raman spectrum shows the presence of scytonemin and carotene in the biological zone

CONCLUSIONS

Raman spectroscopy is a useful analytical technique for planetary exploration because it does not require sample manipulation, and macroscopic and microscopic analyses are possible when identifying components in the specimen. It is sensitive to organic and inorganic compounds and is not destructive to samples that are strictly limited in quantity or accessibility and so they can be used afterwards for other, perhaps more destructive, analyses. In this respect the situation of a Raman spectrometer on a planetary rover or on a fixed lander, to which specimens are brought by a rover, are both acceptable space mission scenarios.

Laser excitation wavelength is an important parameter to consider, not only because fluorescence emission can swamp the weaker Raman scattering, but also because of the different sensitivities shown towards compound identification by different excitation wavelengths. In our assessment 785 nm laser excitation gives a good balance between organic and inorganic compounds, but 514 nm excitation induces a resonance Raman effect for carotenes whose identification is reasonably straightforward, even in small concentrations, although the higher fluorescence emission from some systems can still mask several weaker Raman bands. The observation of more than two spectral biomarker bands is necessary normally for the unambiguous identification of the components of a specimen, particularly when there is a complex mixture of organics and inorganics present.

Spectral resolution is another critical parameter that can affect observed band wavenumbers due to the overlapping of close bands, which can cause erroneous identifications of the material when comparing with standard wavenumber tables and spectral databases. The technical requirements for the construction of a miniaturized Raman spectrometer suitable for extraterrestrial planetary exploration could force a rather low spectral resolution, and so new databases adapted to low resolution spectral data would then be necessary. Nevertheless, good quality spectra are now being obtained from miniaturised Raman spectrometers such as that shown in Figure 1-18, which weighs one kilo. An example of the Raman spectrum of an extremophile obtained from this instrument in our laboratories is shown in Figure 1-19.

The ability to identify organic molecules and minerals present in admixture in a specimen at the same time is an advantage of the Raman analytical technique in astrobiology. Extremophiles are closely related to rock substrata, and biological and geological spectral signatures are shown together in the Raman spectrum; these are distinguishable and can be unambiguously assigned, even if some appear in a relatively low proportion. A list of some of the most important biogeological Raman spectral markers is provided in Table 1-2; this can be regarded as a prototype database for the spectral recognition of indicators from remote life detection instrumentation on planetary landers and rovers.

A detailed knowledge of the different chemical strategies adopted by terrestrial extremophiles, which are considered Martian analogues, is a prerequisite for planetary exploration. Geo- and bio-markers must be studied together to provide a better understanding of the adaptation abilities of the organisms. Geo- and bio-strategies are complementary and the survival of the organisms depends critically upon their being able to modify their microniche environments.

University of Montana Raman Prototype

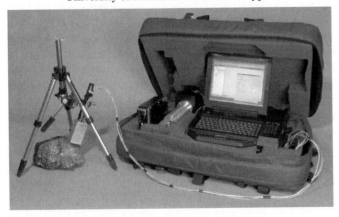

Figure 1-18. Miniaturised Raman spectrometer prototype for space mission, mass 1 kg

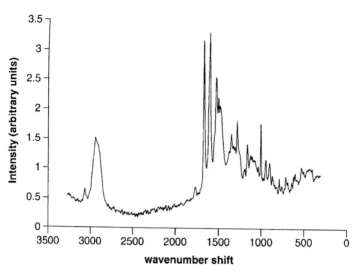

Figure 1-19. Raman spectrum of *Acarospora chlorophana*, an epilithic extremophile from Antarctica, recorded with the miniature Raman spectrometer shown in Figure 1-18

THE FUTURE

In May 2006, the European Space Agency announced that significant funding had been allocated for the forthcoming ExoMars mission on Project AURORA – the most ambitious scientific analytical programme yet devised to detect the signatures of extinct or extant life on Mars (Figure 1-20). A miniaturised scientific instrumental payload on a planetary rover will use a sophisticated battery of analytical techniques, some of which have never before been used in space missions. The scientific instrumentation suite on ExoMars (Pasteur) has to fulfil three specific functions, namely, the characterisation of the geological context of the selected landing site, the detection of life signals and organic molecules in the samples from that site and the evaluation of the hazards to the planned human exploration of Mars. An integral part of this instrumentation will be a novel combined miniaturised Raman/LIBS spectrometer which will be designated the specific task of searching for biogeological marker signatures in the Martian surface and subsurface regolith (Figure 1-21). The Raman spectroscopic data presented in this paper from terrestrial Mars analogues and models will form an important part of the analytical scientific preparation for Mars exploration involving the search for life in the ExoMars mission.

The ExoMars mission and others that follow it, including the manned lunar landings, will culminate in a knowledge base that is essential for the planned human landings and exploration of Mars scheduled for 2033 – the first time that human beings will have set foot on another planet in our Solar System. A symbolism of this leap forward in analytical scientific astrobiology is depicted in Figure 1-22.

Table 1-2. Raman bands of the most common and geo- and biomarkers in extremophile examplars, and their chemical formulae

Mineral / compound	Formula	Raman bands
Calcite	$CaCO_3$	**1086** **712** **282** 156
Aragonite	$CaCO_3$	**1086** **704** **208** 154
Dolomite	$CaMg(CO_3)_2$	**1098** **725** **300** 177
Magnesite	$MgCO_3$	**1094** **738** **330** **213** 119 **141**
Hydromagnesite	$Mg_5(CO_3)_4(OH)_2 \cdot 4H_2O$	1119 728 232 202 184 147 270
Gypsum	$CaSO_4 \cdot 2H_2O$	**1133** **1007** 669 628 **492** **413**
Anhydrite	$CaSO_4$	**1015** 674 628 500 416
Quartz	SiO_2	1081 1064 808 796 696 542 500 128
Haematite	Fe_2O_3	**610** 500 **411** **293** 245 **226** **203**
Limonite	$FeO(OH)nH_2O$	693 **555** 481 **393** **299** **203**
Apatite	$Ca_5(PO_4)_3(F,Cl,OH)$	1034 **963** **586** **428**
Weddellite	$Ca(C_2O_4)2H_2O$	1630 **1475** 1411 910 597 **506** **188** 223 521 **185**
Whewellite	$Ca(C_2O_4)H_2O$	1629 **1490** **1463** 1396 942 **896** 865 596 **504** **517** 207 351 207
Chlorophyll	$C_{55}H_{72}O_5N_4Mg$	1438 1387 **1326** 1287 1067 1048 **988** **916** **744** **665** 499
c-phycocyanin	$C_{36}H_{38}O_6N_4$	1655 **1638** 1582 1463 **1369** 1338 1272 1241 1109 1054 **815**
Beta-carotene	$C_{40}H_{56}$	**1515** **1155** **1006**
Rhizocarpic acid	$C_{26}H_{23}O_6$	**1665** 1610 **1595** **1518** **1496** 1477 1347 1303 **1172** 1002 944 902 768 574 448
Scytonemin	$C_{36}H_{20}N_2O_4$	1605 **1590** **1549** 1444 **1323** 1283 1245 1240 1163 984 752 675 574
Calycin	$C_{18}H_{10}O_5$	1653 **1635** **1611** **1595** **1380** 1344 1240 1155 1034 **960** 878 498 484
Paretin	$C_{16}H_{12}O_5$	**1671** 1631 1613 1387 1370 **1277** 1255 **926** 571 519 467
Usnic acid	$C_{18}H_{16}O_7$	**1694** 1627 **1607** **1322** **1289** 1192 1119 **992** 959 846 602 540 **458**
Emodin	$C_{15}H_{10}O_5$	**1659** 1607 1577 1557 1298 **1281** 1266 942 **565** 467
Atranorin	$C_{19}H_{18}O_8$	**1666** **1658** 1632 **1303** **1294** **588** 504
Pulvinic dilactone	$C_{18}H_{10}O_4$	**1672** **1603** 1455 **1405** 1311 **981** **504**
Gyrophoric acid	$C_{24}H_{20}O_{10}$	**1662** 1628 1612 1334 **1291** **1235** 1304 **1138** 561

Corroborative bands appear in bold

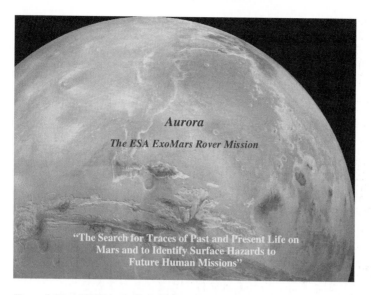

The authors record their appreciation of funding over many years for the basic analytical Raman spectroscopic studies of terrestrial extremophiles and Mars models from the European Space Agency and UK organisations (EPSRC, NERC and PPARC/STFC) which have enabled the preliminary work to be carried out to establish the viability of the Raman technique for incorporation into extraterrestrial life-detection instrumentation and for the opportunity to take novel analytical chemical science to the "last frontier". Professor Edwards is especially grateful for

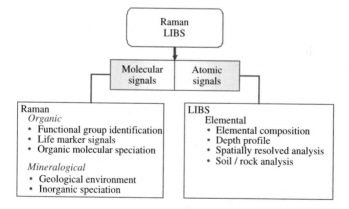

Figure 1-21. The principle of the novel miniaturised Raman/LIBS combined spectrometer for the ExoMars mission to Mars

Figure 1-22. Symbolic depiction of the Mars mission phases, 2007−2033, involving a revisitation of manned lunar exploration before embarking upon the manned missions to Mars (see color plate section)

his scientific collaboration with the late Dr David Wynn-Williams, Head of Antarctic Astrobiology at the British Antarctic Survey, Cambridge, and for the ongoing support of Professors Charles Cockell, John Parnell, Monica Grady, Fernando Rull, Susana Jorge, and Drs Cynan Ellis- Evans, Roger Worland, Andre Brack, Frances Westall, Derek Pullan, Beda Hoffmann, Jan Jehlicka and Liane Benning—a range of expertise encompassing geology, microbiology, geochemistry, meteoritics, earth sciences and biomolecular sciences that is so critical for the effective study of *astrobiology*, the quest for life in the Solar System and beyond.

REFERENCES

1. W.K. Hartman, "A Traveller's Guide to Mars", Workman Publ., New York (2003).
2. W. Thomson, *Nature*, **4**, 262 (1871).
3. C.P. McKay, *Origins of Life & Evolution of the Biosphere*, **27**, 263 (1997).

4. M.C. Malin and K.S. Edgett, *Science*, **288**, 2330 (2000).
5. M.H. Carr, "Water on Mars", Oxford University Press, New York (1996).
6. D.D. Wynn-Williams and H.G.M. Edwards, *Environmental UV Radiation:Biological Strategies for Protection and Avoidance*, in "Astrobiology: The Quest for Life in the Solar System", eds. G. Horneck and C. Baumstarck-Khan, Springer-Verlag, Berlin 244–260 (2000).
7. C.S. Cockell, *Planetary and Space Sciences*, **48**, 203 (2000).
8. R. Caricchidi, *Astrobiology*, **2**, 281 (2002).
9. C.S. Cockell and J. Knowland, *Biol. Revs.*, **74**, 311 (1999).
10. D.D. Wynn-Williams and H.G.M. Edwards, *Icarus*, **144**, 486 (2000).
11. A. Brack, B. Fitton, F. Raulin and A. Wilson, "Exobiology in the Solar System and the Search for Life on Mars", ESA Special Publication (SP-1231), ESA Publications Division, Noordwijk, The Netherlands (1999).
12. S.E. Jorge Villar and H.G.M. Edwards, *Analytical & Bioanalytical Chemistry*, **384**, 100–113 (2006).
13. I.R. Lewis and H.G.M. Edwards, "Handbook of Raman Spectroscopy: From the Process Line to the Laboratory", Marcel Dekker, New York (2000).
14. A. Ellery and D.D. Wynn-Williams, *Astrobiology*, **3**, 565 (2003).
15. H.G.M. Edwards, E.M. Newton, D.L. Dickensheets, D.D. Wynn-Williams, C. Schoen and C. Crowder, *International J.Astrobiology*, **1**, 333, (2003).
16. H.G.M. Edwards, E.M. Newton, D.L. Dickensheets and D.D. Wynn-Williams, *Spectrochimica Acta, Part A*, **59**, 2277 (2003).
17. D.L. Dickensheets, D.D. Wynn-Williams, H.G.M. Edwards, C. Schoen, C. Crowder and E.M. Newton, *J.Raman Spectroscopy*, **31**, 633 (2000).
18. A. Wang, L.A. Haskin, A.L. Lane, T.J. Wdowiak, S.W. Squyres, R.J. Wilson, L.E. Hovland, K.S. Manatt, N. Raouf and C.D. Smith, *J. Geophys. Res. Planets*, **108**, (E1), Art. No.5005 (2003).
19. A. Wang and L.A. Haskin, *Microbeam Analysis 2000, Proc. Inst. Phys. Conf. Series*, **165**, 103 (2000).
20. "ESA Progress Letter 4: Pasteur Instrument Payload for the Exo Mars Rover Mission", ESA Publications Division, Noordwijk, The Netherlands (2004).
21. D.A. Long, "The Raman Effect: A Unified Treatment of the Theory of Raman Scattering by Molecules", John Wiley & Sons Ltd, Chichester, UK. (2002).
22. H.G.M. Edwards, *Origins of Life & Evolution of the Biosphere*, **34**, 3 (2004).
23. S.J. Wentworth and J.L. Gooding, *Parent Planet Meteorites*, **29**, 860–863 (1994).
24. M.D. Lane and P.R. Christensen, *Icarus*, **135**, 528 (1998).
25. D.C. Catling, *J. Geophys. Res-Planets*, **104**, 16453–16469 (1999).
26. J.C. Bridges, D.C. Catling, J.M. Saxton, T.D. Swindle, I.C. Lyon and M.M. Grady, *Space Sci. Revs.*, **96**, 365–392 (2001).
27. M.E.E. Madden, R.J. Bodnar and J.D. Rimstidt, *Nature*, **431**, 821–823 (2004).
28. D.T. Vaniman, D.L. Bish, S.J. Chipera, C.I. Fialips, J.W. Carey and W.C. Feldman, *Nature*, **431**, 663–665 (2004).
29. R.E. Arvidson, F. Poulet, J.P. Bibring, M. Wolff, A. Gendrin, R.V. Morris, J.J. Freeman, Y. Langevin, N. Mangold and G. Bellucci, *Science*, **307**, 1587–1591 (2005).
30. Y. Langevin, F. Poulet, J.P. Bibring and B. Gondet, *Science*, **307**, 1584–1586 (2005).
31. J.B. Murray, J.P. Muller, G. Neukum, S.C. Werner, S. van Gasselt, E. Hauber, W.J. Markiewicz, J.W. Head, B.W. Fong, D. Page, K.L. Mitchell and G. Portyankina, *Nature*, **434**, 352–356 (2005).
32. N.A. Cabrol, E.A. Grin and W.H. Pollard, *Icarus*, **145**, 91–207 (2000).
33. J.M. Moore and D.E. Willhelms, *Icarus*, **154**, 258–276 (2001).
34. D.A. Paige, *Science*, **307**, 1575–1576 (2005).
35. H. Hiesinger and J.W. Head, *Planetary and Space Sciences*, **50**, 939–981 (2002).
36. S.W. Ruff, P.R. Christensen, R.N. Clark, H.H. Kieffer, M.C. Malin, J.L. Bandfield, B.M. Jakosky, M.D. Lane, M.T. Mellon and M.A. Presley, *J. Geophys. Res-Planets*, **106**, 23921–23927 (2001).
37. H.G.M. Edwards, S.E. Jorge Villar, J. Parnell, C.S. Cockell and P. Lee, *Analyst*, **130**, 917–923 (2005).
38. S.E. Jorge Villar, H.G.M. Edwards and L.G. Benning, *Astrobiology*, Icarus, **184**, 158–169 (2006).

CHAPTER 2

X-RAY DIFFRACTION OF PHOTOLYTICALLY INDUCED MOLECULAR SPECIES IN SINGLE CRYSTALS

JACQUELINE M. COLE*

Abstract: We review developments in X-ray diffraction of single crystals that begin to enable one to quantify directly the nature of electronic perturbations induced by light in chemical structures. Such structural information is key to understanding many chemical processes and physical properties activated with light, and the scientific impetus behind this incipient area of structural science is described from academic and industrial perspectives. Photoisomerisation, photochemical reactions in the solid state and spin-crossing magnetic transitions that have enduring or irreversible states induced with light are best understood by unravelling their three-dimensional structure measured in situ in their states converted by light. Investigations conducted with single-crystal X-ray diffraction of structures in a laser-induced *steady state* and the experimental methods used to realise such structures are reviewed. The structural characterisation of transient photo-induced species (down to picosecond lifetime) is paramount to improve understanding of materials that undergo rapid electronic switching, which make operative much of the electronic and optical industry, as there exists an inherent relationship between the structure of the excited state and the physical properties exhibited. Prime instances include structures of molecular conductors and luminescent materials in their excited states with prospective applications as molecular wires, light-emitting diodes, non-linear optical components, triboluminescent and electroluminescent devices. Only indirect and qualitative interpretations of the nature of these excited states were formerly formulated with spectrometric techniques, but the developments in ms-ps *time-resolved* (laser)-pump (X-ray)-probe single-crystal diffraction techniques, described herein, are overcoming this barrier, affording results that are quantitative via a three-dimensional structural representation. Structures of transient species are reviewed and the key experimental parameters that are required for a successful experiment, in terms of characteristics of the X-rays, laser and sample are discussed. The importance of auxiliary spectroscopic experiments is also described. A future outlook on possible X-ray sources to facilitate such work and to extend it to structural studies on even more ephemeral species concludes this review

*Jacqueline M. Cole, Chem. Soc. Rev., 2004, (8), 501–513, DOI: 10.1039/b205339j, http://xlink. rsc.org/?DOI=b205339j, – Adapted by permission of The Royal Society of Chemistry

J.C.A. Boeyens and J.F. Ogilvie (eds.), Models, Mysteries and Magic of Molecules, pp. 29–61.
© 2008 *Springer.*

INTRODUCTION

Photo-induced species form in many compounds, ranging from those occurring in nature, e.g. chlorophyll, to technologically important materials such as photo-darkening semiconductors and light-emitting diodes. On photolysis a molecule becomes excited into otherwise inaccessible states, thus inducing a redistribution of electrons within that molecule. In turn, this condition can produce atomic displacement within the structure of the material or, more severely, isomerisation or chemical reactions in the solid state. Diverse photophysical properties unfold from these structural perturbations, as these cause, for instance, associated changes in dipole moments, luminescence or formation of radicals.

Although photophysical properties are readily measured, their structural origins remain generally elusive as the photo-induced states are typically transient (lifetime: μs or less). Recent developments in temporally resolved experiments have yielded routine methods to determine indirect structural information, particularly from temporally resolved optical (including infrared) spectra. Most indirect structural information existing from temporally resolved spectra emanates from measurements of solutions, but the physical property commonly sought is aimed at solid-state devices. The resulting dearth of comparisons between structure and property in the same phase hampers implicitly their interpretation and use in understanding the role of structure in a given physical phenomenon.

Crystallography is the ultimate technique for determining the bond geometry of a molecule with great accuracy and in a three-dimensional representation. Given that such structural characterisation is also undertaken using solid-state samples, crystallography would seem therefore to be the ideal technique with which to embrace the fourth dimension of time within its capabilities, such that one can understand photophysical processes. Because many processes in all scientific disciplines evolve temporally, a temporally inclusive investigative power of crystallography is attractive to many other areas of structural science. Temporally resolved developments have occurred in all ranges of science, but temporally resolved chemistry is evolving particularly rapidly, revealing results from numerous experiments of varied qualitative and quantitative nature and concerning processes with temporal variations spanning many time domains [1]. Implicit within this evolution is the development of temporally resolved crystallography.

The typically transient nature of photolytically induced processes makes this area of temporally resolved crystallographic development particularly challenging, as do the many practical complications surrounding the optical excitation (see §4). High atomic resolution is commonly a prerequisite: in most cases, the redistribution of electrons, following photo-induction, manifests itself in an associated alteration of bond distance between two or more atoms within a molecule. For example, a luminescent organometallic material classically becomes photo-excited into a state that effects a transfer of charge between metal and ligand (MLCT). The primary structural perturbation in such a case is an altered length of a bond between the metal and ligand(s) concerned. An oxidation might similarly be caused on photo-induction in a material; in such a case the coordination sphere about the metal ion contracts or expands according to an increase or decrease in oxidation number, respectively. Variations 0.1–0.2 Å of bond length are typical, but this extent might be much smaller, particularly in organic compounds: as lengths of C—C bonds are smaller than those of metal-containing bonds, any C—C bond length *changes* are inherently difficult to detect. Obtaining sufficient atomic resolution to observe a structural perturbation sought thus presents a major experimental challenge in this area. One wishes ultimately to be able to collect data with sufficient resolution to perform mapping of the distribution of electron density in three dimensions via a full multipole refinement of charge density, although such activities are likely to remain beyond our capabilities for some time.

Several timely advances in synchrotron and crystallographic instruments have helped overcome some experimental challenges, and have made feasible the subject area of research. New synchrotrons across the world have attained much enhanced intensity, stability, reliability and efficiency of X-ray beams, all critical for this work. Charge-coupled devices (CCD) as area detectors have revolutionised crystallography, not least temporally resolved crystallography for which a substantial decrease in duration of data collection has been critical for the developments described herein. For photolytically induced processes that occur or can be stabilised at low temperatures, the development of open-flow helium-based cryostreams has been instrumental in these advances. The recent revolution in laser technology has greatly increased the feasibility of this area of structural science, with lasers with nanosecond or picosecond pulses becoming standard and all commercial lasers being designed for use as modular bench-top apparatus of type "turn-key operation". The field of optical spectrometry has also aided substantially the progress of this field as much of its highly advanced electronic-timing technology is readily transferable to the subject area.

What follows is a review of photolytically induced processes in materials of the types that are of interest, experimental methods being developed for photolytically induced X-ray diffraction to study these processes, the results obtained from this work, experimental challenges and developments in this area of research, and the future prospects of this technology. We focus on the use of monochromatic oscillatory diffraction of single crystals, necessarily restricting ourselves to chemical results, although recognising, in places, important related Laue-based diffraction work from the biological sphere. For important work similarly aimed at affording direct and quantitative photolytically induced structural information and being attempted using EXAFS spectroscopy, one is referred to reviews by Chen [2] and Saes et al [3] and references therein.

PHOTOLYTICALLY INDUCED PROCESSES OF INTEREST IN MATERIAL SCIENCE

A profound knowledge of the structure-originating functionality of chemical processes driven by light is important on account of the associated technological implication of such materials. The world has experienced an electro-optical revolution: digital light displays that illuminate car dashboards, mobile phones and stereo sound systems; fluorescent and phosphorescent screens, for instance in portable computers; equipment for underwater lighting and apparatus for working at night; optical switches and shock-wave triggers for use within electronic circuitry. The controlling radiative phenomena all result from transfer of electronic charge within a molecule or ionic displacement, and occur in various manifestations, such as photo-, electro-, tribo-, chemi- or sonoluminescence – the application of optical field, electric field, mechanical force, chemical reaction or sound waves, respectively, to produce light (see, for example, Figure 2-1).

An implicit reason that these phenomena are usable in electric and electro-optical devices is the brief interval during which these luminescent processes occur, as electronics require such rapid reactions for high operating frequencies to ensure, for example, constant communication and screen updating. Most fluorescent and phosphorescent processes occur on a time scale of ns to ps or ms to μs, respectively.

There exist also many non-luminescent phenomena driven by light with ephemeral photo-active states that are of technological importance in the electrical or optical industry. In particular, transitions between conductor and insulator in molecular species are of enormous interest in solid-state physics. These phenomena, which occur typically on a timescale of order fs to ns, are led inherently by structural alteration that results from perturbations through transfer of electronic charge in a molecule. Non-linear optical phenomena (see, for example, Figure 2-2) occur on similar time scales, and are governed by molecular charge transfer; it is desirable to comprehend fully their structural manifestations in situ.

Figure 2-1. The application of electroluminescence in lighting the display in this watch. An electric signal, caused by pressing the "light" button on the watch, excites the molecules in the display medium that cause luminescence upon their relaxation (see color plate section)

Figure 2-2. A non-linear optical material, ammonium dihydrogen phosphate, displaying second-harmonic generation, the frequency doubling of light (infrared to blue). The origin of this physical phenomenon is entirely dependent on ionic displacement or molecular charge-transfer (see color plate section)

Materials with light-driven lifetimes ms to ∞ are generally non-luminescent but are also highly desirable industrially. Enduring photodarkening properties have been long exploited in glasses. This area of structural research sustained early attempts at quantifying light-driven structural perturbations [4]. Improved photodarkening materials are sought for device application, as are long-lived photo-active compounds for prospective use as sunscreens for protection against ultraviolet light, for optical storage and magnetic switches. Current photodarkening materials are typically inorganic; their application is typically based on exploiting the low-energy band gaps in these materials to absorb light.

Sunscreens exploit light in various ways. Enduring light-absorbing states are a key physical property in the application of one such variant [5]: a suitable active material of a sunscreen operates by absorbing light in an upper layer of the skin, which undergoes a photochemical change, that in turn blocks farther ultraviolet light penetrating to the more sensitive lower epidermis cells. The longevity of the light-absorbing state ensures that this skin remains protected for several hours or until the sunscreen is removed by washing or skin abrasion. Photoisomerisation is typically the photochemical change involved. Most other sunscreens function by absorbing light in the harmful band of ultraviolet radiation, and emitting it at a lower wavelength that is unharmful to the skin; in these cases, the light-driven lifetime is typically on the transient time scale of ns to μs.

Photoisomerisation, reversible or irreversible, occurs in many areas of organic and organometallic chemistry. Beyond its applied nature as described above, photoisomerisation is important also for the development of fundamental research. For example, such a process might comprise part of a reaction mechanism in organic chemistry that is being developed for a given application, or perhaps induction with light is used to form a new chemical product. Photoisomerisation occurs also with great propensity in nature, for example, in signal transduction of photoreceptor proteins and photorepair processes and in photosystems one and two in plants. The structures of biological species are generally more difficult to characterise than chemical materials, especially in their in vitro state, and the structural characterisation of their photoisomerised states and the associated mechanisms are necessarily even more complicated. Even so, success has been achieved in this direction, for example, on photoactive yellow protein (PYP) [6] and a flavin-bound plant photoreceptor domain [7]. Improved structural knowledge of all processes and mechanisms of chemical photoisomerisation will, however, undoubtedly provide useful insight into many biological processes; it is therefore a valuable goal to strive for the structural characterisation of all species undergoing photolytic isomerisation. Time scales of photoisomerisation vary greatly, but many chemical photoisomerisations occur via metastable states, which are stable under conditions such as decreased temperature, or irreversible transitions. In these cases, their crystal structure can be readily probed, provided that a sample is not destroyed during isomerisation.

Given its importance in computing and memory hardware, the importance of satisfactory materials for optical storage cannot be understated. In particular, there is at present a great commercial interest in developing holographic storage, hailed as the next-generation revolutionary storage medium [8]. Holographic storage allows three-fold data encryption onto a single point, thus exploiting the volume of a material rather than just its surface area, as used on a compact disc (CD) or digital versatile disk (DVD) for instance. Hologram memory can be created on focusing successive interference patterns of two light beams into a specific volume of a material. Each pattern relates to one page of data, coded as light and dark boxes on a screen, through which one incoming beam passes (the beam carrying data). As the two beams have been split from a single laser source, the second beam, which passes through no screen, acts as a reference signal (no data information) upon their coalescence. With the proposed ability to store over 100 times the data capacity of a DVD in a material having the size of a sugar cube (Figure 2-3), the potential of this new medium is incredible, but developments are largely hampered by two factors. One problem lies in the stringent requirements of data retrieval: data are reconstructed on diffracting a further light beam, at *exactly* the same angle as the reference beam enters the material, onto a detector; an error $1 \mu m$ renders the data irrecoverable. The other problem is the dearth of suitable materials for application in enduring devices. Given the rapid industrial drive in this area, it is critical that materials chemistry keeps pace with the technological developments, which is achievable only through understanding the structural manifestations of the radiative processes ensuing in the application. The optical etching created in the

Figure 2-3. A photograph of 100 stacked DVD and a sugar cube. If holographic storage achieves its expectation, one will be able to store a terabyte (1000 Gb) of data in a crystal of size a sugar cube. Employing current commercial optical storage devices, a terabyte of data would fill more than 1000 compact disks, or over a 100 DVD. The use of a sugar cube as a placebo here is a deliberate attempt to emphasise the fact that, although optical technology is developing rapidly, scientists are still struggling to find suitable materials for the application

material inevitably needs to endure, else data become lost irretrievably. Photo-active materials with metastable lifetimes have therefore enormous potential in this area.

Light-induced magnetic spin transitions have shown potential in electronic switching devices and applications in magnetic storage of data [9]. At a given temperature, the magnetic metal ion in the material (typically Fe) converts from the low- to the high-spin state upon excitation with light of a particular wavelength. How sudden this phase transition is and how the hysteresis transpires dictates the level and utility of the "bistability" of the material: a bistable material exhibits a sudden phase transition and hysteresis that allows the material to exist in temperature phase space as either the high- or low-spin magnetic state, for temperature over a wide range. The sudden phase transition at a given temperature also renders materials of these types possibly useful as a temperature sensor. The lifetime of the bistable state corresponds directly to the lifetime of any possible device in which it might be applied: any loss in definition of hysteresis, for example, might cause an error in electronic switching or might corrupt data irrecoverably if such materials are ultimately used in magnetic storage media. It is therefore important that materials with enduring light-induced magnetic spin transitions are understood so that one can design suitable candidates, with more commercially viable operating temperatures (at present transitions occur typically below 50 K), and with greater longevity of bistability. Light-induced X-ray diffraction experiments would be useful to achieve this goal as the crystal structure of a magnetic material in a given spin state can be characterised by X-ray diffraction of a single crystal.

Although there are undoubtedly many other photo-induced phenomena that are omitted here, those described above might provide a representative view of the diverse materials for which X-ray diffraction of a single crystal might generate key advances in these areas, thus underlying the importance of such crystallographic developments.

FOUR TECHNOLOGICAL TIME SCALES OF X-RAY DIFFRACTION EXPERIMENTS

The lifetime of the photo-induced state in a given material dictates the instruments required for the determination of its structure. The smaller is the lifetime of such a species, τ_o, the more challenging the experiment becomes. These instrumental challenges are divisible into four categories: steady-state methods ($\tau_o >$ min), pseudo-steady-state methods (ms $\leq \tau_o \leq$ min), and stroboscopic methods using a pulsed X-ray source generated by means of either a mechanical chopper (μs $\leq \tau_o \leq$ ms) or accelerator-physics (ps $\leq \tau_o \leq \mu$s) methods. Figure 2-4 summarises the instrumental requirements as a function of time.

Steady-State Methods

For materials exhibiting long-lived or irreversible photo-structural changes, *steady-state* methods are used whereby the photo-induced structure is obtained in three experimental steps: one first determines the ground-state structure of the material using conventional methods for X-ray diffraction of small molecules in single crystals; next, the crystal, still mounted on the X-ray diffractometer, is optically pumped for several hours with a suitable lamp, typically either a flash lamp emitting light at wavelengths in a broad band or a laser (monochromatic source); the lamp is then removed and the resulting structure is determined, again using conventional methods and under the same experimental conditions as before. The resulting structure from this second data collection comprises contributions from both the original ground state and the light-induced structure. This is because the light-induced structure is unobtainable experimentally in isolation, as in any photo-induction one never achieves complete conversion in a single crystal. One generally aims to obtain no more than 20–30% photo-conversion, as beyond this point the integrity of the crystal is endangered: if structural perturbations within a crystal lattice are too great or numerous, the crystal fractures or explodes. One can, however, effectively isolate the light-induced structure by analytical means. This achievement requires first importing the refined ground-state structural model,

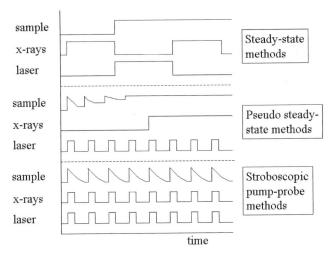

Figure 2-4. A schematic diagram of the relative timing systems between laser, x-ray and sample photo-conversion lifetime. The nature of steady-state, pseudo-steady-state (in its simplest form) and stroboscopic pump-probe methods are illustrated. The absolute timing for each method is on a separate scale: the entire experiment is shown for the steady-state methods; the pseudo-steady-state representation shows up to the beginnings of the first data-collection frame; the stroboscopic representation illustrates a regular pattern that occurs throughout the experiment. Stroboscopic pseudo-steady-state methods are not represented here per se, but they essentially represent a combination of the basic pseudo-steady-state and stroboscopic methods shown here

obtained using the data from the first experimental step, into the "ground plus light-induced" refinement model as a fixed entity, except that all atomic coordinates are normalised to the unit-cell parameters of the "ground plus light-induced" state, to take into account their slight change that is expected on account of a small perturbation in the overall molecular environment from having two similar but not exact molecules present rather than one type. A new scale factor is then refined against the "ground plus light-induced" data and a "photodifference" map thence obtained: this map is essentially a Fourier-difference map that reveals the photo-induced structure exclusively, because account of the ground-state structure has already been taken through the normalised fixed coordinates (see, for example, Figure 2-5). The atomic positions of the photo-induced structure are therefore locatable from this map using standard procedures for interpreting Fourier-difference maps, and are refined to produce the final combined ground-plus-photo-induced model. The occupancy factor of all atoms in each component should be refined as a common factor with both factors summing to 100% – exactly as one would model molecular disorder – so that one can realise the fraction of photo-conversion achieved: that value is the occupancy factor of the light-induced structure.

As steady-state methods rely on the light-induced effect in a single-crystal sample lasting for the duration of collection of X-ray diffraction data that is sufficiently

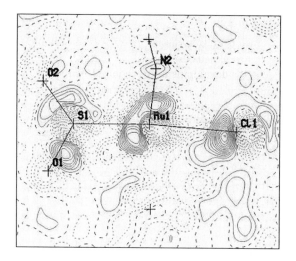

Figure 2-5. A "photo-difference" map showing the ground state (already modelled) depicted by the black lines together with the electron density associated with the light-induced $[Ru(SO_2)(NH_2)_4Cl]Cl$ complex [10, 11]. In this example, SO_2 is the photo-active ligand, undergoing a η^1-SO_2 (end-on) to η^2-SO_2 (side-bound) photoisomerisation. The sulfur atom and one oxygen of the η^2-SO_2 bound ligand are evident in this Fourier-difference map as the green feature and more diffuse green area, respectively, on the left of the figure (the other oxygen is not visible here as it lies out of the plane shown) (see color plate section)

nearly complete to achieve a full structural determination, the intrinsic minimum light-induced lifetime of a sample that can be probed with steady-state methods depends critically on the rate at which data can be collected. In turn, this rate is inherently dependent upon several factors: (i) the crystal symmetry of the sample. The greater is the crystal symmetry, the more reflections that are symmetry equivalents are afforded, and so smaller volumes of reciprocal space must be covered to obtain a complete data set of unique (not symmetry-equivalent) reflections compared with a sample with lower crystal symmetry, e.g. a complete data set of unique reflections requires one quadrant of reciprocal space for an orthorhombic sample whereas a hemisphere is required for a monoclinic crystal. The smaller the area of reciprocal space that must be covered, the more rapid the data collection through the smaller total number of frames of data that must be collected, as the diffractometer must move through fewer angular steps. (ii) Detector coverage, sensitivity and read-out duration. It is generally agreed that without the recent advances in the technology of CCD area detectors for X-ray diffractometers research in this field would not be viable as the rate of data collection is so critical to these experiments. Developments in detector design are continuing rapidly. Detectors with enlarged active areas are being developed that will afford increased coverage of reciprocal space per unit time. Strategies to increase the sensitivity of detector response are being investigated; achieving greater sensitivity inevitably results in more rapid collection of data and accessibility to expanded data (previously too weak to observe above the detector background). The duration of detector read-out is a major draw-back of area-detector technology as not only is it almost invariably the limiting factor of the rate of data collection but also it is so by a large margin. Substantial improvements in this area of detector development are required, without loss of detector sensitivity, before samples with much smaller intrinsic light-induced lifetimes than the present minimum can be studied using steady-state methods. (iii) X-ray diffraction intensity. The controlling factors here are the X-ray source and the size of the sample crystal. Because the brightest X-ray source inevitably affords the most intense diffraction practicable, one should consider performing the experiment at a synchrotron source if intensity from a laboratory source (either a sealed tube or, more intense, a rotating anode) be insufficient. Using a larger crystal also yields more intense diffraction, but one must ensure that the crystal is not so large that it becomes impenetrable to light. Given that the depth of optical penetration of a sample can be small – typically of order μm - one must generally compromise strongly the size of crystal to provide intensity satisfactory to ensure successful photo-conversion, without which the experiment is rendered untenable. Those experiments in which restrictions of crystal size to μm are dictated optically are viable only at synchrotron sources. (iv) X-ray wavelength considerations. Of two competing factors, a small wavelength is desirable as this effect improves the resolution of data collected to a given maximum Bragg angle, 2 θ, whereas the smaller is the X-ray wavelength, the greater is the rate that the X-ray atomic form factor decays as a function of 2 θ for a given element. Given the importance of resolution in these experiments due to the subtle structural changes sought, one should aim for as small a wavelength as avoids diminishing the X-ray form factor such that the X-ray intensity falls beyond feasibility. These arguments assume a completely free choice of X-ray wavelength band, but that condition is available only at a synchrotron source. In a laboratory one is typically restricted to only one or two wavelengths, depending on the X-ray sealed-tube sources available, commonly Mo ($\lambda = 0.71$ Å) or Cu ($\lambda = 1.54$ Å).

Based on the above considerations, in favourable cases – a large and strongly diffracting crystal of a sample with great depth of optical penetration, high symmetry of a crystal, and collecting data at a synchrotron source using a moderately small wavelength – the present minimum light-induced lifetime for experiments in a steady state is estimated to be 20 to 30 min.

Pseudo-Steady-State Methods

Below this steady-state limit of time scale, one can employ *pseudo-steady-state* methods, down to a light-induced lifetime of a sample of order ms. These methods allow a photo-induced state to be activated and maintained on continuously pumping the sample with an optical source that has a pulse frequency (rate of repetition) that repeats more rapidly than the photo-induced lifetime. After an initial cycling period of optical pumping, photo-saturation is achieved, which is the maximum possible fraction of photo-conversion within a sample, for the given optical pumping source.

The sample remains in this "pseudo-steady-state" throughout the experiment as long as the optical source remains pumping the sample at its initial pulse rate. As with the steady-state method, there are several distinct steps to the experiment. In its simplest form, ground-state data of the sample can be first collected conventionally. Optical pumping then commences and the same routine of data collection is repeated when the sample has attained its pseudo-steady state. Once the data have been collected, these can be analysed in a manner identical to that described above for steady-state methods.

For experiments of these types one must consider carefully the effects of laser heating on the sample during the data collection as thermal effects might themselves effect a structural perturbation. Although such a change would be generally small, typically manifesting itself predominantly within the anisotropic displacement parameters of atoms in the structure, the light-induced effect sought might also be subtle. If unaccounted, laser heating effects might therefore readily disguise the light-induced effect sought, or, worse, allow one to attribute erroneously a structural distortion to a photo-effect.

Measurements of heat capacity of a sample might serve for the calculation of the temperature increase expected due to laser heating. One further data collection, post optical-pumping, could then be performed at a sample temperature corresponding to the sum of the temperature of the original data collection and the expected increase in temperature. Any structural differences due to laser heating might therefore be ascertained on comparing the structures relating to the original ground-state experiment and this final one, thus to distinguish any photolytic effects. The accuracy of such temperature calculations might, however, be compromised by the fact that the efficiency of the sample environment to dispel the laser heat away from the crystal is unknown, as it depends on many factors. For example, the temperature gradient between the laser and the sample environment, the level of thermal contact, flow, and heat capacity of the crystal-cooling gas are important considerations. The associated changes in unit-cell parameters can be taken into account on normalisation in a way similar to that described above for the steady-state method.

An alternative method of data collection that circumvents some complications surrounding effects of laser heating involves collecting each frame of data from the ground and photo-induced state in an alternating fashion. Here, the standard strategy of data collection is arranged to optimise the angular coverage of reciprocal space, but, at each angular setting, a data frame is collected twice, first without optical pumping, followed by a delay while the pseudo-steady state is obtained on optical pumping, and then again while the sample remains in its pseudo-steady state. A mechanism of electronic timing is established between the laser shutter and the software to control the diffractometer such that this process can be repeated automatically. This procedure precludes the possibility of a gradual accumulation of laser-heating effects. Given the long duration to read out a data frame from a detector, there is negligible loss of experimental time by this method, if the delay to await the pseudo-steady state can be programmed to act in parallel operation to the detector read-out command. At present, such parallel operation is not standard in most diffractometer software, although it is practicable.

Data-interleaving strategies are also highly advantageous to minimise the danger of sample decay. In photo-induced crystallographic experiments, crystal fracture is a common occurrence due to the severe conditions to which the crystal samples are exposed: laser heating, sample cooling (perhaps to liquid-helium-based temperatures), and intense X-ray sources (commonly synchrotron radiation). Moreover, if the depth of optical penetration of the laser is such that the laser is unable to penetrate through the entire crystal sample, local heating effects within the crystal accumulate that can be dissipated only with a crystal explosion. As mentioned earlier, if too much movement in the crystal lattice occurs as a result of the photo-conversion, this effect can also cause crystal fracture. If the crystal were to fracture halfway through an experiment in which this alternating "light-dark" strategy was in use, at least half the data collected could be analysed, but there is little use in analysing a full ground-state data collection together with data collection from the light-induced crystal for ten minutes, at which point the crystal fractured, as insufficient data would be available to perform a "light-dark" comparative analysis of data. This alternating strategy allows also for the effects of gradual sample decay to be corrected for much more easily than if the data had been collected on the ground state exclusively first, because such decay occurs at the same rate in collecting both data using interleaving data frames. Other time-dependent systematic effects can likewise be eradicated, particularly the effects of synchrotron beam decay if X-rays from this source are used for the experiment. One additional feature of the design of such an experiment that removes systematic effects is the collection of ground and "ground+light-induced" data on the same detector image frame, by shifting the detector by a small and known amount between the collection of data from each of the two states. Systematic errors such as varying ratios of signal to noise between frames are thereby avoided. Such an approach has been demonstrated successfully using image-plate detectors [12], but such a method is suitable only where diffracted spots present on the image are not so many that a detector shift is impossible without reflections from one of these images overlapping with some of those from the second image.

Irrespective of the type of experiment of X-ray diffraction undertaken on a single crystal in a pseudo-steady state with monochromatic light, the necessity to fire the laser onto the sample at varied crystal orientations, owing to the required angular movement of the goniometer for each data frame, can cause anisotropic sample photo-conversion, because a laser beam is inherently polarised. Unless one wishes to exploit this polarisation in some way, it is generally considered best to detune the polarisation to create an incident laser beam with circularly polarised light. In a steady-state experiment, polarisation effects generally cause no problem as they can be averted by either optically pumping a stationary crystal between data collections or, as is common practice, continuously rotating the sample during the optical-pumping period.

The millisecond lower limit of a photo-active lifetime of a sample for pseudo-steady-state experiments is dictated primarily by the maximum rates of repetition of pulses available in most commercial lasers, these being typically of order kHz. Such repetition rates derive from diode lasers, and lasers are more commonly pumped with flash-lamps that operate typically at a repetition rate 10 Hz.

Below the millisecond photo-active lifetime regime, one must collect data in a stroboscopic pump-probe manner in which each frame of data derives from the integration of multiple pump-probe excitations. This condition is distinct from the stroboscopic pump-probe pseudo-steady-state experimental method just described, as each frame of data in that case emanated from a single continuous X-ray (probe) exposure of the pseudo-steady state being captured by the (pump) laser over the period of data acquisition.

A pulsed X-ray source is therefore required to study all processes driven by light with a sub-millisecond photo-active lifetime. If one is to realise the structural perturbations that result from the technologically important high-speed electrical and optical properties described in the second section, one needs an experimental method that provides access to structures enduring ms to ps.

Pulsed X-Rays via a Mechanical Chopper

For samples with photo-active lifetimes in the time regime μs to ms, a mechanical chopper can be used to afford X-ray pulses. The length of each X-ray pulse is designed to be in accordance with the photo-active lifetime of the sample. Ozawa et al [13] gave an example of one such chopper design, in which a rotating chopper operates a wheel comprising slits of varying width radially (see Figure 2-6).

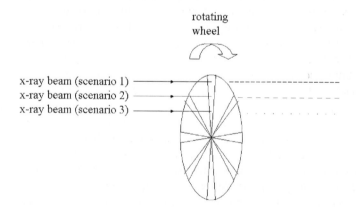

Figure 2-6. A schematic diagram of a mechanical chopper used to create pulses of X-rays of varying widths, according to the position of the chopper relative to the incoming x-ray beam. The chopper axle is moved upwards or downwards to afford the varied pulse widths. Three options are illustrated

To access the greatest possible range of X-ray pulse lengths, this wheel is removable from the chopper such that one of several wheels, each with a separate range of slit apertures, can be chosen for a given experiment. An adaptation of this chopper with a greater speed and rate of repetition has been developed [14].

An alternative chopper design that requires no removal of a wheel has been developed by Cole et al [15]. This chopper has two blades comprising two circular discs, each of identical design; the disc interior (Figure 2-7a) is constructed from aluminium, and the outer rim (Figure 2-7b) of steel. This bimetallic combination is required to optimise lightness (aluminium) to maximise the possible speed of revolution of the chopper, whilst providing a (steel) rim that blocks X-rays in the outer part of the rim. The two disc components for each disc are glued together while ensuring careful disc balancing. The two discs are aligned parallel to each other, and positioned perpendicular to the incident continuous beam of X-rays such that the X-rays meet each chopper disc at its outer perimeter. Half the outer perimeter on the rim of each disc is etched away, thus permitting X-rays to pass freely through a half revolution of each disc. Each disc is held individually in an aluminium protective casing and coupled to a motor that allows each disc to revolve independently at a frequency 10–30 Hz. The mount base and one open casing with a motor attached are shown in Figure 2-7c. Figure 2-3c shows the casing of one disc. A hole at the top of each casing allows the X-rays to enter them (Figure 2-7d). The two discs operate together to achieve a variable length of X-ray pulse. This variability is achieved on rotating both discs in the same sense to each other and at the same frequency, but allowing the relative positions of the open (outer) part of each rim to be adjusted independently. For example, if the relative positions of each disc are almost identical, a large X-ray aperture (longer X-ray pulse) results (Figure 2-7e); conversely, if the relative positions of each disc are almost counter to each other, a small X-ray aperture per disc revolution (short X-ray pulse) is obtained (see Figure 2-7g). The relative positions of the two discs might also lie anywhere between these two extremes, which are 180° apart (e.g. Figure 2-7f), thus yielding a continuous range of X-ray pulses via this design. The limits of X-ray pulses in this range are determined by the speed of revolution of the chopper and the circumference of each disc. The design of this particular chopper allows synchrotron-based time resolution from ms to μs.

Field programmable gate array and timing logic allow the relative positions of each disc to be adjusted independently and in a controllable way. A light-emitting diode and photodiode are positioned, one on either side of each disc casing, pre-X-ray, to monitor and to maintain the relative position of each disc. Post-X-ray, there exists a light-emitting diode on the first casing, but no photodiode. The photodiode detector lies on the second disc, which in contrast houses no light-emitting diode. The comparison of the pre-X-ray signal with that from this post-X-ray arrangement allows a measure of the jitter in the desired X-ray pulse timing. Figure 2-7h shows the holes in each casing in which the light-emitting diodes and photodiodes are located, relative to the (larger central) hole through which the incoming X-ray beam passes.

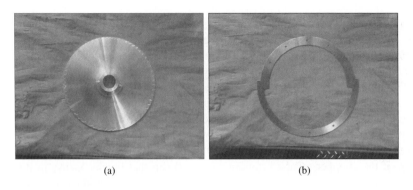

(a)   (b)

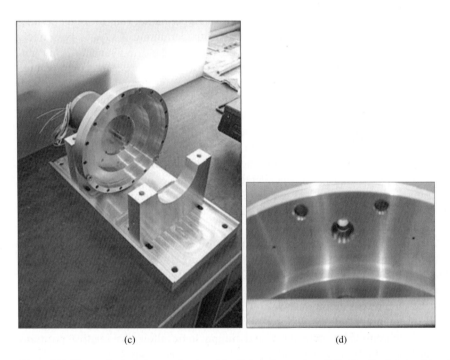

(c)   (d)

Figure 2-7. The mechanical chopper developed by Cole and coworkers [15]. The disc interior (a) and exterior (b) are shown together with the wheel casing and mount (c) and holes for the X-ray and light-emitting diodes (LED) (d). Parts (e), (f) and (g) resemble X-ray pulse lengths that arise from the relative positioning of wheels in phase (e), one rotated 90° to the other (f), and completely out of phase to each other (g). Part (h) is a schematic diagram of the chopper jitter and timing control via LED lying in holes coloured black, and photodiodes lying behind the small hollow hole post-X-ray (photodiodes are behind each LED pre-X-ray)

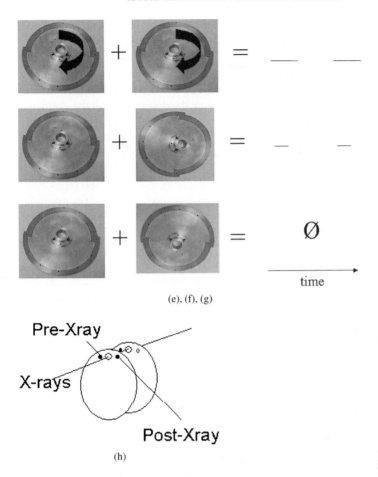

(e), (f), (g)

Pre-Xray

X-rays

Post-Xray

(h)

Figure 2-7. (*Continued*)

For each chopper design featured above, the rate of repetition of the X-ray pulses is matched with that of the pulsed laser, and the timing of the opening of the chopper is synchronised to the shutter opening of the laser such that both the optical and X-ray pulses are either "on" or "off" at a given instant. The "on" periods thereby correspond to the structure of the $x\%$ photo-converted state combined with that of the $100 - x\%$ non-photo-absorbed ground state, whereas the "off" periods relate to the fully recovered ground state. Each frame of data thus acquires Bragg intensity exclusively from the "on" periods; during the "off" periods only background is accrued. The preclusion of any contamination of Bragg intensity due to the exclusive ground-state structure avoids heavy dilution of the contrast that is sought between the ground and light-induced structures: such contamination would present an insurmountable problem. Dark time in a significant amount is introduced in this experimental method. Consequently, although a laboratory X-ray source might be used for many steady-state experiments (even though the greater flux of synchrotron radiation would invariably increase the contrast) synchrotron radiation is mandatory for experiments of these types in which X-ray pulses are created – the much greater X-ray flux available at a synchrotron is required to compensate for the highly depleted time-averaged flux (low "duty-cycle") caused on pulsing the X-rays.

In terms of strategies to collect data, interleaving the data frames for the exclusive ground-state and "ground+light-induced" data sets is important when employing this method, for reasons akin to those described for the pseudo-steady-state pump-probe method, and because such a process allows also the implementation of time-delay strategies that minimise heating problems. Unlike in pseudo-steady-state experiments, the pulsed nature of the X-rays in these experiments affords the freedom to offset temporally the laser and X-ray pulses from one another. Hence, for example, an additional reference data set can be collected in an interleaved fashion, with the collection of the ground and light-activated data sets, which resemble the ground state that has been affected by laser heating but not laser excitation. This effect is achieved by offsetting the timing of the laser and X-ray pulses such that data are not collected (X-ray shutter closed) until the light-induced structure has had time to relax to its ground state, but before it has had a chance to dissipate laser heating effects. Exploiting offsets in laser and X-ray synchronisation can be useful in other ways: for example, one can use it for time-slicing the light-induced state to allow one to follow the evolution of a light-induced state, by observing variations of Bragg intensities of selected signals.

The measurement and analysis of the data proceeds in the same manner as that described for the stroboscopic pseudo-steady-state experiments with the exception that the fraction of photo-conversion in a sample is small. As the duty cycle is small when pulsed X-rays are used, one must collect a frame of data for a considerable duration to afford Bragg signals of adequate intensity. This condition contrasts with a steady-state experiment in which one has generally the luxury of collecting data with satisfactory statistics because, here, the detector read-out time, rather than the rate of data collection, is the time-limiting factor of an experiment. Consequently, owing to practical time constraints, one must generally compromise the accuracy of the Bragg intensities over a realistic time frame to perform an experiment that utilises X-ray pulses. Such a compromise most affects the viability of observing the results sought where the fraction of photo-conversion is small. In these cases, it is considered better to analyse the *difference* between two intensities collected in rapid succession, as this condition yields results more accurate than the parent intensities themselves due to the avoidance of various systematic effects. This effect has been found in similar crystallographic difference experiments in which the differences sought are similarly small [16]. In such cases, the reflection intensities are derived from the response ratio, η, according to

$$\eta_{hkl} = \frac{I_{on}(hkl) - I_{off}(hkl)}{I_{off}(hkl)} = \frac{I_{on}(hkl)}{I_{off}(hkl)} - 1$$

in which I_{on} is the intensity of the light-activated state and I_{off} represents the Bragg intensity of the conventional ground-state structure factor. The response ratios can be refined by least-squares methods using derivative expressions of η [13].

Pulsed X-Rays via the Temporal Structure of a Synchrotron

For samples with photo-active lifetimes from μs to ps, the temporal structure of a synchrotron is exploited to realise matching μs-ps X-ray pulses. This temporal structure derives from the fact that the electrons that are accelerated around a synchrotron ring, before being agitated magnetically to emit X-rays, do so in discreet bunches. The "bunch width" is described in terms of the duration that the full length of an electron bunch takes to pass a given point in the synchrotron, a time scale of order ns to ps. Accelerator physics can be used to harness these bunches individually to provide pulsed X-rays on a ns-ps timescale, or they can be grouped together with other bunches to create a train of X-ray pulses of, in principle, any time scale up to the limiting orbit speed of the synchrotron (of order μs) that would correspond to the head of this train of pulses joining its own tail in the synchrotron circuit. Thus, X-ray pulses on a timescale from μs to ps are available. In practice, there are restrictions on the exact time scales available as a synchrotron operates in one from a small selection of "bunch modes", because the X-ray source serves simultaneously many instruments, some of which require no such time structure but, in contrast, require as great an X-ray intensity (thus high duty cycle) as possible. Certain bunch modes are thereby available in a synchrotron at various times during the year. As an illustration, Figure 2-8 shows the 16-bunch mode that is available at the ESRF, Grenoble, France. One can fine-tune the time-frame of the X-ray pulse

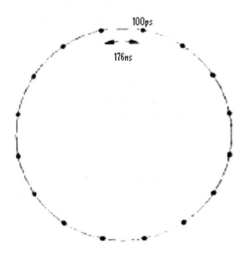

Figure 2-8. One of the (16-bunch) timestructures available in the synchrotron at the ESRF, Grenoble, France

desired for a given experiment by combining this inherent timing with a chopper that isolates one of these bunches from the 16 available. One such chopper, used on the ID9 beamline at the ESRF, is described by Bourgeois et al [17].

To utilise the pump-probe method for such experiments, one can use a laser of ns, ps or fs pulse width, any of which is commercially available, with high-speed electronics such that the required time-gated synchronisation between X-ray and laser pulses for the experiment is achieved.

The much shorter X-ray pulses, used in these experiments, result in such a low duty-cycle that obtaining sufficient X-ray flux is challenging at even the most powerful synchrotrons. Only in year 1998 were the first results achieving nanosecond temporal resolution published, nearly two decades after the first feasibility studies of the technique. The first attempt at time-gating a synchrotron X-ray pulse to a laser pulse was conducted in 1980–1 with an EXAFS experiment, [18] but all attempts failed primarily due to instabilities of the beam accelerator. A reinvestigation of this method in 1984 achieved success on the millisecond time scale, again employing EXAFS techniques [19]. Significant improvements in synchrotron technology made during these two decades were required to enable these ideas to be rejuvenated with success and applied to diffraction; even now there are only a few publications of such results arising from nanosecond temporally resolved crystallography of photo-active species (see §5). Of these studies most have comprised the characterisation of transient biological species on the nanosecond time scale, and have utilised Laue diffraction techniques (cf. Bragg's law: varying λ, fixed θ) rather than monochromatic oscillatorily based methods (cf. Bragg's law: varying θ, fixed λ), which are the principal subject of this review. Laue diffraction has a great advantage of rapid acquisition of data, as the Bragg condition is met many times simultaneously when λ is the variable parameter.

Although oscillatory methods afford data much more slowly than Laue diffraction, they have the intrinsic advantage over Laue diffraction that they preclude the overlap of reflections, as one can assume that the Bragg condition is met only once at a given time in experiments with diffraction of monochromatic X-rays. There is thus no need to deconvolve reflections that commonly hamper the resolution of the data in Laue diffraction, in which reflection overlap is prevalent because the Bragg condition is met many times simultaneously. As the light-induced chemical changes in these experiments are generally on a molecular scale and are typically subtle, great atomic resolution is a prerequisite, thus making Laue diffraction problematic. In the aforementioned biological studies, distinct areas of electron density, which appear in Fourier-difference maps, upon creation of the transient species were revealed rather than individual perturbations on an atomic scale. In addition, instrumentally, Laue diffraction requires X-rays in a white beam that comprises X-ray wavelengths in a continuous range acting simultaneously on the sample. All Laue-based experiments are therefore restricted instrumentally to synchrotron sources. Large strides in the development of Laue-based X-ray crystallography have been made [20], and one can likely look forward to an exciting future in this regard. Furthermore, Laue diffraction is at present the only viable method to use in experiments in which the crystals deteriorate quickly upon photo-activation.

COMPLEMENTARY MEASUREMENTS IN LASER SPECTROSCOPY

Thus far, we have considered primarily only the X-ray requirements for these experiments. Entirely underpinning the diffraction experiments are many laser spectral data, which should be a prerequisite of a diffraction experiment on light-activated species. Without these data one is unable to decide whether a sample is viable for such an experiment, and if so, these spectral data are necessary so that one can be informed about the required laser and X-ray conditions for the diffraction experiment.

The photophysical properties of a compound are generally sensitive to the sample phase; it is therefore crucial that the parameters to match and to optimise X-ray diffraction characteristics on photo-active species derive from measurements based on crystalline samples, to the extent practicable. Before such an X-ray diffraction experiment is performed, each sample candidate should therefore be screened by conducting appropriate laser experiments to establish the spectral parameters associated with the nature of the photo-active structure sought. This screening typically comprises laser experiments, each providing important information *a priori* for the subsequent X-ray diffraction experiments:

(i) *The ultraviolet and visible absorption spectrum in both ground and photoexcited states.* To calculate accurately the depth of optical penetration of light from a laser through a single crystal, absorption spectra must be recorded for both ground and photo-activated states, at a known sample concentration, c, such that the extinction coefficient, ε, can be calculated in each case.

cf. $1/\mu = 1/\varepsilon\ c$, in which $1/\mu$ is the depth of optical penetration ($\mu = $ absorption coefficient)

Regarding sample concentration, c, solid-state measurements are preferable to solution-based data, particularly to avoid solvent effects. Although experimentally more difficult, UV/vis measurements, taken while the compound is in its photo-activated state, are also important as the absorption of the photo-activated state might differ from that of the ground state, thence effecting a varied depth of optical penetration.

(ii) *The temporally resolved emission spectrum of the compound in the solid state at various excitation wavelengths.* These data show which wavelength range should be utilised to achieve maximum photo-conversion and the appropriate timing window for observation of this photo-induced state independent of other possible states.

(iii) *The lifetime of the photo-induced state measured in the crystalline phase, at the temperature at which the X-ray diffraction experiment is to be performed.* This lifetime must be known exactly, as the choice of experimental method for X-ray diffraction derives entirely from the lifetime of the sample. Solid-state requirements are important as solvent effects alter radically the photo-active lifetime of a sample. The temperature aspects are similarly important: decreasing the temperature, for example, can cause the onset of quantum-mechanical stabilisation of the photo-induced state. In more extreme cases, increasing or decreasing the sample temperature can effect a complete change of the structural nature of the dominant emissive state, e.g. in a luminescent material that commonly converts from metal-to-ligand to intraligand charge transfer on a decrease of temperature. From a purely crystallographic point of view, the temperature should be as low as possible, to minimise thermal effects, as libration effects might occlude structural perturbations due to the photo-induced state. A compromise must sometimes be made between laser experiments and crystallographic ideals; in any case, conventional laboratory-based X-ray diffraction should be performed at various temperatures, as part of the testing before a photo-induced X-ray diffraction experiment, to assess the ground-state structure including any libration effects.

(iv) *Tests of laser ablation as a function of laser wavelength, output power, repetition rate, and crystal size.* To ensure that the crystal can withstand laser irradiation for at least the expected duration of the X-ray diffraction experiment, a laser is fired at a crystal sample for various known durations. The laser parameters used are also systematically varied such that one finds the optimal laser conditions that provide a maximal photo-conversion in the X-ray diffraction experiment without destroying the crystal.

(v) *Quantum yield at the proposed wavelength of laser excitation in the X-ray diffraction experiment.* This parameter provides a quantifiable measure of the population level of the photo-converted state that is independent of that obtained through the structural refinement of the X-ray diffraction data.

In practice, not all these laser tests are viable with available resources, feasible preparatory time scales, or due to physical or instrumental restrictions, but at least the basics of tests (i)–(iv) described above are mandatory in some form as the X-ray diffraction parameters derive directly from these measurements. For a comprehensive description of laser spectroscopy experiments, one is referred to Lakowicz [21].

STRUCTURAL RESULTS OBTAINED SO FAR

The chronology of instrumental developments in this area of chemical crystallography has reflected directly the relative levels of these challenges, the structure of metastable photo-induced states being first realised in 1991 [22], whereas the first report of a temporally resolved experiment with X-ray crystallography of a small molecule was published only in 2002 [23]. The chemical literature is therefore some-what skewed at present towards results on materials with enduring photo-induced states as this section illustrates. With continuing developments in crystallographic instrumentation, the proportion of diffraction experiments reported on enduring versus transient photo-induced molecular species is likely to balance, given the myriad of important applications dependent upon ephemeral light-induced structural perturbations.

Although the concern of this review is X-ray diffraction of single crystals, it is pertinent to remark that the first major diffraction experiments on light-induced structural changes arose from work on glasses. Most noteworthy within this con-text is the extensive work on the photodarkening effects of chalcogenide glasses performed in the 1980s [4], using steady-state methods. The EXAFS community was also on the scene earlier than crystallography, for example in biology with the EXAFS-based work on the various geminate states of Myoglobin [24] and the temporally resolved contraction of the Pt-Pt distance in the $[Pt_2(POP)_4]^{4-}$ anion (POP = pyrophosphate, $[H_2P_2O_5]^{2-}$) [25].

The first reported crystallographic investigation in which *steady-state* photo-induced methods were applied involved a putative light-induced structural change in sodium nitroprusside (SNP), $Na_2[Fe(CN)_5(NO)] \cdot 2H_2O$, undertaken by Woike and coworkers [22, 26]. Neutron diffraction was used to reveal the structures of two metastable states, MS_I and MS_{II}, known to exist in this compound under par-ticular conditions of light exposure (wavelength and polarisation direction) and temperature. The structure obtained for MS_I showed small but significant alter-ations in the Fe—N, Fe—C and N—O bonds along the *trans*-O-N-Fe-CN axis, relative to the ground state, also measured. Such changes are also possibly reflected in the anisotropic displacement parameters of the oxygen and nitrogen atoms (see Table 2b in ref. [24]). The structure of MS_{II} indicated more marginal perturba-tions of the lengths of Fe—N and N—O bonds within the same molecular axis. Part of the incentive behind this structural work was the declared potential of SNP for application as an optical-storage device. Given that the ground state and two light-induced states of SNP display varied optical properties [22], the structural characterisation of each state was pertinent from a materials-centred perspective as well as being innovative in an academic sense. This structural work led Coppens and coworkers to initiate X-ray diffraction experiments on SNP [27]; their results revealed that the MS_I state undergoes a light-induced structural perturbation along the *trans*-NC-Fe-N-O axis, although their final interpretation of the data indicates ligand photoisomerisation Fe—NO → Fe—ON rather than a simple variation of bond length. Their structural characterisation of MS_{II} yielded a further photoisomerisa-tion corresponding to the formation of a disordered η^2-bound NO moiety ligated to the Fe ion.

These results became the first in a series of pioneering X-ray diffraction studies of light-induced structural alterations, these being primarily of materials under-going metastable ligand photoisomerisation. Complexes containing η^2-N_2-, NO- and SO_2-based ligands became the subsequent photo-active focus of this work, as described further in their comprehensive review of ligand photoisomerisation [28] and other work [29]. Such investigations have been described to have potential for application to sunscreens and relevance to the understanding of mechanisms in NO- and N_2-based biological processes.

In our work, we have used one of these photoisomers, $[Ru(SO_2)(NH_3)_4(H_2O)]$. [tosylate]$_2$, as a test material in the development of photo-induced X-ray crystallographic techniques at the synchrotron radiation source (SRS) at Daresbury UK. Coppens and coworkers had shown previously that the η^2-SO_2 group photo-isomerised to the metastable η^1-SO_2 species upon application of light at 355 nm, with the crystal held at 100 K [10, 11]. During our tests, we first reproduced this result, and then varied several key experimental parameters to observe and to understand better their effect on the resultant refinement of the light-activated crystal structure.

Noteworthy findings are that (a) one obtains a smaller population of the light-induced crystal structure when a monochromatic (laser) lamp is used relative to a broad-band lamp; this effect is expected as one would envisage the possibility of probing several electronic states with a broad-band source whereas the access to electronic states via a laser is much more restrictive; the use of a broad-band lamp still produces the desired photo-isomer without significant contamination of another light-induced species. Furthermore, this particular sample is so sensitive to light that the background fluorescent lighting in the experimental hutch of the crystallography beamline at SRS had to be blocked as it photo-activated the crystal of its own accord. (b) The results depend strongly on the temperature of the crystal during data collection; the effect is severe for this compound as data collected at 13 K revealed a photo-isomer of a novel type that involves an end-on η^1-O bound Ru-OSO coordination [30]. This example is the first of this type in ruthenium-based organometallic coordination chemistry whose structure has been published, as determined via the Cambridge Structural Database [31]. At temperatures above those where Coppens and coworkers were investigating (100 K), the photo-isomerisation remains metastable to about 240 K, above which its lifetime deteriorates progressively, attaining a lifetime of order microseconds near 295 K. Collections of pseudo-steady-state data on this compound in a series were undertaken in the temperature range 250–280 K to investigate the associated progressive decrease in population of light-activated molecules as a function of temperature, which was revealed to be nearly linear. Measurements at multiple temperatures revealed also that the integrity of the crystal of the compound was most stable when flash-cooled or flash-warmed, i.e. freezing in or heat-shocking each photo-isomer, whilst incremental warming or cooling generally destroyed the crystal. This effect presumably stems from lattice strain being overcome via a sudden variation of the energy of the lattice, which allows one to transcend an otherwise insurmountable barrier of activation energy to reveal a more energetically stable structural form. Correspondingly, a custom-built flash-warming and cooling auxiliary device was constructed for this work. (c) The many refinements undertaken during these tests also revealed satisfactory reproducibility of the structural results. All these noteworthy findings are discussed in detail in a forthcoming paper [32].

Having conducted extensive tests on this compound with monochromatic X-ray irradiation, we sought to investigate the relative merits of monochromatic versus Laue-based photo-crystallographic experiments, by performing analogous steady-state Laue-based X-ray diffraction measurements on this compound. We conducted this work on ID9 at the ESRF, Grenoble, France, in collaboration with Wulff and coworkers [33], following encouraging preliminary tests conducted on Station 7.6 at Daresbury [34]. The profile of the ID9 X-ray beam was optimised to emulate a configuration of bending-magnet type that possessed a suitable *pink-beam* wave-length band, necessary for Laue work on a small molecule. This configuration was achieved by tapering the U46 undulator available to ID9 so as to expand the har-monics to create a pseudo-continuous and broad-bandwidth profile. The millisecond shutter of the ID9 beamline was used to decrease the heavy beam load onto the sample by allowing five 1 ms pulses per diffraction image.

The crystal sample was held stationary while images of three diffraction patterns were obtained, resembling (i) the ground state at 100 K, (ii) the light-activated state at 100 K, after exposure to broadband white light for 5 min; (iii) the ground state after irradiation (light-deactivated on flash warming to room temperature and flash cooling to 100 K using a custom-designed cryostream shutter).

Figure 2-9 shows a striking difference between images for the ground state (left) and light-activated state (middle). The nature of this difference indicates the expected $\eta^1 \rightarrow \eta^2$-$SO_2$ photoisomerisation: the diminution of diffraction seen as a function of resolution in the middle image represents disorder, as one should expect in this case as 10–20% of the molecules are being excited. The first and third images below are identical within experimental error, thus revealing that the sample returns wholly to its ground state in this experimental procedure, and that the sample has not decayed significantly because of exposure to X-rays or light.

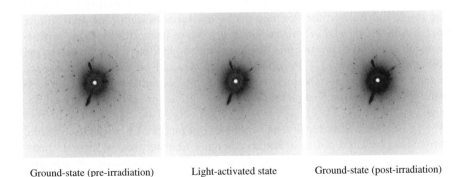

Ground-state (pre-irradiation)   Light-activated state   Ground-state (post-irradiation)

Figure 2-9. $\eta^1 \rightarrow \eta^2$-$SO_2$ photoisomerisation of $[Ru(SO_2)(NH_2)_4(H_2O)].[Tosylate]_2$ by Laue diffraction

That one can use both monochromatic and Laue methods, each to independently provide firm evidence for the photo-structural change in this test compound, is highly encouraging. One would expect that monochromatic methods are likely to yield real-space resolution better than Laue-based results, but this test shows that it is viable to observe photo-structural differences with Laue diffraction on small molecules *and* to undertake full data collections much more quickly than monochromatic methods would allow. Given that the Laue method is the only viable method in some cases, it is important to pursue the development of Laue diffraction in this area concurrent with the more major monochromatic developments.

In parallel with this work, other photoisomerisation experiments have been pursued by Ohashi and coworkers. The impetus behind their work has been photo-reaction and a mechanistic enquiry of reaction intermediates. X-ray diffraction of single crystals has centred about analysing crystalline-state photo-reactions before and after irradiation, and has involved proton transfer [35], formation of a hetero-cyclic ring [36], asymmetry generation and chiral inversion [37], and formation of a radical pair [38]. Many such photo-reactions have been shown to be reversible upon heating, although this condition much depends on the level of conformational change afforded in the photo-reaction, which might be substantial: if the change is too large, crystal fracture occurs as the crystal lattice can withstand molecular movement up to only a certain level before losing its integrity.

As an integral part of this work, the structural manifestations of some conforma-tional changes have afforded useful insight into the topological nature of the reaction cavity in reaction types in a given series. In some cobaloxime complexes, among Ohashi's featured compounds, the nature of reaction cavities has been further probed on photoismerisation of cobaloxime complexes as guests within cyclohexylamine hosts [39]. Coppens and coworkers have also synthesized host-guest complexes in which the guest is a photoactive species; the primary interest is to find a suitable method to "dilute" the photoactive species so that the crystal lattice is stabilised even in the case of large structural changes [40], so that one can augment the maximum fraction of excitation possible above which a crystal fractures.

Ohashi and coworkers deduced mechanistic information about the reaction in some cases, and isolated the structure of the intermediate itself [36]. In a similar vein, the structure of photo-excited triplet states can be deduced with X-ray diffrac-tion, using stabilising cryo-trapping methods, with supporting infrared spectra and theoretical calculations [41]. Reaction intermediates and photo-excited states are generally too ephemeral to be harnessed in this fashion, but developments in tem-porally resolved stroboscopic pump-probe X-ray diffraction methods will facilitate this goal.

The evident feasibility for determinations of light-induced metastable structure, hitherto described, has led to a further emerging field in photo-structural chem-istry: that concerning the characterisation of spin-crossing transitions in magnetic materials. Spiering and coworkers reported the first direct observation of the well known light-induced excited-spin-state-transition (LIESST) effect by photo-induced single-crystal X-ray diffraction [42]; the subject of this work was the complex

[Fe(mtz)$_6$](BF$_4$)$_2$, (mtz = methyltetrazole), which undergoes a photo-induced transition from low spin to high spin (LS $\rightarrow$ HS) at low temperature, and which is sufficiently stable at 10 K for data to be collected to determine the crystal structure corresponding to both the HS state, after irradiation ($\lambda = 514$ nm), and the LS state, obtained before photo-irradiation. The LIESST effect is evident via an extension of the Fe-N bond by $\sim$0.2 Å which corresponds to the electronic transition $^1A_1$ (LS) $\rightarrow$ $^5T_2$ (HS). Complementary measurements of Mössbauer spectra indicate that complete photo-conversion occurs. Other steady-state single-crystal X-ray-diffraction experiments have subsequently realised crystal structures of similarly photo-trapped Fe-containing complexes in a high-spin state, notably Fe(phen)$_2$(NCS)$_2$ (phen = 1,10-phenanthroline) [43] and Fe-based (pyrazol-1-yl)pyridine derivatives [44–46]. In these cases, changes $\sim$0.2 Å in Fe–N bond lengths again afford evidence of the spin transition.

Even more direct evidence for LIESST has been afforded in the pioneering work by Gillon and coworkers in which single-crystal polarised neutron diffraction has been exploited to observe the LIESST effect via the spatial distribution of the magnetisation density in [Fe(ptz)$_6$](BF$_4$)$_2$, (ptz = 1-propyltetrazole) [47]. The topological description of the magnetisation density around the iron position corresponds well to that which one would expect from an Fe^{2+} magnetic form factor. Moreover, the magnetic moment on the iron site was refined (to 4.05(7) μ_B) and compared with the theoretical value of the Fe^{2+} moment at saturation. Such analysis yields a direct and quantitative measure of the fraction of photo-conversion, which, in this case, was practically complete. The power of this technique is great as it stands to reveal unprecedented information about photo-induced magnetic effects that is key to the understanding of their physical properties. The great potential of bistable materials for memory devices and the great effort at present being exerted in industry to find materials for such applications reinforce the importance of this work.

In the microsecond time-resolved (non-steady-state) regime, thus far, there exist results on only a few complexes. The first recorded result concerns the anion Pt$_2$(pop)$_4$$^{4-}$ (pop = pyrophosphate, [H$_2$P$_2$O$_5$]$^{2-}$). The groups of Coppens and Ohashi have independently deduced significant light-induced contraction of the Pt–Pt bond [$\Delta = -0.23$ to $-0.28(9)$Å] in this ion according to single-crystal X-ray diffraction. Separate counter-ions were used: Coppens used [(Et$_4$N)$_4$H]$^{4+}$ whereas Ohashi used [N(C$_5$H$_{11}$)$_4$]$_2$(H)$_2$. Although the results are similar, one would thus expect minor variation due to the slightly varied molecular environments surrounding the anion in the crystal lattice. The two groups also used separate experimental techniques: Coppens utilised stroboscopic pump-probe methods that exploit the aforementioned mechanical-chopper design on the X3 beamline at the NSLS, Brookhaven, USA, with 50 μs resolution and at liquid-helium temperature [23]; Ohashi and coworkers used pseudo-steady-state methods, both in the laboratory [48] and on the BL02B1 beamline at the Spring-8 synchrotron, Japan [12]. The extent of Pt–Pt bond contraction in each case is identical within experimental error, and is consistent also with results from EXAFS, and electronic and optical spectral measurements [25, 49, 50] despite the use of distinct counter-ions. The results are all compatible with a rationalisation that the promotion of a Pt—Pt antibonding $d\sigma^*$ electron into a weakly bonding p-orbital is responsible for this structural change. The work by Coppens and coworkers shows that a concurrent rotation by 3° of the Pt—Pt axis in the plane that bisects two Pt—P vectors occurs in this photo-excitation. Subsequent complementary density-functional calculations have been performed [51]; their results are consistent with the experimental results, except that they show some discrepancy with the EXAFS results in terms of the associated Pt—P bond changes. This effect is unsurprising given the expected propensity of multiple scattering in the EXAFS spectra that might occlude the true Pt—P distance, and the challenging nature of DFT calculations to deduce M—M bond distances in directly bound third-row transition-metal excited states.

Given the highly developmental nature of such pump-probe diffraction experiments coupled with the subtlety of the structural variations sought, obtaining consistency with analogous spectral and theoretical results is important for a quantitative assessment of the reliability of the results and thus the progress of this field. In this regard, Coppens and coworkers performed DFT calculations on a Cu(I)bis(2,9-dimethyl-1,10-phenanthroline) ion [52], of which the metal-ligand charge-transfer (MLCT) excited state had been characterised by Chen and coworkers [53], using stroboscopic pump-probe EXAFS measurements at the Advanced Photon Source, Argonne, USA, which employs the temporal structure of the synchrotron. A mechanical-chopper-based temporally resolved pump-probe X-ray diffraction experiment was subsequently performed by Coppens and coworkers on a similar Cu complex, the sole difference being the substitution of one phenanthroline ligand for a 1,2-bis(diphenylphosphino)ethane ligand [54]. In the X-ray diffraction experiment, the lowest triplet excited state, with a lifetime 85 μs at 16 K, was probed. The structure was refined using response ratios, as described earlier in §3.3, and marked displacements of the Cu atom in each of two independent molecules in the unit cell were observed. In a similar experiment, a significant contraction ($\sim$0.85 Å) of a Rh—Rh bond was observed in the ion, $[Rh_2(1,8\text{-diisocyano-p-menthane})_4]^{2+}$ in its excited triplet state that possesses a lifetime 11.7 μs at 23 K [55].

Diffraction experiments that exploit such technology, thus making practicable the investigation of species that might be as transient as 100 ps, have featured mostly biological molecules and have thus been Laue-based in design. Although the subject of this review is restricted to X-ray diffraction of single crystals containing small molecules, there are tantalising results from this biological research sphere (see for example all references herein on work by the groups of Wulff and Moffat); these have strongly aided technical progress in monochromatic oscillatorily based pump-probe diffraction experiments at the level ps to ns of temporal structure, through the continuing improvements of the instrument ID9 at ESRF, Grenoble, France, developed by Wulff and coworkers.

Several small-molecule single-crystal diffraction experiments have thus been conducted using the ID9 beamline, the most celebrated of which is the photo-induced paraelectric (neutral) to ferroelectric (ionic) structural phase transition in tetrathiafulvalene-p-chloranil (TTF-CA) [56]. Here, a 300-fs pulsed-laser pump and a 100-ps X-ray probe were used to deduce the structure of its ground state and photo-induced state, data being collected stroboscopically 2 ns before and 1 ns after each laser pulse. Before data collection, the phase transition proceeds via cooperative accumulation of each photo-induced molecule such that there is a lead time $\sim$500 ps required for the effect to convert from a molecular to a macroscopic phenomenon. Accordingly, the evolution of the photo-induced state was monitored before the data collection on recording the changes in relative intensity of certain Bragg reflections over the timescale 0–500 ps. Other experiments have involved similar experimental arrangements to investigate transient structures with lifetimes from ps to ns. For instance, Cole and coworkers used ID9 at the ESRF to probe the ephemeral ^{3}MLCT state of a rhenium carbene complex, [HNCH$_2$CH$_2$NHCRe (2,2'bipyridine)(CO)$_3$]Br, which has a lifetime $\sim$230 ns [57]; the feasibility of the experiment was confirmed in terms of the experimental arrangement and structural simulations. Accordingly, data were collected and initial indications from data analysis show that the experiment has been successful. Further data analysis is in progress. Although this review focuses on single-crystal X-ray diffraction, we mention that powder X-ray diffraction has been used with this ps-time-resolved monochromatic oscillatorily based diffraction experiment at the ID9 facility to deduce photo-induced structural changes in the organic molecule N,N-dimethylamino-benzonitrile (DMABN) [58].

FURTHER INSTRUMENTAL DEVELOPMENTS: PROSPECTS FOR THE FUTURE

Given the high regard for the ID9 beamline at the ESRF, there is much impetus to export this technical knowledge to create similar experiments at synchrotron sources such as APS and Spring-8. If such developments prevail, not only will experiments similar to those presently possible at the ESRF be possible at these sites but also one will be able to access entirely new frontiers of temporally resolved X-ray diffraction of small molecules in single crystals. At APS, even more ephemeral light-induced structures might be probed, given the temporal structure of the APS synchrotron, 20 ps. As Spring-8 is the brightest X-ray source in the world, at 8 GeV, it might be used to overcome the X-ray duty-cycle limitations that at present render some structures elusive. The new wave of forthcoming medium-energy synchrotrons, notably Diamond in the UK and Soleil in France, might also afford unique opportunities for temporally resolved single-crystal X-ray diffraction. In these cases, much will depend on the final specifications regarding the available modes of electron bunches in the synchrotron and the single-crystal X-ray diffraction beamline specifications that will be implemented. At the Diamond facility, pump-probe time-resolved small-molecule crystallography is one of six key areas of structural science that have been targeted for the approved single-crystal diffraction beamline [59]. Moreover, the design of this beamline has incorporated specifications for anomalous X-ray scattering (AXS) experiments as this area is also one target of structural science. AXS might be exploited usefully in photo-induced experiments to increase the contrast of the light-induced structure over that of the dominant ground state, in the area of expected change within a molecule. In luminescent organometallic materials, for example, the alteration expected commonly manifests itself in a variation of the M—X bond distance. A diffraction experiment could be performed readily on collecting structural data first using X-rays of a wavelength near that of the absorption edge of this metal and then at a wavelength far from the edge. The difference between the two experiments would reveal a change in the atomic form factor explicitly due to the metal. The use of AXS in this fashion, proposed by Cole et al [57], has yet to be exploited in practice, but it is prospectively a powerful tool for analysis of the commonly subtle light-induced structural changes in metal-containing compounds. The beamline design at the Diamond synchrotron would be well suited in this regard.

Aside from synchrotrons, fs X-ray pulses will also be obtainable from free-electron lasers, and several so-called X-FEL facilities are currently planned world-wide. In Europe, the TESLA facility in Hamburg, Germany, has received much attention. The viability of fs temporally resolved X-ray diffraction experiments might be hampered by the fact that the X-rays resulting from an X-FEL have far too great a flux for a crystal to survive the irradiation, although it has to survive only a few fs as it is a "one-shot" experiment; i.e. there is no possibility for stroboscopic pump-probe experimentation as the X-rays arrive in a single flash. The "one-shot" nature of an X-FEL limits inherently also the use of the technique for X-ray diffraction experiments, given the intrinsic requirements for reciprocal space coverage so that a crystal structure can be afforded.

Fs-laser-induced ps X-ray generation has also been demonstrated in the laboratory by exploiting the high spatial and spectral brightness of plasmas that, once created, emit X-ray pulses on a time scale of order ps and typically with a kHz rate of repetition [60, 61]. The development of these instrumental advances for affording laboratory-based, quantitative X-ray-diffraction results would prove highly beneficial to this area of structural endeavour.

CONCLUDING REMARKS

The excitement shared by researchers in this new area of chemical crystallography has been conveyed here, through the major technological and experimental advances and key results that have been achieved already. The great scientific impact that small-molecule crystallography stands to deliver in studying photo-induced species has been demonstrated, and one can expect an exciting future, particularly in temporally resolved stroboscopic pump-probe diffraction experiments when further results unfold from the instrumental developments, still strongly under way. The great rate at which detector technology is currently advancing might yield detectors that are more appropriate to photo-induced structural studies, especially in terms of data-read-out time, background levels and sensitivity, and such advances would strongly influence continuing experimental developments across all regimes of sample lifetime. One also envisages much more advanced Laue-based methods for small-molecule chemical crystallography, which would lead to experiments that are otherwise impossible due to rapid crystal fracture upon laser irradiation. The combined use of other facilities at present under development and the increased use of complementary techniques such as EXAFS, powder diffraction, temporally resolved infrared and Raman spectrometry, and density-functional calculations, will also ease the way ahead. We have seen that complementary laser spectra provide crucial information *a priori* to all X-ray-diffraction experiments on photo-induced species, and it appears to be the continuing goal of the laser industry to market lasers that are reliable, fully tunable and truly of "turn-key operation", so this development will greatly aid the crystallographer. However things unfold, one can be certain that the future of determination of light-induced structure looks bright.

ACKNOWLEDGEMENTS

I thank the Royal Society for a University Research Fellowship and St. Catharine's College, Cambridge, for a Senior Research Fellowship.

REFERENCES

1. J. R. Helliwell, "Faraday Discussions: Time-Resolved Chemistry: fromstructure to function", 2003, Volume **122**.
2. L. X. Chen, *Angew. Chem. Int. Ed.*, 2004, **43**, 2886–2905.
3. M. Saes, C. Bressler, F. van Mourik, W. Gawelda, M. Kaiser, M. Chergui, C. Bressler, D. Grolimund, R. Abela, T. E. Glover, P. A. Heimann, R. W. Schoenlein, S. L. Johnson, A. M. Lindenberg and R. W. Falcone, *Rev. Sci. Instruments*, 2004, **75**, 24–30, and references therein.
4. S. R. Elliott, *J. Non Cryst. Solids*, 1986, **81**, 71–98 and references therein.
5. D. Kimbrough, *Journal of Chemical Education*, 1997, **74(1)**, 51–53.
6. B. Perman, V. Srajer, Z. Ren, T. Teng, C. Pradervand, T. Ursby, D. Bourgeois, F. Schotte, M. Wulff, R. Kort, K. Hellingwerf and K. Moffat, *Science*, 1998, **279**, 1946–1950.
7. S. Crosson and K. Moffat, *The Plant Cell*, 2002, **14**, 1067–1075.
8. W. F. Brinkman and M. R. Pinto, *Bell Labs Technical Journal*, 1997, 57–75.
9. O. Kahn, "Molecular Magnetism", VCH, New York, 1993.
10. A. Y. Kovalevsky, K. A. Bagley and P. Coppens, *J. Am. Chem. Soc.*, 2002, **124**, 9241–9248.
11. A. Y. Kovalevsky, K. A. Bagley, J. M. Cole and P. Coppens, *Inorg. Chem.*, 2003, **42**, 140–147.
12. Y. Ozawa, M. Terashima, M. Mitsumi, K. Toriumi, N. Yasuda, H. Uekusa and Y. Ohashi, *Chem. Lett.*, 2003, **32**, 62–63.
13. Y. Ozawa, M. R. Pressprich and P. Coppens, *J. Appl. Cryst.*, 1998, **31**, 128–135.
14. M. Gembicky, D. Oss, R. Fuchs and P. Coppens, *J. Syn. Rad.*, 2005, **12**,665.
15. J. M. Cole, S. L. G. Husheer, S. J. Teat and G. Bushnell-Wye, to be published.
16. A. Paturle, H. Graafsma, H.-S. Sheu and P. Coppens, *Phys. Rev. B*, 1991, **43**, 14683–14691.
17. D. Bourgeois, T. Ursby, M. Wulff, C. Pradervand, A. Legrand, W. Schildkamp, S. Labouré, V. Srajer, T. Y. Teng, M. Roth and K. Moffat, *J. Synchrotron Rad.*, 1996, **3**, 65–74.
18. D. R. Sandstrom, S. C. Pyke and F. W. Lytle, *Stanford Synchrotron Radiat. Lab. Rep. 80/01* (1980) and *Stanford Synchroton Radiat. Lab. Activities Rep. April 1980 to March 1981* (1981).
19. D. M. Mills, A. Lewis, A. Harootunian, J. Huang and B. Smith, *Science*, 1984, **223**, 811–813.
20. Z. Ren, D. Bourgeois, J. R. Helliwell, K. Moffat, V. Srajer and B. L. Stoddard, *J. Synchrotron Rad*, 1999, **6**, 891–917.
21. J. R. Lakowicz, "Principles of fluorescence spectroscopy", 2nd Edition, 1999, Plenum Press.

22. M. Rüdlinger, J. Schefer, G. Chevrier, N. Furer, H. U. Güdel, S. Hassühl, G. Heger, P. Schweiss, T. Vogt, T. Woike and H. Zöllner, *Z. Phys. B: Cond. Matt.*, 1991, **83**, 125–130.

23. C. D. Kim, S. Pillet, G. Wu, W. K. Fullagar and P. Coppens, *Acta Crystallogr. A*, 2002, **58**, 133–137.

24. L. Powers, B. Chance, M. Chance, B. Campell, J. Friedman, S. Khalid, C. Kumar, A. Naqui, K. S. Reddy and Y. Zhou, *Biochemistry*, 1987, **26**, 4785–4796 and references therein.

25. D. L. Thiel, P. Livins, E. A. Stern and A. Lewis, *Nature*, 1993, **362**, 40–43 (corrigendum: **363**, 565).

26. M. Rüdlinger, J. Schefer, T. Vogt, T. Woike, S. Haussühl and H. Zöllner, *Physica B*, 1992, **180/181**, 293–298.

27. M. D. Carducci, M. R. Pressprich and P. Coppens, *J. Am. Chem. Soc.*, 1997, **119**, 2669–2678 and references therein.

28. P. Coppens, I. Novozhilova and A. Kovalevsky, *Chem. Rev.*, 2002, **102**, 861–883.

29. A. Y. Kovalevsky, G. King, K. A. Bagley and P. Coppens, *Chem. Eur. J.*, 2005, **11**, 7254.

30. K. F. Bowes, J. M. Cole, S. L. G. Husheer, P. R. Raithby, T. L. Savarese, H. A. Sparkes, S. J. Teat and J. E. Warren, *Chem. Commun.*, 2006, 2448–2450.

31. I. J. Bruno, J. C. Cole, P. R. Edgington, M. Kessler, C. F. Macrae, P. McCabe, J. Pearson and R. Taylor, *Acta Crystallogr.*, 2002, **B58**, 389–397.

32. K. F. Bowes, J. M. Cole, S. L. G. Husheer, P. R. Raithby, H. A. Sparkes, S. J. Teat and J. E. Warren, *J. Syn. Rad.*, in preparation.

33. J. M. Cole, S. L. G. Husheer, M. Lorenc, Q. Kong and M. Wulff, *Acta Cryst. A*, in preparation.

34. S. L. G. Husheer, J. M. Cole, D. Laundy and S. J. Teat, to be published.

35. T. Ohhara, J. Harada, Y. Ohashi, I. Tanaka, S. Kumazawa and N. Niimura, *Acta Crystallogr. B*, 2000, **56**, 245–253.

36. T. Takayama, M. Kawano, H. Uekusa, Y. Ohashi and T. Sugawara, *Helv. Chim. Acta*, 2003, **86**, 1352–1358.

37. A. Sekine, H. Tatsuki and Y. Ohashi, *J. Organomet. Chem.*, 1997, **536–7**, 389–398.

38. M. Kawano, Y. Ozawa, K. Matsubara, H. Imabayashi, M. Mitsumi, K. Toriumi and Y. Ohashi, *Chem. Lett.*, 2002, 1130–1131.

39. D. Hashizume and Y. Ohashi, *J. Chem. Soc., Perkin Trans.*, 1999, **2(8)**, 1689–1694.

40. P. Coppens, B. Ma, O. Gerlits, Y. Zhang and P. Kulshrestha, *Cryst. Eng. Comm.*, 2002, **4**, 302–309.

41. M. Kawano, K. Hirai, H. Tomioka and Y. Ohashi, *J. Am. Chem. Soc.*, 2001, **123**, 6904–6908.

42. J. Kusz, H. Spiering and P. Gütlich, *J. Appl. Cryst.*, 2001, **34**, 229–238.

43. M. Marchivie, P. Guionneau, J. A. K. Howard, G. Chastanet, J.-F. Letard, A. E. Goeta and D. Chasseau, *J. Am. Chem. Soc.*, 2002, **124**, 194.

44. J. Elhaïk, V. A. Money, S. A. Barrett, C. A. Kilner, I. Radosavljevic Evans and M. A. Halcrow, *J. Chem. Soc., Dalton Trans.*, 2003, **10**, 2053–2060, and references therein.

45. C. Carbonera, J. Sanchez Costa, V. A. Money, J. Elhaik, J. A. K. Howard, M. A. Halcrow and J.-F. Letard, *Dalton Trans.*, 2006, **25**, 3058.

46. A. L. Thompson, V. A. Money, A. E. Goeta, J. A. K. Howard and C. R. Chimie, 2005, **8**, 1365.

47. A. Goujon, B. Gillon, A. Gukasov, J. Jeftic, Q. Nau, E. Codjovi and F. Varret, *Phys. Rev. B*, 2003, **67**, 220401–4.

48. N. Yasuda, M. Kanazawa, H. Uekusa and Y. Ohashi, *Chem. Lett.*, 2002, **11**, 1132–1133.

49. W. A. Fordyce, J. G. Brummer and G. A. Crosby, *J. Am. Chem. Soc.*, 1981, **103**, 7061–7064.

50. P. Stein, M. K. Dickson and D. M. Roundhill, *J. Am. Chem. Soc.*, 1983, **105**, 3489–3494.

51. I. V. Novozhilova, A. V. Volkov and P. Coppens, *J. Am. Chem. Soc.*, 2003, **125**, 1079–1087.

52. L. X. Chen, G. B. Shaw, I. Novozhilova, T. Liu, G. Jennings, K. Attenkofer, G. J. Meyer and P. Coppens, *J. Am. Chem. Soc.*, 2003, **125**, 7022–7034.

53. L. X. Chen, G. Jennings, T. Liu, D. J. Gosztola, J. P. Hessler, D. V. Schaltrito and G. J. Meyer, *J. Am. Chem. Soc.*, 2002, **124**, 10861–10867.

54. P. Coppens, I. I. Vorontsov, T. Graber, A. Y. Kovalevsky, Y.-S. Chen, G. Wu, M. Gembicky and I. V. Norozhilova, *J. Am. Chem. Soc.*, 2004, **126**, 5980.

55. P. Coppens, O. Gerlits, I. I. Vorontsov, A. Y. Kovalevsky, Y.-S. Chen, T. Graber, M. Gembicky and I. V. Norozhilova, *Chem. Commun.*, 2004, **19**, 2144.

56. E. Collet, M.-H. Lemée-Cailleau, M. Buron-Le Cointe, H. Cailleau, M. Wulff, T. Luty, S.-Y. Koshihara, M. Meyer, L. Toupet, P. Rabiller and S. Techert, *Science*, 2003, **300**, 612–615.
57. J. M. Cole, P. R. Raithby, M. Wulff, F. Schotte, A. Plech, S. J. Teat and G. Bushnell-Wye, *Faraday Discuss.*, 2003, **122**, 119–129.
58. S. Techert, F. Schotte and M. Wulff, *Phys. Rev. Lett.*, 2001, **86**, 2030–2033.
59. W. Clegg, J. M. Cole, R. Morris, P. R. Raithby, S. J. Teat, C. C. Wilson, C. Wilson, J. Evans and M. Smith, *Diamond Light Source Technical Report*, SCI-BLP-028-0101 (for ease of availability, see also: http://www.diamond.ac.uk/Publications/1987/sci-blp-028-0101.pdf).
60. C. Rischel, A. Rousse, I. Uschmann, P.-A. Albouy, J.-P. Geindre, P. Audebert, J.-C. Gauthier, E. Forster, J.-L. Martin and A. Antonetti, *Nature*, 1997, **390**, 490–492.
61. R. J. Tompkins, I. P. Mercer, M. Fettweis, C. J. Barnett, D. R. Klug, Lord G. Porter, I. Clark, S. Jackson, P. Matousek, A. W. Parker and M. Towrie, *Rev. Sci. Instruments*, 1998, **69**, 3113–3117.

CHAPTER 3

MOLECULAR CONFORMATION AND CRYSTAL LATTICE ENERGY FACTORS IN CONFORMATIONAL POLYMORPHS

ASHWINI NANGIA

Abstract: In crystal structures of flexible molecules the total energy is a summation of the molecular conformer and crystal lattice energy contribution. These two energy factors are of comparable magnitude in organic solids because bond torsions and intermolecular interactions have similar energies, worth a few $kcal\,mol^{-1}$. The two contributions may be additive or cancel one another. Polymorphism is likely in molecular systems wherein molecular conformer and crystal lattice energy effects compensate each other, i.e. a metastable conformer resides in a stable packing arrangement or a stable rotamer is assembled in a metastable crystal environment. Consequently, conformational polymorph energy differences occur in a small window of $<3\,kcal\,mol^{-1}$. Several organic conformational polymorph clusters that highlight this principle are discussed in this chapter

INTRODUCTION

McCrone [1] defined polymorphism as the existence of 'a solid crystalline phase of a given compound resulting from the possibility of at least two different arrangements of the molecules of that compound in the solid state' over 40 years ago. This broad definition is widely accepted today in crystal engineering, materials science and pharmaceutical development [2]. The existence of polymorphism implies that free energy differences between various forms are small $(0.5–8\,kcal\,mol^{-1})$ and that kinetic factors are important during crystal nucleation and growth. Molecular conformations, hydrogen bonding, packing arrangements, and lattice energies of the same molecule in different supramolecular environments may be compared in polymorphic structures [3]. Polymorphs are ideal systems to study molecular structure–crystal structure–crystal energy relationships with a minimum number of variables, because differences arise due to molecular conformations, hydrogen bonding, and crystal packing effects but not due to a different chemical species. There

63

J.C.A. Boeyens and J.F. Ogilvie (eds.), Models, Mysteries and Magic of Molecules, pp. 63–86.
© 2008 *Springer.*

is keen interest in understanding polymorphism, the mechanism of crystal nucle-ation from solution, growing new crystal forms, controlling the selective growth of one form, transformations between polymorphs, and high-throughput crystalliza-tion of drugs [4]. Polymorphism is more widespread in pharmaceutical solids, with estimates of 30–50% in drug-like molecules, [5] compared to 4–5% polymorphic crystals [6] in the Cambridge Structural Database (CSD) [7]. Table 3-1 lists the

Table 3-1. CSD refcodes of organic polymorph clusters up to the recent update of the Cambridge Structural Database (August 2006 update)[a]

Heptamorph			
QAXMEH			

Pentamorph			
GLYCIN	IFULUQ	SUTHAZ	

Tetramorphs			
ADULEQ	BISMEV	KAXHAS	STARAC
AMBACO	CBMZPN	MABZNA	SLFNMB
BEWKUJ	CILHIO	PYRZIN	VISKAJ
BIXGIY	HEYHUO	RUWYIR	WUWTOX

Trimorphs					
AMNTPY	DIYJUQ	GEHBAX	MBYINO	PUBMUU	UDAYUT
AWAKIS	DLABUT	GISRIJ	MBZYAN	PUPBAD	UJORIU
AZADAG	DLMSUC	HADKIG	MCHTEP	QOGNEF	UNEWUF
BALWEQ	DMANTL	HIMWIJ	METHOL	RBTCNQ	WEFKIC
BANHOO	DMFUSC	HNIABZ	MEZKEH	SAMPYM	WIRXAW
BIMYAX	DMMTCN	HYQUIN	MNIAAN	SIFLOI	WUWTIR
BIYSEH	DOBTUJ	IJETOG	NADQAL	SIKLIH	XINBEB
BOPKOG	DPYRAM	IMDIAC	NAGHOT	SILTOW	YACTEC
BZCHOL	DUCKOB	IVADUE	NAPYMA	SLFNMA	YERRUI
CENRIW	DUVFUV	JATFUF	NAZLAC	SOBPEE	YUYHIJ
COMXAD	DUVZOJ	JIBCIG	NIMFOE	SULAMD	ZEPFAB
DATREV	ESTRON	JUSBUU	NOJHEZ	TAWRIT	ZEXREZ
DBEZLM	FACRIK	KTCYQM	OCHTET	TELYAK	ZOGQAN
DCBFRO	FAFWIS	LAURAC	PARQUI	TEPHTH	ZZZHWI
DCLANT	FAWFOY	LAVMOK	PATVEM	THIOUR	ZZZIYE
DCLBEN	FOMNEB	LCYSTN	PCBZAM	TNBENZ	ZZZVTY
DEGGEB	FEGWAP	LILXIN	PDABZA	TORSEM	
DETBAA	FESKAP	MACCID	PEFTIE	TPEPHO	
DHNAPH	FIDYIA	MALEHY	PHBARB	TUHBAZ	
DIMETH	FILGEM	MALOAM	PHTHCY	TURPYB	
DIWWEL	GADSIO	MBPHOL	PNEOSI	UCECAG	

[a] This list does not include polymorphs reported in recent papers. Benzidine (tetramorphs): M. Rafilovich and J. Bernstein, J. Am. Chem. Soc., 128 (2006) 12189. Oxalyl dihydrazide (pentamorphs): S. Ahn, F. Goo, B. M. Kariuki and K. D. M. Harris, J. Am. Chem. Soc., 128 (2006) 8441.

number of polymorphs of small organic compounds up to the recent update of the CSD. These statistics are highly under-representative on the prevalence of polymorphism because drug molecules are often not archived in the CSD for proprietary reasons. Among organic crystal structures, there is one example of a compound with nine polymorphs, 5-methyl-2-[(2-nitrophenyl)amino]-3-thiophenecarbonitrile, common name ROY [8] because of its red, orange and yellow colored polymorphic forms. Single crystal X-ray structures are reported for seven of these forms (QAXMEH), followed by 3 pentamorphs, 16 tetramorph clusters, and 121 trimorphic systems (Table 3-1). Interest in polymorphism is growing because different solid-state modifications have different physical, chemical and functional properties such as melting point, stability, color, bioavailability, toxicity, pharmacological activity, nonlinear optical response, etc. Polymorph screening is now regarded as an important and routine step in the development of specialty chemicals, drugs and pharmaceuticals.

Polymorphs are classified according to the following terminologies. Concomitant polymorphs crystallize simultaneously from the same solvent and crystallization flask under identical crystal growth conditions. They may be viewed as supramolecular isomers in a chemical reaction. Conformational polymorphs occur for flexible molecules, i.e. these molecules can adopt more than one conformation under ambient conditions. When different conformers of the same molecule are present in the same crystal structure the situation represents conformational isomorphs. Conformational isomorphism, the existence of multiple conformations in the same crystal structure, is closely related to the presence of more than one molecule in the asymmetric unit, i.e. $Z'>1$. The exact reasons why some crystals have $Z'>1$ are still not properly understood even as several research groups are working to seek answers to this enigma [9]. Pseudopolymorphism, [10] the occurrence of the same molecule with different solvent molecules in the crystal lattice or the same solvent in a different stoichiometry, is closely related to polymorphism.

In this chapter I will discuss conformational polymorphs in relation to (1) energy of the molecular conformer, (2) crystal lattice energy, (3) comparable magnitude of these energies in organic solids, (4) compensation of one by the other leading to a delicate balance, (5) benchmarking of calculated crystal structure energies with the experimental stability of polymorphs. Scheme 3.1 lists some molecular solids that exhibit conformational polymorphism.

MOLECULAR CONFORMER AND CRYSTAL LATTICE ENERGIES

A molecule is defined by three different parameters: bond distances, bond angles and torsion angles. Of these three parameters, bond stretching and compression is insignificant to cause structural changes because of the high bond energies of single, double and triple bonds (about 80, 150 and 200 kcal mol^{-1}). The distortion of a single bond by 0.03 Å is worth 0.3 kcal mol^{-1} while the values for double and triple bonds are proportionately higher (0.6–1.0 kcal mol^{-1}). Distortion of bond angle by 6–10° has the same energy penalty as bond distance changes of

0.03–0.05 Å. Rotation about C—C single bonds, or bond torsions have energy barrier of 0.5–3.0 kcal mol^{-1} depending on steric factors. Thus, bond angle and bond torsion deviations have approximately one and two orders of magnitude less energy penalty than bond stretching.

Among intermolecular interactions, the energy scale is roughly of the order: van der Waals interactions $= 0.5–1.0$ kcal mol^{-1} $\sim RT$ at ambient temperature, weak C—H $\cdots$ O interactions $= 1.0–4.0$ kcal mol^{-1}, strong O—H $\cdots$ O, N—H $\cdots$ O hydrogen bonds $= 4.0–10.0$ kcal mol^{-1} [11]. Thus, a torsion angle deformation is about the energy of a weak C—H $\cdots$ O or van der Waals interaction and several deformations may add up to the energy of a strong H bond. This leads to the situation that bond torsion changes, which determine molecular shape, are of comparable energy to intermolecular interactions that direct crystal packing. Thus, a molecule may adopt a metastable conformation if it can form a stronger interaction that makes up for this lost energy in crystal packing, or may reside in the stable conformation but engage in less stable intermolecular interactions. Joel Bernstein [12] discussed several examples of this phenomenon over two decades ago, and I cite only two cases from his exhaustive list. The observed conformation of butaclamol in the crystal structure of hydrobromide salt **1.HBr** is 1.4 kcal mol^{-1} higher in energy than the stable rotamer, which is over-compensated by the dominant van der Waals interactions. The molecular conformation of **1.HCl** is similar. The most important torsion angle in adenosine-5'-mono-phosphate monohydrate **2** is the rotation of the phosphate group with respect to the furanose ring. The *gg* conformation observed in the monoclinic crystal structure is $\sim$4 kcal/mol higher in energy that the stable *tg* conformer. The torsion angle differs about the base–sugar bond by 47° and the furanose ring pucker is also different in the two polymorphs of **2** (space group $P2_1$ and $P2_12_12_1$).

We have observed several examples of conformational polymorphs in my group during the last few years, which I will discuss in this article (Scheme 3.1). Molecular conformer energies were calculated in Gaussian 03 or Spartan 04 and crystal lattice energies were computed in Cerius^2 [13]. These systems illustrate the delicate interplay and mutual influence of molecular conformation, crystal packing effects, strong and weak hydrogen bonds, and lattice energies in conformational polymorphs of organic solids.

WEAK C—H $\cdots$ O HYDROGEN BONDS IN POLYMORPH CLUSTERS

4,4-Diphenyl-2,5-cyclohexadienone **3** exists as four polymorphs, labeled A–D, whose crystallographic details are summarized in Table 3-2 [14]. Polymorphs A–D are conformational polymorphs since they have different molecular conformers in their crystal structures. They are also concomitant polymorphs because they crystallized simultaneously from the same flask and under identical crystal nucleation and growth conditions. Forms B–D with multiple molecules in the asymmetric unit ($Z' > 1$) may also be classified as conformational isomorphs. Such polymorph clusters with different molecular conformations, crystal packing and C—H $\cdots$ O

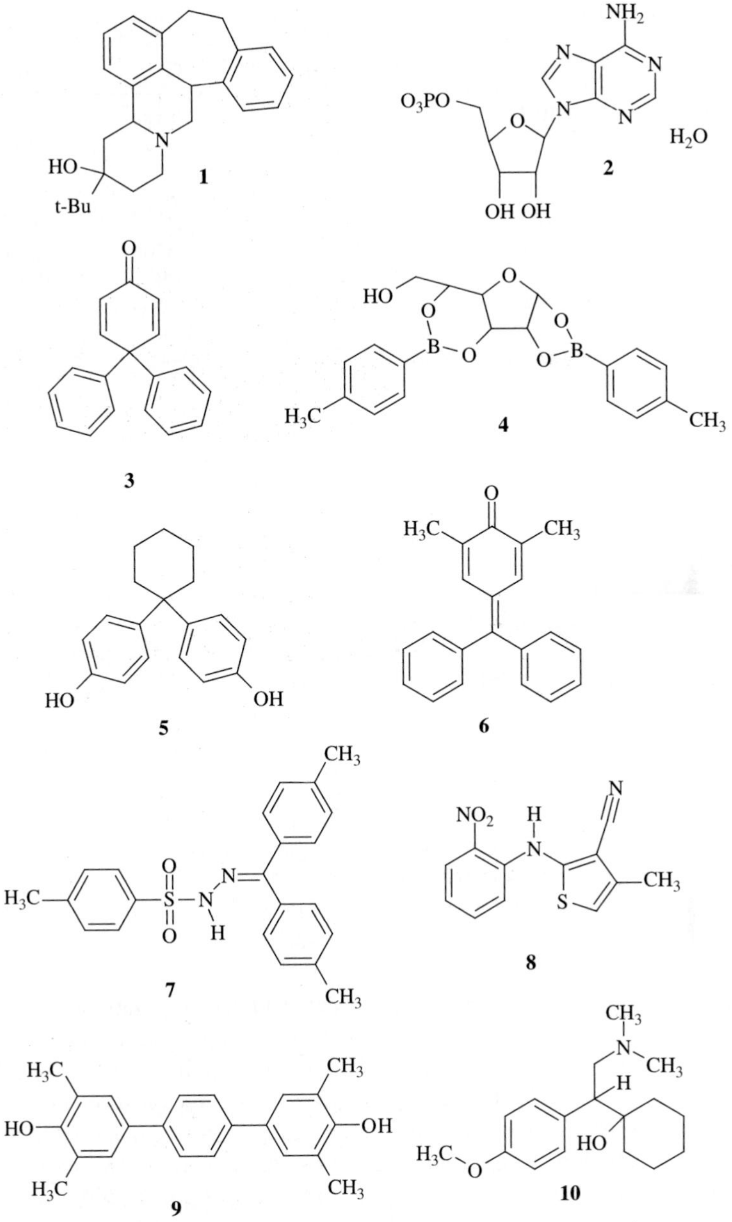

Scheme 3.1 Chemical structure of some conformational polymorph systems

Table 3-2. Crystallographic data on polymorphs A–D of diphenyl quinone **3** [14]

	Form A	Form B	Form C	Form D
CSD refcode	HEYHUO	HEYHUO01	HEYHUO02	HEYHUO03
Space group	$P2_1$	$P\bar{1}$	$P\bar{1}$	$Pbca$
Z', Z	1, 2	4, 8	12, 24	2, 16
a [Å]	7.9170(6)	10.0939(2)	18.3788(4)	10.7921(6)
b [Å]	8.4455(6)	16.2592(3)	19.9701(4)	17.4749(12)
c [Å]	10.3086(9)	16.2921(4)	24.4423(5)	27.9344(19)
α[°]	90	88.2570(10)	95.008(1)	90
β[°]	105.758(2)	85.3380(10)	111.688(1)	90
γ[°]	90	83.6450(10)	105.218(1)	90
V [Å^3]	663.36(9)	2648.00(10)	7871.8(3)	5268.2(6)
R-factor	0.050	0.068	0.112	0.059

Table 3-3. Relative energies[a] [per molecule, kcal mol^{-1}] of crystal forms A, B and D of **3**.[b] Lattice energies, U_{latt}, were calculated in both COMPASS and DREIDING force field of Cerius^2 but only COMPASS numbers are discussed in text. Molecular conformer energies were computed in Spartan (HF/6-31G**)

Polymorph	U_{latt}		E_{conf} of each conformer	E_{conf} average[c]	$E_{total} = U_{latt} + E_{conf}$	
	COMPASS	DREIDING			COMPASS	DREIDING
A	0.00	0.30	1.22	1.22	1.22	1.52
B	1.03	2.76	0.00, 0.06, 0.66, 1.12	0.46	1.49	3.22
D	0.82	0.00	1.08, 1.25	1.16	1.98	1.16

[a] Values are taken from ref. 15. Both U_{latt} and E_{conf} are relative energies.
[b] Form C is excluded because of high crystal structure R factor.
[c] Average E_{conf} value was calculated for multiple conformers as $\Sigma\ E_{conf} \div Z'$.

interactions offer ideal chemical systems to better understand the stated objectives in the introduction. The energies of the molecular conformers in forms A, B, D of **3** and crystal lattice energies of these polymorphs are listed in Table 3-3 [15]. We consider only forms A, B, and D for further discussion because (1) the experimental accuracy of X-ray crystal structure C coordinates is modest due to poor crystal quality and low data-to-parameter ratio of reflections ($R_1 = 0.112$), (2) it is a "disappeared polymorph" after our initial discovery in 2002 and further confirmation of any data is not possible. Conformer energies (E_{conf}) of the 19 rotamers and the most stable conformer in the gas phase have the energy order: gas phase ($-2.78\,\text{kcal mol}^{-1}$) < form B < form D < form A. Crystal lattice energies (U_{latt}, COMPASS force field, Cerius^2) follow the order form A < form D < form B. The most stable gas phase rotamer is not observed in any crystal structure so far. Notably, conformer and lattice energies follow a different order and their energy differences are comparable. A consideration of both the lattice energy and the conformer energy takes into account energy penalty from a metastable rotamer in a stable crystal structure and vice versa. The total energy (E_{total}) order in crystal forms A, B and D is A < B < D (COMPASS values), which is consistent with phase transformation experiments. However, consideration of U_{latt} alone, whether in COMPASS or DREIDING force field, is not consistent with experiments. The COMPASS force field gives accurate results for typical organic molecules.

Variable-temperature powder X-ray diffraction on a mixture of forms A–D [15] of **3** at room temperature gave essentially pure form A upon heating to 70°C (Figure 3-1), indicating that polymorph A is the thermodynamic (stable) modification in the enantiotropic cluster A–D in the 30–70°C range. Melt crystallization at 115°C gave the kinetic form B in pure yield (Figure 3-2). The observed powder diffractions patterns were matched with the calculated profiles from the respective single crystal structures in Powder Cell 2.3 [16]. Both the metastable and the stable polymorph was identified through phase transitions and heating experiments. The observed phase relationships are consistent with calculations when both the conformer penalty and the lattice energy stabilization are considered together under the E_{total} column of Table 3-3. The energy difference of $0.3\,\mathrm{kcal\,mol^{-1}}$ between forms A and B (1.22 vs. $1.49\,\mathrm{kcal\,mol^{-1}}$) serves as a benchmark to calibrate the force fields used, given the excellent match between computation and experiment.

An important and novel observation about multiple Z' in polymorphs A–D is the correlation between the strength of weak C—H$\cdots$O hydrogen bonds in their crystal structures and the value of Z'. Thus, polymorph C with the highest Z' has the shortest interaction, polymorph B with next lower Z' has longer interaction, and form A with $Z' = 1$ has the longest interaction (Figure 3-3) [15]. Whereas strong O—H$\cdots$O hydrogen bonds have been ascribed to multiple Z' in the literature, [9] we show that even weak C—H$\cdots$O H bonds show the same trend and in a graded manner.

A confirmation of the above idea is borne out from our own work [17] as well as that of William Clegg [18]. α-D-Glucofuranose-1,2:3,5-bis(p-tolyl)boronate **4** exists in two modulated crystalline forms at 298 K and 100 K. When reflections were collected at room temperature, data were solved in orthorhombic space group $P2_12_12_1$ with $Z' = 1$. However, data collection at low temperature gave a structure in monoclinic space group $P2_1$ but now with $Z' = 2$ [17]. The three symmetry independent molecules have similar conformations except for small differences in the C1—C2 boronate ester portion (Figure 3-4), which are significant for the present discussion. C—H$\cdots$O interactions in the low T, high Z' crystal structure are shorter (stronger) than the low Z', high T form (Figure 3-5). Crystal packing in the two structures is identical as conformed by their unit cell similarity index, $\Pi = 0.005$. Differential scanning calorimetry suggests that phase transition temperature between the monoclinic and orthorhombic forms of **4** is between 200–235 K. Carolyn Brock [19] has defined modulated structures as those in which $Z' > 1$ and relatively small displacements or orientation of molecules would make the molecules crystallographically identical. According to Jonathan Steed [9][b] the origin behind modulation is the

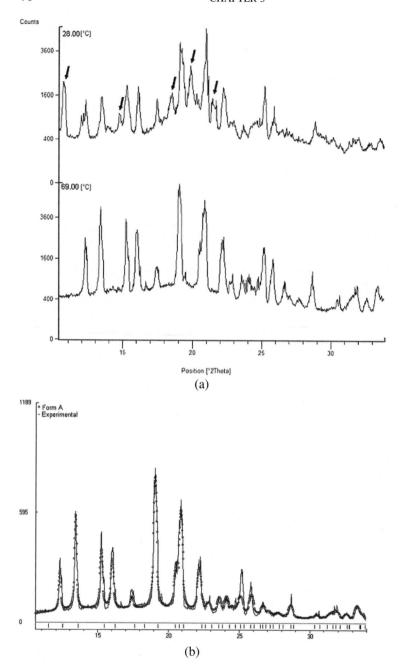

Figure 3-1. (a) Powder XRD of **1** at 28°C (top) and 69°C (bottom). Peaks that disappear upon heating are marked with an arrow. Note the increase in intensity of peaks and overall simplification of profile at higher temperature. (b) Experimental powder XRD of **3** at 69°C (black line) matches well with the calculated powder pattern of polymorph A (dotted line). Least squares refinement in Powder Cell 2.3: $R_p = 14.39$, $R_{wp} = 18.81$. The starting solid in (a) was a mixture of forms A–D

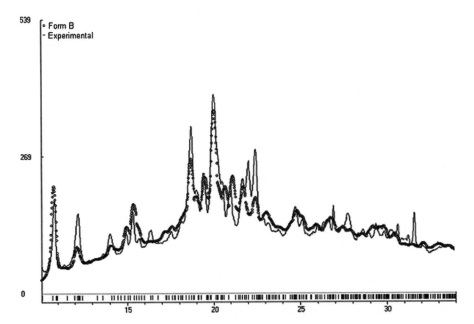

Figure 3-2. Experimental powder XRD of **3** from melt crystallization at 115°C (black line) shows good agreement with the calculated powder pattern of polymorph B (dotted line). Least squares refinement in Powder Cell 2.3: R_p=11.78, R_{wp}=15.86. The starting solid was a mixture of forms A–D

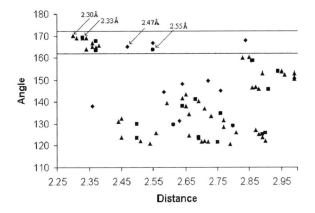

Figure 3-3. H···O distance (2.2–3.0 Å) vs. C—H···O angle (120–180°) scatter plot of interactions in tetramorphs A–D. A= ● (Z'=1), B = ■ (Z'=4), C= ▲ (Z'=12), D = ♦ ($Z' = 2$). The shortest H···O distance (marked with an arrow in the linear band) is inversely related to Z' (number of symmetry-independent conformations)

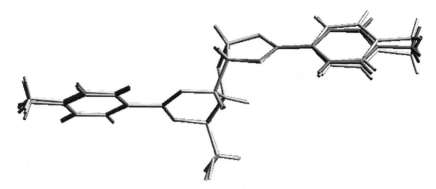

Figure 3-4. Overlay of symmetry-independent molecules in **4** by keeping the furanose ring as the fixed fragment (Cerius²). The C3—C5 boronate ester overlays nicely (left side) but the C1—C2 portion has different orientations (right side). The orthorhombic form is shown in black and the two symmetry-independent molecules of the monoclinic form are shown in grey

presence of intermolecular interactions that do not fit conventional space group operations, which results in multiple Z' due to pseudo-symmetry.

A distinction should be made between what one should call modulated structures and polymorphs. Two crystal structures of the same molecule having different Z'

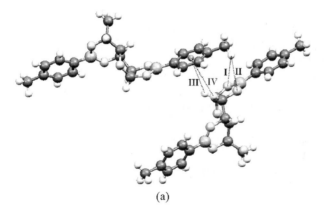

(a)

Figure 3-5. (a) C—H⋯O and C—H⋯π contacts between symmetry-related molecules ($Z' = 1$) in the orthorhombic form at 298 K. H⋯O, C⋯O (Å), ∠C—H⋯O (°); i = 3.84, 4.03, 93.9; ii = 2.94, 3.68, 135.4. H⋯π, C⋯π (Å), ∠C—H⋯π(centroid) (°); iii = 3.43, 4.03, 121.9; iv = 3.68, 4.18, 118.0. (b) C—H⋯O and C—H⋯π interactions between symmetry-independent molecules ($Z' = 2$) in the monoclinic form at 100 K: i = 2.88, 3.45, 118.7; ii = 3.05, 3.65, 122.1; iii = 3.91, 4.44, 122.1; iv = 3.48, 4.15, 131.9. Note that as the C—H⋯O interactions become shorter in (b) C—H⋯π distances lengthen due to the relative displacement of molecules. The C3—C5 tolyl group is not involved in weak interactions and this portion overlays nicely in Figure 3-4. Typical e.s.u. in heavy atom positions are 0.004–0.006 Å (100 K) and 0.006–0.009 Å (298 K)

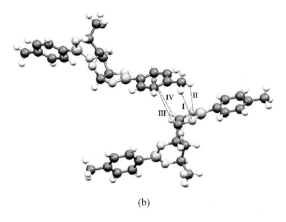

(b)

Figure 3-5. (Continued)

should be called modulated structures if there is little change in the overall packing of molecules from visual inspection, similarity indices, and powder XRD match. On the other hand, if molecular packing and/or intermolecular interactions in the two crystal structures are different then they are defined as polymorphs [17].

GUEST-FREE POLYMORPHS OF A PURE HOST

We recently crystallized two polymorphs of the pure host compound 1,1-bis(4-hydroxyphenyl)cyclohexane **5** by melt crystallization and sublimation [20]. The experimental conditions for each polymorph were slightly different. Sublimation of pure **5** at 150–175°C gave thin plate and fine needle shaped crystals on the cold finger. When the same starting material was heated up to 180–190° C and the melt liquid flash-cooled, crystals of plate and block morphology appeared. These crystals, designated as melt and sublimed forms respectively, were confirmed to be polymorphs by single crystal X-ray diffraction (**5s**: $P\bar{1}$, $Z' = 1$; **5m**, *Pbca*, $Z' = 2$). The OH groups in the bis-phenol are oriented syn in **5s** and in one of the symmetry-independent molecules of **5m** (A) whereas they are anti in the second molecule (B) of **5m** (Figure 3-6). In triclinic form **5s**, one of the OH groups engages in O—H $\cdots$ O H bonds along [010] and the second OH is bonded in an O—H $\cdots$ π interaction [21]. In the orthorhombic form **5m**, the syn diol molecule A has identical H bonding to the above polymorph but the anti diol B uses both its OH groups in making O—H $\cdots$ O H bonds in cooperative chains along [100] (Figure 3-7). The reason for multiple Z' in **5m** is ascribed to stronger H bonding in oligomers crystallized from the neat liquid under fast cooling (kinetic) conditions [9].

Molecular conformer and crystal lattice energies were computed using Gaussian 03 and Cerius^2 software (Table 3-4). The conformer as well as the lattice energy of the sublimed polymorph is stable relative to the melt form. Whereas there is no rotamer and lattice energy compensation here, crystal energy differences are very

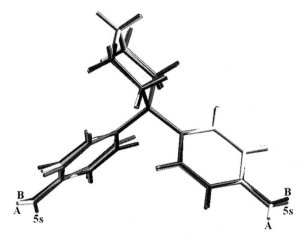

Figure 3-6. Overlay of conformer in sublimed polymorph **5**s, and symmetry-independent molecules A and B in melt phase **5**m. The diol OH groups are syn in **5**s and A molecule whereas they are anti in B molecule

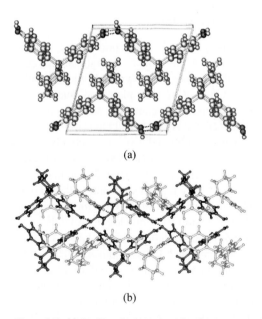

(a)

(b)

Figure 3-7. (a) O—H···O chain along [010] in the triclinic polymorph **5**s (OH groups of diol are syn). (b) Symmetry independent molecules in **5**m. The rectangular voids in the molecular ladders along [010] of the black B molecules (OH groups are anti) have a cross-section of 3.6 × 2.2 Å. The O—H···O chain down [100] is not shown. The grey A molecules have similar H bonding to **5**s

Table 3-4. Relative energies[a] [per molecule, kcal mol^{-1}] of melt and sublimed phases **5m** and **5s**.[b] U_{latt} were calculated using COMPASS force field in Cerius^2 and E_{conf} was computed in Gaussian 03 at the DFT, B3LYP/6-31G level

Polymorph	U_{latt}	E_{conf} of each conformer	E_{conf} average[b]	$E_{total} = U_{latt} + E_{conf}$
5s ($Z' = 1$)	−37.11	0.00	0.00	−37.11
5m ($Z' = 2$)	−36.23	0.23, 0.29	0.26	−35.97

[a] Values are taken from ref. 20. E_{conf} are relative energies.
[b] Average E_{conf} value was calculated for multiple conformers as $\Sigma\,E_{conf} \div Z'$.

small and surely indicative of polymorphism. Moreover, melting point and packing fraction of kinetic phase **5m** are lower than that of the thermodynamic form **5s** (**5m** 1.261 g/cm^3, 69.9%; **5s** 1.275 g/cm^3; 71.1%). Differential scanning calorimetry (Figure 3-8) shows that these conformational polymorphs do not undergo phase transition up to 160° C (monotropic cluster) but the metastable form **5m** transforms to the stable phase **5s** between 180–200° C (enantiotropic system).

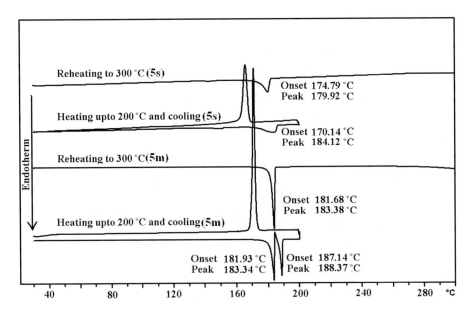

Figure 3-8. DSC of **5m** and **5s** polymorphs. Metastable phase **5m** shows phase transition to the sublimation polymorph **5s** and transformation to the thermodynamic form upon heating up to 200°C. Polymorph **5s** does not show phase changes under the same conditions except the sublimation endotherm. The reheating cycle endotherm is shifted to ∼5°C lower T than the first heating cycle due to better contact of the melted solids with the sample holder

Table 3-5. Occurrence of Z′ in organic crystals crystallized 'from the melt' (83 hits), 'by sublimation' (334 hits), and 'overall statistics' in the CSD.[a] Overall percentages are values from 160 850 organic crystal structures in the CSD

Z′	Sublimation % (# hits)	From melt % (# hits)	Overall Organic age	S+M ÷ 2×O %age values
< 1	29.04 (97)	22.89(19)	17.64	1.47
1	53.89 (180)	59.04(49)	71.88	0.78
> 1	17.06 (57)	18.07(15)	11.54	1.52
2	12.28 (41)	10.84(9)	10.04	1.15
≥ 3	3.89 (13)	7.23(6)	1.24	4.48
> 3	2.99 (10)	3.61(3)	0.69	4.78
4	2.69 (9)	2.40(2)	0.45	5.65

[a] Taken from ref. 20.

In addition to showing that solvent-free melt and sublimation crystallization conditions offer an attractive route to new polymorphs, a CSD survey of these methods of crystallization and the frequency of Z' was performed (Table 3-5). There is a dramatic increase in the occurrence of $Z' \geq 3$ crystal structures when melt or sublimation crystallization conditions are used [20]. The occurrence of high Z' in melt crystallization and sublimation methods is ascribed to the rapid cooling of the hot liquid or vapor (100–300°C) in the open flask or on the cold finger (kinetic phase), conditions under which hydrogen-bonded clusters are likely to condense in a pseudo-symmetric crystalline arrangement. On the other hand, the slower nucleation process of solution crystallization gives the frequent situation of $Z' \leq 1$ (88% hits).

Differences in the unit cells, hydrogen bonding, X-ray crystal structures, and powder diffraction patterns confirm that bis-phenol **5** is a pair of polymorphs with different Z' whereas crystal structures of **4** at different temperatures exhibiting near identical powder diffraction peaks are modulated structures [17].

REVISITING CLASSICAL POLYMORPHIC SYSTEMS

Terry Lewis, Ian Paul and David Curtin reported polymorphs of dimethyl fuschone **6** over two decades ago [22]. It exists in three crystalline modifications with one molecule in the asymmetric unit of each polymorph, α, β and γ forms ($P2_1/c$, $P2_12_12_1$, $Pna2_1$). They appeared as concomitant polymorphs in our hands, [23] affording a mixture of dominant α (80%), minor β (18%) and trace amount of γ form (2%) at room temperature in benzene. Crystal lattice energy and molecular conformation energy show that the α form is the most stable, followed by β, and γ form is the least stable. Heating the polymorphic mixture to 170°C gave β polymorph as the sole product, indicating that both α and γ forms convert to it (Figure 3-9). This is an example wherein the crystal structure predicted to be the most stable polymorph from calculations (Table 3-6) is different from the experimental results. This discrepancy could be due to two reasons: (1) inaccuracy in the

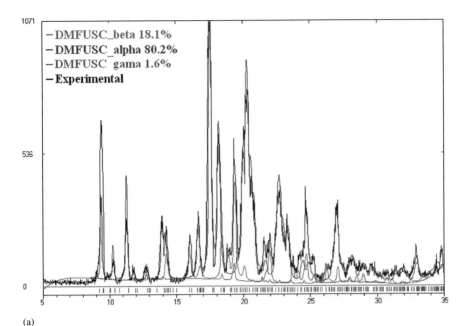

(a)

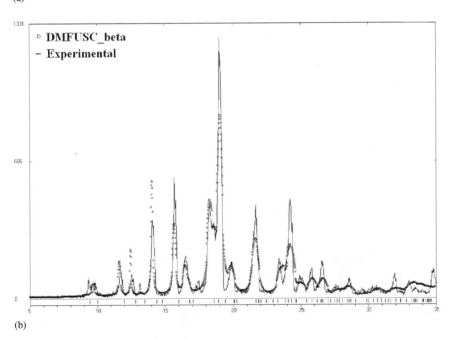

(b)

Figure 3-9. (a) Powder XRD of crystalline dimethyl fuschone **6** shows a mixture α (80%), β (18%) and γ forms (2%) at 30°C. (b) Heating the concomitant mixture to 170°C shows pure β form in >97% polymorphic purity from PXRD match with the simulated powder profile from the X-ray crystal structure (see color plate section)

Table 3-6. Relative energies [per molecule, kcal mol^{-1}] of α, β and γ forms of fuschone **6**. U_{latt}, COMPASS, Cerius^2 and E_{conf}, Spartan, HF/6-31G**

Polymorph	U_{latt}	E_{conf}	E_{total}
$\alpha(P2_1/c)$	0.00	0.00	0.00
$\beta(P2_12_12_1)$	0.41	0.61	1.02
$\gamma(Pna2_1)$	0.44	9.48	9.92

calculation of molecular conformer or crystal lattice energy or both, and/or (2) difference in the energy order at 0 K, assumed to be the temperature for computations in the gas phase, and the energy rankings at 300–450 K, the temperature regime for crystallizations and PXRD measurements.

Sulfonamide molecules are known to exhibit polymorphism in 50% cases. Sulfapyridine exists in 5 crystalline forms. We examined polymorphism in tolyl sulfonamide **7** and found it to be trimorphic in preliminary experiments [24]. Form 1 (*C2/c*), form 2 (*C2/c*) and form 3 (*P2$_1$/c*) have one symmetry-independent molecule but the conformations are different in the orientation of tolyl rings (Figure 3-10). Whereas full structural details and thermodynamic relationships will be discussed

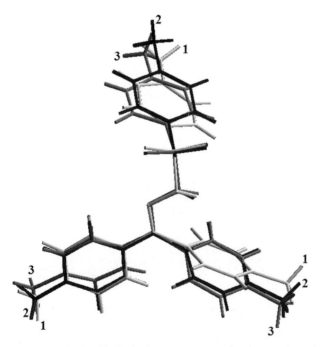

Figure 3-10. Different conformations of symmetry-independent molecules in polymorphs 1–3 of sulfonamide **7**. The tolyl groups are oriented differently in the three forms

Table 3-7. Lattice energy (U_{latt}, Cerius^2, COMPASS, kcal mol^{-1}), conformation energy (E_{conf}, Gaussian 03, B3LYP/6-31G(d,p), kcal mol^{-1}) and total energy ($E_{total} = U_{latt} + E_{conf}$) of tolyl sulfonamide **7**. All values are scaled to the lowest U_{latt} and E_{conf} as 0.00

Polymorph	U_{latt}	E_{conf}	E_{total}
form 1	3.39	0.00	3.39
form 2	2.42	6.29	8.71
form 3	0.00	0.85	0.85

in a forthcoming article, suffice to say for now that there is intramolecular (conformer) and intermolecular (lattice) energy compensation in **7** (Table 3-7). Form 1 is the kinetically controlled crystal structure, form 3 is the stable, thermodynamic phase, and form 2 is a disappeared polymorph, consistent with computed energies of these forms. The high-energy rotamer makes form 2 an elusive polymorph, an observation that is derived only from a consideration of E_{conf} but not U_{latt} alone. The melting point of stable form 3 is 17°C higher than metastable form 1 (Figure 3-11). The trimorph cluster of **7** once again validates the use of both conformer and lattice energy summation to calculate "real" energy differences. A consideration of lattice energy alone would have led to the erroneous conclusion that form 3 < form 2 < form 1.

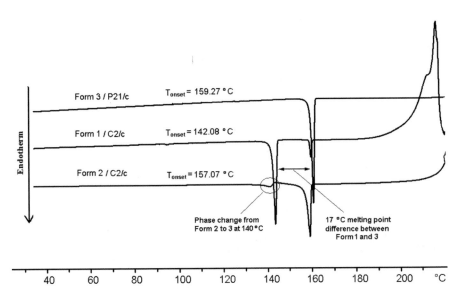

Figure 3-11. DSC of forms 1–3 of sulfonamide **7** polymorphs. Form 1 crystallizes concomitantly with a minor amount of thermodynamic form 3. Form 2 undergoes phase transition to form 3 at ca. 140°C. Form 3 exhibits a sharp melting endotherm at 159°C as the only phase transition

ROY – AN EXCEPTION TO THE ENERGY COMPENSATION PRINCIPLE

ROY **8** is the archetype of polymorph clusters with a record nine polymorphs as of date [8]. Of these 9 polymorphs, single crystal X-ray structures are reported for 7 forms. Gavezzotti and Dunitz [25] computed conformer and lattice energy values of 6 ROY polymorphs (Table 3-8, $Z' = 1$ in each structure). However, ROY is a classic example of a conformational polymorph cluster wherein there is no conformer and lattice energy compensation. As a matter of fact, the two effects are additive. The most stable crystal structure, Y, also has the stable rotamer of this flexible molecule. The difference between the lowest and highest energy polymorphs, Y and ORP, is 5.2 kcal mol^{-1} when both molecular and crystal energies are considered together. The absence of energy balance and the large difference between polymorph energies leads to the thought: will metastable polymorphs of ROY one day transform to the win-win thermodynamic state of the stable molecule and crystal lattice in the yellow form?

PHENYLOGUE POLYMORPH SERIES

Many examples of polymorphs are discovered by chance. There are as yet no "rules" to predict what combination of molecular features or functional groups will give a new polymorph system. Notwithstanding, the percentage of polymorphs in drugs and pharmaceuticals is about 10 times higher than that in the global crystal structure archive. Of course, this is only a validation of McCrone's age-old dictum that, "... every compound has different polymorphic forms and that, in general, the number of forms known for a given compound is proportional to the time and money spent in research on that compound". Drug molecules are crystallized several times (possibly hundreds) and this exercise often leads to new modifications. Further, there is deliberate effort to find new drug polymorphs for their improved

Table 3-8. Relative energies of the molecular conformation (E_{conf}), crystal lattice (U_{latt}) and total energy (E_{total}, kcal mol^{-1}) of the six polymorphs of ROY **8**[a]

Polymorph	U_{latt}	E_{conf}	E_{total}	Torsion ϕ[b]
Y	0.00	0.00	0.00	75
R	2.86	−1.19	1.67	22
ON	1.90	0.00	1.90	53
OP	2.86	0.95	3.81	46
YN	2.86	1.43	4.29	76
ORP	3.09	2.14	5.23	39

[a] Taken from ref. 25. U_{latt} were calculated by the Pixel method from UNI atom–atom force field and E_{conf} using MP2/6-31G** calculations. Energy values are converted from kJ mol^{-1} in the original paper to kcal mol^{-1} for comparison.
[b] Torsion angle between the mean planes of the phenyl and thiophene rings.

properties and patenting. An exhaustive search for all possible polymorphs of an organic compound [26] has now become a routine exercise in chemists' laboratories.

The terphenyl diol **9** was synthesized with the idea of reproducing the elusive β-quinol network via crystal engineering principles [27]. To our pleasant surprise, the extended diol **9** crystallized in two modifications, a rhombohedral polymorph **9r** that is topologically identical to the β-hydroquinone network, [28] and a monoclinic form **9m** resembling γ-hydroquinone (Figure 3-12). The central terphenyl ring adopts different conformations in the two polymorphs of **9** (conformational polymorphs). The designed molecule **9** as a phenylogue extension of hydroquinone is the first example of structural mimicry in a polymorph series to our knowledge.

NOVEL API POLYMORPH

The last example is of conformational polymorphs and multiple Z' in an active pharmaceutical ingredient, commonly referred to as API. Venlafaxine **10** is a serotonin–norepinephrine reuptake inhibitor drug (SNRI) for treating anxiety and depression, and marketed as the hydrochloride salt formulation in forms 1 and/or 2.

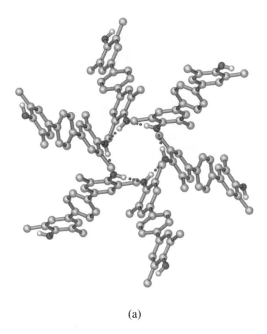

(a)

Figure 3-12. (a) Chair cyclohexane ring of O—H · · · O hydrogen bonds in the rhombohedral polymorph **9r**. (b) Hydrogen bonding of the axially oriented terphenyl groups to six different O—H · · · O hexamers forms the super cube of the β-quinol network. (c) Infinite chain of O—H · · · O hydrogen bonds along [010] in monoclinic polymorph **9m**. The middle phenyl ring is disordered over two orientations (s.o.f. 0.69, 0.31). H atoms are omitted except for OH groups

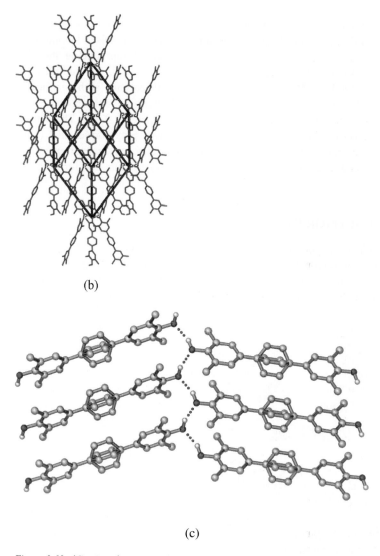

(b)

(c)

Figure 3-12. (Continued)

We recently discovered a novel polymorph, designated form 6, with unit cell parameters significantly different from both forms [29]. Single crystal X-ray structures of
form 3 (melt phase), form 4 (hydrate/solvate) and form 5 (sublimed phase) [30] are
not known and they are therefore not relevant to the present discussion on molecular conformations and multiple Z'. The conformation of molecule **10** in form 1
and 2 is different only in the opposite orientation of OCH_3 group, while the rest
of the tricyclic skeleton overlays very nicely. Interestingly, form 6 has both these

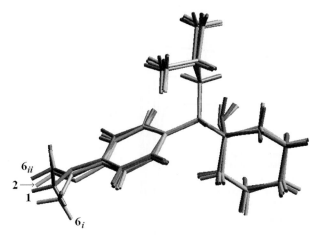

Figure 3-13. Overlay of symmetry-independent venlafaxine molecules in form 1, form 2 and form 6 (molecules i and ii). Chloride counter ion is omitted

molecular conformations of **10** in the crystal structure asymmetric unit ($Z' = 2$). Forms 1, 2 and 6 of venlafaxine are conformational polymorphs (Figure 3-13) and form 6 with multiple conformers is also an example of conformational isomorphism. Form 1 rotamer is more stable than form 2 conformer by $0.2\,\mathrm{kcal\,mol^{-1}}$ at the B3LYP/6-31G (d,p) level in Gaussian 03. Full structural details of this novel form 6 will be discussed in a forthcoming article. Suffice to say for now that a likely reason for two molecules in crystal form 6 could be (1) that it was crystallized at high temperature (170–190°C), (2) the presence of a second molecular conformation lead to better packing, and (3) the structure has ionic $\mathrm{O-H\cdots Cl^-}$ and $\mathrm{N^+-H\cdots Cl^-}$ hydrogen bonds. When one polymorph has $Z' = 1$ and another has $Z' > 1$, the high Z' structure was referred to as a metastable crystal nucleus of the low Z' stable modification [9]^b, [9]^e, [14]. In contrast, the high Z' form 6 of venlafaxine hydrochloride has higher melting point (by ~10°C) than forms 1 and 2 and it is the most stable modification known (Figure 3-14).

The exact reasons why molecules crystallize with multiple molecules in the asymmetric unit are becoming clear from a flurry of papers in the recent literature [9]. According to Steed, Brock and our results there are several reasons for multiple Z'. (1) The molecule has a packing problem because of its awkward shape, which is reconciled by having two or more molecules in different conformations. Such molecule have $Z' > 1$ crystal structures. (2) The molecules organize in stable clusters prior to reaching the highest symmetry arrangement in strong $\mathrm{O-H\cdots O}$ hydrogen-bonded structures because of the enthalpic advantage from σ-cooperative chains, e.g. as in alcohols, phenols, steroids, nucleotides, nucleosides. Molecule **5** (melt phase) would fall in this category. (3) There are several low-lying molecular conformations inter-converting in solution and more than one molecule may crystallize simultaneously because of kinetic factors. Molecule **3** (polymorphs B, C and D) and **10** (form 6) are in this category.

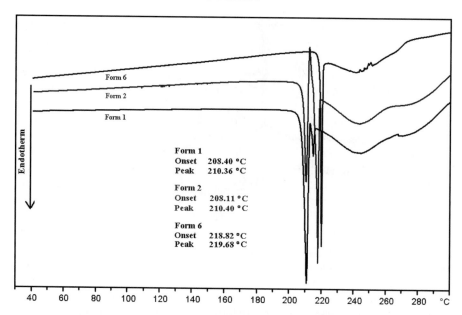

Figure 3-14. DSC of form 1, form 2 and form 6 at heating rate of 5°C min^{-1}. Form 6 has the highest melting point of 219–220°C compared to form 1 and 2 of 210–211°C assigned by us [30]

According to Desiraju, [31] these different reasons are simply different ways of saying the same thing. The proportion of high Z' crystal structures (> 1) is relatively constant at $11\pm1\%$ between 1970 to 2006 even as the CSD has grown 43 times in the same period. The fundamental physical basis for multiple Z' is the difference between $\Delta G^{\ddagger}_T$ and $\Delta G^{\ddagger}_K$ (T = thermodynamic, K = kinetic) for the crystallization of a given molecule. These energy differences are related to the modest energy of intermolecular interactions in organic crystals (0.5–8 kcal mol^{-1}) and so the appearance of high Z' will depend on the temperature of crystallization. Since most crystallization experiments are carried out between 10–30°C, the proportion of high Z' crystal structures will stay the same. His argument is that chemical or geometric or symmetry factors are different manifestations of the constancy of temperature range for crystallization and the energy range of intermolecular interactions in organic solids [32].

CONCLUSIONS

It is very difficult to make definitive conclusions in a review article on polymorphism because the subject is still evolving. Even as our understanding of this phenomenon improves with more structural data pouring in the complexity of systems being studied is also increasing. Experimental techniques, procedures and automated protocols are being optimized to carrying out crystallization screens for the discovery of new polymorphs using solution crystallization, solid-state grinding, solvent drop grinding, cocrystal former, crystal structure prediction, functionalized polymer support, cross-nucleation, etc [26]. As is always true in science, serendipity favors the prepared mind.[33] Whereas crystal engineers have developed strategies to control supramolecular organization using strong O—H$\cdots$O/ N—H$\cdots$O H bonds, weak C—H$\cdots$O and C—H$\cdots$O H bonds, I$\cdots$N and I$\cdots$O$_2$N halogen bonds, π-stacking in the last decade, a proper understanding of polymorphism is not equally advanced. We still have no way of predicting whether a molecule will be polymorphic, how many forms will it have, their crystal packing, when polymorphism will strike an API manufacturing process, and so on. This article is a "status update" on conformational polymorphs at the end of 2006. The results in this article show that polymorphism is likely when there is a balance of intramolecular and intermolecular energies in flexible molecules. Conformational polymorphism and multiple Z' are related issues and we have explanations, at least after crystal structure determination and analysis, about why some polymorphs have multiple molecules in the asymmetric unit. High Z' polymorph structures are generally metastable relative to their low Z' cousins but there are exceptions to this trend.

ACKNOWLEDGMENT

I thank the Department of Science and Technology, Government of India for financial support through Project No. SR/S5/OC-02/2002 entitled "Polymorphism in molecular crystals". Travel funds to attend the INDABA5 conference were provided by the UPE grant at UH. The variable-temperature powder XRD experiments were carried out in collaboration with Prof. Gert J. Kruger, University of Johannesburg, South Africa.

REFERENCES

1. W. C. McCrone, in Physics and Chemistry of the Organic Solid State, Vol. 2, D. Fox, M. M. Labes and A. Weissberger (Eds.), Wiley Interscience: New York, 1965, pp. 725–767.

2. J. Bernstein, Polymorphism in Molecular Crystals, Clarendon, Oxford, 2002.

3. a) A. Kálmán, L. Fábián, G. Argay, G. BernPth and Z. Gyarmati, J. Am. Chem. Soc., 125 (2003) 34; b) I. Weissbuch, V. Y. Torbeev, L. Leiserowitz and M. Lahav, Angew. Chem. Int. Ed., 44 (2005) 3226; c) M. Morimoto, S. Kobatake and M. Irie, Chem. Eur. J., 9 (2003) 621; d) P. Raiteri, R. Martoňák and M. Parrinello, Angew. Chem. Int. Ed., 44 (2005) 3769; e) P. K. Thallapally, R. K. R. Jetti, A. K. Katz, H. L. Carrell, K. Singh, K. Lahiri, S. Kotha, R. Boese and G. R. Desiraju, Angew. Chem. Int. Ed., 43 (2004) 1149; f) H. Chow, P. A. W. Dean, D. C. Craig, N. T. Lucas, M. L. Scudder and I. G. Dance, New J. Chem., 27 (2003) 704; g) R. G. Gonnade, M. M. Bhadbhade and M. S. Shashidhar, Chem. Commun., (2004) 2530; h) C. Guo, M. B. Hickey, E. R. Guggenheim, V. Enkelmann and B. M. Foxman, Chem. Commun., (2005) 2220; i) W. I. F. David, K. Shankland, C. R. Pulham, N. Bladgen, R. J. Davey and M. Song, Angew. Chem. Int. Ed., 44 (2005) 7032.

4. a) R. J. Davey, Chem. Commun. (2003) 1463; b) N. Blagden and R. J. Davey, Cryst. Growth Des., 3 (2003) 873; c) P. Erk, H. Hengelsberg, M. F. Haddow and R. van Gelder, CrystEngComm, 6 (2004) 474; d) J. Bernstein, Chem. Commun., (2005) 5007; e) L. Yu, J. Am. Chem. Soc., 125 (2003) 6380; f) P. Vishweshwar, J. A. McMahon, M. Oliveira, M. L. Peterson and M. J. Zaworotko, J. Am.

Chem. Soc., 127 (2005) 16802; g) C. P. Price, A. L. Grzesiak and A. J. Matzger, J. Am. Chem. Soc., 127 (2005) 5512; h) R. J. Davey, G. Dent, R. K. Mughal and S. Praveen, Cryst. Growth Des., 6 (2006) 1788.

5. a) S. R. Bryn, R. R. Pfeiffer and J. G. Stowell, Solid-State Chemistry of Drugs, SSCI, West Lafayette, IN, 1999; b) R. Hilfiker, F. Blatter and M. von Raumer, in Polymorphism in the Pharmaceutical Industry, R. Hilfiker (Ed.), Wiley–VCH, Weinheim, 2006, 1–19.

6. J. van de Streek and S. Motherwell, Acta Crystallogr., B61 (2005) 504.

7. J. Chisholm, E. Pidcock, J. van de Streek, L. Infantes, S. Motherwell and F. H. Allen, CrystEngComm, 8 (2006) 11.

8. S. Chen, I. A. Guzei and L. Yu, J. Am. Chem. Soc., 127 (2005) 9881.

9. a) C. P. Brock and L. L. Duncan, Chem. Mater., 6 (1994) 1307; b) J. W. Steed, CrystEngComm, 5 (2003) 169; c) S. Aitipamula and A. Nangia, Chem. Eur. J., 11 (2005) 6727; d) X. Hao, J. Chen, A. Cammers, S. Perkin and C. P. Brock, Acta Crystallogr., B61 (2005) 218; e) D. Das, R. Banerjee, R. Mondal, J. A. K. Howard, R. Boese and G. R. Desiraju, Chem. Commun., (2006) 555; f) K. M. Anderson, A. E. Goeta, K. S. B. Hancock and J. W. Steed, Chem. Commun., (2006) 2138; g) N. J. Babu and A. Nangia, Cryst. Growth Des., 6 (2006) 1995.

10. A. Nangia, Cryst. Growth Des., 6 (2006) 2.

11. G. R. Desiraju and T. Steiner, The Weak Hydrogen Bond in Structural Chemistry and Biology, IUCR Monograph, Oxford University Press, Oxford, 1999.

12. J. Bernstein, in Organic Solid State Chemistry, G. R. Desiraju (Ed.), Elsevier: Amsterdam, 1987, pp. 471–518.

13. Gaussian 03: www.gaussian.com; Spartan 04: www.wavefun.com; Cerius^2: www.accelrys.com.

14. V. S. S. Kumar, A. Addlagatta, A. Nangia, W. T. Robinson, C. K. Broder, R. Mondal, I. R. Evans, J. A. K. Howard and F. H. Allen, Angew. Chem. Int. Ed., 41 (2002) 3848.

15. S. Roy, R. Banerjee, A. Nangia and G. J. Kruger, Chem. Eur. J., 12 (2006) 3777.

16. N. Krauss and G. Nolze, Federal Institute for Materials Research and Testing, Berlin, Germany, 2000.

17. S. K. Chandran and A. Nangia, CrystEngComm, 8 (2006) 581.

18. G. S. Nichol and W. Clegg, Cryst. Growth Des., 6 (2006) 451.

19. X. Hao, M. A. Siegler, S. Parkin and C. P. Brock, Cryst. Growth Des., 5 (2005) 2225.

20. B. Sarma, S. Roy and A. Nangia, Chem. Commun., (2006) 4918.

21. E. Weber, K. Skobridis, A. Wierig, L. R. Nassimbeni and L. Johnson, J. Chem. Soc., Perkin Trans 2, (1992) 2123.

22. a) T. W. Lewis, I. C. Paul and D. Y. Curtin, Acta Crystallogr., B36 (1980) 70; b) E. N. Duesler, T. W. Lewis, I. C. Paul and D. Y. Curtin, Acta Crystallogr., B36 (1980) 166.

23. S. K. Chandran, S. Roy and A. Nangia, unpublished results.

24. S. Roy and A. Nangia, unpublished results.

25. J. D. Dunitz and A. Gavezzotti, Cryst. Growth Des., 5 (2005) 2180.

26. a) B. Rodríguez-Spong, C. P. Price, A. Jayashankar, A. J. Matzger and N. Rodríguez-Hornendo, Adv. Drug. Del. Rev., 56 (2004) 241; b) C. R. Gardner, C. T. Walsh and Ö. Almarsson, Nature Reviews, 3 (2004) 926; c) J. Bernstein, Chem. Commun., (2005) 5007; d) A.V. Trask and W. Jones, Top. Curr. Chem., 254 (2005) 41.

27. S. Aitipamula and A. Nangia, Chem. Commun., (2005) 3159.

28. T. C. W. Mak and C.-K. Lam, in Encyclopedia of Supramolecular Chemistry, Vol. 1, J. L. Atwood and J. W. Steed (Eds.), Marcel Dekker, 2004, pp. 679–685.

29. S. Roy, P. M. Bhatt, A. Nangia and G. J. Kruger, Cryst. Growth Des., 7 (2007) 476.

30. S. Roy, S. Aitipamula and A. Nangia, Cryst. Growth Des., 5 (2005) 2268.

31. G. R. Desiraju, CrystEngComm, 9 (2007) 91.

32. K. M. Anderson and J. W. Steed, CrystEngComm, 9 (2007) 328.

33. M. Rafilovich and J. Bernstein, J. Am. Chem. Soc., 128 (2006) 12189.

CHAPTER 4

HOW GUESTS SETTLE AND PLAN THEIR ESCAPE
ROUTE IN A CRYSTAL
Structural metrics of solvation and desolvation for inorganic diols

ALESSIA BACCHI

SUMMARY

Wheel-and-axle organic diols are host systems well known in the solid state. The introduction of a metal into the wheel-and-axle molecular frame opens the way for the design of materials with multiple functions, joining chemical and structural properties of the metal centre with the steric and supramolecular attitudes of the organic matrix. This chapter treats the rationalization of host-guest properties of wheel-and-axle inorganic diols.

   A reversible dynamic reorganization between the pure host (apohost) and the host-guest phases requires two conditions: a low-cost structural rearrangement between two states represented by the close apohost and the final open host framework, and an easily accessible path of migration for the outcoming and incoming guest molecules.

   The first requirement is tackled by realizing a bistable system, capable of reversibly switching between networks based on host-host (self-mediated) and host-guest (guest-mediated) interactions. The use of a simple geometric model for the analysis of the basic interactions responsible of the cohesion in the structures of the apohosts and in those of the host-guest complexes shows that the 'Venetian-blind' mechanism allows switching between two networks with modest structural changes. The comparison of the crystal structures of the initial and final states of the guest removal for inorganic wheel-and-axle diols allows an understanding of the effects of guest inclusion on the host framework, and provides clues to postulate a mechanism for the structural rearrangements accompanying the process.

   The preferential mechanism of escape of guests from the crystal bulk of inorganic wheel-and-axle diols is studied by using the knowledge base Isostar for intermolecular interactions. Two preferential directions of coordination to guest molecules

J.C.A. Boeyens and J.F. Ogilvie (eds.), Models, Mysteries and Magic of Molecules, pp. 87–108.
© 2008 *Springer*.

are discovered, related to two possible solvation sites within the guest-mediated networks. Application of the principles of structure correlation to supramolecular host-guest interactions indicates that, for inorganic wheel-and-axle diols, the preferential way for guests to escape from the crystal bulk corresponds to a migration of a guest along an array of alternate guest sites.

INTRODUCTION: CRYSTAL ENGINEERING OF INCLUSION COMPOUNDS

Interest in Crystalline Host-Guest Systems

During the past twenty years much effort has been devoted to the design of crystalline species capable of incorporating small molecules into the lattice through weak interactions, which might be broken at will, producing solid materials with host-guest properties [1]. The challenge was initially directed to develop organic counterparts to zeolites and clays [2, 3], which have been extensively used in the petrochemical industry to bind small molecules to a robust and porous inorganic framework. The interest in potential applications of host-guest materials has rapidly extended to the fields of separation, storage, heterogeneous catalysis, sensor devices and solid-state 'green' chemistry [4].

The formation and decomposition of a host-guest compound is described [5] with this equation,

$$H(s, \alpha) + n \ G(l \ or \ v) \rightleftarrows H \cdot G_n(s, \beta)$$

representing the equilibrium according to which a pure crystalline host H in phase α includes guest molecules (G) *via* recrystallization (liquid G) or absorption of vapour (vapour G), producing a crystalline host-guest β phase $H \cdot G_n$, with ratio n of host to guest. The pure host H is called an apohost; it might present pre-formed cavities, or it might be non-porous.

According to topological considerations, crystalline host-guest compounds are described as channel (tubulate), layer (intercalate), and cage (cryptate) clathrates [6]. General rules have been formulated to relate a topology to a crystallization temperature [7]: the ratio of guest to host decreases with increasing temperature of crystallization, and the topology varies in the order intercalate $\rightarrow$ tubulate $\rightarrow$ cryptate $\rightarrow$ apohost on proceeding from low to high temperature. The relations between crystal packing and the kinetics of enclathration and desorption of inclusion compounds have been reviewed [8].

Crystalline materials with inclusion attitudes towards small guests have been classified according to the behaviour of the host framework when the guests are removed [9]: first-generation compounds have microporous frameworks that collapse on the removal of guest molecules, second-generation compounds exhibit a permanent porosity also in an absence of included guests, and third-generation compounds have flexible and dynamic frameworks, which rearrange reversibly in response to the presence of guests. The latter are the subject of this chapter.

Porous Networks

The simple existence of cavities, or pores, or prospective void volumes in a crystal structure does not imply that the material can be an apohost with permanent porosity. Porosity must be assessed experimentally to demonstrate that a guest can be introduced and removed from the lattice without structural alterations [10]. Many inorganic materials constructed with covalent bonds, such as zeolites, show permanent porosity. The realization of synthetic porous covalent three-dimensional networks has been successfully achieved with the design of coordination polymers with thermally stable, robust and rigid open frameworks [11], [10]^b. Non-covalent porous networks, in which the molecular building blocks constituting the apohost framework are held together by weak interactions (typically hydrogen bonds), have also been designed using the tools and strategies of crystal engineering and molecular tectonics [12]. By analogy with their covalent counterparts, in these compounds the removal and inclusion of a guest does not affect the apohost structure [13].

Soft Networks

Organic and metallo-organic molecular crystals typically undergo remarkable structural modifications upon inclusion of a guest [8]. The lattice rearrangements induced by the entrance of a guest generally produce crystal fracturing or loss of crystallinity. Some compounds, however, possess a flexible apohost structure capable of dynamically accommodating guest molecules by converting to a clathrate phase through a process of single crystal to single crystal [14], [4]^b. Such dynamic porosity has been described for cholic acid [15], t-butyl-calix [4] arenes [16] and for the hybrid organic-inorganic systems that are the subject of the present chapter [17]. The dynamic pores might result from a 'bistability' of the soft apohost framework [9], capable of converting from a closed phase to an open phase in response to guest molecules.

Solvate Systems

The guests involved in clathration are generally the solvent molecules employed in the crystallization of the host-guest compounds. A survey of the Cambridge Structural Database (CSD) [18] has shown that, for organic and organometallic compounds, solvation occurs with more than 220 organic solvents [19]. The greatest occurrence is observed for 15 molecules, which alone cover more than 82 percent of all solvates (methanol, dichloromethane, benzene, ethanol, chloroform, acetonitrile, acetone, toluene, tetrahydrofuran, ethyl acetate, diethyl ether, dioxane, dimethylsulfoxide, dimethylformamide, hexane). Host molecules can be divided into two classes: those that possess an inherent concavity and form molecular complexes by adapting the guest inside the cavity, as in the case of cyclodextrins, and those that leave cavities to accommodate the guest in the lattice during self assembly, on which this chapter is primarily focused. In some cases the crystallization of solvates followed by controlled desolvation on heating or evacuation might be the only way to afford a pure apohost phase, or might be one way to obtain polymorphs [20].

WHEEL-AND-AXLE DIOLS

The design of molecular compounds that can potentially generate dynamic porous networks is guided by Weber's specifications: an effective host should be rigid and bulky and contain hydroxyl groups, which facilitate enclathration by coordination of the guest *via* hydrogen bonding [21]. Characteristic examples are the scissors and ring structures [22], and the wheel-and-axle diols [23] (*waad*). (Scheme 4.1)

In particular, *waad* are constituted with two bulky termini (wheels), comprising aromatic rings (found in some cases with ortho-substituents or as fluorenol groups, Scheme 4.1 and Table 4-1), carrying two hydroxyl groups separated by a central, rigid and linear organic spacer (axle). The great inclusion ability of these hosts is likely due to steric crowding around the hydroxy groups. Through this steric factor, the host molecules cannot associate directly by host…host hydrogen bonds. When a guest molecule is able to connect the host molecules by forming a hydrogen-bond network, the inclusion compound can form a stable guest-mediated crystalline lattice [23]. In general, *waad* coordinate acceptor guest molecules (G) through OH…G hydrogen bonds, and the stoichiometry *waad*·2G is generally

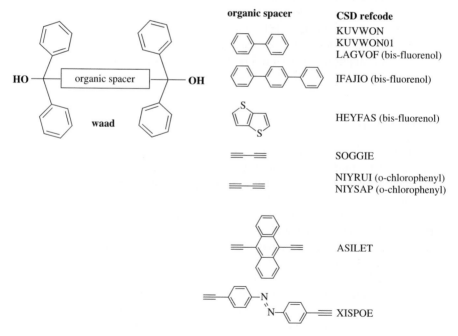

Scheme 4.1

Table 4-1. Occurrence of solvates crystal structures for organic and inorganic *waad* for which the apohost structure is known. For each apohost the number of known solvates is reported (N), together with the list of the corresponding CSD [18] refcodes. Structures *ASILIX*, *ASILOD*, and *ASILUJ* refer to the bis-fluorenil apohost, whose crystal structure is not known. Inorganic *waad* solvates have been grouped together

Apohost, CSD refcode	N	Solvates, CSD refcode	Apohost, CSD refcode	N	Solvates, CSD refcode
KUVWON KUVWON01	7	IVOFEE, IVOFOO, IVOFOO01, KUVWUT, KUVXAA, KUVXEE, KUVXII	SOGGIE	51	BABKAQ, BABKAQ10, BUSQIP, BUSQOV, BUSQUB, BUTNUZ, CAJNEH, CUMQAC, CUMQEG, CUMQIK, DEGPUR, DIMZUU, DIMZUU10, EROKAX, EROKEB, EROKIF, EROKOL, GAXFEQ, GIKYEE, GIKYOO, GUMDIB, GUMDUM, GUMFAV, GUMFAV02, HUDHIX, HUDHOD, HUDHUJ, HUMHOM, JODYIK, JULKIK, JULKUW, JULLAD, KERSAB, PODRUV, PODSAC, PODSEG, PODSIK, SOGHAX, VAJYAH, VAJYEL, VEMKUT, VEMLAA, WUPSUV, XELKII, XELKOO, XELKUU, XIWDEM, XOMLOA, XOMLOA01, XOMLOA02, YEXQAT
LAGVOF	6	LAGVUL, LAGWAS, LAGWEW, LAGWIA, DEBLAY, DEBLEC			

Table 4-1. (Continued)

Apohost, CSD refcode	N	Solvates, CSD refcode	Apohost, CSD refcode	N	Solvates, CSD refcode
 IFAJOO	1	IFAJEE	 NIYRUI NIYSAP	9	GIKZAB, GIMKES, GIMKIW, KOMRUZ, SAPLAW, TAFDIN, TAFDOT, VAXFUV, VONPET
 HEYFAS	8	HEYFIA, JUNDOL, KUMHIJ, KUMHIJ10, POKSAJ, POKSAJ10, PONPUD, PONPUD10	 ASILET	4	ASILIX, ASILOD, ASILUJ, MIFPAS
 XISPOE	4	AWUHEF, XISPUK, UHEMUP, UHEAW	 FIFHAE, FIFHUY, FIFJIO, JAWZ	10	FIFHEI, FIFHIM, FIFHOS, FIFJAG, FIFJEK, FIFJOU, FIKCUY, JAWYIQ, JAWYOW, JAWYUE

found. Beginning with the basic skeleton of the *waad* molecule, several derivatives have been designed by engineering the structure of the axle [23–25]. Scheme 4.1 and Table 4-1 report the diagrams of organic *waad* apohosts of which the crystal structure has been determined.

Occurrence of Wheel-and-Axle Diols

Based on the axle nature, several organic and inorganic *waad* have been synthesized. In all cases the length of the axle ranges approximately from 6 to 18 Å. The organic *waad* can be subdivided into two families, according to the nature of the axle: one family contains derivatives similar to 4,4′-bis(diphenylhydroxy-methyl)biphenyl [25][a] (CSD refcode: KUVWON), in which the HO-C is linked to an aromatic ring; another family is related to 1,1,6,6-tetraphenylhexa-2,4-diyne-1,6-diol [25][b] (CSD refcode: SOGGIE), in which the HO-C is linked to an alkyne moiety (Scheme 4.1). Inorganic *waad* are analogues of the 'KUVWON' family, obtained through the insertion of a metal atom into the middle of the axle, on coordinating α-(4-pyridyl)benzhydrol (LOH) or similar derivatives to various transition metals (Scheme 4.2).

The structural and clathrating behaviour of several *trans*-M(LOH)$_2$X$_n$ (M = Ag$^+$, Pd^{2+}; X = anion) complexes has been reported [17, 25][c–e]. The introduction of a metal into the wheel-and-axle molecular frame opens the way for the design of materials with multiple functions, joining chemical and structural properties of the metal centre with the steric and supramolecular attitudes of the organic matrix. The final structural and clathrating properties of the inorganic diol depend on the coordination number of the metal, on its geometry and configuration, on the steric and supramolecular characteristics of the counteranions, and on the hydrogen-bond network involving the carbinol —OH groups. Pd(LOH)$_2$X$_2$ complexes (X = Cl, CH$_3$, CH$_3$ COO and I) have been found to behave as dynamic porous networks, as they are capable of reversibly altering their solid-state organization in response to solvent inclusion or release, without losing crystallinity.

Many crystalline solvates of organic and inorganic *waad* are known in the crystallographic literature [24, 25]; for most of them the structure of the corresponding apohost phase has been also reported (Table 4-1). The comparison of the crystal

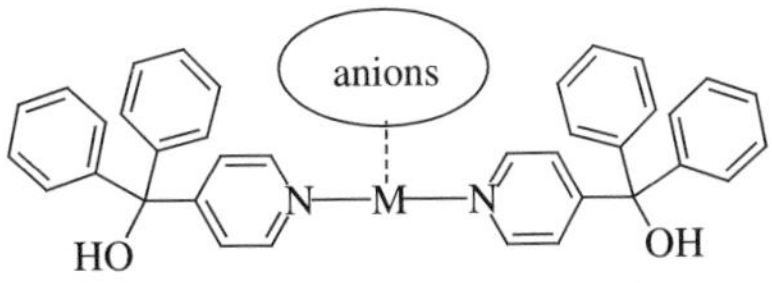

M	anion	CSD refcode
Pd^{2+}	Cl$^-$	FIFHAE
Pd^{2+}	Cl$^-$, CH$_3^-$	FIFHUY
Pd^{2+}	CH$_3$COO$^-$	FIFJIO
Pd^{2+}	I$^-$	JAWZAJ
Ag$^+$	BF$_4^-$	AYOGUQ
Ag$^+$	PF$_6^-$	AYOHAX

Scheme 4.2

structures of the initial and final states of the guest removal allows one to analyse the effects of guest inclusion on the host framework, and might provide clues about a mechanism for the structural rearrangements accompanying the process.

A GEOMETRIC MODEL OF GUEST INCLUSION BY WHEEL-AND-AXLE DIOLS

The basic interactions responsible for the cohesion both in the structures of the apohosts and in those of the host-guest complexes are describable with a simple geometric model that takes into account the necessity to achieve the maximal close packing for molecules with an awkward shape such as *waad* (Scheme 4.3). The most sensible way to optimize the space occupation for a pair of *waad* is to match approximately the protuberance of the terminal aryl moieties (the wheels) with the hollow of the molecular axle. The spatial relation between pairs of parallel adjacent host molecules is described by the distance between the axles (**d**) and the offset of the axle midpoints (Δ, expressed as a fraction of the axle length). Through this offset, the vector representing the translation of one molecule relative to the other (a′) is tilted through an angle θ with respect to the molecular axle. Molecular pairs choose the best **d** and Δ to optimize the contacts; a′ and θ are determined by the resulting geometry. In most cases, a′ coincides with one cell parameter.

Analysis of Apohost Structures

The analysis of all organic and inorganic apohost crystal structures shows that in many cases *waad* aggregate in pairs of parallel, displaced molecules, according to the above model. The aggregation of *waad* pairs is characterized by various inter-actions involving the —OH moieties both as hydrogen-bond donor OH...X (X=N, anion, π), and as CH...O hydrogen-bond acceptor (Figure 4-1).

The **d** values span from 3.6 Å to 7.2 Å, corresponding to a range of contacts from wheel...axle to wheel...wheel (Figure 4-2). The offset parameter Δ is well cor-related with the **d** value (Figure 4-3), showing that wheel...wheel contacts occur at extreme values of Δ, whereas wheel...axle contacts cause displacements about one half the molecular length.

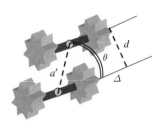

Scheme 4.3

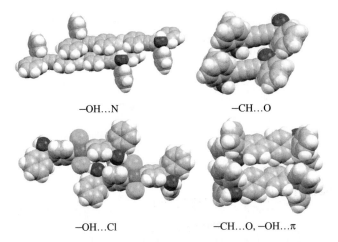

−OH...N   −CH...O

−OH...Cl   −CH...O, −OH...π

Figure 4-1. Typical interactions between *waad* pairs in the crystal structures of apohost phases (from top left, clockwise: XISPOE, SOGGIE, KUVWON, FIFHAE)

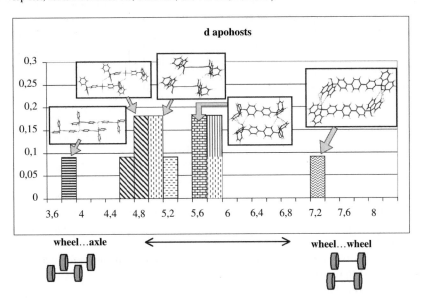

Figure 4-2. Occurrence of **d** values for the interactions between pairs of *waad* molecules in the apohost structures. **d** values span from 3.6Å to 7.2 Å, corresponding to a range of contacts from wheel...axle to wheel...wheel. Analogue apohost molecules and polymorphs are identified by bar meshing, and it can be seen that they are grouped around similar **d** values. The insets show the collocation of some of the structures discussed in the paper (from left: XISPOE, FIFHAE, SOGGIE, KUVWON, IFAJIO)

Effect of Guest Inclusion on d and Δ

In general, solvates of the same parent apohost tend to cluster about host-dependent **d** values; these are larger than those found in the corresponding apohost structures

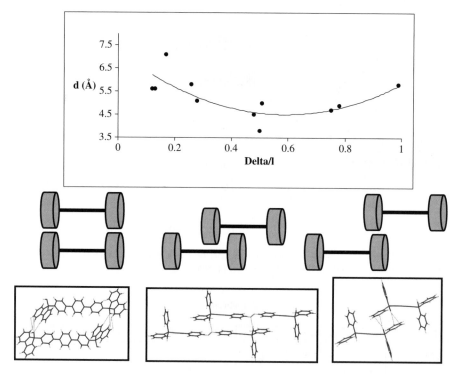

Figure 4-3. Correlation between **d** and Δ for pairs of *waad* molecules in the apohost structures. Shortest axle...axle distances **d** correspond to displacements of about one half of the molecule, which optimize the packing between the partners. Some cases are exemplified in the insets, from left: IFAJIO, XISPOE, SOGGIE (in the direction perpendicular to the one shown in Figure 4-2)

(Figure 4-4), indicating that, upon guest inclusion, wheel...wheel and wheel...axle interactions are replaced by wheel...guest interactions, which induce a slight expansion of **d**. The molecular offset Δ is confined between 1/3 and 2/3 the molecular length, and the correlation between **d** and Δ disappears.

The conversion from apohost to clathrate structure requires that the diol molecules can reversibly switch from a situation in which the host molecules interact with themselves (self-mediated network) to another in which they interact with the guests (guest-mediated network), with a concomitant rearrangement of geometric parameters **d** and Δ.

DESIGN OF BISTABLE NETWORKS

A reversible dynamic reorganization between the apohost and the host-guest phases requires two conditions: a low-cost structural rearrangement between two states represented by the close apohost and the final open-host framework, and an easily accessible path of migration for the outgoing and incoming guest molecules [26].

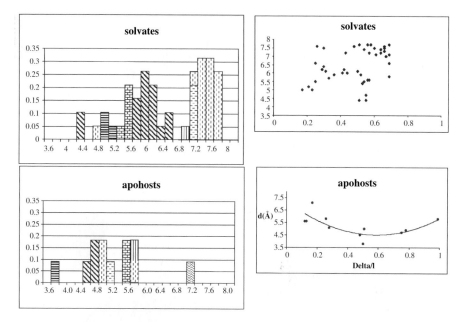

Figure 4-4. Effect of guest inclusion on the geometry of the pair of *waad* molecules. Solvates of the same parent host (indicated by bar meshing as in Figure 4-2) tend to cluster around host dependent **d** values, which are larger than those found in the corresponding apohost structure. Δ values for solvates are grouped between 0.2 and 0.7. The correlation between **d** and **Δ** disappears

Focusing on the first condition, the design of a bistable network requires the evaluation of the intermolecular association.

Self-Mediated and Guest-Mediated Networks

As described earlier under '*Analysis of apohost structures*', the assembly of *waad* molecules is based on OH...X (X=N, anion, π) or hydrogen bonds between host molecules, or can be mediated by the guest (G) via OH...G interactions. In the former case we define the structure as a self-mediated network, in the latter as a guest-mediated network (Scheme 4.4).

Of particular interest is the case in which the same structure might switch between a self-mediated network and a guest-mediated network, as this might facilitate a reversible exchange of a guest *via* a solid-gas process. In the case of *waad* pairs, this result can be achieved with a modest structural reorganization involving the rotation of both molecules about their centres of mass (Scheme 4.5). A simple oscillation of a pair of wheel-and-axle molecules about the midpoint of the axles accounts for large variations in parameters **d** and Δ without displacing the centres of mass.

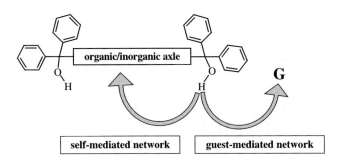

Scheme 4.4

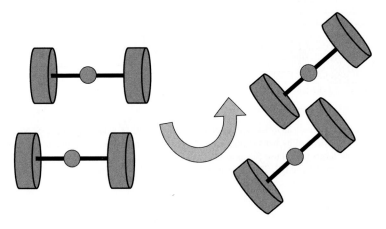

Scheme 4.5

The "Venetian-Blind" Mechanism

The comparison between apohost and solvate structures shows that, in many cases, the *waad* molecules are arranged in columns with similar spacings in both the presence and absence of a guest (Figure 4-5). The columns respond to guest inclusion or removal by altering the inclination of the molecular axes relative to the column axis. This condition corresponds to a modest concerted rotation of all molecules about their centres of mass, accompanied by a transition between the self-mediated state and the guest-mediated state, with consequent breaking and forming of the related OH . . . X and OH . . . G hydrogen bonds (Scheme 4.6). The network responds to the presence of a hydrogen-bond acceptor guest by converting between self-mediated and guest-mediated arrangements.

Inorganic *waad trans*-$[M(LOH)_2X_2]$ (M=Pd, Pt, X=Cl, I, CH_3COO, CH_3) exploit this "Venetian-blind" mechanism to convert between the anion-mediated network of the apohost, sustained by OH . . . X hydrogen bonds, and the guest-mediated network based on OH . . . G hydrogen bonds [17], [25][e].

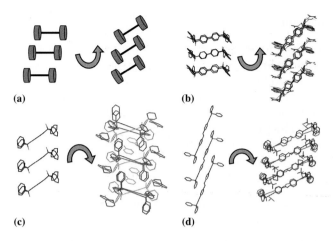

Figure 4-5. The comparison between apohost (left) and solvate structures (right) for families (b–d) of similar organic *waad* shows that in many cases the molecules are arranged in columns with similar spacings and different inclination. The columns respond to guest inclusion or removal by changing the inclination of the molecular axles relatively to the column axis, according to the 'Venetian blinds' machinery(a)

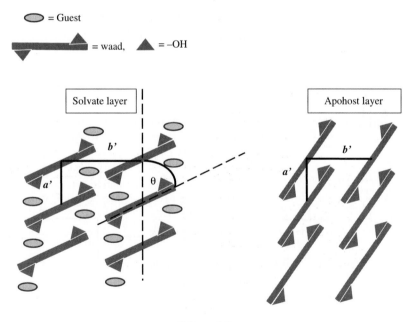

Scheme 4.6

Network Metrics for Inorganic Wheel-and-Axle Diols

The family of inorganic *waad trans*-$Pd(LOH)_2X_2$ (X=Cl, CH_3, CH_3COO, I) apohosts and several corresponding solvates containing guest molecules *trans*-$Pd(LOH)_2X_2 \cdot nG$ ($n = 2$, G=acetone, tetrahydrofurane THF, dimethylsulfoxide DMSO, dimethylformamide DMF; $n = 3/2$, G=1,4-dioxane) as hydrogen-bond acceptors have been investigated [17]. Solvation occurs on hydrogen bonding between the terminal —OH groups of the complex diol and the acceptor atom of the guest. All these solvate forms are organized in layers with practically invariant metrics. The layers consist of parallel columns in arrays formed by stacked $Pd(LOH)_2X_2 \cdot 2G$ units, with average $\mathbf{d} = 5.9\,\text{Å}$ and $\Delta = 0.34$. The stacking of complex units within a column occurs through —OH...G and π...G interactions (Figure 4-6, left). The structures of apohosts $Pd(LOH)_2X_2$ are based on —OH...X hydrogen bonds that generate columns of parallel units, arranged in layers (Figure 4-6, right). These are related to the metrics of the solvate forms by rotation through 28° of the complex molecules within the layer plane, giving $\mathbf{d} = 4.7\,\text{Å}$ and $\Delta = 0.69$. Thanks to the 'Venetian-blinds' mechanism, the transition between solvate and apohost structures does not alter significantly the location of palladium centres, as the separations respectively within a column (a') and between adjacent columns (b') pass from a' = 7.6, b' = 18.1 Å for the host-guest systems to a' = 10.7, b' = 17.4 Å, for the apohosts (Figure 4-6).

In all cases the non-solvate form is completely converted into the corresponding crystalline solvate forms on exposure to the vapour of the guest; conversely it is quantitatively recovered from the solvate upon removal of the guest under mild conditions (Scheme 4.7), without observing any transient amorphous phase during desolvation.

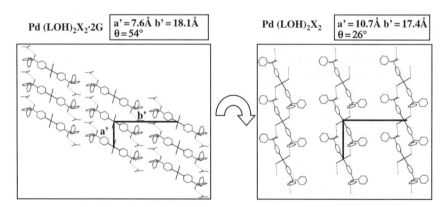

Figure 4-6. Layer organization and metrics for $Pd(LOH)_2X_2$ and $Pd(LOH)_2X_2 \cdot 2G$ (in the picture X=Cl, G=acetone). According to the 'Venetian blinds' machinery, the transition between solvate and apohost structures does not alter significantly the location of palladium centers

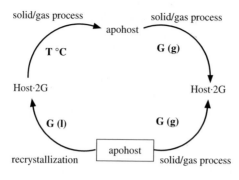

Scheme 4.7

The process $[Pd(LOH)_2Cl_2] \rightleftharpoons [Pd(LOH)_2Cl_2].2(\text{acetone})$ has been recycled three times with only a modest loss of crystallinity of the final non-solvate product. On the basis of the structural data, the solvation and desolvation is proposed to be based on a concerted rotation of the complex molecules by 28° in the layer plane, concomitantly with the migration of solvent through the lattice (Scheme 4.6).

The desolvation and solvation might be reversible because only minimum displacements are required for the single host molecules, and because only a modest reorientation is necessary for their molecular long axes.

GUEST MIGRATION PATH

As reported in the previous paragraph, when *waad* are organized in layers, the 'Venetian-blinds' rearrangement allows switching between self-mediated and guest-mediated networks without substantial alteration of the crystal structure. This condition realizes a bistable network, which is the first condition required for a reversible dynamic reorganization between the apohost and the host-guest phases. The second condition that must be fulfilled is the existence of a readily accessible path of migration for the outgoing and incoming guest molecules. A preliminary idea of the most accessible locations for the guest molecules within the layers of *waad* can be achieved on applying the general principles of structure correlation [27] to the collection of *waad* crystal structures that present the potentially bistable layer organization. All inorganic and many organic *waad* are organized in layers with a common host framework, with metrics that depend on the dimensions of the host molecules, but that in general comprise columns of parallel molecules, generated by translation through a', inclined by θ relative to the a' vector. The columns are separated by the b' translation, which is generally comparable with the molecular length (Scheme 4.6). The analysis of the distribution of guest positions over these structures might indicate the locations of the most accessible halts during guest displacement during solvation and desolvation.

The application of the utility *isogen*, belonging to the package of the *knowledge base of intermolecular interactions* Isostar [28], to the structural data of all organic and inorganic *waad* solvates demonstrates that the host molecules present two preferential directions of coordination to G, indicated as *exo* and *endo* (Figure 4-7). These directions are related to two possible sites of solvation within the *waad* guest-mediated networks: a site between the columns within a layer (*exo*), and a site between the aromatic groups and the anions of two adjacent layers (*endo*), as shown in Scheme 4.8.

Analysis of Guest Substructures for Inorganic *Waad*

The substructures presented by the guest molecules in the inorganic *waad* solvates have been analyzed by estimating the extension and shape of the crystal volume enclosed by the Hirshfeld surfaces (HS) of the guests [17]. The HS of a molecule in a crystal is defined such that for every point on the HS exactly half the electron density is due to the spherically averaged non-interacting atoms comprising the

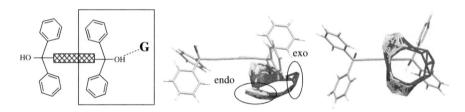

Figure 4-7. Location of preferential directions of coordination of *waad* to G according to the Isogen [28] analysis of the fragment reported in the box, over all the *waad* known in the CSD. Exo and endo sites are indicated

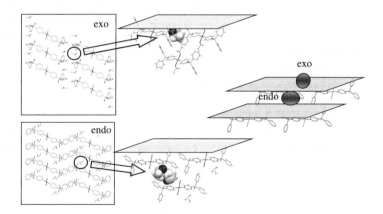

Scheme 4.8

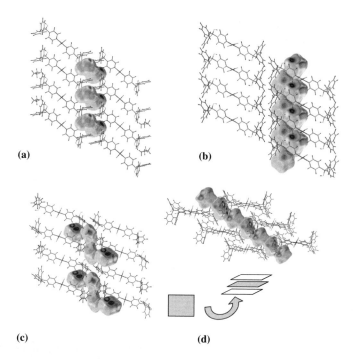

(a)   (b)

(c)   (d)

Figure 4-8. Hirshfeld surfaces [29] of guest molecules showing the occupation of *exo* and *endo* sites in the Pd(LOH)$_2$X$_2$·nG inorganic *waad*. (a) isolated pairs of *exo* sites in FIFHIM. (b) continuum of *endo* sites in JAWYIQ. (c) communication of *exo* and *endo* sites creating channels running across the layers JAWYUC. (d) edge-on view of the layers, showing the extension of the channels

molecule, and the other half is due to the atoms comprising the rest of the crystal; the HS essentially describes how much space a molecule occupies in a crystal [29].

Diverse situations are found for the Pd(LOH)$_2$X$_2$·nG inorganic *waad* (Figure 4-8): the occupation of site *exo* is generally based on pairs of G molecules belonging to adjacent columns of which the volumes touch through the intracolumnar space within a layer (Figure 4-8a); consecutive pairs are separated by aromatic rings. Conversely, the occupation of site *endo* generates a continuum of guest domains producing guest channels between the layers (Figure 4-8b). The spatial communication between sites *exo* and *endo* is governed by the position of the aromatic rings of the host molecules; this condition is attained in [Pd(LOH)$_2$I$_2$]·3/2(1,4-dioxane), in which a solvent chain is generated through the contact between G units that run diagonally across the layers (Figure 4-8c, d). We might regard the collection of solvent substructures (Figure 4-9) as snapshots of a possible path for solvent migration.

Metrics of Guest Migration

Guest substructure analysis has shown that the preferential mechanism of escape from the crystal bulk might correspond to a guest migration along an array of

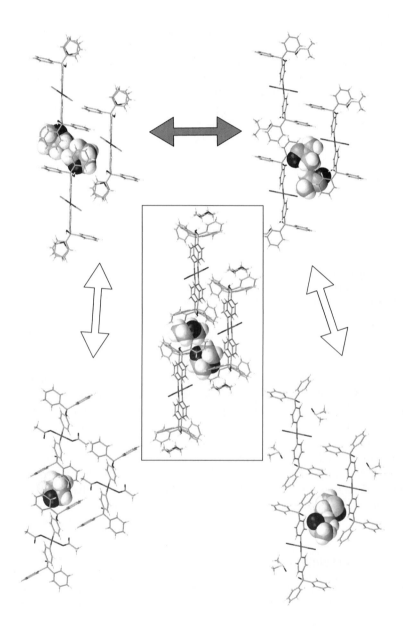

Figure 4-9. Immaginary snapshot sequence of guest migration through the solvate lattice, by sliding between *exo* and *endo* sites, with the assistance of conformational rearrangement of aromatic rings. The structure of [Pd(LOH)₂I₂]·3/2(1,4-dioxane) (JAWYUC, central inset) suggests a structural continuity for the process, as it can be viewed as the combination of [Pd(LOH)₂(CH₃COO)₂]·THF (FIFJOU, top left) and [Pd(LOH)₂I₂]·2DMSO (JAWYIQ, bottom left), through intermediates represented by [Pd(LOH)₂Cl₂]·2THF (FIFHIM, top right) and [Pd(LOH)₂I₂]·2DMF (JAWYOW, bottom right)

alternate *exo* and *endo* sites (Figure 4-9). The assistance of aromatic rings is required to facilitate the translational motion of the guests inside the lattice.

To relate the solvent release to the structural features of these systems, the morphologic alterations associated with the initial steps of desolvation have been analysed [17]. After having its faces indexed, a well shaped crystal of [Pd(LOH)$_2$I$_2$]·2DMF was heated. At a temperature 120°C, a preferential alteration was observed with an optical microscope on the face (001), indicating that desolvation can proceed on preferential sliding of the guest across the layers (Figure 4-10a), following exactly the path occupied by 1,4-dioxane molecules in [Pd(LOH)$_2$I$_2$]·3/2(1,4-dioxane) (Figure 4-10b). A similar observation was made for the organic *waad* H·2ETHYL [30] (H = 9,9′-(biphenyl-4,4′-diyl)difluoren-9-ol, ETHYL = ethylamine, CSD refcode = DEBLAY), for which desolvation occurs along [010] channels, corresponding to the array of *exo* and *endo* sites described in the preceding paragraph (Figure 4-10c). In this organic bis-fluorenol host, the

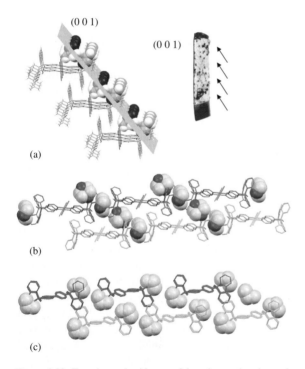

(a)

(b)

(c)

Figure 4-10. Experimental evidences of the solvent migration path across the layers. (a) After 15 minutes at 100°C defects appear selectively as crackings on face (0 0 1) of [Pd(LOH)$_2$I$_2$]·2DMF (JAWYOW). (b) The direction of solvent escape corresponds to the channels created by structural continuity of sites *exo* and *endo* in [Pd(LOH)$_2$I$_2$]·3/2(1,4-dioxane) (JAWYIQ) [17]. (c) Migration along equivalent channels has been detected for the organic analogue H·2ETHYL (DEBLAY) [30]

covalent link between the aromatic rings facilitates communication between the sites along the channel.

CONCLUSIONS

Waad are molecules with an elongated central spacer between large, rigid substituents at both ends, carrying also two OH groups. Numerous *waad* have been characterized, spanning diverse geometries and molecular dimensions, depending on the nature of the central spacer [23, 24]. Their propensity to include small guest molecules (G) in the crystal lattice derives from the combination of the hydrogen-bond donor capability of the OH groups and the steric bulkiness of the terminal diaryl moieties [23]. Inorganic *waad* have been synthesized on coordinating α-(4-pyridyl)benzhydrol (LOH) to various transition metals [17], [25]$^{c-e}$: the final structural and clathrating properties of the inorganic diol depend on the coordination number of the metal, its geometry and configuration, on the steric and supramolecular characteristics of the counteranions, and on the hydrogen-bond network involving the carbinol —OH groups. The possibility of a reversible dynamic reorganization between the apohost and solvate phases requires two conditions: a low-cost structural rearrangement between two states represented by the closed apohost and the final open host framework, and a readily accessible path of migration for the outgoing and incoming guest molecules.

The conversion from non-solvated to solvated structures requires that the diol molecules can switch from a situation in which the host molecules interact with themselves (self-mediated network) to another in which they interact with the guests (guest-mediated network). The design of a crystal framework capable of switching from a self-mediated to a guest-mediated state is supported by the analysis of the association patterns of pairs of *waad*, characterized by the distance between the axles (**d**) and the offset of the axle midpoints (Δ). The analysis of known crystal structures shows that the dimer geometry in the non-solvate structures is a consequence of wheel...wheel or wheel...axle interactions, with a correlation between **d** and Δ. With the inclusion of guest molecules, wheel...G contacts induce a slight expansion of **d**, whereas the offset values Δ are regularized and cluster between 1/3 and 2/3 the molecular length. This effect can be achieved through a modest structural reorganization involving the rotation of both molecules about their centres of mass, according to the 'Venetian-blinds' mechanism [25]^e. In most structures considered here, the *waad* pairs are organized in layers, and a similar concerted mechanism might convert a non-solvate to a solvate state. Inorganic *waad* *trans*-[M(LOH)$_2$X$_2$] (M = Pd, Pt, X = Cl, I, CH$_3$COO, CH$_3$) might exploit this mechanism to convert from the anion-mediated network of the apohost, sustained by OH...X hydrogen bonds, to the guest-mediated network based on OH...G hydrogen bonds [17].

In the inorganic *waad*, solvation occurs with a ratio 1:2, host to guest, by a OH...G hydrogen bond [17]. All these solvate forms are organized in layers with practically invariant metrics. The layers consist of parallel columns in arrays formed from stacked M(LOH)$_2$X$_2$·2G units, with average $d = 5.9$ Å and $\Delta = 0.34$.

The guest molecules are included in the space between the chains within the layers. The structures of the corresponding apohosts are based on —OH...X hydrogen bonds that again generate columns of parallel units, arranged in layers. These structures are related to the metrics of the solvate forms by a rotation 28° of the complex molecules within the layer plane, giving $d = 4.7\text{Å}$ and $\Delta = 0.69$. In all cases the apohost is completely converted into the corresponding crystalline solvate forms on exposure to the vapour of the guest; conversely it is quantitatively recovered from the solvate upon removal of the guest under mild conditions, without observing any transient amorphous phase during desolvation.

The use of the *knowledge base of intermolecular interactions* Isostar [28] shows that *waad* present two preferential directions of coordination to G molecules, named *exo* and *endo*. These directions are related to two possible sites for solvation within the *waad* guest-mediated networks. Experiments on desolvation have shown that for some organic [30] and inorganic [17] *waad* the preferential mechanism of escape of guests from the crystal bulk corresponds to a migration of a guest along an array of alternate *exo* and *endo* sites. An assistance of aromatic rings is required to facilitate the translational motion of the guests inside the lattice.

REFERENCES

1. J. W. Steed and J. L. Atwood (eds), Supramolecular Chemistry, Ch.1, Wiley, Chichester, 2000.
2. (a) A. K. Cheetham, G. Férey and T. Loiseau, Angew. Chem. Int. Ed. 38 (1999) 3268–3292. (b) Y. Aoyama, Topics in Current Chemistry, 198 (Design of Organic Solids) (1998) 131–161.
3. J. L. Atwood, J. E. Davies, D. D. MacNicol and F. Voegtle (eds), Comprehensive Supramolecular Chemistry, Vol. 7, Pergamon, Oxford, 1997.
4. (a) S. Kitagawa, R. Kitaura and S.-I. Noro, Angew. Chem., Int. Ed., 43 (2004) 2334–2375. (b) M. Albrecht, M. Lutz, A. L. Spek and G. van Koten, Nature, 406 (2000) 970–974. (c) M. Eddaoudi, J. Kim, N. Rosi, D. Vodak, J. Watcher, M. O'Keeffe and O. M. Yaghi, Science, 295 (2002) 469–472. (d) D. Meinhold, W. Seichter, K. Koehnke, J. Seidel and E. Weber, Adv. Mater., 9 (1997) 958–961. (e) S.-K. Yoo, J. Y. Ryu, J. Y. Lee, C. Kim, S. J. Kim and Y. Kim, Dalton Trans., (2003) 1454–1456. (f) G. Kaupp, CrystEngComm, 5 (2003) 117–133.
5. L. R. Nassimbeni, CrystEngComm, 5 (2003) 200–203.
6. E. Weber and H.-P. Josel, J. Inclusion Phenom., 1 (1983) 79–85.
7. (a) B. T. A. Ibragimov, J. Inclusion Phenom. Macrocyclic Chem., 34 (1999) 345–353. (b) L. R. Nassimbeni and H. Su, Acta Crystallogr., Sect. B., B58 (2002) 251–259.
8. M. R. Caira and L. R. Nassimbeni, Comprehensive Supramolecular Chemistry, D. D. MacNicol, F. Toda, R. Bishop (eds), Vol. 6, Ch. 25, Pergamon, Oxford, 1996.
9. S. Kitagawa and K. Uemura, Chem. Soc. Rev., 34 (2005) 109–119.
10. (a) L. J. Barbour, Chem. Commun., (11) (2006) 1163–1168. (b) O. M. Yaghi, M. O'Keeffe, N. W. Ockwig, H. K. Chae, M. Eddaoudi and J. Kim, Nature, 423 (2003) 705–714.
11. (a) B. F. Hoskins and R. Robson, J. Am. Chem. Soc., 112 (1990) 1546–1554. (b) R. Kitaura S. Kitagawa, Y. Kubota, T. C. Kobayashi, K. Kindo, Y. Mita, A. Matsuo, M. Kobayashi, H.-C. Chang, T. C. Ozawa, M. Suzuki, M. Sakata and M. Takata, Science, 298 (2002) 2358–2361. (c) N. L. Rosi, J. Eckert, M. Eddaoudi, D. T. Vodak, J. Kim, M. O'Keeffe, O. M. Yaghi, Science, 300 (2003) 1127–1130. (d) S.-I. Noro, S. Kitagawa, M. Kondo and K. Seki, Angew. Chem. Int. Ed., 39 (2000) 2082–2084.
12. (a) D. Braga, F. Grepioni and A. G. Orpen (eds), Crystal Engineering: From Molecules and Crystals to Materials, Kluwer, Dordecht, Netherlands, 1999. (b) M. Simard, D. Su and J. D. Wuest, J. Am. Chem. Soc., 113 (1991) 4696–4698. (c) M. W. Hosseini, Acc. Chem. Res., 38 (2005) 313–323.

13. P. Brunet, M. Simard and J. D. Wuest, J. Am. Chem. Soc., 119 (1997) 2737–2738.
14. K. T. Holman, A. M. Pivovar, J. A. Swift and M. D. Ward, Acc. Chem. Res., 34 (2001) 107–118.
15. M. Miyata and K. Miki, React. Mol. Cryst. 173–5 (1993) 153–164.
16. J. L. Atwood, L. J. Barbour, A. Jerga and B. L. Schottel, Science, 298 (2002) 1000–1002.
17. A. Bacchi, E. Bosetti and M. Carcelli, CrystEngComm, 7 (2005) 527–537.
18. (a) F. H. Allen, Acta Crystallogr., B58 (2002) 380–388. (b) F. H. Allen, O. Kennard and R. Taylor, Acc. Chem. Res., 16 (1983) 146–153. (c) I. J. Bruno, J. C. Cole, P. R. Edgington, M. Kessler, C. F. Macrae, P. McCabe, J. Pearson and R. Taylor, Acta Cryst. B58 (2002) 389–397.
19. H. Gorbitz and P. Hersleth, Acta Crystallogr., B56 (2000) 526–534.
20. L. R. Nassimbeni, Acc. Chem. Res., 36 (2003) 631–637.
21. E. Weber, Inclusion Compounds, J. L. Atwood, J. E. D. Davies and D. D. MacNicol (eds), Vol. 4, Ch.5, Oxford University Press, Oxford, 1991.
22. (a) E. Weber and M. Czugler, Topics in Current Chemistry, 149 (Mol. Inclusion Mol. Recognit. – Clathrates 2) (1988) 45–135. (b) E. Weber, M. Heckerl, I. Csoregh and M. Czugler, J. Am. Chem. Soc., 111 (1989) 7866–7872.
23. F. Toda, Comprehensive Supramolecular Chemistry, D. D. MacNicol, F. Toda, R. Bishop (eds). Vol. 6, Ch. 15, Pergamon, Oxford, 1996, 465–516.
24. For recent examples see: (a) M. R. Caira, A. Jacobs, L. R. Nassimbeni and F. Toda, Cryst. Eng. Comm., 5 (2003) 150–153. (b) I. Csoregh, T. Brehmer, S. I. Nitsche, W. Seichter and E. Weber, J. Inclus. Phen. Macr. Chem., 47 (2003) 113–121.
25. (a) E. Weber, K. Skobridis, A. Wierig, L. R. Nassimbeni and L. Johnson, J. Chem. Soc., Perkin Trans. 2, (12) (1992) 2123–2130. (b) D. R. Bond, L. Johnson, L. R. Nassimbeni and F. Toda, J.Solid State Chem., 92 (1991) 68–79. (c) A. Bacchi, E. Bosetti, M. Carcelli, P. Pelagatti and D. Rogolino, CrystEngComm, 6 (2004) 177–183. (d) A. Bacchi, E. Bosetti, M. Carcelli, P. Pelagatti and D. Rogolino, Eur. J. Inorg. Chem., (10) (2004) 1985–1991. (e) A. Bacchi, E. Bosetti, M. Carcelli, P. Pelagatti, D. Rogolino, G. Pelizzi, Inorg. Chem. 44 (2005) 431–442.
26. (a) F. H. Herbstein, Acta Cryst., B62 (2006) 341–383. (b) F. Mallet, S. Petit, S. Lafont, P. Billot, D. Lemarchand and G. Coquerel, J. Thermal Anal. Cal., 73 (2003) 459–471. (c) Galwey, A. K. Thermochim. Acta, 355 (2000) 181–238. (d) S. Petit and G. Coquerel, Chem. Mater., 8 (1996) 2247–2258. (e) S. K. Makinen, N. J. Melcer, M. Parvez and G. H. K. Shimizu, Chem. Eur. J., 7 (2001) 5176–5182.
27. H.-B. Buergi, J. D. Dunitz (eds), Structure Correlation, VCH, Weinheim, 1994.
28. I. J. Bruno, J. C. Cole, J. P. M. Lommerse, R. S. Rowland, R. Taylor and M. L. Verdonk, J. Comput.-Aided Mol. Des., 11 (1997) 525–537.
29. J. J. McKinnon, A. S. Mitchell and M. A. Spackman, Chem. Eur. J., 4 (1998) 2136–2141.
30. M. R. Caira, T. le Roex, L. R. Nassimbeni and E. Weber, Cryst. Growth. Des., 6 (2006) 127–131.

CHAPTER 5

STRUCTURAL DETERMINATION OF UNSTABLE SPECIES

YUJI OHASHI

INTRODUCTION

According to an enduring belief, an analysis of crystal structure with X-rays can show only the static structure of a molecule and the dynamic nature of the molecule should be obtained with spectral methods. The reasons are that application of a diffraction method depends on a periodic structure of molecules and that collecting diffraction data requires an extensive duration. It was therefore thought that a molecule can not alter its structure without destroying the periodic structure, that is, the single-crystalline form. We found, however, that, when a crystal was exposed to X-ray or visible light, chiral molecules of a cobaloxime complex gradually altered to racemic ones in a crystal whilst it retained a single-crystalline form [1]. Several related cobaloxime complex crystals showed a transformation from chiral to racemic without decomposition of the single-crystalline form [2]. Moreover, the chiral enrichment of a crystal was observed only with irradiation from a xenon lamp [3]. The chirality of a cobaloxime molecule was inverted in a crystal only on photoirradiation [4]. Such reactions were observed with not only X-ray but also neutron diffraction [5]. As the reactions that maintain the single-crystalline form are important for analysis of reaction mechanisms, we defined such a reaction as a 'crystalline-state reaction' to distinguish it from conventional solid-state reactions [6]. The crystalline-state reaction differs from a reaction from single crystal to single crystal as the latter includes the reactions in which vapour or solution state or dislocation appears in the transformation from a single crystal to a single crystal. The mechanisms of such reactions are difficult to deduce from the initial and final structures, as the molecular structure might alter much in the intermediate or solution state.

Among crystalline-state reactions, some have multiple steps. On photoirradiation or heating, a reactant molecule becomes excited to a transition structure and then converts to product. Analysis of the crystal structure during the crystalline-state reaction reveals the disordered structure of the reactant and product molecules. The concentration of the product increases gradually from 0 to 100% in the disordered structure when the crystal structures are analyzed at constant intervals during the

109

J.C.A. Boeyens and J.F. Ogilvie (eds.), Models, Mysteries and Magic of Molecules, pp. 109–135.
© 2008 *Springer*.

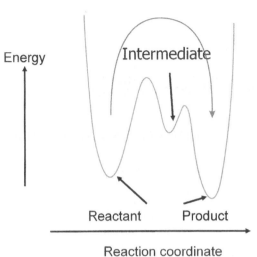

Reaction coordinate

Figure 5-1. Reaction coordinate including the intermediate stage

reaction. If the metastable intermediate exists as shown in Figure 5-1, the disordered structure including three molecules – reactant, intermediate and product – would be observed during the reaction. The concentration of the intermediate molecule increases gradually during the initial stage but gradually decreases during the final stage. In contrast, the concentrations of reactant and product molecules monotonically decrease and increase, respectively, during the reaction.

To observe the intermediate structures one must collect the intensity data rapidly. We designed and made a diffractometer, R-AXIS IIcs, for rapid collection [7] of data, which has been marketed in a revised form by Rigaku Company as R-AXIS RAPID since 1998. The two-step crystalline-state reaction was first found in the inversion of a chiral alkyl group bonded to the cobalt atom in a cobaloxime complex when the crystal on the diffractometer was irradiated with a xenon lamp [8]. Figure 5-2 shows the two-step structural change of the complex, ((R)-1-methoxy-carbonylethyl) [(S)-1-phenylethylamine] cobaloxime. After exposure for 2 h, the structure (Figure 5-2b) showed the disordered structure of the chiral (R)-1-methoxycarbonylethyl group; this disorder was caused by rotation of the chiral group about the C—C bond, not by inversion of the chiral group. After exposure for 5 h (Figure 5-2c), most of the chiral group rotated about 100° about the C—C bond and a small portion of the groups was inverted to the opposite configuration. After exposure for 10 h (Figure 5-2d), most groups were inverted to the S configuration. This result clearly indicates that the inversion proceeded in two steps, that is, first rotation about the C—C bond and then inversion of the entire group. In the intermediate stage the chiral group is expected to have a structure shown in Figure 5-1e. The rotation about the C—C bond might make a void space between neighbouring molecules; such a void space might cause inversion of the chiral group of the

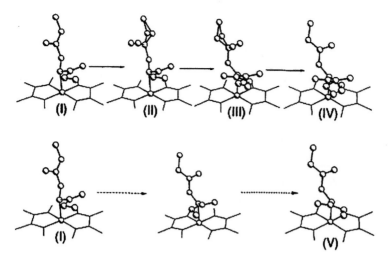

Figure 5-2. Two-step inversion of (R)-1-methoxycarbonylethyl group bonded to the cobalt atom in a cobaloxime complex. The upper figure shows the change of the chiral group in the four stages, (I) initial, (II) after 2 h, (III) after 5 h, (IV) after 10 h and (V) after 24 h. The lower figure indicates the intermediate structure between (I) and (V) stages

neighbouring molecule. By this mechanism such a bulky group can be inverted to the opposite configuration whilst maintaining the single-crystalline form.

The two-step inversion prompted an idea that various metastable intermediate structures might be observed if the disordered structures at the intermediate stages were analyzed carefully. We review here several examples of metastable reaction intermediates and unstable excited-state molecules.

PHOTOCHROMISM OF SALICYLIDENEANILINE DERIVATIVE

The first example is the structure of the metastable red species of a salicylideneaniline derivative, *N*-3,5-di-*tert*-butylsalicylidene-3-nitroaniline, which, when the pale yellow crystal is irradiated with a mercury lamp, shows photochromism with the largest lifetime among various salicylideneaniline derivatives [9]. Much research on this subject has been undertaken but the structure of the photo-coloured red species remains controversial [10].

The most important point for crystalline-state photoreactions is the selection of the wavelength of the light. If light with a wavelength at the absorption maximum were irradiated, only surface molecules of the crystal might react and the crystallinity of the surface might be broken, because the light cannot penetrate into the crystal. The inner part of the crystal remains unreacted. Light with much larger wavelength than that of the absorption maximum was proposed to penetrate readily into the crystals [11]. Absorption spectra of this crystal appear in Figure 5-3; the solid and dashed curves are absorption spectra of the stable pale yellow and

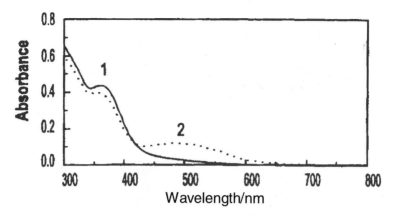

Figure 5-3. Absorption spectra of initial pale yellow crystal (**1**, solid curve) and red-colored crystal (**2**, dotted curve)

metastable red species, respectively. Although wavelengths at the absorption edge of the pale yellow species, 450–500 nm, are effective to form the red species within the crystal, these wavelengths would be absorbed effectively by the metastable red species so that it can revert to the pale yellow species. For this reason the structure of the red species has not been analyzed. We used light from a laser with wavelength 730 nm; this light has half the energy of that at the absorption maximum of the pale yellow species, 365 nm. This laser light is not effectively absorbed by the red species but is effectively absorbed by the pale yellow species in a two-photon excitation [12]. Irradiated with laser light at 730 nm for 4 h at room temperature, the crystal turned dark red.

The crystal structures before and after laser irradiation at 90 K are shown in Figures. 5-4a and 5-4b. The disordered structure was observed after irradiation. The minor part of the disordered structure, of which the occupancy factor is 0.104(2), is significantly different from the major part, which is the same as the original pale yellow species. The minor part should reflect the structure of the metastable red species. When the crystal was irradiated with light at 530 nm, the red colour of the crystal disappeared, and the structure analyzed was the same as that of the pale yellow crystal.

The structure before irradiation indicated that the pale yellow species exists in the *enol* form, that is, the proton is attached to the O1 atom as shown in Figure 5-4c. In the red species, the proton is translocated to the imine nitrogen atom and the *cis-keto* form produced is transformed to the *trans-keto* form as shown in Figure 5-4d. An interconversion between the *enol* and *trans-keto* forms is clearly responsible for the colour changes.

Although the structural alteration between the *enol* and *trans-keto* forms is large, the peripheral parts of both forms are similar, as shown in Figure 5-4b. For this reason such a large motion might occur reversibly without degradation of the crystallinity.

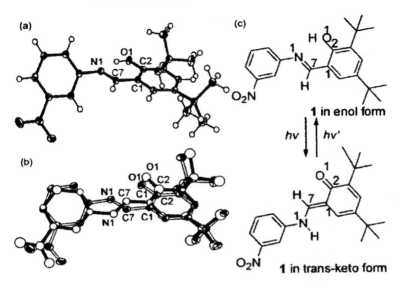

Figure 5-4. Photochromism of N-3,5-di-tert-Butylsalicylidene-3-nitoroaniline. (a) molecular structure before irradiation, (b) molecular structure after irradiation, (c) structural change between the enol form to the trans-keto form in the photochromism

The reason that the coloured species of this crystal has the largest lifetime was explained on comparing the photochromism of the polymorphic crystal. When the molecule adopts a trans-keto form, the N1-H group forms a hydrogen bond with the neighbouring molecule in the crystal; the trans-keto form becomes stabilized by the hydrogen bond.

RADICAL-PAIR FORMATION FROM HEXAARYLBIIMIDAZOLYL DERIVATIVES

The second example involves the formation of a radical pair of hexaarylbiimidazolyl derivatives. A crystallographic investigation of an energetically unstable reactive radical pair seems never to have been reported, although stable radicals such as radical ion pairs have been analyzed with X-rays. When the crystal of 2,2′-di(o-chlorophenyl) 4,4′5,5′-tetraphenyl-biimidazole (o-Cl-HABI) is irradiated with UV light, the central N—C bond is cleaved and the radical pair is produced as shown in Figure 5-5. The radical pair returns readily to the original dimer molecule when kept at room temperature. The HABI derivatives have been used as polymerization photo-initiators in imaging materials and holographic photopolymers [13].

The molecular structure before irradiation is shown in Figure 5-6 [14]. A crystal in the diffractometer was cooled below 103 K and irradiated with a high-pressure mercury lamp for 20 min. The intensity data were collected at 103 K. The analyzed structure is disordered as shown in Figure 5-7.

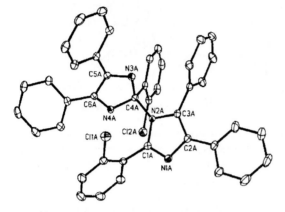

Figure 5-5. Radical pair formation from 2,2′-di(o-chlorophenyl)-4, 4′5, 5′- tetraphenylbiimidazole

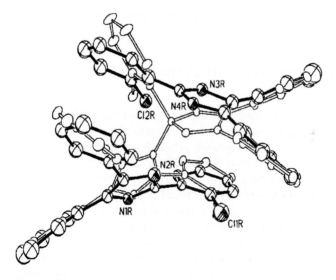

Figure 5-6. Molecular structure before irradiation. Two five-membered rings are connected by an N—C bond

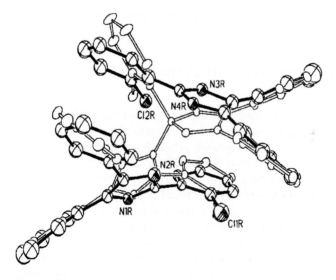

Figure 5-7. Molecular structure after irradiation. The molecule with white bonds is the original dimer molecule and the two molecules with black bonds are produced radical pair. The ellipsoids are drawn at 30% probability level

The minor part is the planar radical pair that has an occupancy factor about 0.10. The small intermolecular distances within the radical pair are in a range between 3.1 A and 3.5 A; such short contacts should enable the coupling of the radical pair to resume the original dimer molecule upon heating. The CW-ESR measurements revealed that, after irradiation at 77 K, the triplet signal appeared in a single-crystalline sample but disappeared when the crystal was warmed to 183 K.

The ESR experiments revealed that a triplet-state radical pair, RP II, was produced at 2 K on exposure to UV light but the signal disappeared at 30–40 K, although another triplet-state radical pair, RP I, with another D value appeared above 40 K [15]. To examine the structural change between species RP I and II, using a X-ray vacuum camera we collected intensity data before and after irradiation at 24 K and 30 K at beamline BL02B1 of SPring-8 [16]. The crystal after irradiation at 30 K was warmed to 40 K and 70 K, and intensity data were collected at both temperatures. Although the structural differences between the radical pairs RP II and RP I are small, as shown in Figure 5-8, the distance between atoms C2R and C5R, 0.05Å, for which spin densities are estimated to be the greatest with density-functional calculation, agrees satisfactorily with an estimated change of distance, 0.06Å, according to ESR measurements.

When the mother compound, HABI, was used, unexpected intermediates were observed [17]. The crystal structure of HABI at 108 K was analyzed with X-rays. The bond distances and angles are identical to the corresponding ones of o-Cl-HABI. A pale yellow crystal of HABI at temperatures under 108 K was irradiated with a high-pressure mercury lamp for 30 min. The crystal turned black; the crystal structure was revealed to be disordered structure as shown in Figure 5-9.

Although the space group P-1 was retained after irradiation, the two molecules across an inversion centre were much altered. The C—N bonds connecting the two

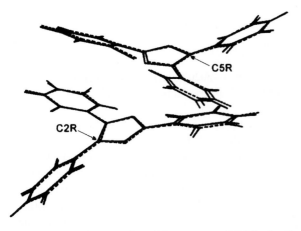

Figure 5-8. Superimposed view of the structures of RPI (broken lines) and RPII (solid lines)

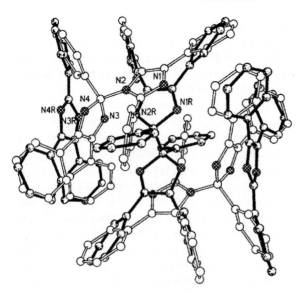

Figure 5-9. Disordered structure after photoirradiation. The two molecules with white bonds are initial 1,2′-dimer molecules whereas the molecules with black bonds are one 2,2′-dimer and two lophyl radicals

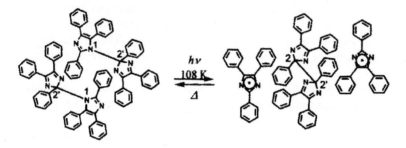

Figure 5-10. Reaction scheme with molecular swapping for HABI

imidazolyl groups were cleaved in both molecules across the inversion centre to produce a 2,2′-dimer and two lophyl radicals in yields 13.9%. The reaction scheme is shown in Figure 5-10.

The 2,2′-dimer, which is known as a piezodimer, has a C2—C2′ bond instead of the N1—C2′ bond of the original 1,2′-dimer. The structure of the 2,2′-dimer was determined for the first time, although the existence of the 2,2′-dimer was deduced spectroscopically. On warming to 25 °C, the black crystal returned to the original pale yellow crystal. The ESR spectra of a single crystal of HABI after irradiation showed a broad doublet signal corresponding to free lophyl radicals. Infrared spectra before and after irradiation indicate the decrease and increase of 1,2′- and 2,2′-dimers, respectively.

The structure after irradiation at 50 K was analyzed at beamline BL02B1 of SPring-8. The 2,2′-dimer was not observed but four lophyl radicals appeared in the disordered structure. On warming to 100 K, the 2,2′-dimer was observed with X-rays. Thermal motion of the lophyl radicals and a void space around the radicals might be important to produce the 2,2′-dimer [18].

FORMATION OF AN UNSTABLE CARBENE

Our third example is the unstable structure of a carbene. The work of Staudinger about 1913 on the decomposition of diazo compounds contributed much to the recognition of carbenes as reaction intermediates [19]. Since then, triplet carbenes stable enough for X-ray analysis have not yet been realized, although singlet carbenes were isolated by Arduengo in 1991 [20]. The reason is partly that a triplet carbene is less readily stabilized thermodynamically than its singlet counterpart. Tomioka et al. reported the formation of a carbene as an intermediate upon photolysis of bis(triaryl)diazomethane but isolation of the carbene was impracticable [21]. We tried to analyze the carbene structure using the 2,4,6-trichlorophenyl derivative as shown in Figure 5-11 [22].

As the produced carbene dimerizes readily, the photolysis and collection of X-ray data were performed for a sample at 80 K. The crystal was irradiated at 365 nm for 2 h using a high-pressure mercury lamp and a band-pass filter in combination. Although the crystal turned from yellow to red, there was no evidence for deterioration of the crystal quality. Figure 5-12 shows the molecular structures before and after irradiation. The dihedral angle between the two phenyl rings before irradiation is 70.2(1)°.

After irradiation new features appeared in the difference electron-density map; these features were assigned to the photoinduced carbene and a nitrogen molecule, which is shown with black bonds. The occupancy factor became 0.197(5). The produced molecule is shown in Figure 5-13. The carbene bonds, C1A–C11A and C1A–C21A are 1.437(15) and 1.423(16) Å, respectively. The carbene angle of C11A–C1A–C21A is 142(2)°. The nitrogen molecule takes an ordered structure; the bond distance is 1.05(3). The dihedral angle between the two phenyl rings, 76(2)°, differs insignificantly from that of the original molecule, 70.2(1)°. Further photoirradiation or warming the crystal resulted in a loss of crystallinity.

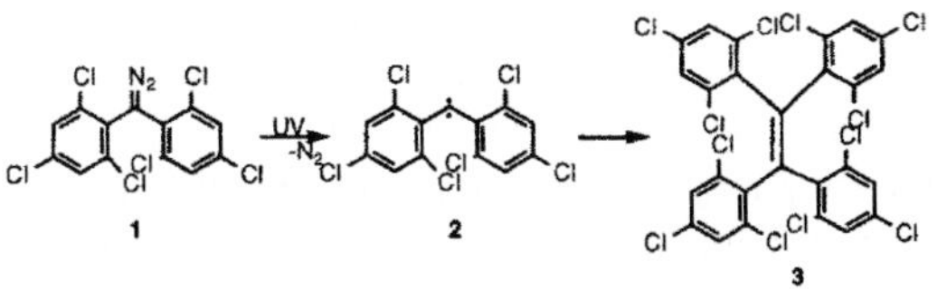

Figure 5-11. Photolysis of bis(2,4,6-trichlorophenyl)diazomethane

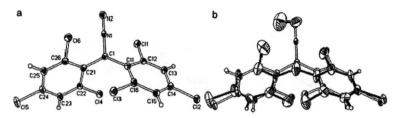

Figure 5-12. Molecular structure of di(2,4,6-trichlorophenyl)diazomethane before (a), after (b) the photoirradiation. The molecules with black bonds after irradiation are the produced carbene and dinitrogen molecule

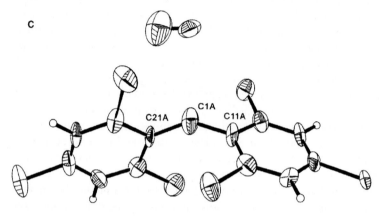

Figure 5-13. Molecular structures of the photo produced ones, dinitrogen molecule and the carbene. The thermal ellipsoids are drawn at 50% probability level

ESR spectra of a single crystal after irradiation at 77 K showed no clear signal ascribed to the triplet carbene but a weak doublet signal. The latter might be caused by a strong exchange coupling of the triplet states and an antiferromagnetic interaction. We therefore performed calculations on the triplet and singlet carbenes. The energy of the triplet carbene is less than that of the singlet carbene if the carbene angle is 142°, and the infrared spectra are consistent with those calculated for the triplet carbene.

Crystals of the related bis(triaryl)diazomethane were examined in the same way, replacing chlorine with bromine or a methyl group, with analogous results [23].

Nitrene Formation

Photochemical reactions of derivatives of phenyl azide have been intensively investigated in diverse fields such as synthetic chemistry, material science and industrial applications, [24] since Bertho proposed the reaction intermediate, nitrene, as shown in Figure 5-14, [25] but the structure of the nitrene has never been analyzed. More

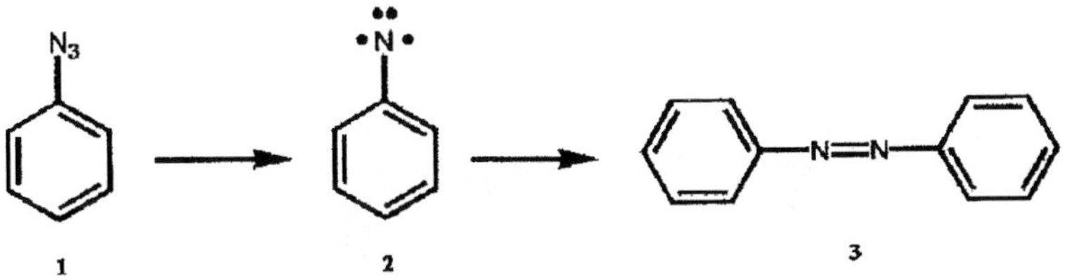

Figure 5-14. A mechanism proposed by Bertho (1924)

than 15 phenyl azides were prepared, and the crystals were exposed to the high-pressure mercury lamp with filters. The crystals decomposed upon irradiation except those of 2-azidebiphenyl and 1-azido-2-nitrobenzene, which showed gradual variations of cell dimensions [26]. For those two crystals, the structures before and after irradiation were analyzed with X-rays, and the reactions were analyzed.

The first example is of the nitrene produced from 2-azidobiphenyl. Figures 5-15a and 5-15b show the structures before and after irradiation, respectively, at 80 K [26]. Before irradiation the short contact of the N1 atom within the molecule is 3.091(1)Å, which is nearly the same as the sum of van der Waals radii (3.07Å). After exposure for 2 h, the features appearing in the difference electron-density map showed the nitrene and dinitrogen, which are shown with black bonds in Figure 5-15b. The structure of dinitrogen is clear and the occupancy is about 0.20. In contrast, the structure of the photo-induced nitrene is unclear because the nitrene is approximately superimposed upon the original 2-azidobiphenyl molecule; it is impossible to estimate the nitrene bond distance, C1—N1.

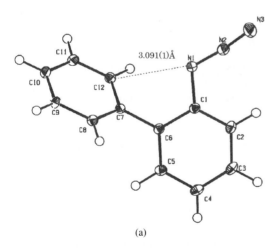

(a)

Figure 5-15. Molecular structures of 2-azidobiphenyl before and after the irradiation

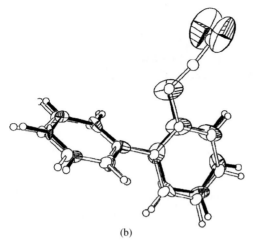

(b)

Figure 5-15. (Continued)

The second example is a photoproduct through a nitrene from 1-azido-2-nitrobenzene, which is a well studied precursor for syntheses of heterocyclic rings via photolysis and pyrolysis in both liquid and gaseous phases [27]. The molecular structure before irradiation at 80 K is shown in Figure 5-16 [28]. The intramolecular

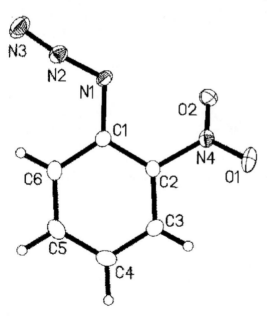

Figure 5-16. Molecular structure of 1-azido-2-nitrobenzene before irradiation. The thermal ellipsoids are drawn at 505 probability level

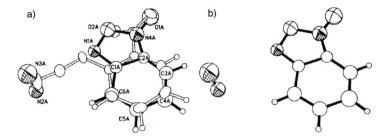

Figure 5-17. Molecular structure after irradiation. (a) disordered structure with the original and produced molecules, (b) the photo produced benzofuroxan molecule

distance between N1 and O_2, 2.710(1)Å, is significantly smaller than the sum of van der Waals radii (3.07Å). The dihedral angle between the nitro group and the phenyl ring is 37.6(1)° to avoid steric repulsion.

The crystal at 80 K was irradiated with light at wavelength 436 nm using a high-pressure mercury lamp and filters in combination for 30 min. The analyzed structure showed disorder as presented in Figure 5-17a. The structures of the photoproduced benzofuroxan (2,1,3-benzoxa-diazole 1-oxide) and dinitrogen shown in Figure 5-17b are clear; the occupancy factor is about 0.10. This structure indicates that the produced nitrene makes a bond with the neighbouring oxygen atom of the nitro group, O_2, to become a five-membered group.

The ESR work indicated that the signal due to the triplet nitrene was observed at 5.4 K but that the signal disappeared at 80 K, consistent with the X-ray result. Infrared spectra indicated that photolysis of this compound at 7 K yielded benzo-furoxan as a major product and 1,2-dinitrosobenzene as a minor one in agreement with the report for a Xe matrix [29]. The latter 1,2-dinitrosobenzene disappeared at 80 K.

The third example is the nitrene produced from an acid-base complex. Although the production of a nitrene was clearly confirmed in the above examples, the nitrene structure remained uncertain because the nitrene is almost overlapped upon the original azide molecule or exists only at a low temperature. To obtain a precise structure of nitrene, we formed the acid-base complex on introducing a carboxyl group onto the phenyl ring as shown in Figure 5-18, in which 3-azidobenzoic acid

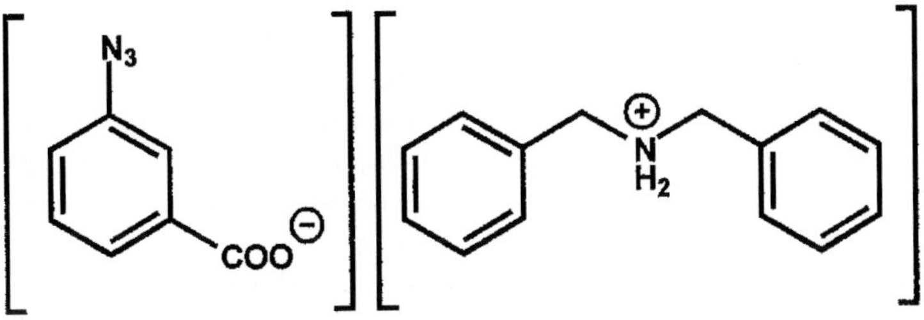

Figure 5-18. Acid-base complex formation between 3-azidobenzoic acid and dibenzylamine

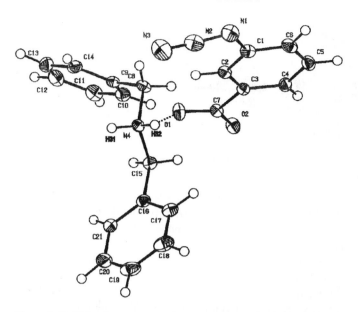

Figure 5-19. Molecular structure before irradiation. The hydrogen atom of the carboxyl group of 3-azidobenzoic acid is transferred to the nitrogen atom of the bibenzylamine and makes a strong hydrogen bond

is hydrogen-bonded with dibenzylamine. To observe unstable intermediates in the crystalline-state reactions, the formation of this acid-base complex, introduced in the photoisomerization of a cobaloxime complex, [30, 31] has the following three merits: the void space around the azide group might be expanded so that the crystal is not decomposed; the reactive molecules are diluted in the complex crystal so that the light can penetrate readily into the crystal, and the crystal is readily formed. Figure 5-19 shows the molecular structure of the complex before irradiation at 25 K at beamline BL02B1 of SPring-8 [32].

The carboxyl group of the azide makes a hydrogen bond with the imino group of the amine, and the proton of the carboxyl group is transferred to the imino group. The crystal at 25 K was irradiated with a high-pressure mercury lamp through a band-pass filter for 2 h. The analyzed structure clearly indicates that a nitrene and a dinitrogen were produced in the crystal although the occupancy factor was only 0.075 as shown in Figure 5-20. The amine molecule occupies almost the same position, and a hydrogen bond connects the two molecules. Because the hydrogen bond fixes the two molecules, the produced nitrene slides significantly to avoid a short contact with the produced dinitrogen. The length of the nitrene bond, C1R-N1R, is 1.34(4)Å, in satisfactory agreement with a calculation, 1.343, at the CASSCF(6,6)/6-31G* level. The other bond distances between the structures of the azide and nitrene are the same within experimental error.

The same combination between 3-azidobenzoic acid and dibenzylamine of the crystal was produced in another form, II [33]. In that crystal, the two azide groups

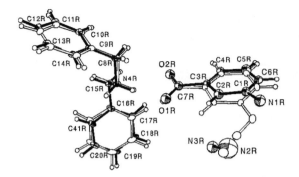

Figure 5-20. Molecular structure after the irradiation. The hydrogen bond is conserved and the structure is not changed. The nitrene bond moved from the original position to a considerable extent, since the carboxyl group is fixed by the hydrogen bond

of neighbouring molecules across an inversion centre closely contact each other, as shown in Figure 5-21. When the crystal was irradiated with the Hg lamp, the two nitrenes might be produced across the inversion centre. Two dinitrogen molecules are confirmed to be produced about the original azide molecule, as shown in Figure 5-21. Across the inversion centre, 3,3-carboxyazobenzene was clearly produced, although the occupancy factor is only 0.078(4). As the occupancy factor of the dinitrogen molecules is 0.189(4), the nitrene should exist about 11%, but the orientation is not fixed; the structure might not appear in the difference electron-density map. Figure 5-21 shows the produced structure. Such a dimer formation is the same as that proposed by Bertho about 80 years ago [25].

When the dibenzylamine was replaced with other amines, several novel reaction intermediates such as a seven-membered ring from a benzene ring or a reaction product with a five-membered ring were observed. These results indicate clearly that the produced nitrene proceeds to the next reaction depending on the environment in the crystalline lattice [33].

Excited Molecules

Several excited-state structures have been analyzed with synchrotron radiation. A combination of excitation by laser light and Laue diffraction using polychromatic synchrotron radiation made it possible to obtain the excited-state structure [34, 35]. Another method proposed is stroboscopic, in which molecules in a crystal were repeatedly excited with a pulsed laser and the structural change was probed in microseconds immediately after the excitation with pulsed monochromatic X-rays from the synchrotron [36]. Although such temporally resolved diffraction is attractive to analyze transient irreversible structural changes, such techniques are unnecessary for the observation of reversible structural changes between the two states, ground and excited, even if the change occurs within a picosecond. The diffraction should be caused by the periodic structure composed of the equilibrium between the ground- and excited-state molecules. If the concentration of the excited molecules

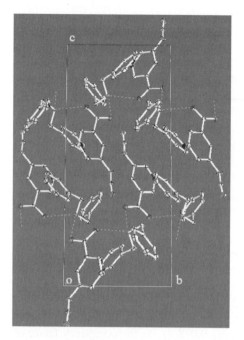

Figure 5-21. (a) Crystal structure of form II. The two azido groups are faced at the inversion center (see color plate section)

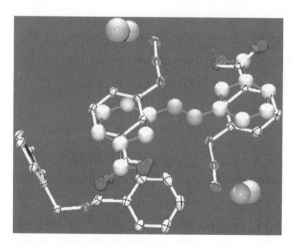

Figure 5-21. (b) molecular structure after the irradiation (see color plate section)

exceeds the threshold value in the equilibrium structure, the structure of the excited molecule can be analyzed with the diffraction data during irradiation. To obtain the structure of the excited molecule, the concentration or occupancy factor of excited molecules is of the utmost importance. The light of wavelength at the absorption

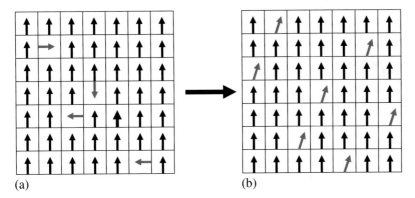

Figure 5-22. Excited molecules produced in the crystalline lattice. (a) at the early stage, (b) at the equilibrium state (see color plate section)

maximum might excite only surface molecules, and be unable to excite molecules within the crystal, as mentioned above. Although light of greater wavelength can penetrate into the crystal more effectively, the number of photons absorbed would be small and an extensive duration would be required for data collection.

The second important point is that the excited molecules are produced in a crystal at random during photoirradiation in early stages as shown in Fig. 5-22a. The excited molecules produced then gradually alter the original lattice to become the equilibrium lattice as shown in Figure 5-22b. With the diffraction method the structure of excited molecules with scattered orientations can not be analyzed, but can be analyzed only if the excited molecules have approximately the same orientation in the crystalline lattice of the equilibrium state.

To observe the excited-state molecules, we sought to develop a conventional photo-crystallographic method utilizing irradiation with continuous wavelengths and continuous X-rays with monochromatic wavelength. The crystal would be in an equilibrium state between the ground and excited states when it was photoirradiated and kept at low temperatures. The light of wavelength greater than that of the absorption maximum was selected to allow the light to penetrate effectively into the crystal. If the proportion of excited molecules exceeds the threshold value (ca. 5%) at the equilibrium state, the reversible structural change, no matter how quickly it occurs, should be observed with this method.

As a diplatinum complex $[Pt_2(pop)_4]^{4-}$ (pop = diphosphite) has an excited state with a long lifetime, ca. 7 μs, we selected it as the first sample [37]. For the cations, five alkylammonium cations were selected, as shown in Figure 5-23. For the tetrabutylammonium (Bu), tetrapentylammonium (Pn) and benzyltriethylammonium (Bzte) cations, there are two crystal forms, named Bu1 and Bu2, Pn1 and Pn2, and Bzte1 and Bzte2, respectively. On analysis of these eight crystal structures, the Pt anions in all crystals except Bzdmp were found to have two extra protons: the Pt anions should be represented as $[Pt_2(pop)_2(popH)_2]^{2-}$; hence only two cations are included for a Pt anion in those crystals. The Bu2 and Bzte2 crystals have a

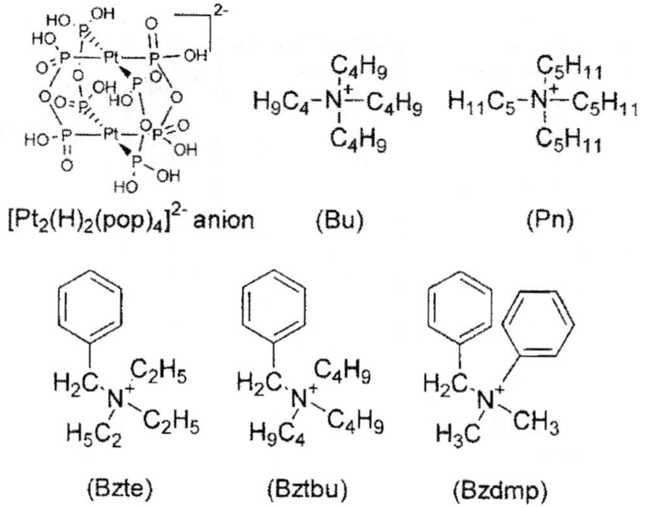

[Pt₂(H)₂(pop)₄]²⁻ anion   (Bu)   (Pn)

(Bzte)   (Bztbu)   (Bzdmp)

Figure 5-23. Structure of the Pt complex anion and five alkylammonium cations

methanol molecule in the asymmetric unit. The Pn2 crystal has a disordered part in the alkyl groups. The remaining four crystals – Bu1, Pn1, Bzte1 and Bztbu – were compared for their photochemical changes as they have similar conditions.

A crystal was mounted on the diffractometer and irradiated with a xenon lamp and a combination of filters to selected wavelengths between 400 and 550 nm. The temperature was 173 K for Bu1 and Pn1 and 103 K for Bzte1 and Bztbu. Such temperatures were selected for which the largest volume change of the unit cell was obtained before and during irradiation. The unit-cell dimensions were significantly altered during photoirradiation. The changes of the unit-cell volumes for the four crystals with dark and light conditions are summarized in Table 5-1. All crystals significantly shrank in light. For Bu1, Pn1 and Bzte1 the ratio of $\Delta|V_{on}-V_{off}|/V_{off}$ became 0.6–0.7%.

To examine whether the unit-cell change is affected by the crystal size, the powder diffraction patterns of Pn1 with light off and on were measured at SPring-8, as shown in Figure 5-24a. The diffraction features shift to angles greater than those with darkness. The shifted pattern returned to the original one with darkness. The sharp diffraction features with light indicate no mixing of the original

Table 5-1. Change of the unit-cell volume (Å) at light-off (V_{off}) and light-on (V_{on}) stages for Bu1, Pn1, Bzte1 and Bztbu crystals

	$V_{off}/\text{Å}^3$	$V_{on}/\text{Å}^3$	$\Delta(V_{on}-V_{off})$	$\Delta|V_{on}-V_{off}|/V_{off}$ (%)
Bu1	1320.18(3)	1312.16(5)	−8.02(4)	0.61
Pn1	3046.45(7)	3026.12(17)	−20.33(12)	0.67
Bzte1	2108.78(12)	2094.92(11)	−13.86(11)	0.66
Bztbu	2696.42(16)	2691.27(12)	−5.15(14)	0.19

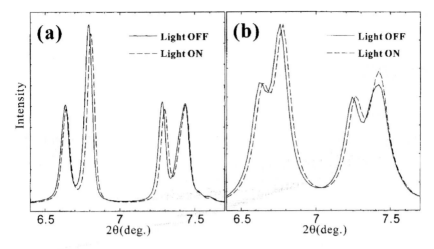

Figure 5-24. (a) The powder diffraction patterns of the Pn1 complex at the light-off and light-on stages at the BL02B2 beamline of SPring-8. (b) The plots of the calculated diffraction angles based on the unit-cell dimensions of the single crystals at the light-off and light-on stages

diffraction features. This condition indicates that photo-excited and ground-state molecules reach an equilibrium state and that a crystalline lattice corresponding to the equilibrium state was produced. The diffraction pattern in Figure 5-24b shows the calculated diffraction pattern based on the single crystal change. Both photo-produced structures at the equilibrium state for the powdered crystals and the single crystal are clearly equivalent, as the two diffraction patterns are essentially the same. The equilibrium structure is independent of the number of incident photons, and the number of excited molecules reaches a maximum at the equilibrium state.

There are no significant structural differences in the cation molecules between dark and light, but the bond distances in the Pt complex anions are significantly altered from the corresponding ones with darkness. We claim safely that the Pt complex has $D_{4h}(4/m)$ symmetry except for the exocyclic P=O or P—OH groups. Table 5-2 shows the averaged Pt—Pt, Pt—P and P—O(—P) bond distances in the Pt anions of the four crystals, assuming D_{4h} symmetry. The Pt—Pt and Pt—P distances become significantly smaller with light relative to darkness. The changes of the other bond distances are within the experimental error. The contractions of the Pt—Pt and Pt—P bonds have ranges 0.0031(3)–0.0127(5)Å and 0.0019(12)–0.0085(14)Å, respectively. Such differences are probably caused by the varied concentrations of excited molecules in each crystal. The difference in transmittance of the incident light and in the packing around the Pt complex might be responsible for the varied concentrations of the excited molecules. The ratios $|\Delta(\text{Pt}—\text{Pt})| / |\Delta(\text{Pt}—\text{P})|$ in the two crystals of Pn1 and Bzte1 are 1.53 and 1.49, respectively, thus being similar. Hence the excited Pt complex anion has the same structure in each crystal, although the concentrations of the excited molecules might differ among the four crystals. The change at the equilibrium stage is schematically drawn in Figure 5-25.

Table 5-2. Averaged bond distances (Å) of Pt—Pt, Pt—P, P—O(—P), P=O and P—O(—H) at light-off (D_{off}) and light-on (D_{on}) stages and their differences (ΔD)

	D_{off}	D_{on}	ΔD		D_{off}	D_{on}	ΔD
Bul				Bztbu			
Pt—Pt	2.9419(3)	2.9381(3)	−0.0038(3)	Pt—Pt	2.9394(3)	2.9363(4)	−0.0031(3)
Pt—P	2.3338(10)	2.3269(10)	−0.0069(10)	Pt—P	2.3254(12)	2.3235(12)	−0.0019(12)
P—O(—P)	1.632(3)	1.629(3)	−0.003(3)	P—O(—P)	1.630(4)	1.628(4)	−0.002(4)
P—O	1.535(3)	1.530(3)	−0.005(3)	P=O	1.527(4)	1.521(4)	−0.006(4)
P—O(—H)	1.563(3)	1.561(3)	−0.002(3)	P—O(—H)	1.562(3)	1.563(4)	0.001(4)
Pnl							
Pt—Pt	2.9546(3)	2.9465(3)	−0.0081(3)				
Pt—P	2.3318(11)	2.3265(11)	−0.0053(11)				
P—O(—P)	1.630(3)	1.627(4)	−0.003(3)				
P=O	1.522(3)	1.518(3)	−0.004(3)				
P—O(—H)	1.565(3)	1.561(4)	−0.004(4)				
Bztel							
Pt—Pt	2.9726(4)	2.9599(6)	−0.0127(5)				
Pt—P	2.3374(13)	2.3289(14)	−0.0085(14)				
P—O(—P)	1.639(4)	1.628(6)	−0.011(5)				
P=O	1.538(4)	1.532(6)	−0.006(5)				
P—O(—H)	1.558(4)	1.553(5)	−0.005(5)				

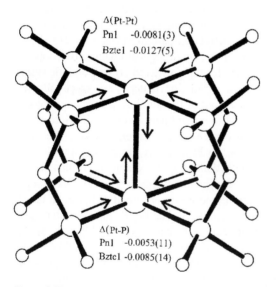

Δ(Pt-Pt)
Pnl   −0.0081(3)
Bztel −0.0127(5)

Δ(Pt-P)
Pnl   −0.0053(11)
Bztel −0.0085(14)

Figure 5-25. A schematic drawing of the structural change of the Pt complex anion at the excited state

The complex anion of $[Pt_2(pop)_4]^{4-}$ has a strong singlet absorption band associated with a weakly luminescent state, $^1A_{2u} \leftarrow {}^1A_{1g}$ ($d\sigma^* \rightarrow p\delta$), which can populate a strongly emissive long-lived triplet state, $^3A_{2u}$ ($d\sigma^* \rightarrow p\delta$), that can undergo many chemical reactions. Much work has been performed to elucidate the spectra, photophysics and photochemistry associated with excited electronic states of $[Pt_2(pop)_4]^{4-}$ in the region 300–450 nm. According to vibronically resolved absorption and emission spectra of low-temperature crystalline samples of $[N(C_4H_9)_4]_4[Pt_2(pop)_4]$, the Pt—Pt bond for the $^3A_{2u}$ state becomes shorter by about 0.21Å than in the $^1A_{1g}$ ground state [38, 39]. From the combination of the EXAFS method with rapid-flow laser spectra in an aqueous solution with glycerol, the $^3A_{2u}$ state with a lifetime about 4 μs is estimated to undergo a contraction of the Pt—Pt distance 0.52 ± 0.13Å and of the Pt—P distance 0.047 ±0.011Å relative to the $^1A_{1g}$ ground state [40]. From a resonance-Raman intensity analysis of the $^1A_{2u} \rightarrow {}^1A_{1g}$ transition at room temperature, the Pt—Pt contraction was about 0.225Å [41]. The temporally resolved stroboscopic method indicated that the Pt—Pt bond distance in the crystal of $[N(C_2H_5)_4]_3[Pt_2(pop)_3(popH)]$ contracted 0.28(9)Å, and the population of the excited state was 2.0% at 17 K [36]. Using the low-temperature vacuum camera installed at BL02B1 of SPring-8 the Pt—Pt bond contraction was reported to be 0.23(4)Å and the population of the excited state was 1.4(2)% [42].

All these results indicate that the Pt—Pt bond becomes shorter by 0.2–0.3Å and the Pt—P bond contracts 0.1–0.2Å. Consistent with these observed bond contractions of Pt—Pt and Pt—P distances, the concentrations of excited molecules might be 4–5% in Pn1 and Bzte1 crystals and 1–2 % in Bu1 and Bztbu crystals in light. This estimate agrees satisfactorily with the populations observed in the above experiments, but the above values are based on an assumption that the structure of the excited molecule in the solid state has nearly the same structure as that in the gaseous or solution state. As the excited molecule in the solid state might suffer from steric repulsion due to strong intermolecular interactions, a direct comparison of the excited molecule in the solid state with that in the gaseous or solution state might be inadequate.

The unit cell of a crystal of VO(acac)$_2$ (acac = acetylacetonato, $C_5H_7O_2^-$) (Figure 5-26) expanded significantly when the single crystal at low temperature was irradiated with the xenon lamp. The VO(acac)$_2$ complex was proposed to have a short-lived excited species due to a d-d transition under irradiation with visible light, [43] but the lifetime is too short to observe the emission. Because the anisotropy in expansion of the unit cell on photoirradiation clearly differed from the thermal one, it is possible to observe the excited structure of the VO(acac)$_2$ complex in the equilibrium state even if the lifetime of the excited state is small.

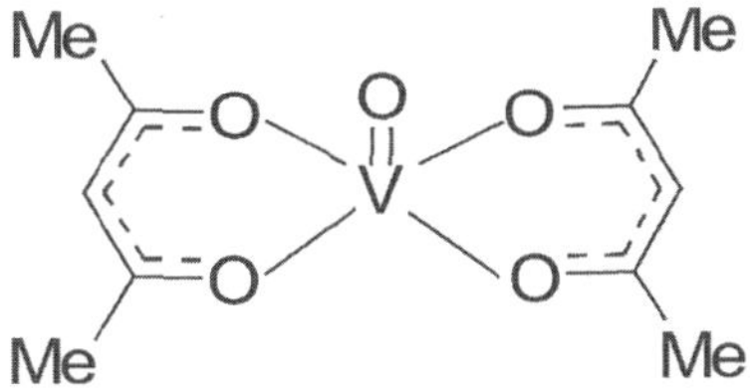

Figure 5-26. Structure of bis(acetylacetonato)oxobanadium(IV)

Table 5-3. Change of unit-cell dimensions due to photoirradiation (Δpc) and thermal expansion (Δtc) and the ratio
of the changes (Δpc/Δtc)

	Δpc	Δtc	Δpc/Δtc
a(Å)	+0.0178(2)	+0.0447(3)	0.398
b(Å)	+0.0202(3)	+0.0322(3)	0.627
c(Å)	+0.0273(4)	+0.0276(4)	0.989
$\alpha(°)$	+0.004(2)	+0.049(2)	0.008
$\beta(°)$	−0.006(2)	−0.072	0.083
$\gamma(°)$	−0.044(2)	−0.105(2)	0.419
V(Å³)	+4.07(3)	+6.66(4)	0.611

Δpc and Δtc are differences between 140K-off and 140K-
on and between 168K-off and 140K-off, respectively.

Intensity data were collected at four temperatures – 168, 160, 150 and 140 K –
with light and dark conditions [44]. At all temperatures the unit cell expanded in
light. To assess whether the change was caused by photoirradiation, the unit-cell
dimensions on photoirradiation at 140 K (photochemical change) were compared
with the corresponding ones caused by a temperature increase (thermal change),
which were obtained from the difference of unit-cell dimensions at 168 K and
140 K. Table 5-3 shows the photochemical and thermal changes and the ratio of
both changes. The expansion of the unit cell on photoirradiation is clearly more
anisotropic than that on thermal treatment.

Figure 5-27 shows the bond elongation at the equilibrium state with the light
on at 140 K. The four V—O bonds are averaged. The observed bond elonga-
tions of V=O and V—O, 0.0043(8) and 0.0045(7)Å respectively, are significant. The changes of the C—O and C—C bonds are within experimental error.
Such bond elongation indicates that the electrons of V—O bonds are transferred to the V=O bond in the d-d transition, as proposed after a theoretical
calculation.

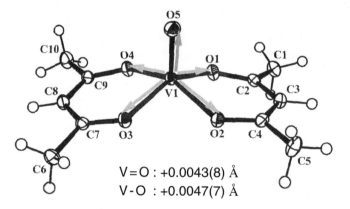

Figure 5-27. Elongation of V=O and V—O bonds at the excited state (see color plate section)

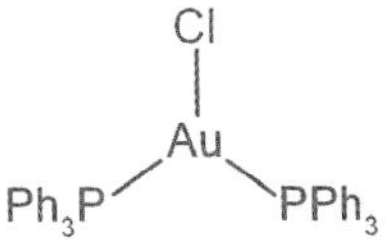

Figure 5-28. Structure of $[AuCl(PPh_3)_2]$

Table 5-4. Selected bond distances (Å) and angles (°) at 84 K and their difference, Δ_{on-off}

	Light-off stage	Light-on stage	Δ_{on-off}
Au1—P1	2.3278(4)	2.3222(4)	−0.0056(4)
Au1—P2	2.3457(4)	2.3403(4)	−0.0054(4)
Au1—Cl1	2.5821(4)	2.5764(4)	−0.0057(4)
P1—C111	1.8253(17)	1.8208(17)	−0.0045(17)
P1—C121	1.8285(17)	1.8245(18)	−0.0040(18)
P1—C131	1.8335(17)	1.8291(18)	−0.0044(18)
P2—C211	1.8342(17)	1.8287(18)	−0.0055(18)
P2—C221	1.8257(17)	1.8221(17)	−0.0036(17)
P2—C231	1.8251(17)	1.8212(18)	−0.0039(18)
P1—Au1—P2	136.446(15)	136.441(15)	−0.005(15)
P1—Au1—Cl1	115.184(14)	115.179(15)	−0.005(15)
P2—Au1—Cl1	108.276(14)	108.286(15)	+0.010(15)
Au1—P1—C111	120.45(6)	120.47(6)	+0.02(6)
Au1—P1—C121	110.10(5)	110.05(6)	−0.05(6)
Au1—P1—C131	109.63(6)	109.69(6)	+0.06(6)
Au1—P2—C211	112.21(6)	112.17(6)	−0.04(6)
Au1—P2—C221	109.25(5)	109.30(6)	+0.05(6)
Au1—P2—C231	119.75(6)	119.79(6)	+0.04(6)

A lattice shrinkage in the equilibrium state was found for the Au(I) complex shown in Figure 5-28. The three-coordinated Au(I) complex of $[AuCl(PPh_3)_2]$ has a long-lived excited state of lifetime 3.7 μs, at room temperature and shows phosphorescence with $\lambda_{max} = 511$ nm, after excitation at $\lambda = 330$ nm [45]. The unit-cell dimensions were measured with dark and light conditions at four temperatures – 175, 156, 136 and 84 K. The decrease of the unit-cell volume, 10.23Å^3, is largest at 84 K. The differences in bond distances and angles between dark and light at 84 K are compared in Table 5-4. The Au—Cl and the average Au—P bonds with light are significantly shorter by 0.0057(4) and 0.0055(4)Å respectively than the corresponding ones with darkness. These contractions correspond to an excitation from $^3E''$ to $^1A_{1'}$ derived from an electronic transition from the d_{xy} and d_{x2-y2} (anti-bonding) orbitals to the p_{z2} (bonding) orbital.[45,46,47] According to calculations involving $[Au(PH_3)]^{3+}$, the Au—P bonds are shortened in the excited state. The Au—P bond contraction was estimated to be 0.05Å from that calculation [47].

In the platinum, vanadium and gold complexes, the structural changes in the excited state were determined when the crystal structures were analyzed accurately with the light on, at the equilibrium state, but the bond shrinkage or elongation is somewhat smaller than those predicted in a calculation. This calculation was performed assuming a vacuum and no intermolecular interactions, whereas in the crystal the molecule is closely packed among other molecules; the structure can thus not alter as if the molecule were in vacuum. As the lifetime of the excited molecule might be several microsecond, the molecule should cycle between the excited state and the ground state rapidly in the equilibrium state. Thus the energy diagrams of the ground and excited states differ between a vacuum and a crystal. As shown in Figure 5-29 with dotted curves, the structure at the energy minimum in the excited state differs significantly from the corresponding one in the ground state in vacuum, but the structures at the energy minimum in the ground and excited states should be similar in the crystal because steric hindrance in the crystalline lattice is large as shown in Figure 5-29 (dotted curves). Thus the amount of the bond elongation and shrinkage is smaller than that predicted by theoretical calculation. Therefore, the concentration predicted in the Pt complexes should be the minimum one if there is no intermolecular interaction in the crystal. The real concentration of the excited molecules in the equilibrium state may exceed 10%. However, we must consider the definition of the excited molecule in such condensed substances as crystals.

In order to detect the diffraction data very rapidly, we have made a new photon-counting X-ray detector, MSGC, whose size is $10 \, \text{cm} \times 10 \, \text{cm}$. The detector can record one-shot data within 1 ms. If the crystal was rotated around the φ-axis on the four–circle diffractometer using the rotating anode X-ray generator and the MSGC detector, all the intensity data of a standard single crystal were collected within

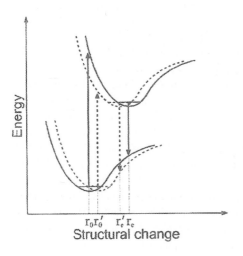

Figure 5-29. Proposed energy diagram under vacuum (solid curves) and in the crystal (dotted curves)

2 s and the structure was analyzed [48, 49]. When the Pn1 crystal was mounted on the diffractometer with the MSGC detector and was irradiated with the xenon lamp at the Beam line BL46XU at SPring-8, the change of the diffraction angle was recorded at 500 ms time intervals at 80 K, which is shown in Figure 5-30. The diffraction angle gradually increased (the unit-cell volume shrank) after the light was put on and it gradually decreased (the unit-cell volume expanded) after the light was off. It was emphasized that the change did not reach the equilibrium state within 20 s [50]. The quantitative discussion may be inadequate since the size of the crystal should effect the rate of the change. Further experiments should be necessary. However, it seems plausible that the time necessary to reach the equilibrium state may be in the order of second.

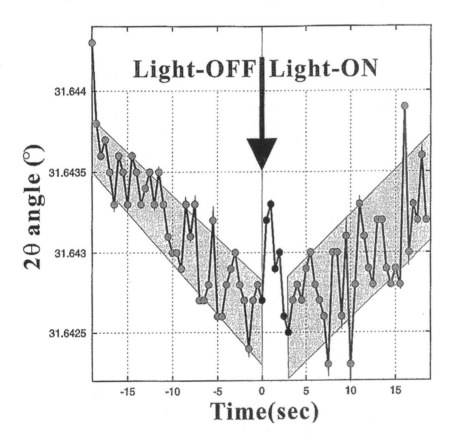

Figure 5-30. Gradual change in diffraction angle of the photoirradiated diplatinum complex crystal. When the light was put on at $t = 0$, the abrupt chnage occurred and then the diffraction angle gradually increased. When the light was put off at $t = 20$, the diffraction angle gradually decreased as shown in left part of the figure after the abrupt change for a short time

CONCLUSION

A variety of reaction intermediates have been analyzed by X-rays. If the chemical reaction due to photoirradiation or thermal change occurs in a single crystal with retention of the single crystal form (which is call crystalline-state reaction), the structure at any stage of the reaction can be obtained by X-ray analysis. The intermediate structure of the two step reaction was observed in the cobaloxime crystal. If the crystal was decomposed in the course of reactions, the reaction was stopped before decomposition and a small portion of the reaction intermediates or unstable species can be obtained as disordered structure with the original molecules by X-ray analysis at low temperatures. The structure of the unstable red species in the photochromic crystals of salicylidenanilines has been analyzed by X-rays. If the intermediates are unstable at room temperature or gas molecules should be produced after reaction, the cryo-method should be introduced. The crystal was mounted on the diffractometer and was cooled less than 100 K using the cold-nitrogen-gas-flow method. The crystal was irradiated with the Hg or Xe lamp and the intensity data were collected just before the diffraction peaks became significantly broader. The unstable radical pair, carbene and nitrene structures have been analyzed by X-rays. The structures of the excited molecules are, in general, very unstable even if the temperature is less than 100 K. Even the longest life-time of the excited molecule may be $10 \mu s$. However, the molecules go up and down between the ground and excited states reversibly. Within a few second, the crystal may reach the equilibrium state between the ground and excited states. The unit-cell dimensions gradually changed, expanded or shrank, from the ground state to the equilibrium state. In the equilibrium state, the structures of the excited molecules were analyzed as disordered structures with the ground state molecules at low temperatures. The shortening of the Pt—Pt and Pt—P bond distances were observed in the Pt complex crystals with different cation molecules. Moreover, the lengthening of the V=O and V—O bonds and the shortening of Au—P and Au—Cl bonds have been observed in the vanadium and the gold complexes, respectively. All of these structures have made clear the processes or routes in the chemical reactions by visualizing the intermediate structures. Such a dynamic analysis of X-ray diffraction will play more and more important role in the analysis of chemical reaction than before.

REFERENCES

1. Y. Ohashi and Y. Sasada, *Nature*, **267**, 142 (1977)
2. Y. Ohashi, *Acc. Chem. Res.*, **21**, 268 (1988)
3. Y. T. Osano, A. Uchida and Y. Ohashi, *Nature*, **352**, 510 (1991)
4. Y. Ohashi, "Chiral Photochemistry", Ed. by A. Ramamurthy and Y. Inoue, John Wiley, 2005
5. Y. Ohashi, T. Hosoya and T. Ohhara, *Crystallography Reviews*, **12**, 83 (2006)
6. Y. Ohashi, K. Yanagi, T. Kurihara, Y. Sasada and Y. Ohgo, *J. Am. Chem. Soc.*, **103**, 5805 (1981); Y. Ohashi, A. Uchida, Y. Sasada and Y. Ohgo, *Acta Cryst*, **B39**, 54 (1983)
7. Y. Ohashi and H. Uekusa, *J. Mol. Struc.*, **374**, 37 (1996)
8. Y. Ohashi, Y. Sakai, A. Sekine, Y. Arai, Y. Ohgo, N. Kamiya and H. Iwasaki, *Bull. Chem. Soc. Jpn.*, **68**, 2517 (1995)

9. J. Harada, H. Uekusa and Y. Ohashi, *J. Am. Chem. Soc.*, **121**, 5809 (1999)

10. M. D. Cohen and G. M. J. Schmidt, *J. Phys. Chem.*, **66**, 2442 (1962); K. Kownacki, A. Mordzinski, R. Wilbrandt and A. Grabowska, *Chem. Phys. Lett.*, **227**, 270 (1994)

11. V. Enkelmann, G. Wagner, K. Novak and K. B. Wagener, *J. Am. Chem. Soc.*, **115**, 10390 (1993)

12. D. A. Parthenopoulos and P. M. Rentzepis, *Science*, **245**, 843 (1989)

13. B. M. Monroe and G. C. Weed, *Chem. Rev.*, **93**, 435 (1993)

14. M. Kawano, T. Sano, J. Abe and Y. Ohashi, *J. Am. Chem. Soc.*, **121**, 8106 (1999)

15. J. Abe, T. Sano, M. Kawano, Y. Ohashi, M. M. Matsushita and T. Iyoda, *Angew. Chem. Intl. Ed.*, **40**, 580 (2001)

16. M. Kawano, Y. Ozawa, K. Matsubara, H. Imaizumi, M. Mistumi, K. Toriumi and Y. Ohashi, *Chem. Lett.*, **2002**, 2213 (2002)

17. M. Kawano, T. Sano, J. Abe and Y. Ohashi, *Chem. Lett.*, **36**, 1372–1373 (2000)

18. M. Kawano, K. Hirai, H. Tomioka and Y. Ohashi, submitted to *J. Am. Chem. Soc.*, (2007)

19. U. H. Brinker, Ed., *Advances in Carbene Chemistry*, JAI Press, Greenwich and Stanford, 1994, 1998, 2000. References are therein.

20. A. J. Arduengo, III, *Acc. Chem. Res.*, **32** 913 (1999)

21. H. Tomioka, *Acc. Chem. Res.*, **30** 315 (1997)

22. M. Kawano, K. Hirai, H. Tomioka and Y. Ohashi, *J. Am. Chem. Soc.*, **123** 6904 (2001)

23. M. Kawano, K. Hirai, H. Tomioka and Y. Ohashi, *J. Am. Chem. Soc.*, 129, 2383 (2007)

24. E. F. V. Scriven (ed.), "Azides and Nitrenes, Reactivity and Utility" Academic Press, New York (1984)

25. Bertho, *Ber.*, **57** 1138 (1924)

26. T. Takayama, *Master Thesis*, Tokyo Institute of Technology (2001)

27. A. R. Katritzky and M. F. Gordeev, *Heterocycles*, **35** 483 (1993)

28. T. Takayama, M. Kawano, H. Uekusa, Y. Ohashi and T. Sugawara, *Helv. Chem. Acta,* **86** 1352 (2003)

29. N. P. Hacker, *J. Org. Chem.*, **56**, 5216 (1991)

30. D. Hashizume and Y. Ohashi, *J. Chem. Soc., Perkin Trans. 2*, **1998**, 1931 (1998)

31. D. Hashizume and Y. Ohashi, *Perkin Trans. 2*, **1999**, 1689 (1999)

32. M. Kawano, T. Takayama, H. Uekusa, Y. Ohashi, Y. Ozawa, K. Matsubara, H. Imabayashi, M. Mistumi and K. Toriumi, *Chem. Lett.*, **32** 922 (2003)

33. T. Mitsumori, *Master Thesis*, Tokyo Institute of Technology (2005)

34. S. Techert, F. Schotte and M. Wulff, *Phys. Rev. Lett.*, **86**, 2030 (2001)

35. K. Moffat, *Chem. Rev.*, **101**, 1569 (2001)

36. C. D. Kim, S. Pillet, G. Wu, W. K. Fullagar and P. Coppens, *Acta Cryst.*, **A58**, 133 (2002)

37. N. Yasuda, H. Uekusa and Y. Ohashi, *Bull. Chem. Soc. Jpn.*, **77**, 933 (2004)

38. S. F. Rice and H. B. Gray, *J. Am. Chem. Soc.*, **105**, 4571 (1983)

39. A. E. Stiegman, S. F. Rice, H. B. Gray and V. M. Miskowski, *Inorg. Chem.*, **26**, 1112 (1987)

40. D. J. Thiel, P. Livins, E. A. Stern and A. Lewis, *Nature*, **362**, 40 (1993)

41. K. H. Leung, D. L. Phillips, C.-M. Che and V. M. Miskowski, *J. Raman Spectro.*, **30**, 987 (1999)

42. Y. Ozawa, M. Terashima, M. Mitsumi, K. Toriumi, N. Yasuda, H. Uekusa and Y. Ohashi, *Chem. Lett.*, **2002**, 62 (2002)

43. S. D. Bella, G. Lenza, A. Gulino and I. Fragala, *Inorg. Chem.*, **35**, 3885 (1996)

44. M. Hoshino, A. Sekine, H. Uekusa and Y. Ohashi, *Chem. Lett.*, **34**, 1228 (2005)

45. C. King, M. N. I. Khan, R. J. Staples and J. P. Fackler, Jr., *Inorg. Chem.*, **31**, 3236 (1992)

46. T. M. McCleskey and H. B. Gray, *Inorg. Chem.*, **31**, 1733 (1992)

47. K. A. Barakat, T. R. Cundari and M. A. Omary, *J. Am. Chem. Soc.*, **125**, 14228 (2003)

48. A. Ochi, H. Uekusa, T. Tanimori, Y. Ohashi, H. Toyokawa, Y. Nishi, Y. Nishi, T. Nagayoshi and S. Koishi, *Nucl. Inst. & Mat. Phys. Res.*, **A467–468**, 1148 (2001)

49. A. Takeda, H. Uekusa, H. Kubo, K. Miuchi, T. Nagayoshi, Y. Ohashi, Y. Okada, R. Orito and Toru Tanimori, *J. Synchrotron Rad.*, **12**, 820 (2005)

50. Y. Ohashi, *J. Cryst. Soc. Jpn.*, **46**, 59–64 (2004)

CHAPTER 6

IS POLYMORPHISM CAUSED BY MOLECULAR
CONFORMATIONAL CHANGES?

IVAN BERNAL

Abstract: The possibility that polymorphism is connected with variations of molecular conformations within a unit cell is explored graphically and numerically. The conclusion is that
no direct relationship exists between one and the other because polymorphic crystals are
found, equally frequently, in which the conformations of their molecules agree almost
exactly, or agree somewhat, or display large differences

INTRODUCTION

An enduring mystery of molecules is polymorphism [1, 2] – their ability to pack
together in repeating units within a crystal in multiple manners. Understanding why
this phenomenon occurs, how it occurs, and how to control it is a fundamental
problem of chemistry, physics, biology, pharmacy etc. A possible origin might
be variations of molecular conformations according to which molecules can adopt
multiple conformations that have approximately the same ground-state energies, or
there might exist a multi-well surface of potential energy on which the molecules
can readily transform from one conformation to another because the barriers are
small. As shown in Figure 6-1, the orientations of the pyridine ligands vary slightly
despite the fact that the pyridine-N—Cu bond is a single bond, but there is a prominent variation in the orientation of the nitro ligands. Analogously for the nickel
compound in Figure 6-2, the NCS ligands point in markedly varied directions,
indicating a conformational origin for the polymorphism relationship of these two
substances. Reference to Figure 6-3 shows that there is virtually no difference in
the conformation of the molecules of this cobalt compound despite its also having
a singly bonded pyridine ligand that might readily rotate about the Co—N bond;
the largest distance between chemically equivalent atoms is associated with C12
that does not belong to the pyridine ligand. In Figure 6-2 some equivalent atoms
are over 2.0 Å apart. Accordingly, a detailed examination of molecules in diverse

J.C.A. Boeyens and J.F. Ogilvie (eds.), Models, Mysteries and Magic of Molecules, pp. 137–165.
© 2008 *Springer.*

Copper pyridine JUCNIE02 vs JUCNIE03	Mean deviation of the atoms		
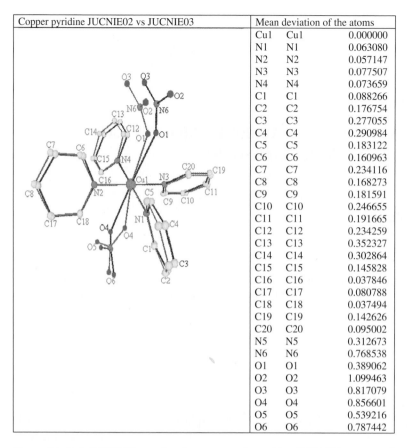	Cu1	Cu1	0.000000
	N1	N1	0.063080
	N2	N2	0.057147
	N3	N3	0.077507
	N4	N4	0.073659
	C1	C1	0.088266
	C2	C2	0.176754
	C3	C3	0.277055
	C4	C4	0.290984
	C5	C5	0.183122
	C6	C6	0.160963
	C7	C7	0.234116
	C8	C8	0.168273
	C9	C9	0.181591
	C10	C10	0.246655
	C11	C11	0.191665
	C12	C12	0.234259
	C13	C13	0.352327
	C14	C14	0.302864
	C15	C15	0.145828
	C16	C16	0.037846
	C17	C17	0.080788
	C18	C18	0.037494
	C19	C19	0.142626
	C20	C20	0.095002
	N5	N5	0.312673
	N6	N6	0.768538
	O1	O1	0.389062
	O2	O2	1.099463
	O3	O3	0.817079
	O4	O4	0.856601
	O5	O5	0.539216
	O6	O6	0.787442

Figure 6-1. Red: JUCNIE02 .... Space group: Pnna. Blue: JUCNIE03 .... Space group: Ccca Ref. [5]

classes should be undertaken to explore the possibilities that there is a systematic explanation for the occurrence of polymorphism and that it might have a molecular origin.

With this aim in mind, we conducted a search of CSD [3] and other sources (see Acknowledgements) to answer the question "Is this phenomenon conformationally driven?".

CONDITIONS FOR ACCEPTING ENTRIES IN THE SAMPLING

We searched CSD [3] according to the following conditions.

(a) There were no restrictions on the space group; the compounds could hence be found in racemic or enantiomorphic space groups, provided that they were reported from the same source; there was no restriction on z or z′.

Ni(en)$_2$(NCS)$_2$		Mean deviation of the atoms		
		Ni1	Ni1	0.000000
		N1	N1	0.020099
		N2	N2	0.071034
		N3	N3	0.137061
		N4	N4	0.094950
		N5	N5	0.197870
		N6	N6	0.116532
		C1	C1	0.163581
		C2	C2	0.355885
		C3	C3	0.817204
		C4	C4	0.205434
		C5	C5	0.108135
		C6	C6	0.831591
		S2	S2	2.020545
		S1	S1	2.046873

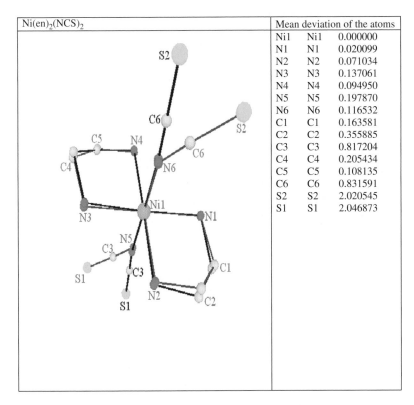

Figure 6-2. Red: EDITCN01 .... Space group: P2$_1$/a. Blue: EDITCN02 .... Space group: P2$_1$ Ref. [6]

(b) The compounds must be polymorphic and have identical composition; no solvation polymorphs were accepted even though CSD [3] fails to filter these species. The entries were mostly filtered by hand, even though there is some interest in solvation polymorphism [1]; some examples are illustrated below, Figures 6-4 and 6-5. In these examples, crystal 2 has two molecules in the asymmetric unit, with the same space group. Comparison of polymorph **1** with molecules **a** and **b** of **2** shows marked variations in the stereochemistry of these species, including **2a** and **2b** of which the conformations are greatly disparate.

(c) No disordered structures were accepted, with the caveat that the occurrence of some disorder in charge-compensating anions, such as ClO$_4^-$, PF$_6^-$ and BF$_4^-$, is common, generally affecting little the desired species. Some such structures might have been accepted provided that the important fragment was sensibly behaved.

(d) When the same polymorph was studied more than once, that with the least *R*-factor was selected.

HEFTAN Vs. HEFTAN01	Mean deviation of the atoms		
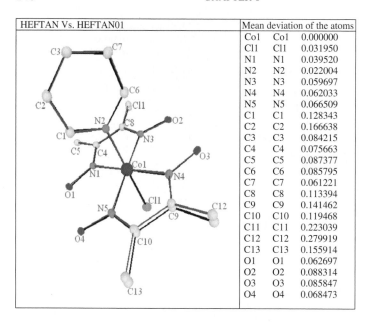	Co1	Co1	0.000000
	Cl1	Cl1	0.031950
	N1	N1	0.039520
	N2	N2	0.022004
	N3	N3	0.059697
	N4	N4	0.062033
	N5	N5	0.066509
	C1	C1	0.128343
	C2	C2	0.166638
	C3	C3	0.084215
	C4	C4	0.075663
	C5	C5	0.087377
	C6	C6	0.085795
	C7	C7	0.061221
	C8	C8	0.113394
	C9	C9	0.141462
	C10	C10	0.119468
	C11	C11	0.223039
	C12	C12	0.279919
	C13	C13	0.155914
	O1	O1	0.062697
	O2	O2	0.088314
	O3	O3	0.085847
	O4	O4	0.068473

Figure 6-3. Red: HEFTAN .... Space group: Cc Ref. [7]. Blue: HEFTAN01 .... Space group: Cmcm Ref. [8]

Hydration polymorphism between **1** (2 H$_2$O) vs. **2a** (4.5 H$_2$O)	Mean deviation of the atoms		
	Co1	Co1	0.000000
	O1	O1	0.102296
	O2	O2	0.072882
	O3	O3	0.150452
	O4	O4	0.434871
	C7	C7	0.147989
	C8	C8	0.209214
	N1	N1	0.493686
	N2	N2	0.312945
	N3	N3	0.357350
	N4	N4	0.456786
	C3	C3	0.051671
	C4	C4	0.771262
	C5	C5	3.927201
	C6	C6	2.009095
	C1	C1	2.513958
	C2	C2	3.964877

Figure 6-4. Red: Complex **1** .... Space group: P-1 Ref. [9]. Blue: Complex **2a** .... Space group: P-1 Ref. [10]

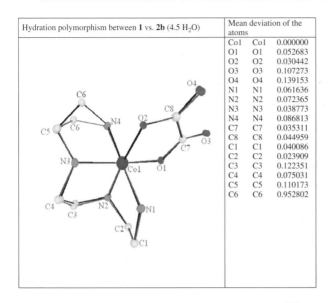

Hydration polymorphism between **1** vs. **2b** (4.5 H_2O)	Mean deviation of the atoms		
	Co1	Co1	0.000000
	O1	O1	0.052683
	O2	O2	0.030442
	O3	O3	0.107273
	O4	O4	0.139153
	N1	N1	0.061636
	N2	N2	0.072365
	N3	N3	0.038773
	N4	N4	0.086813
	C7	C7	0.035311
	C8	C8	0.044959
	C1	C1	0.040086
	C2	C2	0.023909
	C3	C3	0.122351
	C4	C4	0.075031
	C5	C5	0.110173
	C6	C6	0.952802

Figure 6-5. Red: Complex **1** .... Space group: P-1 Ref. [9]. Blue: Complex **2b** .... Space group: P-1 Ref. [11]. Glycine, below, crystallizes in three space groups – $P2/n$, $P3_2$ and $P2_1$. Despite these variations in packing, the molecules are essentially identical, as seen in Figures 6-6 and 6-7

Under only these conditions we performed the search. We made the final selection of examples by first examining the molecular structures using the Mercury routine of CSD [3].

PROGRAM MATCHIT

This program [4] accepts structural data from *CIF* documents and from *ins* files. The atomic coordinates are converted into orthogonal Å coordinates and a selected atom is placed at a common origin, as seen below. Other atoms are added that define a set of 3-D orientations. Least-square fitting of some, or all, atoms is performed. The choice is dictated by the extent of disagreement (or agreement) in stereochemistry of the molecules in question. To distinguish the molecules, a pair of colours is selected for the two molecular bonds, and atoms are coded by colour. Finally, the molecule is rotatable in 3-D space to locate the most convenient view; the picture is labeled and saved. For the latter purpose, of several format options that are available, PAINT or WORD formats are most useful.

In numerous examples to follow, we explore the extent of match or mismatch between atomic centres in molecules of diverse types that crystallize in polymorphic structures, both illustrating the compared molecular structures in the figures and tabulating the mean deviations of corresponding atoms between two unit cells of polymorphic crystals; the caption of each figure explains the comparison.

Glycine, compounds **1** vs. **2**	Mean deviation of the atoms in Ångstroms.		
	C2	C2	0.000000
	C1	C1	0.027900
	N1	N1	0.028803
	O1	O1	0.025706
	O2	O2	0.105993

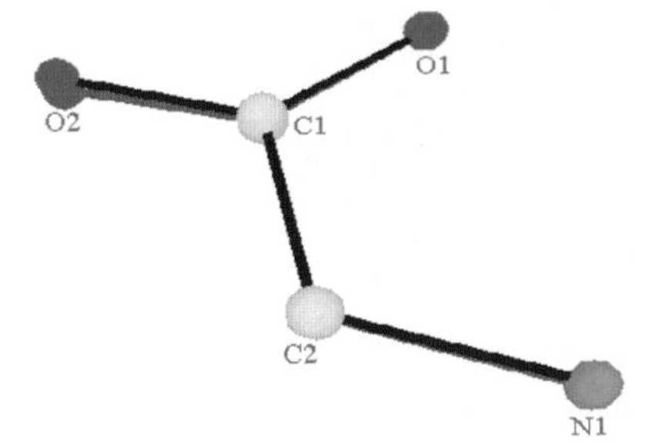

Figure 6-6. Red: Space group: $P2_1/n$. Ref. [12]. Blue: Space group: $P3_2$. Ref. [13]. The differences in stereochemistry of glycine between these two polymorphs are incredibly small despite the great variation of space groups. **2** crystallizes in an enantiomorphic space group (γ-form), whereas **1** crystallizes in a centrosymmetric space group (α-form). As the molecular skeleton is virtually identical, the physical properties of these two crystalline forms must reflect packing differences and not molecular stereochemical variations. If the molecules are so nearly identical, why, and how, do they pack so disparately? The minor deviations in atomic positions are amplified enormously by the fact that a crystal containing 10^{-5} mol still has 2×10^{18} molecules. This is a giant amplification factor for the tiny differences between those molecules. The same condition is applicable to subsequent examples

Glycine compounds **1** vs. **2**	Mean deviation of the atoms		
	C1	C1	0.000000
	C2	C2	0.019063
	O1	O1	0.015086
	O2	O2	0.012297
	N1	N1	0.124053

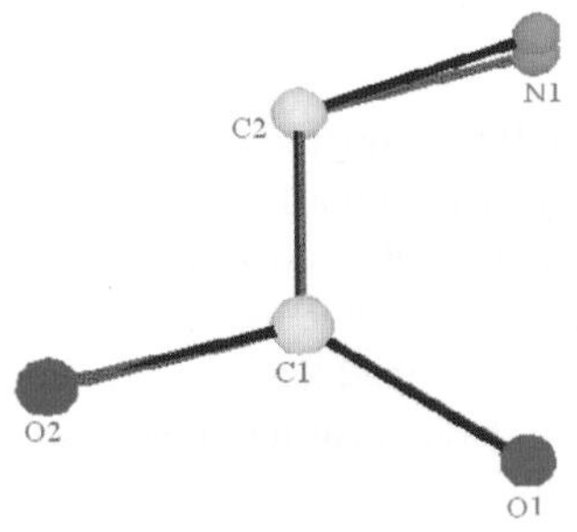

Figure 6-7. Red: Compound **1** .... Space group: $P2_1/n$. Ref. [12]. Blue: Compound **2** .... Space group: $P2_1$. Ref. [14]. The distances between chemically equivalent atoms are still satisfactory, even if N1 deviates considerably more than other atoms in this pair and in the preceding one. We again match a molecule in an enantiomorphic lattice (**2**; b-form) with another molecule in a centrosymmetric space group (**1**; a-form). The geometry at C2 shows that the configuration at C2 is properly preserved, as is the case in Figure 6-6. Matching of the β and γ forms is superfluous, for obvious reasons

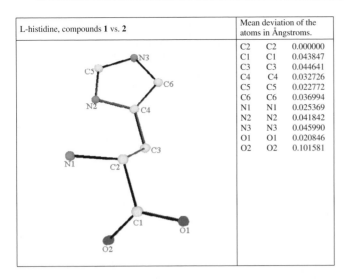

L-histidine, compounds **1** vs. **2**	Mean deviation of the atoms in Ångstroms.		
	C2	C2	0.000000
	C1	C1	0.043847
	C3	C3	0.044641
	C4	C4	0.032726
	C5	C5	0.022772
	C6	C6	0.036994
	N1	N1	0.025369
	N2	N2	0.041842
	N3	N3	0.045990
	O1	O1	0.020846
	O2	O2	0.101581

Figure 6-8. Red: Compound **1**............Space group: P2$_1$. Ref. [15]. Blue: Compound **2**............ Space group: P2$_1$2$_1$2$_1$. Ref. [16]. With the minor exception of O2, the deviations from a nearly perfect match do not exceed 0.046 Å. For this pair of polymorphs at least, the implication is therefore that the skeleton of this amino acid is little susceptible to conformational variation, possibly as a result of a hydrogen-bonding "lock" between N1 and N2. Further work on this compound would be desirable, including NMR measurements

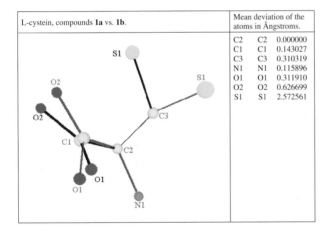

L-cystein, compounds **1a** vs. **1b**.	Mean deviation of the atoms in Ångstroms.		
	C2	C2	0.000000
	C1	C1	0.143027
	C3	C3	0.310319
	N1	N1	0.115896
	O1	O1	0.311910
	O2	O2	0.626699
	S1	S1	2.572561

Figure 6-9. Red: Compound **1a**............ Space group: P2$_1$. Ref. [17]. Blue: Compound **1b**............ Space group: P2$_1$. Same r Ref. [17]. The asymmetric unit of this polymorph contains two independent molecules (**1a** and **1b**). The match is poor in general. The variation of the S1 atoms is particularly puzzling because the two molecules deposited from the solution under identical conditions and at the same time; no environmental origin is thus available to rationalize the disparity observed

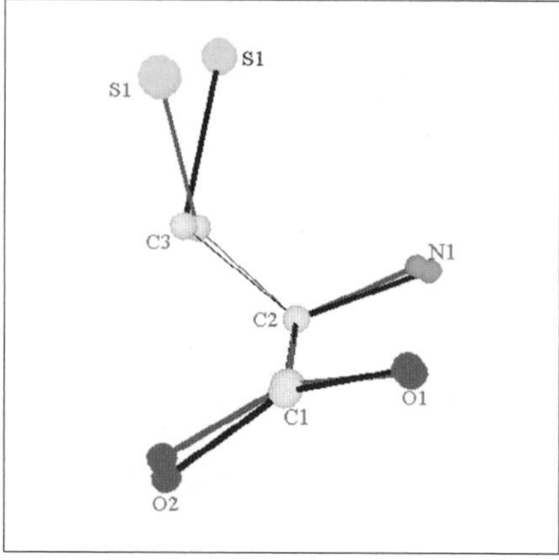

L-cystein, compound **1a** vs. compound **2**	Mean deviation of the atoms in Ångstroms.		
	C2	C2	0.000000
	N1	N1	0.148776
	C1	C1	0.066771
	C3	C3	0.200806
	O1	O1	0.157810
	O2	O2	0.243774
	S1	S1	1.808490

Figure 6-10. Red: Compound **1a** .... Space group: P2₁. Ref. [17]. Blue: Compound **2** .... Space group: P2₁2₁2₁. Ref. [18]. The stereochemistry of **1a** [23] differs less with respect to **2** than with its twin **1b** (see Figure 6-11).This condition is astonishing if one takes into account that the space groups differ, whereas **1a** and **1b** are both in the asymmetric unit of the same polymorphic crystal. Moreover, whereas for **1a** and **1b** the crystallization conditions, solvents (if any) etc. were identical, between **1a** and **2** they were unlikely the same; the authors [24] provided no details as to the origin of their crystal. In this match, the sulfur atoms are close, but the match is still poor

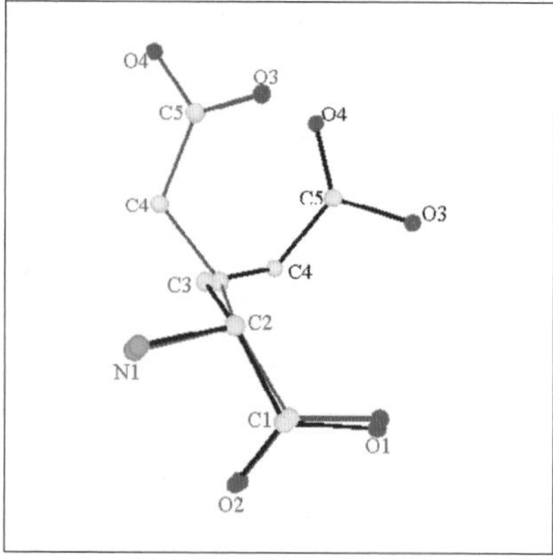

L-glutamic acid, compounds **1** vs. **2**	Mean deviation of the atoms		
	C2	C2	0.000000
	C1	C1	0.100887
	C3	C3	0.224446
	O1	O1	0.188542
	O2	O2	0.092278
	N1	N1	0.176287
	O3	O3	3.142765
	O4	O4	2.764956
	C4	C4	2.260690
	C5	C5	2.624132

Figure 6-11. Red: Compound **1** .... Space group: P2₁2₁2₁. Ref. [19]. Blue: Compound **2** .... Space group: P2₁2₁2₁ Ref. [20]. The space group is the same for both crystalline forms, but the cell parameters are differ appreciably. The match in the lower part of the figure not poor, but the upper part is markedly mismatched. Distances between the carboxylic oxygens range as great as 3.143 Å. If such conformational variation can occur in polypeptides, conformations might differ radically at various sites of the polypeptide at which glutamic acid might occur in multiple locations

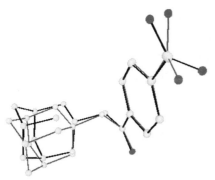

Red-Color-bond: JUHLUT10.search1.cif
Blue-Color-bond: JUHMAA10.MOL-1.cif

Figure 6-12. This adamantane derivative contains a benzene ring and a keto group that agree in stereo-chemistry, but discrepancies occur in the carboxylate and the adamantane ring. Of two molecules in the asymmetric unit of JUHMAA, this is molecule 1. JUHLUT10—Ref. [21], JUHMAA10—Ref. [22]

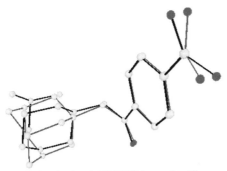

Red-Color-bond: JUHLUT10.search1.cif
Blue-Color-bond: JUHMAA10.MOL-2.cif

Figure 6-13. The same adamantane derivative contains a benzene ring and a keto group that agree in stereochemistry, but discrepancies occur again in the carboxylate and the adamantane ring. Of two molecules in the asymmetric unit of JUHMAA10, this is molecule 2. Refs. are, respectively [21] and [22]

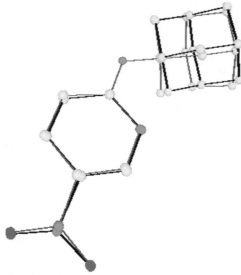

Red-Color-bond: QOTGOV01.search1.cif
Blue-Color-bond: QOTGOV02.search1.cif

Figure 6-14. For this adamantane derivative, the two polymorphs agree closely. Within molecules of the same class both molecular matching and mismatching can occur within crystalline polymorphs. QOTGOV01—Ref. is same for both [23]

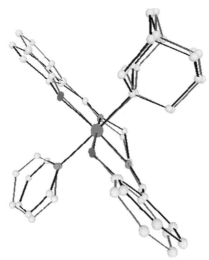

Red-Color-bond: YOCTAL01.YOCTAL01.cif
Blue-Color-bond: YOCTAL.YOCTAL.cif

Figure 6-15. For this Schiff base derivative of a rhodium coordination compound with axial pyridine and adamantane ligands, the molecules of the two polymorphic crystals agree satisfactorily, except for the torsional angle of the pyridine ligands. YOCTAL01 Ref. same for both [24]

Red-Color-bond: AMYTAL10.search1.cif
Blue-Color-bond: AMYTAL11.mol-1.cif

Figure 6-16. In this match between two polymorphic barbiturates, a small discrepancy in stereochemistry in the iso-butyl chain is minor but, apparently, significant. That AMYTAL 11 crystallized with two molecules in the asymmetric unit is significant for crystalline behavior. This is molecule 1. AMYTAL10 Ref. same for both [25]

Red-Color-bond: AMYTAL10.search1.cif
Blue-Color-bond: AMITAL11.mol-2.cif

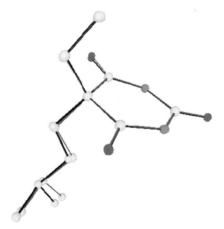

Figure 6-17. In this match between AMYTAL10 and molecule 2 of AMYTAL11, there is closer agreement between the stereochemistry of the two molecules, particularly for the is-butyl chain. References are also the same [25]

Red-Color-bond: AMYTAL10.search1.cif
Blue-Color-bond: AMYTAL11.mol-1.cif

Figure 6-18. For another pair of polymorphs of a barbiturate, the match is satisfactory. DETBAA01 Ref. [26] DETBAA04 Ref. [27]

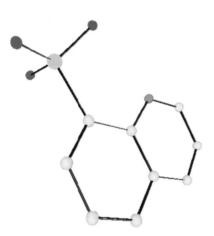

Red-Color-bond: BEYZIO.search2.cif
Blue-Color-bond: BEYZIO02.search2.cif

Figure 6-19. Here the match is nearly perfect despite the sulfonate groups being readily rotatable about their C—S bonds. BEYZIO Ref. [28] BEYZIO02 Ref. [29]

Red-Color-bond: TBCHTS01.search2.cif
Blue-Color-bond: TBCHTS11.search2.cif

Figure 6-20. For this cyclohexyl derivative of toluene sulfonate, the mismatch of the cyclohexyl rings is marked. TBCHTS01 Ref. [30] TBCHTS11 Ref. [31]

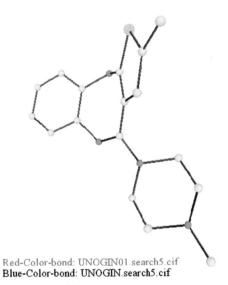

Red-Color-bond: UNOGIN01.search5.cif
Blue-Color-bond: UNOGIN.search5.cif

Figure 6-21. Polymorphs in this pair that are diazepines used to treat psychic disorders such as schizophrenia exhibit a perfect match. UNOGIN01 Ref. [32] UNOGIN Ref. [33]

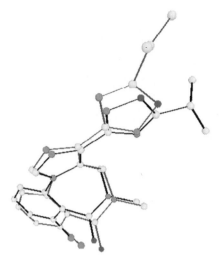

Red-Color-bond: SEBFAG.search5.cif
Blue-Color-bond: SEBFAH.mol-2.cif

Figure 6-22. The mismatch in another pair of polymorphs of diazepines is appreciable in most places, specially in the five-membered ring (upper right)

Red-Color-bond: FEROCA03.mol-1.cif
Blue-Color-bond: FEROCA.search1.cif

Figure 6-23. Even thought these carboxylates would be expected to repel each other coulombically and to rotate away from each other, possibly into trans positions (see Figure 6-24), they are exactly matched. FEROCA03 Ref. [35] FEROCA Ref. [36]

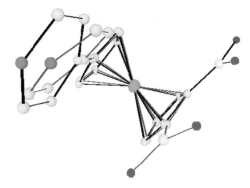

Red-Color-bond: BATFEI01.search1.cif
Blue-Color-bond: BATFEI.mol-2.cif

Figure 6-24. Here the carboxylates behave as expected and point away from one another; the phenyls likewise do not match well. Same reference for both Ref. [37]

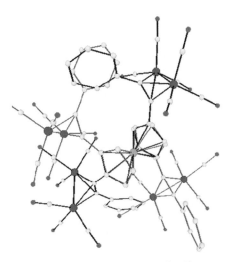

Red-Color-bond: WIJDIC.search1.cif
Blue-Color-bond: WIJDIC01.search1.cif

Figure 6-25. The match here is poor except at the central $(Cp)_2Fe$ moiety. WIJDIC Ref. [38] WIDJIC01 Ref. [39]

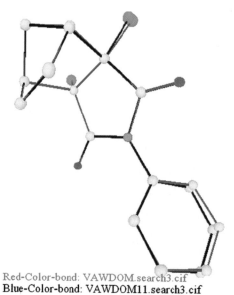

Red-Color-bond: VAWDOM.search3.cif
Blue-Color-bond: VAWDOM11.search3.cif

Figure 6-26. For this case of appreciable match, there is a single bond between the phenyl ring and the imide nitrogen; there is a slight torsional difference. VAWDOM Ref. [40] VAWDOM11 Ref. [41]

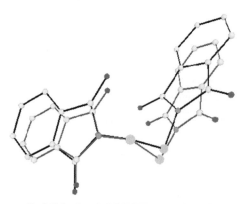

Red-Color-bond: QOGNEF01.search3.cif
Blue-Color-bond: QOGNEF.MOL-1.cif

Figure 6-27. This pair of maleimides shows much mismatch. QOGNEF01 Ref. [42] QOGNEF Ref. [43]

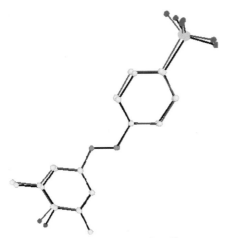

Red-Color-bond: ZAYKOZ01.search1.cif
Blue-Color-bond: ZAYKOZ.search1.cif

Figure 6-28. This pair of diazo compounds shows much match despite the phenyl rings being readily rotatable about the phenyl—N bond; the fit is worst at the sulfonate groups. Same reference for both Ref. [44]

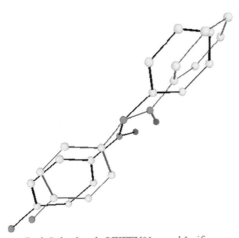

Red-Color-bond: QIKFEV01.search1.cif
Blue-Color-bond: QIKFIZ01.search1.cif

Figure 6-29. The orientation of the NO moieties and the phenyl rings are poorly matched, especially at the upper right: they are almost orthogonal to one another. Same reference for both Ref. [45]

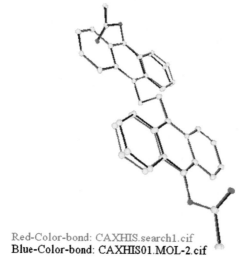

Red-Color-bond: CAXHIS.search1.cif
Blue-Color-bond: CAXHIS01.MOL-2.cif

Figure 6-30. Despite the number of flexible groups present, the match is appreciable. Same reference for both Ref. [46]

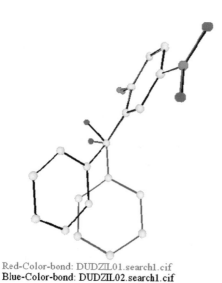

Red-Color-bond: DUDZIL01.search1.cif
Blue-Color-bond: DUDZIL02.search1.cif

Figure 6-31. The match in this case was made with only the atoms of the p-nitroaniline, for obvious reasons; the orientation of fragments in the lower half is notable. DUDZIL01 is Ref. [47] and DUDZIL02 is Ref. [48]

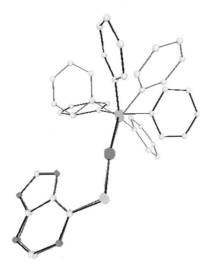

Red-Color-bond: HEGLEK.search1.cif
Blue-Color-bond: HEGLEK01.search1.cif

Figure 6-32. In the upper half the match is poor; in this pyrimidine compound the phosphine phenyl rings do not match at all. HEGLEK is Ref. [49] and HEGLEK01 is Ref. [50]

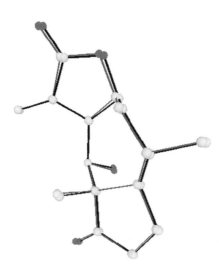

Red-Color-bond: FEGWAP.FEGWAP.cif
Blue-Color-bond: FEGWAP02.FEGWAP02.cif

Figure 6-33. For this polymorphic pair of sesquiterpenes the match is satisfactory. FEGWAP is Ref. [51] and FEGWAP02 is Ref. [52]

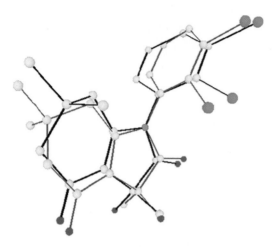

Red-Color-bond: CEWSIG10.search2.cif
Blue-Color-bond: CEWSOM10.search2.cif

Figure 6-34. Terpenes in this pair match poorly at any ring, particularly the six-membered rings. CEWSIG10 and CEWSOM10 are both Ref. [53]

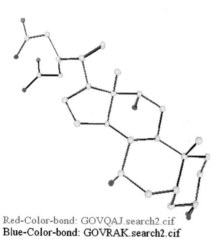

Red-Color-bond: GOVQAJ.search2.cif
Blue-Color-bond: GOVRAK.search2.cif

Figure 6-35. Terpenes in this pair match well except the terminal amide fragment. GOVOAJ and GOVRAK are both Ref. [54]

Red-Color-bond: DEKDAY.search1.cif
Blue-Color-bond: DEKDAY01.search1.cif

Figure 6-36. For this pair, a mono-protonated hexamethylenetetramine, the match is outstanding. DEKDAY and DEKDAY01 are both Ref. [55]

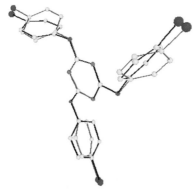

Red-Color-bond: HEXWIQ01.search1.cif
Blue-Color-bond: HEXWIQ.search1.cif

Figure 6-37. Although the match is imperfect, there are several flexible groups to consider. HEXWIQ01 Ref. is [56] and HEXWIQ is Ref. [57]

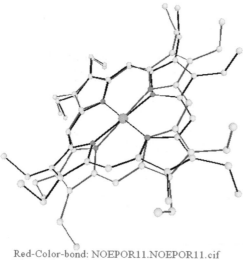

Red-Color-bond: NOEPOR11.NOEPOR11.cif
Blue-Color-bond: NOEPOR.NOEPOR.cif

Figure 6-38. In this octaethylporphine two ethyl chains match well and the others poorly. The porphyrin rings seem to disagree in their deviation from planarity. NOEPOR11 Ref. is [58] and NOEPOR Ref. is [59]

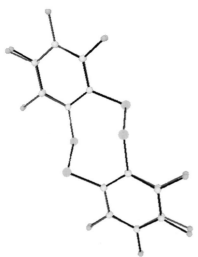

Red-Color-bond: BIBKUS.search1.cif
Blue-Color-bond: BIBKUS01.search1.cif

Figure 6-39. This complex polysulfide exhibits a nearly perfect fit, and indeed a perfect match of the torsional angles of the two disulfide chains although they are from polymorphic crystals – both monoclinic, $P2_1/c$ and $P2_1/a$. BIBKUS and BIBKUS01 Ref. is [60]

3,2am and N-Methyl En **1** vs. **2**	Mean deviation of the atoms		
	Co1	Co1	0.000000
	Cl1	Cl1	0.452822
	N1	N1	0.321385
	N2	N2	0.242438
	N3	N3	0.467313
	N4	N4	0.350583
	N5	N5	0.270656
	C1	C1	0.404884
	C2	C2	0.661458
	C3	C3	0.942061
	C4	C4	0.313852
	C5	C5	0.668304
	C6	C6	0.395083
	C7	C7	0.670395
	C8	C8	0.539234

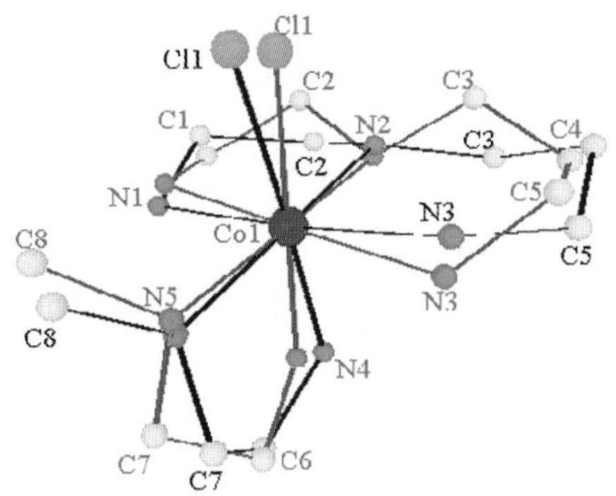

Figure 6-40. Red: Complex **1** .... Space group: $P2_1/c$. Blue: Complex **2** .... Space group: $P2_12_12_1$. Mismatch occurs everywhere despite the rigidity of the 1,5,8-triazooctane ligand defined by N1...... N3; the torsional angles of the N-Me-ethylenediamine are opposite and even the chloride ligands disagree; both Ref. [61]

N-methyl tren **1** vs. **2**	Mean deviation of the atoms		
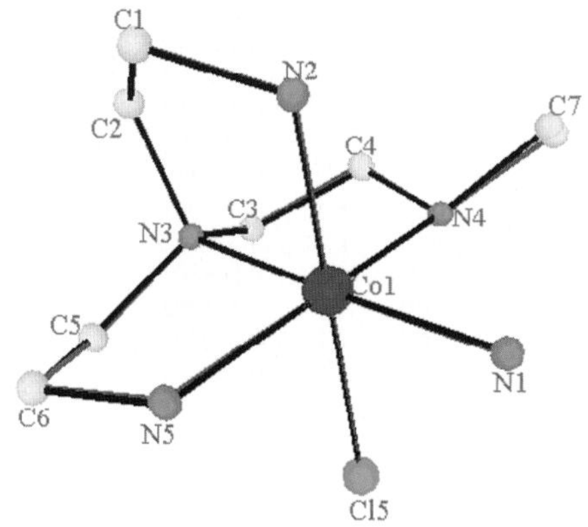	Co1	Co1	0.000000
	Cl5	Cl5	0.033581
	N1	N1	0.028373
	N2	N2	0.015701
	N3	N3	0.019237
	N4	N4	0.030827
	N5	N5	0.037109
	C1	C1	0.029849
	C2	C2	0.022473
	C3	C3	0.030286
	C4	C4	0.056709
	C5	C5	0.035759
	C6	C6	0.015012
	C7	C7	0.102250

Figure 6-41. Red: Complex **1** .... Space group: $P2_1/c$. Blue: Complex **2** .... Space group: $P2_12_12_1$. Both have Ref [62]. The match here is remarkable; the space groups of the compounds in Figure 6-40 are the same as those found here, but the mismatch there is pronounced. The packing forces in both cases are the same because the symmetry operations are identical; thus packing forces seem not to be the cause of mismatch. These prototypical cases of polymorphism are taken from the work of Yu et al., Ref. [63]. QUAXMEH was arbitrarily selected as the standard for comparison

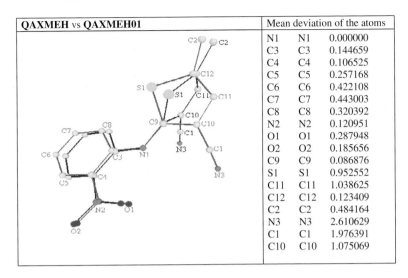

QAXMEH vs QAXMEH01	Mean deviation of the atoms		
	N1	N1	0.000000
	C3	C3	0.144659
	C4	C4	0.106525
	C5	C5	0.257168
	C6	C6	0.422108
	C7	C7	0.443003
	C8	C8	0.320392
	N2	N2	0.120951
	O1	O1	0.287948
	O2	O2	0.185656
	C9	C9	0.086876
	S1	S1	0.952552
	C11	C11	1.038625
	C12	C12	0.123409
	C2	C2	0.484164
	N3	N3	2.610629
	C1	C1	1.976391
	C10	C10	1.075069

Figure 6-42. Red: **QAXMEH**, Space group: $P2_1/c$. Blue: **QAXMEH01**, Space group: $P2_1/n$. Both rings diagree with a maximum discrepancy at the five-membered ring despite sharing the same space group; packing forces must thus be similar if not identical

QAXMEH vs QAXMEH02	Mean deviation of the atoms		
	N1	N1	0.000000
	C3	C3	0.091797
	C4	C4	0.125830
	C5	C5	0.214782
	C6	C6	0.269283
	C7	C7	0.261812
	C8	C8	0.186651
	N2	N2	0.098541
	O1	O1	0.407795
	O2	O2	0.442628
	C9	C9	0.726705
	C10	C10	0.435884
	C11	C11	1.072216
	S1	S1	3.290350
	N3	N3	4.309128
	C1	C1	2.521531
	C2	C2	4.720814
	C12	C12	2.970582

Figure 6-43. Red: **QAXMEH**, Space group: $P2_1/c$. Blue: **QAXMEH01**, Space group: $P^{-}1$. The discrepancy in the five-membered ring is here worse than in the previous case, whereas the agreement in the six-membered ring is better; the space groups are distinct

QAXMEH vs QAXMEH03	Mean deviation of the atoms		
	N1	N1	0.000000
	C3	C3	0.140850
	C4	C4	0.241014
	C5	C5	0.317192
	C6	C6	0.351107
	C7	C7	0.302141
	C8	C8	0.220008
	C9	C9	0.485065
	N2	N2	0.247620
	O1	O1	0.592493
	O2	O2	0.113218
	C10	C10	0.803148
	C11	C11	0.301140
	S1	S1	3.060593
	C1	C1	3.032679
	C2	C2	3.990851
	N3	N3	4.841734
	C12	C12	2.379899

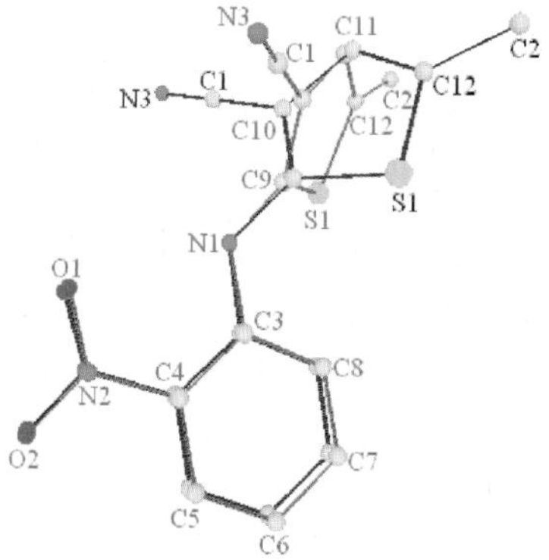

Figure 6-44. Red: **QAXMEH**, Space group: $P2_1/c$. Blue: **QAXMEH03**, Space group: $P2_1/n$. The space groups are again the same, but the disagreement in the five-membered ring is marked; some distances between chemically related atoms are nearly 5.0 Å

QAXMEH vs QAXMEH04	Mean deviation of the atoms		
	N1	N1	0.000000
	C3	C3	0.072061
	C4	C4	0.082540
	C5	C5	0.100146
	C6	C6	0.145825
	C7	C7	0.164392
	C8	C8	0.129088
	C9	C9	0.129001
	N2	N2	0.145031
	O1	O1	0.192223
	O2	O2	0.153878
	C2	C2	0.427247
	C11	C11	1.309894
	C1	C1	2.439683
	C10	C10	1.334204
	C12	C12	0.102097
	N3	N3	3.266544
	S1	S1	1.061295

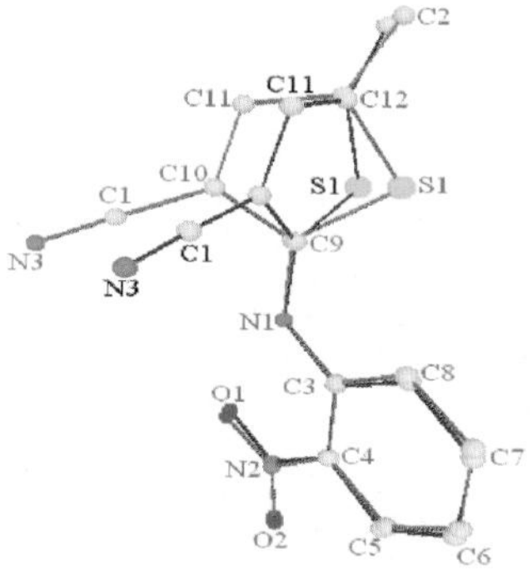

Figure 6-45. Red: **QAXMEH**, Space group: $P2_1/c$. Blue: **QAXMEH04**, Space group: P^-1. The space groups are again monoclinic vs. triclinic; the discrepancies arise between the positions of the chemically related atoms of the five-membered rings, between which the distances are up to 3.25 Å

QAXMEH vs QAXMEH05		Mean deviation of the atoms	
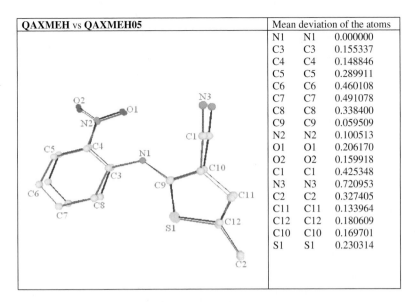	N1	N1	0.000000
	C3	C3	0.155337
	C4	C4	0.148846
	C5	C5	0.289911
	C6	C6	0.460108
	C7	C7	0.491078
	C8	C8	0.338400
	C9	C9	0.059509
	N2	N2	0.100513
	O1	O1	0.206170
	O2	O2	0.159918
	C1	C1	0.425348
	N3	N3	0.720953
	C2	C2	0.327405
	C11	C11	0.133964
	C12	C12	0.180609
	C10	C10	0.169701
	S1	S1	0.230314

Figure 6-46. Red: **QAXMEH**, Space group: $P2_1/c$. Blue: **QAXMEH05**, Space group: *Pbca*. This pair matches fairly well, with the maximum deviation between chemically related atoms ca. 0.5 Å. One space group is monoclinic whereas the second one is orthorhombic. Thus, packing forces must be quite different as the orthorhombic space group has three mutually perpendicular mirror planes whereas the monoclinic has only one

Note that there were six of them and, therefore, five matches, all using QAXMEH as the standard for comparison. The best match was with QAXMEH05, as seen in Figure 6-46, the worst one is shown in Figure 6-44.

ACKNOWLEDGEMENTS

The Robert A. Welch Foundation supported this research through grant E-592. I thank the organizers of the Indaba for their kind invitation to give this lecture, and the journal Mendeleev for permitting reproduction of some figures from my paper published therein [Mendeleev, **6**, 270–276 (2004)].

REFERENCES

1. J. Bernstein, Polymorphism in Molecular Crystals, Oxford, Oxford University Press, 2002.
2. G. R. Desiraju, Crystal Engineering—the design of organic solids, *Material Science Monographs*, Vol. 54, Elsevier. Amsterdam, The Netherlands, 1989.
3. Cambridge Structural Database, Cambridge Crystallographic Data Centre, 12 Union Road, Cambridge CB2 1EZ UK, Tel, = +44-1223-336408. Available from http://www.ccdc.cam.ac.uk Released by Wavefunction, Inc. 18401 von Karman Ave., Suite 370, Irvine CA 92612, Tel. = 949-955-2120. Fax = 949-955-2118. http://www.wavefun.com. Version 1.8, 2005.
4. MATCHIT is a program for the graphical and mathematical superposition of molecules or fragments that are chemically related, performing a least-square fit of designated atoms and calculating the

distances between related pairs of atoms. It was written at the University of Houston Chemistry Department by Rathnakumar Ramanujam, under the guidance of Ivan Bernal, 2002 and 2003.

5. M. Lutz, A. L. Spek, P. de Hoog, P. Gamez, W. L. Driessen and J. Reedjik, Private Communication to CSD, 2002.

6. N. V. Podberezskaya, T. P. Shakhshneider, A. V. Virovets and P. A. Stabnikov, Zh. Strukt. Khim. (Russ.) (J. Struct. Chem.), **32**, 96 (1991).

7. S. Geremia, R. Dreos, L. Randaccio, G. Tauzher and L. Antolini, Inorg. Chim. Acta, **216**, 125 (1994).

8. Yu. M. Chumakov, V. N. Biyushkin, V. I. Tsapkov, M. V. Gandzii, N. M. Samus' and T. I. Malinovskii, Koord. Khim.(Russ.) (Coord. Chem.), **20**, 381 (1994).

9. H. Chun, B. J. Salinas and I. Bernal, Eur. J. Inorg. Chem., 723 (1999).

10. I. Bernal, J. Cetrullo and J. Cai, Trans. Met. Chem., **19**, 221 (1994). This crystal contains two molecules in the symmetric unit—see the next reference.

11. I. Bernal, J. Cetrullo and J. Cai, Trans. Met. Chem., **19**, 221 (1994). Of two molecules in the asymmetric unit, this is the second.

12. L. F. Power, K. E. Turner and F. H. Moore, Acta Crystallogr., **B32**, 11 (1974).

13. A. Kvick, W. M. Canning and T. F. Koetzle, Acta Crystallogr., **B36**, 115 (1980).

14. T. N. Debrushchak, E. V. Boldyreva and E. S. Shutova, Acta Crystallogr., **E58**, o634 (2002).

15. M. T. Averbuch-Puchot, Z. Kristallogr., **207**, 111 (1993).

16. J. J. Madden. E. L. McGandy and N. C. Seeman, Acta Crystallogr., **B28**, 2377 (1972).

17. C. H. Gorbitz and B. Dalhus, Acta Crystallogr., **C31**, 2022 (1975).

18. K. A. Kerr, J. P. Ashmore and T. F. Koetzle, Acta Crystallogr., **B51**, 1059 (1995).

19. W. Marcoin, H. Dude, J. Kusz and B. Bzowski, Applied Crytallogr. Conference, p. 40 (1999).

20. N. Hiyayama. K. Shibata. Y. Ohashi and Y. Sasada, Bull. Chem. Soc. Japan, **53**, 30 (1980).

21. R. Jones, J. R. Scheffer, J. Trotter and Jie Yang, Crystallogr., Sect. B: Struct. Sci., **50**, 601 (1994).

22. JUHMAA10 No reference is available in CSD.

23. M. Yu. Antipin, T. V. Timofeeva, R. D. Clark, V. N. Nesterov, F. M. Dolgushin, J. Wu, and A. Leyderman, J. Mater. Chem., **11**, 351 (2001).

24. R. Dreos-Garlatti, S. Geremia, L. Randaccio, S. Ruffini and G. Tauzher, J. Organomet. Chem., **487**, C24 (1995).

25. B. M. Craven and E. A. Vizzini, ActaCrystallogr., Sect. B: Struct. Crystallogr., Cryst. Chem., **25**, 1993 (1969).

26. B. M. Craven, E. A. Vizzini and M. M. Rodrigues, Acta Crystallogr., Sect. B: Struct. Crystallogr. Cryst. Chem., **25**, 1978 (1969).

27. R. K. McMullan, R. O. Fox Jr and B. M. Craven, Acta Crystallogr.,Sect. B: Struct. Crystallogr. Cryst. Chem., **34**, 3719 (1978).

28. B. M. Gatehouse, Cryst. Struct. Commun., **11**, 493 (1982).

29. E. Silina, Yu. Bankovsky, V. Belsky and J. Lejejs, Latv. Khim. Z. (Latvian J. Chem.), 155 (1994).

30. P. L. Johnson, C. J. Cheer, J. P. Schaefer, V. J. James and F. H. Moor, Tetrahedron, **28**, 2893 (1972).

31. M. Spiniello and J. M. White, Org. Biomol. Chem, **1**, 3094 (2003).

32. S. M. Reutzel-Edens, J. K. Bush, P. A. Magee, G. A. Stephenson and S. R. Byrn, Cryst. Growth Des., **3**, 897 (2003).

33. I. Wawrzycka-Gorczyca, A. E. Koziol and M. Glice, J. Cybulski, Acta Crystallogr., Sect. E: Struct. Rep. Onlin, **60**, o66 (2004).

34. F. Watjen, R. Baker, M. Engelstoff, R. Herbert, A. MacLeod, A. Knight, K. Merchant, J. Moseley, J. Saunders, C. J. Swain, E. Wong and J. P. Springer, J. Med. Chem., **32**, 2282 (1989).

35. F. Takusagawa and T. F. Koetzle, Acta Crystallogr., Sect. B, Struct. Crystallogr. Cryst. Chem., **35**, 2888 (1979).

36. G. J. Palenik, Inorg. Chem., **8**, 2744 (1969).

37. D. Braga, M. Polito, M. Bracaccini, D. D'Addario, E. Tagliavini, L. Sturba and F. Grepioni, Organometallics, **22**, 2142 (2003).
38. K. Onitsuka, Xin-Qung Tao, Wen-Qing Wang, Y. Otsuka, K. Sonogashira, T. Adachi and T. Yoshida, J. Organomet. Chem., **473**, 195 (1994).
39. E. Champeil and S. M. Draper, J. Chem. Soc., Dalton Trans., 1440 (2001).
40. Ya. M. Nagiev, A. N. Shnulin, Sh. T. Bagirov, V. A. Adigezalov, V. I. Shil'nikov and I. M. Mamedov, Zh. Strukt. Khim. (Russ.) (J. Struct. Chem.), **29**, 105 (1988).
41. A. N. Shnulin, Ya. M. Nagiev, Sh. T. Bagirov, V. A. Adigezalov, V. I. Shil'nikov and I. M. Mamedov, Kristallografiya (Russ.) (Crystallogr. Rep.), **34**, 96 (1989).
42. D. M. M. Farrell, C. Glidewell, J. N. Low, J. M. S. Skakle and C. M. Zakaria, Acta Crystallogr., Sect. B: Struct. Sci., **58**, 289 (2002).
43. J. M. S. Skakle, J. L. Wardell, J. N. Low and C. Glidewell, Acta Crystallogr.,Sect. C: Cryst. Struct. Commun., **57**, 742 (2001).
44. N. Ehlinger and M. Perrin, Acta Crystallogr., Sect. C: Cryst. Struct. Commun., **51**, 1846 (1995).
45. K. Ejsmont, M. Broda, A. Domanski, J. B. Kyziol and J. Zaleski, Acta Crystallogr., Sect. C: Cryst. Struct. Commun., **58**, o545 (2002).
46. H.-D. Becker, S. R. Hall, B. W. Skelton and A. H. White, Aust. J. Chem., **35**, 2357 (1982).
47. P. J. Cox, A. T. Md. Anisuzzaman, R. H. Pryce-Jones, G. G. Skellern, A. J. Florence and N. Shankland, Acta Crystallogr., Sect. C: Cryst. Struct. Commun., **54**, 856 (1998).
48. P. J. Cox and J. L. Wardell, Int. J. Pharmaceutics, **194**, 147 (2000).
49. P. D. Cookson, E. R. T. Tiekink and M. W. Whitehouse, Aust. J. Chem., **47**, 577 (1994).
50. B. R. Vincent, D. J. Clarke, D. R. Smyth, D. de Vos and E. R. T. Tiekink, Met.-Based Drugs, **8**, 79 (2001).
51. F. R. Fronczek, A. G. Ober and N. H. Fischer, Acta Crystallogr., Sect. C: Cryst. Struct. Commun., **43**, 358 (1987).
52. U. Rychlewska, M. Budesinsky, B. Szczepanska, E. Bloszyk and M. Holub, Collect. Czech. Chem. Commun., **60**, 276 (1995).
53. K. Go, G. Kartha and N. Viswanathan, Acta Crystallogr., Sect. C: Cryst. Struct. Commun., **41**, 417 (1985).
54. K. Sada, T. Kondo, M. Ushioda, Y. Matsuura, K. Nakano, M. Miyata and K. Miki, Bull. Chem. Soc. Jpn., **71**, 1931 (1998).
55. R. D. Gilardi and R. J. Butcher, J. Chem. Cryst., **28**, 673 (1998).
56. K. Reichenbacher, H. I. Suss, H. Stoeckli-Evans, S. Bracco, P. Sozzani, E. Weber, J. Hulliger, New J. Chem. (Nouv. J. Chim.), **28**, 393 (2004).
57. A. Anthony, G. R. Desiraju, R. K. R. Jetti, S. S. Kuduva, N. N. L. Madhavi, A. Nangia, R. Thaimattam, V. R. Thalladi, Crystal Engineering, **1**, 1 (1998).
58. D. L. Cullen, E. F. Meyer Jr, J. Am. Chem. Soc., **96**, 2095 (1974).
59. E. F. Meyer Jr, Acta Crystallogr., Sect. B: Struct. Crystallogr. Cryst. Chem., **28**, 2162 (1972).
60. T. Chivers, M. Parvez, I. Vargas-Baca and G. Schatte, Can. J. Chem., **76**, 1093 (1998).
61. Z. Tao, Q.-J. Zhu, S.-F. Xie, X.-Q. Luo, G.-Y. Zhang, Z.-Y. Zhou and X.-G. Zhou, Chin., J. Inorg. Chem., **18**, 147, (2002). Ref. [61]
62. D. A. Buckingham, J. D. Edwards and G. McLaughlin, Inorg. Chem., **21**, 2770, (1982).
63. L. Yu, G. A. Stephenson, C. A. Mitchel, C. A. Burrell, S. V. Norek, J. J. Bowyer, T. B. Borchardt, J. G. Stowell and S. R. Bryn, J. Amer. Chem. Soc., **122**, 585 (2000).

CHAPTER 7

CRYSTALLINE AMINO ACIDS
A link between chemistry, materials science and biology

ELENA BOLDYREVA

INTRODUCTION

Amino acids belong to systems that are replete with *mystery* and *magic* and that serve as beautiful *models*. These small *molecules* form the elementary basis of peptides and proteins, which are crucial for life. This condition alone would suffice to make the study of amino acids exciting and important, but studies of amino acids are not restricted to biological problems. Amino acids can form not only variable peptide chains but also three-dimensional crystalline structures, which are no less variable and beautiful than biopolymers. Polymorphism of crystalline amino acids is a challenge for crystal engineers, and also for those researchers who are involved in studying intermolecular interactions, in particular – hydrogen bonds. Crystalline amino acids can serve as biomimetics modeling interactions in biopolymers. Phase transitions in crystalline amino acids are related to the folding of peptides. Crystalline amino acids serve as drugs and molecular materials, and the control of their structures and properties finds important applications in industry. Biomolecular assemblies – polyaminoacids, two-dimensional peptide-layered structures, peptide nanoporous structures – form a link between proteins and crystalline amino acids. Comparative studies of the sequence: crystalline amino acids – polyaminoacids – peptide layers – proteins provide insight into the properties of matter in its diverse forms.

CRYSTAL STRUCTURES

Amino acids are small organic molecules that have both carboxylic amino groups (e.g. of general formula NH_2-CH(R)-COOH). Their side chains R can vary – small or bulky, hydrophobic or hydrophilic, polar, charged or neutral. In both

167

J.C.A. Boeyens and J.F. Ogilvie (eds.), Models, Mysteries and Magic of Molecules, pp. 167–192.
© 2008 *Springer*.

aqueous solution and the crystalline state amino acids exist as zwitter ions, in which both the amino and carboxylic groups are ionized ($^+NH_3$-CH(R)-COO$^-$). In matrices of noble gases they can be obtained in their non-ionized form.

At the onset of X-ray structural analysis, the structures of crystalline amino acids were studied to obtain information about bond lengths, valence and torsion angles of the primary building blocks of proteins. These data were used in refining the first crystal structures of proteins, as diffraction data then collected for biopolymers suffered from poor resolution [1].

After direct high-resolution structural measurements on proteins became practicable, interest in crystalline amino acids decreased, but became renewed in attempts to simulate the distribution of electronic charge density in biopolymers using amino acids and simple peptides as model systems, to derive some transferable parameters and potentials [2–15]. Another research direction involved these systems and their packing patterns to mimic selected folds and interaction patterns of biopolymers. Typical molecular conformations and patterns of hydrogen bonds and packing in the structures of crystalline amino acids have been reviewed [16–19]. Similar analyses were performed for small peptides [18, 20–22].

Scrutiny of the crystal structures of amino acids, their racemates and complexes, reveals that each hydrogen bond connecting the α-amino and α-carboxylate groups and its symmetry equivalents generally produces an infinite sequence head to tail in which the two groups are periodically brought into close proximity (see examples in Figure 7-1). Such sequences, which are suggested to have relevance to prebiotic polymerisation, appear to be an almost universal feature of amino-acid aggregation

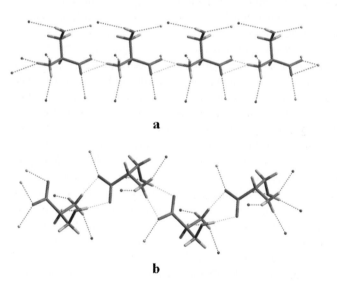

a

b

Figure 7-1. Examples of chains head to tail in crystalline amino acids: (a) – linear, L-serine, (b) – helical, DL-serine, the chain formed by zwitter ions of the same chirality [23]

in the solid state. These sequences belong to two main categories in terms of the geometric arrangement of their amino-acid molecules. The sequences in the first category comprise straight chains of molecules related mostly by the shortest cell translation in the crystals. The sequences of the second category form hydrogen-bonded two-fold helices centred around crystallographic 2_1 screw axes. The sequences are further divisible on the basis of geometrical features of their hydrogen bonds. A few sequences involving both L and D isomers have been observed in the crystal structures of some DL-amino acids. The shortest cell translation in most crystals under consideration has a value near 5.3 Å that corresponds to the periodicity of a straight head to tail sequence or, less frequently, that of a helical sequence or both. The crystal structures of amino acids and their complexes are classified in terms of the occurrence and the geometric disposition of their head to tail chains of varied types [17].

The "head-to-tail chains" are linked via hydrogen bonds with each other, producing either stacked layers (double centrosymmetric layers resulting in alternating hydrophobic and hydrophilic regions in the crystal structure, or parallel layers in polar structures), three-dimensional frameworks, triple helices linked with each other, or nano-porous three-dimensional structures with hydrophobic or hydrophilic cavities of tunable size and shape that can incorporate various guest molecules (see examples in Figure 7-2). Formation of a hydrogen bond is interrelated with the conformational diversity of amino acids, which arises due to not only rotation around main-chain covalent bonds but also the existence of various side-chain rotamers [18, 19].

Crystalline amino acids thus provide models that allow one to simulate various features of biopolymers in a wide range – from those of the main backbone frameworks in fibrils and rigid robust amyloid structures to those of cavities in globular proteins. Through the presence of the head to tail chains as the main structure-forming unit, the structures of crystalline amino acids resemble in many respects the structures of polyaminoacids and of the fragments of peptides, although the zwitter ions within a chain in the crystals of amino acids are linked by dipole-dipole interactions and hydrogen bonds, not by peptide bonds. As examples, compare the structures of polymorphs of glycine with those of polymorphs of polyglycine: layered (α- [24], β- [28], δ- [29]) polymorphs can be related to polyglycine I [30, 31], whereas helical γ-polymorph [25]–to polyglycine II [32] (Figure 7-3).

Polyamino acids can be considered as models for conformational studies, providing an atomistic description of the secondary structural motifs typically found in proteins [30–39]. Two-dimensional hydrogen-bonded layers and columns in the structures of crystalline amino acids can mimic β-sheets and helices in proteins and amyloids [40–45], and can be compared with two-dimensional crystalline layers at interfaces [46–58]. Nano-porous structures of small peptides can mimic cavities in proteins [24, 59–63]. One can also prepare crystals in which selected functional groups and side chains are located with respect to each other in the same way, as at recognition sites of substrate-receptor complexes, and use the systems to simulate the mutual adaptation of components of the complex responsible for recognition.

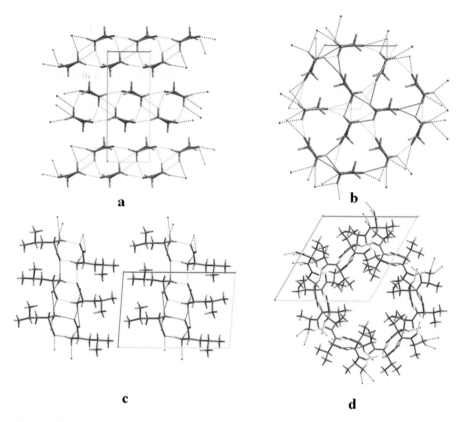

Figure 7-2. Fragments of crystal structures of selected amino acids and small peptides: (a) – α-glycine [24], (b) – γ-glycine [25], (c) – L-leucine [26], (d) – L-Ala-L-Val [27]

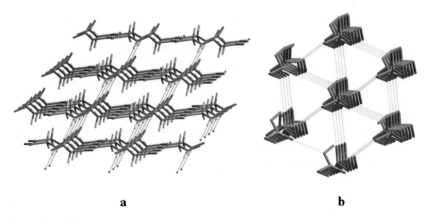

Figure 7-3. (a) – Fragments of structures of polyglycine I, (b) polyglycine II (data from [30–32])

It is important to mimic not only the static structures but also their dynamic properties. Conformational transitions, changes of folds, denaturation, and renaturation of biopolymers can be understood better if lattice dynamics, phase transitions, amorphization of crystalline amino acids and small peptides are studied and compared with those in synthetic polyaminoacids and in two-dimensional layers at the interfaces. Variable-temperature [44, 64–84] and variable-pressure [29, 81, 82, 85–134] IR- and Raman spectroscopy, inelastic neutron scattering, SAXS, NMR, X-ray and neutron diffraction, DSC are applied to study the structure and dynamics of crystalline amino acids, small peptides, synthetic polymers, interface layers and biopolymers [73–153].

POLYMORPHS

The problem of polymorphism of crystalline amino acids is important for practical applications of these compounds as non-linear optical materials [135–141], piezo-electrics [25, 28, 142] and drugs, and is also related to the self-organization and self-assembling of biomolecules. The most numerous polymorphs were found for the smallest and simplest non-chiral glycine. Three polymorphs (α-,β-,γ-) were obtained on crystallization from solutions at ambient conditions by Bernal [143]; their structures were solved and refined by Kvick and Iitaka [24, 25, 28], and then re-refined by others under ambient conditions, at low temperatures, and at high pressures. Heating, exposure to moisture or to wet NH_3 and storage in solution result in transitions between these three polymorphs (Figure 7-4) [144–148]. Cooling of β-glycine results in a second-order phase transition [80]. New polymorphs (β'- [123, 129], δ- [29, 122, 129], ζ- [125]) are formed on increasing the hydrostatic pressure and on subsequent decompression. All polymorphs but the γ-form are polytypes, having 2D-layers of similar structure (formed by head to tail chains) stacked variously. In the structure of γ-glycine the head to tail chains form triple helices resembling those in polyglycine II, or in collagen.

Three polymorphs were observed for L-serine, of which two were formed at large hydrostatic pressures and not quenchable to ambient conditions [124, 126, 127, 131, 134]. All polymorphs have similar 3D-structures described by the same space-group symmetry $P2_12_12_1$, and differ in the hydrogen-bonding patterns and conformations of zwitter ions (Figure 7-5). Four polymorphs were reported for L-cysteine, of which some were obtained at low temperature or increased pressure, or on decompression [133, 149–153]. They differ also in patterns of hydrogen bonding and molecular conformations.

Two polymorphs of DL-methionine, the α- and β-forms, have been known since 1950 as concomitant conformational polymorphs [154–157]. The crystals have almost identical shape, for which reason the second form was discovered by chance. The two forms are almost equally stable, and their crystal structures are also similar (Figure 7-6).

L-glutamic acid provides an example of the existence of two concomitant polymorphs, of which the crystal habits differ greatly – needles and plates. Their crystal structures look very different (Figure 7-7) [158–162], but, despite the distinct

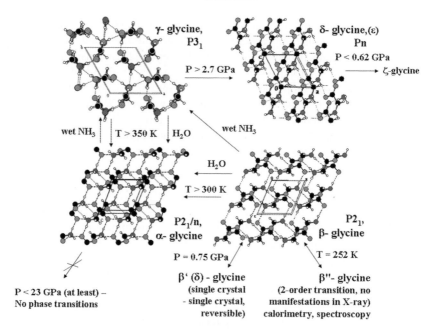

Figure 7-4. Schematic presentation of transitions between polymorphs of glycine [29, 80, 122, 123, 128, 144, 145]. Notation in brackets was suggested later [129]

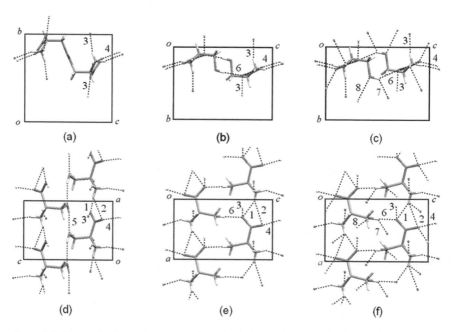

Figure 7-5. Fragments of crystal structures of L-serine in three polymorphs: (a, d) – L-serine-I at ambient pressure, (b, e) – L-serine-II at 5.4 GPa, (c, f) – L-serine-III at 8.0 GPa [127]

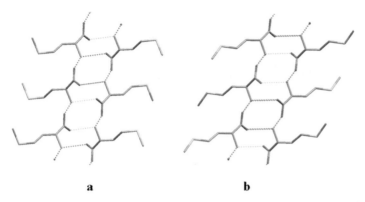

Figure 7-6. Fragments of crystal structures of polymorphs of DL-methionine: (a) – α- , (b) – β- [154–157]

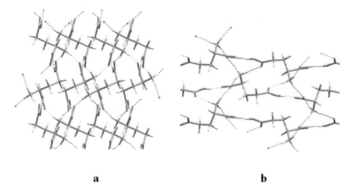

Figure 7-7. Fragments of crystal structures of L-glutamic acid in: (a) – α – (b) β – polymorphs [158–162]

molecular conformations, considerable similarities between the two structures were decoded via graph-set analysis; only the higher-order graph sets reveal the differences [163]. When kept in solution, the metastable α-polymorph transforms into the stable β-form, and this transformation is mediated by the solvent [164–168].

Beyond proper polymorphs, one can consider and compare the crystal structures of pairs of chiral (L-/D-), and racemic (DL-) amino acids. This practice is justified because in many DL-compounds L- and D-zwitter ions are not mixed at random, but zwitter ions of the same chirality form chains and even two-dimensional layers; L- and DL-serine provide such an example (Figure 7-1 and 7-8) [23, 82]. In such cases, one can consider these chains or layers as "polymorphs", and apply to them correlations between structure and property.

Crystalline amino acids form readily hydrates, salts and various molecular complexes and co-crystals. The molecular conformations, the hydrogen bond geometric parameters and patterns, dipole moments and the charge distributions can be tuned finely in these systems, and many structure-forming units and structural patterns

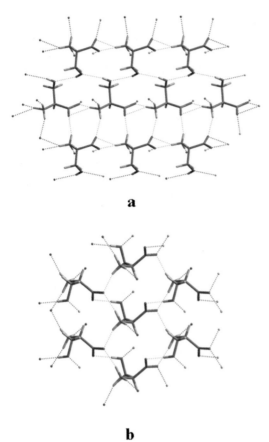

a

b

Figure 7-8. Layers formed by L-zwitter ions in the crystal structures of (a) L –, (b) DL – serine [82]

are preserved. These systems are thus used to extend correlations of structure and properties, in addition to true polymorphs.

CRYSTAL GROWTH

The crystal growth of amino acids is puzzling. As already mentioned, they commonly grow as concomitant polymorphs. An altered rate of crystallization results readily in the formation of a hydrate instead of an anhydrous form. For example, attempts to grow L-serine by slow evaporation of its aqueous solutions yield a monohydrate, unstable in air [81, 169]. If a powder sample of L-serine is added to the solution (ethanol and H_2O, 1:1), large single crystals of the stable anhydrous L-serine precipitate immediately [81]. The latter process is not observed when D-serine is taken instead of L-serine, likely because of the presence of other impurities in the D-serine powder [170]. The dissolution of DL-serine powder produces a subsequent crystallization of DL-serine crystals; dissolution of D- and L-serine powders in an equimolar mixture results in the crystallization of D- and L-serine monohydrates.

Crystallization is highly sensitive to conditions, seeding, aging of solution before and after crystallization has begun. Exotic conditions of crystallization – in an electric field, microemulsions, droplets of particular size and gels, at the surface of metal nano-particles, etc. – are required to obtain a desirable polymorph [25, 28, 144, 145, 171–184]. Although the "algorithms of reproducible growth of a desirable polymorph" seem to be more or less known, the mechanisms governing formation of polymorphs are incompletely understood, despite much work and several hypotheses published. The control of the growth of a desirable crystal polymorph is complicated additionally by the fact that polymorphs might have similar crystal habits, and be therefore not readily recognized as distinct polymorphs (see example of DL-methionine above). Crystallization conditions also affect strongly the habit of a crystal, the crystal structure and the polymorphic form remaining the same. This effect was demonstrated, for example, for glycine polymorphs [182, 184].

The control of growth of selected polymorphs is related to the ability to tune the fundamental unit of crystallization. A solution is inhabited by clusters packed as various polymorphs, and the experimental conditions determine which of these clusters yield the crystalline seeds with a desirable structure. On varying pH or the rate of precipitation (anti-solvent crystallization, or crystallization in gels/microemulsions), or adding impurities in small proportions, applying an electric field etc., one can control the occurrence also of desirable clusters in the solution. In case of polymorphs of glycine, the main ideas are either to destroy dimers present in solution and thus to favour the growth of the α-polymorph, in which double centrosymmetric layers are present, or to obstruct the growth of particular faces. Selected additives can prevent growth of some clusters and facilitate the growth of another – principle of key and lock. For example, selected amino acids added in small proportions to solutions of *other* amino acids control the growth of selected polymorphs: additives of DL-phenylalanine, DL-methionine, DL-hexafluorovaline etc. serve to control the crystal habit or polymorphism of glycine [182, 184]; additives of L-phenylalanine inhibit the growth of stable β-L-glutamic acid, and thus increase the yield of the metastable α-L-glutamic acid polymorph [185, 186].

Seeding and aging of solutions are also important to control the polymorph. For example, pure β-glycine is obtainable on spontaneous anti-solvent precipitation by acetone from acetic-acid solution, only if the solution was sufficiently aged; otherwise impurities of other polymorphs (α- and γ-) are present in the precipitated sample [144, 145, 181]. The nature of this aging ("solution memory") effect requires investigation. According to recent literature, the structures of glycine solutions vary depending on which polymorph was dissolved, and also evolve temporally [187]. AFM studies have shown that dimers present in the crystals of α-glycine are preserved on dissolution [188]. Computer simulations also show that clusters of molecules existing in a molecular crystal are preserved long after dissolution; vice versa, clusters present in solution are preserved in the crystals formed [189].

If left in solution, crystals of a polymorph can convert into other polymorphs (a solvent-mediated transformation). The products of these transformations might be unpredictable and irreproducible. Through kinetic factors in the crystallization, the so-called "slurry experiments" might mislead in attempts to estimate the relative stability of polymorphs from observations of their interconversion in the presence of solvents. In "slurry experiments" not the most stable polymorph is formed, but the polymorph growing best under these particular conditions. Preferential crystallization of a polymorph does not directly depend on its relative stability.

For example, glycine in its α-form is most readily crystallized from aqueous solutions under ambient conditions. When obtained, this form is preserved indefinitely, also at high pressures [128], on heating [144, 145], or on cooling [78], but this form is not the most thermodynamically stable under ambient conditions. The most stable polymorph under ambient conditions is the γ-form [144, 145], which crystallizes only under special conditions [144, 145, 175–180, 184]. Solvent-mediated transformations of β-glycine to α-glycine [148], and from α-glycine to γ-glycine [146] in aqueous solutions are reported. Exposure of glycine polymorphs to water vapour resulted in a transformation into the α-form; exposure of the same polymorphs to wet NH_3 gave the γ-form, whereas dry NH_3 had no effect on glycine polymorphs [144, 145]. Through the interplay of kinetic and thermodynamic factors, and permanently varying conditions of supersaturation during crystal growth, the sequence of the growth of polymorphs can be complicated. According to published data on crystallization from aqueous solutions, on cooling first the stable needle-like β-form of L-glutamic form is formed, which becomes then incorporated into plate-like crystals of the more rapidly growing metastable α-form, which, in turn, subsequently converts into the stable β-form in large crystals [165, 166].

Transformations between the polymorphs of crystalline amino acids often require the reconstruction of the extended hydrogen-bond network in the crystal, are kinetically controlled and related to recrystallization. Both the crystal growth and the polymorphic transitions in the crystals are therefore extremely sensitive to the interaction with molecules of gases or liquids that can be involved in the formation of hydrogen bonds with molecules at the surface.

The importance of hydrogen bonds, the proximity in energies between polymorphs – varied spatial organizations of the same molecules or zwitter ions, and the lability of the systems make crystalline amino acids resemble biopolymers, which can also adopt conformations similar in energy, and which are extremely sensitive to minor variations in the environment and P-T conditions. Many experimental techniques to control the growth of selected polymorphs have been successfully applied to control also the supramolecular self-assembly of peptides. For example, self-assembly on the surface of Au nano-particles was proposed both as a method to control crystal size and to generate polymorphs – preferential crystallization of β-glycine [176], and as a method of self-assembling glycine-rich peptide nanostructures [190]. Crystallization of glycine polymorphs from emulsions, microemulsions and lamellar phases served to control polymorphism, β- or γ-forms being stabilized in the presence of surfactants, and the crystal habit and assembly of crystals being strongly modified – spheres of β-glycine needles were obtained [175]. This phenomenon can have relevance for the understanding of peptide-membrane interactions. Tuning self-assembly via additives, successfully used to control the habit and polymorphism of glycine crystals [182, 184], works also to control the formation of ordered peptide assemblies at interfaces [57]. The well established fact that minor additives of one amino acid can control self assembly of another amino acid (polymorph control) provokes a question: how can the presence of a particular amino acid in solution affect the folding or unfolding of peptides or proteins? The ordered self assembly of glycine-rich peptides, formation of linear fibril versus cyclic spherrulite-like constructs from the same amino-acid sequence attract much attention due to their importance for both materials science and biology [39]. Crystallization of amino acids, the interrelation between the structure of solutions and temporal evolution and the polymorph formed can model many important features of these processes. Self-aggregative propensities in peptides are manifested by amino acids lacking interactive side chains; the morphologies generated are a function of conformational ensembles dominating in solution. This effect is especially true for peptides with highly flexible backbone geometries, for which thermodynamic aging causes the stabilization of preferred conformation in solution, via intra- and intermolecular hydrogen bonds, eliminating disfavoured structures [191].

CRYSTAL PROPERTIES

Investigated properties of crystalline amino acids and their complexes include those important practically (piezoelectric [25, 28, 142] or non-linear optical [135–141] characteristics), relative stability (with a special emphasis on the comparison of D-, L-, and DL-forms), lattice dynamics by Raman and IR-spectroscopy or neutron scattering, and structural distortions or transformations induced by variations in temperature or pressure. The objective of such work is to improve the understanding of interactions between zwitter ions, their conformational energies and the factors determining formation and reorganization of their structures. These studies are related to the problems of the design of self-assembled peptide-based materials, and of the understanding of the peptides dynamics, folding/ unfolding/misfolding.

Experiments at high pressure – a just emerged research direction – provides an example. Raman spectra of amino acids, mainly of those with non-linear optical and piezoelectric properties, under varied pressures were first measured by a group in Brazil [111–113]. Somewhat later, systematic X-ray diffraction and spectroscopic experiments of crystalline amino acids and small peptides under non-ambient conditions began as joint efforts in Novosibirsk and Marburg Universities, using facilities of both laboratories, and instruments available at the Swiss-Norwegian beam line of ESRF [29, 81, 82, 114–127]. Shortly afterwards, other groups have joined research in the same field [128–134].

High-pressure studies of crystalline amino acids and small peptides facilitate comparison of the bulk compressibility of various structures, the compressibility of hydrogen bonds and structural synthones of selected types, the conformational flexibility of selected molecular fragments and structural synthones, the stability of selected structures with respect to phase transitions, including the effects of medium (solvent), and mechanisms of phase transitions.

Bulk compressibility is not only a basic quantitative characteristic of the response of a structure to pressure but also provides insight into dynamical properties. For example, compressibility measurements are widely applied to study protein dynamics in solutions, which contributes to the biological function of protein molecules [85–103, 192, 193]. For crystals, the compressibility is calculable from X-ray diffraction data on the variation of cell parameters with pressure. For the same phase within its range of stability, compressibility was measured for α-, β-, and γ-polymorphs of glycine [115, 122, 129, 194], for L-serine [81, 82, 127, 131, 134], DL-serine [82], L-cysteine [133], L-cystine [130], α-glycylglycine [132], L-alanine, DL-alanine [195] and β-alanine [196]. Despite disparate crystal structures, these systems have similar values of bulk compressibility, the volume change being about –5 %/GPa. L-cysteine is likely the only pronounced outlet up to now – its compressibility is noticeably greater [133]. The values are comparable with the reported compressibilities of proteins: –1%/GPa [106–110, 192, 193].

The study of not merely the bulk compressibility but also of the anisotropy of structural strain gives additional information. Measured variations of cell parameters with pressure give linear strain in the directions of the principal axes of strain tensors, and also in any other selected direction in the structure [117, 197, 198]. The analysis of the anisotropy of strain reveals the directions in which the structure is rigid and those in which it is soft. Linear strain calculations are complemented by data on the shortening or stretching of hydrogen bonds with pressure, if the structure is refined and coordinates of at least non-hydrogen atoms are known [29, 82, 114–122, 126, 127, 129–134, 199, 200]. One can analyze the variation of other interatomic distances with pressure, for example, the S-S intramolecular distances within a covalent bond and of the S—S intermolecular non-covalent contacts in L-cystine [130], or the CH—CH contacts in the hydrophobic parts of double layers present in many crystal structures of amino acids. Attempts were made even to correlate pressure-induced structural variation with the shortening of "CH…O" hydrogen bonds [129–134]. The analysis of geometric parameters from diffraction data should be completed with spectroscopic experiments [201, 202]. In some cases, spectroscopy did not confirm the presence of hypothetical CH…O bonds, although the corresponding interatomic C—O distances meet commonly accepted criteria [123–125, 128, 203, 204].

We take as an example the anisotropy of pressure-induced compression of the structures of glycine in its α- and γ-polymorphs [115]. The bulk compressibility of γ-glycine is about 0.9 times that of α-glycine. Both polymorphs have a head to tail chain formed by zwitter ions linked by NH…O hydrogen bonds as a common structural unit [24, 25]. These chains are considered structural analogues of peptide chains, although lacking peptide bonds [17]; they are preserved in all polymorphs of glycine [122, 125]. γ-Glycine has a polar structure, in which head to tail chains of zwitter ions form triple helices, which are linked with each other by NH...O hydrogen bonds in a three-dimensional framework (Figure 7-2). The structure of γ-glycine is about 0.4 times as compressible along those chains as in the plane normal to these chains (Figure 7-9). In the structure of α-glycine, the same head to tail chains are linked with each other via NH...O hydrogen bonds into layers, which, in turn, form double centrosymmetric layers, also with NH...O hydrogen bonds. The inner part of a double layer is hydrophilic; the outer parts of the double layers in contact with each other are hydrophobic. No hydrogen bonds occur between the double layers (Figure 7-2). One could expect the structure of α-glycine to be most compressible in the direction normal to the double layers, but it is not – the structure is about 1.2 times as compressible along the direction of hydrogen bonds linking the head to tail chains within a layer. As in γ-glycine, the structure is the most rigid in the direction along the head to tail chains [115].

The structures of other crystalline amino acids are also shown to be most rigid in a direction along the head to tail chains [81, 82, 127, 129–131, 134]. The compressibility of shorter NH...O hydrogen bonds linking zwitter ions along these chains is affected only slightly even by jump-wise structural rearrangements in the course of phase transitions. For example, in L-serine, the N—O distance in this hydrogen bond decreases practically linearly about –0.01 Å/GPa in the entire range of pressure from ambient to 10 GPa, although the crystal structure undergoes two phase transitions, at about 5 and about 8 GPa, which are accompanied by a jump-wise *increase* in the cell parameter along the same head to tail chain (Figure 7-9) [127]. The compressibility of the shorter NH...O hydrogen bonds in the chains head to tail remains almost unaffected by a structural arrangement of the triple helices formed by these chains in γ-glycine into a layer in δ-glycine in the course of an irreversible extended single-crystal – powder phase transition beginning about 3.5 GPa [121, 122]. Other hydrogen bonds in the structures of crystalline amino acids are more compressible than the short NH...O bonds within the head to tail chains, the variations of the N—O distances being typically about ±0.02–0.05 Å/GPa [82, 121, 122, 127, 129–134]. Similar values were measured for the compressibility of hydrogen bonds in other organic crystals [105, 114, 116–121, 199, 200]. Hydrogen bonds are compressible or extensible with increasing pressure; the altered interatomic distance in a hydrogen bond does not invariably correlate with the value of linear strain of the entire crystal structure in the same direction, because of the conformational alterations and rotation of molecules [82, 116–121, 127, 200]. For comparison, measured typical values for proteins are about ±0.1–0.01 Å/GPa [88, 107–110].

Analysis of the linear strain in crystalline amino acids, in particular – the analysis of the compressibility of selected structural elements – molecular chains, layers, intermolecular hydrogen bonds, voids – is important for an understanding of the compressibilities of structural fragments of peptides and proteins – helices, sheets, turns, non-structured fragments and cavities. For example, in proteins loops are often more compressible than helices, which are in turn more compressible than β-sheets [9, 88, 107, 108, 110, 192, 193]. The compressibility of main chains is comparable with the strain in crystalline amino acids. This comparison should

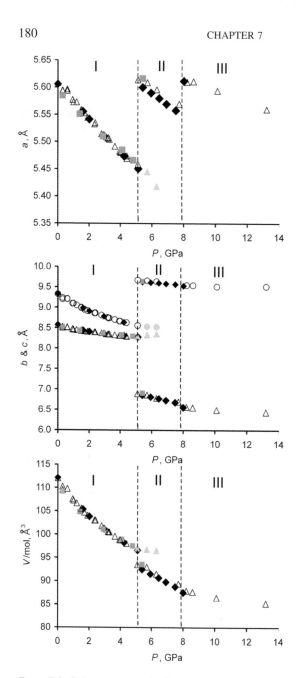

Figure 7-9. Cell parameters and volume versus pressure in L-serine in three polymorphs [127]

be expected to be more informative for fibrillar proteins and amyloid structures than for globular proteins. Whereas the globular native forms of proteins are side-dominated compact structures evolved on pursuing a unique fold with optimal packing of amino-acid residues, the structure of amyloid fibrils is dominated by a main chain, with an extensive hydrogen-bond network [191]. Pressure-induced phase transitions in crystalline amino acids, for example, the transformation of triple-helix (γ-glycine) to layer (δ-glycine) [29, 122, 125], can mimic conformational changes in amyloid structures and in fibrilar proteins such as collagen [205]. Amyloids are shown to undergo pressure-induced structural transformations during which contrasting conformational changes occur consecutively – first a pressure-induced reorganization of fibrils, and then a pressure-induced unfolding [206]. The crystal structures of simple dipeptides can mimic those of the amyloid polypeptides [40]; detailed high-pressure studies of these simpler systems could thus provide valuable information on the pressure effects on the amyloids themselves. The anisotropy of lattice strain in the crystals of amino acids with layered structures bears comparison with the measured elastic properties of two-dimensional layers of oligopeptide films [207].

The compressibility of cavities of biopolymers and the contribution of rigidity of the cavity to the conformational stability of the biopolymer can be also mimicked by studying structures of smaller molecules. Attempts were made to describe the anisotropic compression of some crystalline amino acids by "closing voids" [129–134]. Although a pressure-induced process is expected to produce a structure with increased density and smaller voids, crystalline amino acids are not the best systems to study compression of cavities as their properties are largely determined by dipole-dipole interactions and strong hydrogen bonds (OH...O and NH...O). Many crystals are piezoelectric; hence pressure induces polarization and noticeable redistribution of electron density that must be taken into account during analysis of the anisotropy of pressure-induced structural strain and the mechanisms of phase transitions. Systematic comparisons of amino acids – their salts and complexes – in series and of polymorphs of the same amino acid and of amino acids with varied side chains would be helpful. Improved mimetic models for the compressibility of cavities are selectable among dipeptides in a family with nano-size cavities and channels [40, 59–63]. One can compare the effect of pressure on layered dipeptides, and on the dipeptide crystal structures that possess large cavities of variable size and hydrophobic or hydrophilic properties, mimic the effect of liquids on the compressibility and the conformational stability of the cavity. One can study compression in various liquids – hydrophilic, hydrophobic, containing special organic additives known to stabilize proteins of ocean piezophiles, using model crystal structures with the cavities of similar size, but with varied – hydrophilic or hydrophobic – properties of the inner and outer walls of the cavities.

Comparison of the compressibilities of polymorphs and of the crystal structures of separate amino acids might be relevant to understand why fragments of proteins built of amino acids in varied sequences compress variously [88, 192, 193].

A knowledge of the elastic properties of the selected fragments of amino-acid crystals is essential when considering muscles, or biopolymers forming silk or spider threads. The strain induced by hydrostatic pressure is pertinent to the conformational transitions induced by substrate-receptor interactions, or by collisions of the biopolymers. Varying side chains, or the length of the main backbone chains of amino acids and peptides forming the crystal structures, one can take control of dipole-dipole interactions, H-bond patterns, the occurrence or absence of an inversion center, and then investigate systematically the effect of the molecular arrangement on the mechanical properties. Hydrates can be compared to anhydrous amino acids, salts to amino-acid molecules, mixed crystals with homomolecular phases, etc. To vary dipole-dipole interactions in a wide range, amino acids can be modified chemically, substituting protons for methyl groups. Selective deuteration can affect the kinematic characteristics of zwitter ions and their H-bonding ability. Biomolecular assemblies – polyaminoacids, peptide two-dimensional layered or nano-porous structures – serve as an important bridge between crystalline amino acids and proteins.

CONCLUSIONS

An interest in the crystals of amino acids and small peptides has reemerged with attention to the dynamic properties of these systems, the subtle kinetic factors determining their crystallization and polymorphism, phase transitions and anisotropic structural response to variation of temperature, pressure and an environment. The systems can serve as a unique interface between biology, chemistry, material science and nanotechnology of immediate and future importance.

ACKNOWLEDGEMENTS

Our research in the field described in this chapter was supported by grants from several foundations (DFG, BMBF, DLR, A. von Humboldt Foundation, RFBR, RAS, BRHE program – CRDF, USA, and Russian Ministry of Education and Science). Experimental results were obtained in collaboration with Dr. H. Sowa, Dr. H. Ahsbahs, Dr. H. Uchtmann, Dr. S.V. Goryainov, Prof. Yu.V. Seryotkin, Dr. S.N. Ivashevskaya, Prof. V.V. Chernyshev, Dr. E.B. Burgina, Dr. T.N. Drebushchak, Dr. V.A. Drebushchak, Dr. Yu.A. Chesalov, Prof. V.P. Dmitriev, Prof. H.-P. Weber, E.N. Kolesnik, E.S. Shutova, G.B. Chernoby, A.F. Achkasov, and A. Zhilin. Experiments using synchrotron radiation were performed at the SNBL station of ESRF (Grenoble), those under laboratory conditions – in Marburg and in Novosibirsk. I am grateful to Dr. T.N. Drebushchak for assistance with preparing figures for this chapter, and to my family and my research group – for their patience and support of my work.

REFERENCES

1. *Gurskaya, G.V.* The molecular structure of amino acids: determination by X-ray diffraction analysis, Consultants Bureau, NewYork USA (1968).
2. *Kvick, A., Al-Karaghouli, A.R., Koetzle, T.F.* Deformation electron density of α-glycylglycine at 82 K. I. The neutron diffraction study, Acta Cryst. B33 (1977) 3796–3801.
3. *Kvick, A., Koetzle, T.F., Stevens, E.D.* Deformation electron density of α-glycylglycine at 82 K. II. The X-ray diffraction study, J. Chem. Phys. 71 (1979) 173–179.
4. *Destro, R., Marsh, R.E., Bianchi, R.* A low-temperature (23 K) study of L-alanine, J. Phys. Chem. 92 (1988) 966–973.
5. *Destro, R., Roversi, P., Barzaghi, M., Marsh, R.E.* Experimental charge density of α-glycine at 23 K, J. Phys. Chem. 104 (2000) 1047–1054.
6. *Bouhmaida, N., Ghermani, N.-E., Lecomte, C., Thalal, A.* Modelling electrostatic potential from experimentally determined charge densities. II. Total potential, Acta Crystallogr. 53A(5) (1997) 556–563.
7. *Jelsch, C., Pichon-Pesme, V., Lecomte, C., Aubry, A.* Transferability of multipole charge-density parameters: application to very high resolution oligopeptide and protein structures, Acta Crystallogr. 54D (1998) 1306–1318.
8. *Lecomte, C., Guillot, B., Jelsch, C., Podjarny, A.* Frontier example in experimental charge density research: experimental electrostatics of proteins, Int. J. Quant. Chem. 101(5) (2005) 624–634.
9. *Jelsch, C., Guillot, B., Lagoutte, A., Lecomte, C.* Advances in protein and small-molecule charge-density refinement methods using MoPro, J. Appl. Crystallogr. 38(1) (2005) 38–54.
10. *Lecomte, C., Aubert, E., Legrand, V., Porcher, F., Pillet, S., Guillot, B., Jelsch, C.* Charge density research: from inorganic and molecular materials to proteins, Z. Kristallogr. 220(4) (2005) 373–384.
11. *Koritsanszky, T., Volkov, A., Coppens, P.* Aspherical-atom scattering factors from molecular wave functions. 1. Transferability and conformation dependence of atomic electron densities of peptides within the multipole formalism, Acta Crystallogr. 58A(5) (2002) 464–472.
12. *Volkov, A., Li, X., Koritsanszky, T., Coppens, P.* Ab initio quality electrostatic atomic and molecular properties including intermolecular energies from a transferable theoretical pseudoatom databank, J. Phys. Chem. 108A(19) (2004) 4283–4300.
13. *Flaig, R., Koritsanszky, T., Dittrich, B., Wagner, A., Luger, P.* Intra and intermolecular topological properties of amino acids: a comparative study of experimental and theoretical results, J. Am. Chem. Soc. 124 (2002) 3407–3417.
14. *Dittrich, B., Koritsanszky, T., Grosche, M., Scherer, W., Flaig, R., Wagner, A., Krane, H.G., Kessler, H., Riemer, C., Schreurs, A.M.M., Luger, P.* Reproducability and transferability of topological properties; experimental charge density of the hexapeptide cyclo-(D,L-Pro) -(L-Ala) monohydrate, Acta Crystallogr. B58 (2002) 721–727.
15. *Dittrich, B., Hubschle, C.B., Luger, P., Spackman, M.A.* Introduction and validation of an invariom database for amino-acid, peptide and protein molecules, Acta Crystallogr. D62 (2006) 1325–1335.
16. *Vinogradov, S.N.* Hydrogen bonds in crystal structures of amino acids, peptides and related molecules, Int. J. Peptide Protein Res. 14(4) (1979) 281–289.
17. *Suresh, C.G., Vijayan, M.* Occurrence and geometrical features of head-to-tail sequences involving amino acids in crystal structures, Int. J. Peptide Protein Res. 22(2) (1983) 129–143.
18. *Görbitz, C.H.* Hydrogen-bond distances and angles in the structures of amino acids and peptides, Acta Crystallogr. B45 (1989) 390–395.
19. *Görbitz, C.H.* Structures and conformational energies of amino acids in the zwitterionic, hydrogen-bonded state, J. Mol. Struct. (Theochem) 775(1–3) (2006) 9–17.
20. *Görbitz, C.H., Etter, M.C.* Hydrogen bond connectivity patterns and hydrophobic interactions in crystal structures of small, acyclic peptides, Int. J. Peptide Protein Res. 39(2) (1992) 93–110.

21. *Görbitz, C.H.* Peptide structures, Current Opinion in Solid State and Materials Science. 6(2) (2002) 109–116.

22. *Karle, I.* Folding, aggregation and molecular recognition in peptides, Acta Crystallogr. B48 (1992) 341–356.

23. *Kistenmacher, T.J., Rand, G.A., Marsh, R.E.* Refinements of the crystal structures of DL-serine and anhydrous L-serine, Acta Crystallogr. B30 (1974) 2573–2578.

24. *Jösson, P.G., Kvick, A.* Precision neutron diffraction structure determination of protein and nucleic acid components. III. The crystal and molecular structure of the amino acid α-glycine, Acta Crystallogr. B28 (1972) 1827–1833.

25. *Iitaka, Y.* The crystal structure of γ-glycine, Acta Crystallogr. 14 (1960) 1–10.

26. *Görbitz, C.H., Dalhus, B.* Redetermination of L-leucine at 120 K, Acta Crystallogr. C52 (1996) 1754–1756.

27. *Görbitz, C.H.* An exceptionally stable peptide nanotube system with flexible pores, Acta Crystallogr. B58(5) (2002) 849–854.

28. *Iitaka, Y.* The crystal structure of β-glycine, Acta Crystallogr. 13 (1960) 35–45.

29. *Boldyreva, E.V., Ivashevskaya, S.N., Sowa, H., Ahsbahs, H., Weber, H.-P.* Effect of high pressure on the crystalline glycine: formation of a new polymorph, Dokl. Akad. Nauk. 396 (2004) 358–361.

30. *Bamford, C.H., Brown, L., Cant, E.M., Elliott, A., Hanby, W.E. Malcolm, B.R.* Structure of polyglycine, Nature. 176 (1955) 396–397.

31. *Kajava, A.V.* Dimorphism of polyglycine I: structural models for crystal modifications, Acta Crystallogr. D55 (1999) 436–442.

32. *Crick, F.H.C., Rich, A.* Structure of polyglycine II, Nature. 176 (1955) 780–781.

33. *Meyer, K., Go, Y.* Observations roentgenographiques sur des polypeptides inferieurs et superiurs, Helv. Chim. Acta. 17 (1934) 1488–1492.

34. *Astbury, W.T., Dagliesh, C.E., Darmon, S.E., Sutherland, G.B.B.M.* Studies of the structure of synthetic polypeptides, Nature. 162 (1948) 596–600.

35. *Bamford, C.H., Elliott, A., Hanby, W.E.* Syntheitic Polypeptides: Preparation, Structure, and Properties, Academic Press, N.Y. (1956).

36. *Frazer, R.D.B., MacRae, T.P.* Conformation in Fibrous Proteins and Related Synthetic Polypeptides, Academic Press, N.Y. (1973).

37. *Kajava, A.V.* Proteins with repeated sequence – structural prediction and modeling, J. Struct. Biol. 134 (2001) 132–144.

38. *MacPhee, C.E., Woolfson, D.N.* Engineered and designed peptide-based fibrous biomaterials, Current Opinion in Solid State and Materials Science 8(2) (2004) 141–149.

39. *Zanuy, D., Nusinov, R., Aleman, C.* From peptide-based material science to protein fibrils: discipline convergence in nanobiology, Phys. Biology 3. (2006) S80–S90.

40. *Görbitz, C.H.* The structure of nanotubes formed by diphenylalanine, the core recognition motif of Alzheimer's β-amyloid polypeptide, ChemComm. 22 (2006) 2332–2334.

41. *Maji, S.K., Haldar, D., Drew, M.G.B., Banerjee, A., Das, A.K., Banerjee, A.* Self-assembly of β-turn forming synthetic tripeptides into supramolecular β-sheets and amyloid-like fibrils in the solid state, Tetrahedron. 60(14) (2004) 3251–3259.

42. *Das, A.K., Banerjee, A., Drew, M.G.B., Haldar, D., Banerjee, A.* Stepwise self-assembly of a tripeptide from molecular dimers to supramolecular β-sheets in crystals and amyloid-like fibrils in the solid state, Supramolecular Chemistry. 16(5) (2004) 331–335.

43. *Dutt, A., Drew, M.G.B., Pramanik, A.* β-Sheet mediated self-assembly of dipeptides of ω-amino acids and remarkable fibrillation in the solid state, Organic Biomol. Chem. 3(12) (2005) 2250–2254.

44. *Drebushchak, T.N., Kolesnik, E.N., Boldyreva, E.V.* Variable temperature (100–295 K) single-crystal X-ray diffraction study of the α-polymorph of glycylglycine and a glycylglycine hydrate, Z. Kristallogr. 221 (2006) 128–138.

45. *Ray, S., Drew, M.G.B., Das, A.K., Banerjee, A.* Supramolecular β-sheet and nanofibril formation by self-assembling tripeptides containing an N-terminally located γ-aminobutyric acid residue, Supramolecular Chemistry. 18(5) (2006) 455–464.

46. *Rapaport, H., Kuzmenko, I., Howes, P.B., Kjaer, K., Als-Nielsen, J., Leiserowitz, L., Lahav, M.* Structural characterization of valinomycin and nonactin at the air-solution interface by grazing incidence X-ray diffraction, J. Am. Chem. Soc., 119 (1997) 11211–11216.

47. *Rapaport, H., Kuzmenko, I., Kjaer, K., Als-Nielsen, J., Weissbuch, I., Lahav, M., Leiserowitz, L.* Crystalline architectures at the air-liquid interface: from nucleation to engineering, Synchrotron Radiation News. 12 (1999) 25–33.

48. *Rapaport, H., Kim, H.S., Kjaer, K., Howes, P.B., Cohen, S., Als-Nielsen, J., Ghadiri, M.R., Leiserowitz, L., Lahav, M.* Crystalline cyclic peptide nanotubes at interfaces, J. Am. Chem. Soc. 121 (1999) 1186–1191.

49. *Rapaport, H., Kuzmenko, I., Berfeld, M., Edgar, R., Popovits-Biro, R., Kjaer, K., Als-Nielsen, J., Weissbuch, I., Leiserowitz L., Lahav M.* From nucleation to engineering of crystalline architectures at air-liquid interfaces, J. Phys. Chem. B. 104 (2000) 1399–1428.

50. *Rapaport, H., Kjaer, K., Jensen, T.R., Leiserowitz, L., Tirrell D.A.* Two-dimensional order in β-sheet peptide monolayers, J. Am. Chem. Soc. 122 (2000) 12523–12529.

51. *Kuzmenko, I., Rapaport, H., Kjaer, K., Als-Nielsen, J., Weissbuch, I., Lahav, M., Leiserowitz L.* Design and characterization of crystalline thin film architectures at the air-liquid intreface: simplicity to complexity, Chem. Rev. 101 (2001) 1659–1696.

52. *Rapaport, H., Möller, G., Knobler, C.M., Jensen, T.R., Kjaer, K., Leiserowitz, L., Tirrell, D.A.* Assembly of triple-stranded β-sheet peptides at interfaces, J. Am. Chem. Soc. 124 (2002) 9342–9343.

53. *Weissbuch, I., Berfeld, M., Bouwman, W., Kjaer, K., Als-Nielsen, J., Lahav, M., Leiserowitz, L.* Separation of enantiomers and racemate formation in two-dimensional crystals at the water surface from racemic – amino acid amphiphiles: design and structure, J. Amer. Chem. Soc. 119 (1997) 933–942.

54. *Weissbuch, I., Rubinstein, I., Weygand, M.J., Kjaer, K., Leiserowitz, L., Lahav, M.* Crystalline phase separation of racemic and nonracemic zwitterionic α -amino acid amphiphiles in a phospholipid environment at the air/water interface: a grazing-incidence X-Ray diffraction study, Helv. Chim. Acta. 86 (2003) 3867–3874.

55. *Weissbuch, I., Bolbach, G., Leiserowitz, L., Lahav, M.* Chiral amplification of oligopeptides via polymerization in two-dimensional crystallites on water, Origins of Life and Evolution of the Biosphere. 34 (2004) 79–92.

56. *Martin, S.M., Kjaer, K., Weygand, M.J., Weissbuch, I., Ward, M.D., Lahav, M.* Hydrogen-bonded monolayers and interdigitated multilayers at the air-water interface, J. Phys. Chem. B. 110(29) (2006) 14292–14299.

57. *Rapaport, H.* Ordered peptide assemblies at interfaces, Supramolecular Chemistry. 18(5) (2006) 445–454.

58. *Davies, R.P.W., Aggeli, A., Beevers, A.J., Boden, N., Carrick, L.M., Fishwick, C.W.G., McLeish, T.C.B., Nyrkova, I., Semenov, A.N.* Self-assembling β-sheet tape forming peptides, Supramolecular Chemistry 18(5) (2006) 435–443.

59. *Görbitz, C.H.* Nanotube formation by hydrophobic dipeptides, Chem. Europ. J. 7(23) (2001) 5153–5159.

60. *Görbitz, C.H.* β Turns, water cage formation and hydrogen bonding in the structures of L-valyl-L-phenylalanine, Acta Crystallogr. B58 (2002) 512–518.

61. *Görbitz, C.H.* Nanotubes from hydrophobic dipeptides: Pore size regulation through side chain substitution, New J. Chem. 27(12) (2003) 1789–1793.

62. *Görbitz, C.H.* Monoclinic nanoporous crystal structures for L-valyl-L-alanine acetonitrile solvate hydrate and L-valyl-L-serine trifluoroethanol solvate, CrystEngComm. 7 (2005) 670–673.

63. *Görbitz, C.H., Nilsen, M., Szeto, K., Tangen, L.W.* Microporous organic crystals: An unusual case for L-leucyl-L-serine, ChemComm. 34 (2005) 4288–4290.

64. *Kosic, T.J., Cline, R.E., Dlott, D.D.* Picosecond coherent Raman investigation of the relaxation of low frequency vibrational modes in amino acids and peptides, J. Chem. Phys. 81(11) (1984) 4932–4949.

65. *Husain, S.K., Hasted, J.B., Rosen, D., Nicol, E., Birch, J.R.* FIR Spectra of amino acids and related molecules, Infrared Phys. 24(2/3) (1984) 201–208.

66. *Krimm, S., Bandekar, J.* Vibrational spectroscopy and conformation of peptides, polypeptides, and proteins, in: Advances Protein Chemistry, Vol. 38, Ed. C.B. Anfinsen, J.T. Edsall, F.M. Richards, Academic Press, N.Y. (1986) 181–364.

67. *Freedman, T.B., Nafie, L.A., Keiderling, T.A.* Vibrational optical activity of oligopeptides, Biopolymers – Peptide Science Section 37(4) (1995) 265–279.

68. *Dovbeshko, G., Berezhinsky, L.* Low frequency vibrational spectra of some amino acids, J. Mol. Struct. 450(1–3) (1998) 121–128.

69. *Jalkanen, K.J., Elstner, M., Suhai, S.* Amino acids and small peptides as building blocks for proteins: comparative theoretical and spectroscopic studies, J. Mol. Struct. (Theochem). 675 (2004) 61–77.

70. *Matei, A., Drichko, N., Gompf, B., Dressel, M.* Far-infrared spectra of amino acids, Chem. Phys. 316 (2005) 61–71.

71. *Pawlukojc, A., Leciejewicz, J., Natkaniec, I.* The INS spectroscopy of amino acids: L-leucine, Spectrochim. Acta. A52 (1996) 29–32.

72. *Pawlukojc, A., Bajdor, K., Dobrowolski, J.Cz., Leciejewicz, J., Natkaniec, I.* The IINS spectroscopy of amino acids: L-isoleucine, Spectrochim. Acta. A53(6) (1997) 927–931.

73. *Pawlukojć, A., Leciejewicz, J., Tomkinson, J., Parker, S.F.* Neutron scattering, infra-red, Raman spectroscopy and ab initio study of l-threonine, Spectrochim. Acta. A57(12) (2001) 2513–2523.

74. *Pawlukojć, A., Leciejewicz, J., Tomkinson, J., Parker, S.F.* Neutron spectroscopic study of hydrogen bonding dynamics in L-serine, Spectrochim. Acta. A58(13) (2002) 2897–2904.

75. *Pawlukojć, A., Leciejewicz, J., Ramirez-Cuesta, A.J., Nowicka-Scheibe, J.* L-Cysteine: Neutron spectroscopy, Raman, IR and ab initio study, Spectrochim. Acta. A61(11–12) (2005) 2474–2481.

76. *Barthes, M., Vik, A.F., Spire, A., Bordallo, H.N., Eckert, J.* Breathers or structural instability in solid L-alanine: a new IR and inelastic neutron scattering vibrational spectroscopic study, J. Phys. Chem. A. 106 (2002) 5230–5241.

77. *Moreno, A.J.D., Freire, P.T.C., Melo, F.E.A., Mendes Filho, J., Nogueira, M.A.M., Almeida, J.M.A., Miranda, M.A.R., Remédios, C.M.R., Sasaki, J.M.* Low-temperature Raman spectra of monohydrated L-asparagine crystals, J. Raman Spectrosc. 35(3) (2004) 236–241.

78. *Boldyreva, E.V., Drebushchak, T.N., Shutova, E.S.* Structural distortion of the α, β, and γ-polymorphs of glycine on cooling, Z. Kristallogr. 218 (2003) 366–376.

79. *Boldyreva, E.V., Drebushchak, V.A., Kovalevskaya, Yu.A., Paukov, I.E.* Low-temperature heat capacity of α and γ polymorphs of glycine, J. Therm. Analys. Calorim. 73 (2003) 109–120.

80. *Drebushchak, V.A., Boldyreva, E.V., Kovalevskaya, Yu.A., Paukov, I.E., Drebushchak, T.N.* Low-temperature heat capacity of β-polymorph of glycine and a phase transition at 252 K, J. Therm. Analys. Calorim. 79 (2005) 65–70.

81. *Boldyreva, E.V., Kolesnik, E. N., Drebushchak, T.N., Ahsbahs, H., Beukes, J.A., Weber, H.-P.* A comparative study of the anisotropy of lattice strain induced in the crystals of L-serine by cooling down to 100 K or by increasing pressure up to 4.4 GPa, Z. Kristallogr. 220 (2005) 58–65.

82. *Boldyreva, E.V., Kolesnik, E.N., Drebushchak, T.N., Sowa, H., Ahsbahs, H., Seryotkin, Yu.V.* A comparative study of the anisotropy of lattice strain induced in the crystals of DL-serine by cooling down to 100 K, or by increasing pressure up to 8.6 GPa, Z. Kristallogr. 221 (2006) 150–161.

83. *Drebushchak, V.A., Kovalevskaya, Yu.A., Paukov, I.E., Boldyreva, E.V.* Heat capacity of D- and DL-serine in a temperature range of 5.5 to 300 K, J. Therm Analys. Calorim. (2006) on-line (doi: 10.1007/S10973-006-7668-1).

84. *Drebushchak, V.A., Kovalevskaya, Yu.A., Paukov, I.E., Boldyreva, E.V.* Heat capacity of α-glycylglycine in a temperature range of 6 to 440 K. Comparison with glycines, J. Therm. Analys. Calorim. 85(2) (2006) 485–490.

85. *Mozhaev, V.V., Heremans, K., Frank, J., Masson, P., Balny, C.* High-pressure effects on protein structure and function, Proteins. 24 (1996) 81–91.
86. *Marchal, S., Torrent, J., Masson, P., Kornblatt, J.M., Tortora, P., Fusi, P., Lange, R., Balny, C.* The powerful high pressure tool for protein conformational studies, Brazilian J. Med. Biol. Res. 38(8) (2005) 1175–1183.
87. *Balny, C., Masson, P., Heremans, K.* High pressure effects on biological macromolecules: From structural changes to alteration of cellular processes, Biochim. Biophys. Acta – Protein Struct. Molec. Enzym. 1595(1–2) (2002) 3–10.
88. *Li, H., Akasaka, K.* Conformational fluctuations of proteins revealed by variable pressure NMR, Biochim. Biophys. Acta. 1764 (2006) 331–345.
89. *Frauenfelder, H., Alberding, N.A., Ansari, A., Braunstein, D., Cowen, B.R., Hong, M.K., Iben, I.E.T., Johnson, J.B., Luck, S., Marden, M.C., Mourant, J.R., Ormos, P., Reinisch, L., Scholl, R., Schulte, A., Shyamsunder, E., Sorensen, L.B., Steinbach, P.J., Xie, A., Young, R.D., Yue, K.T.* Proteins and pressure, J. Phys. Chem. 94 (1990) 1024–1037.
90. *Panick, G., Malessa, R., Winter, R., Rapp, G., Frye, K.J., Royer, C.A.* Structural characterization of the pressure-denatured state and unfolding/refolding kinetics of staphylococcal nuclease by synchrotron small-angle X-ray scattering and Fourier-transform infrared spectroscopy, J. Mol. Biol. 275 (1998) 389–402.
91. *Panick, G., Winter, R.* Pressure-induced unfolding/refolding of ribonuclease A: static and kinetic Fourier transform infrared spectroscopy study, Biochem. 39 (2000) 1862–1869.
92. *Vogtt, K., Winter, R.* Pressure-assisted cold denaturation of hen egg white lysozyme: the influence of cosolvents probed by hydrogen exchange NMR, Braz. J. Med. Biol. Res. 38 (2005) 1185–1193.
93. *Daniel, I., Oger, P., Winter, R.* Origins of life and biochemistry under high-pressure conditions, Chem. Soc. Rev. 35 (2006) 858–875.
94. *Dzwolak, W., Jansen, R., Smirnovas, V., Loksztejn, A., Porowski, S., Winter, R.* Template-controlled conformational patterns of insulin fibrillar self-assembly reflect history of solvation of the amyloid nuclei, Phys. Chem. Chem. Phys. 7 (2005) 1349–1351.
95. *Gabke, A., Kraineva, J., Köhling, R., Winter, R.* Using pressure in combination with X-ray and neutron scattering techniques for studying the structure, stability and phase behaviour of soft condensed matter and biomolecular systems, J. Phys.: Condens. Matter. 17 (2005) S3077–S3092.
96. *Jansen, R., Dzwolak, W., Winter, R.* Amyloidogenic self-assembly of insulin aggregates probed by high resolution atomic force microscopy, Biophys. J. 88 (2005) 1344–1353.
97. *Winter, R.* High pressure effects in molecular bioscience, in: Chemistry at Extreme Conditions (Edited by M.R. Manaa), Elsevier B.V., 29–82 (2005).
98. *Smirnovas, V., Winter, R., Funck, T., Dzwolack, W.* Thermodynamic properties underlying the α-helix-to-β-sheet transition, aggregation and amyloidogenesis of polylysine as probed by calorimetry, densimetry and ultrasound velocimetry, J. Phys. Chem. B. 109 (2005) 19043–19045.
99. *Grudzielanek, S., Jansen, R., Winter, R.* Solvational tuning of the unfolding, aggregation and amyloidogenesis of insulin, J. Mol. Biol. 351 (2005) 879–894.
100. *Eisenblätter, J., Winter, R.* Pressure effects on the structure and phase behavior of DMPC-gramicidin lipid bilayers – a synchrotron SAXS and ^{2}H-NMR spectroscopy study, Biophys. J. 90 (2006) 956–966.
101. *Mitra, L., Smolin, N., Ravindra, R., Royer, C., Winter, R.* Pressure perturbation calorimetric studies of the solvation properties and the thermal unfolding of proteins in solution – experiments and theoretical interpretation, Phys. Chem. Chem. Phys. 8 (2006) 1249–1265.
102. *Smirnovas, V., Winter, R., Funck, T., Dzwolak, W.* Protein amyloidogenesis in the context of volume fluctuations: a case study on insulin, Phys. Chem. Chem. Phys. 7 (2006) 1046–1049.
103. *Grudzielanek, S., Smirnovas, V., Winter, R.* Solvation-assisted pressure tuning of insulin fibrillation: from novel aggregation pathways to biotechnological applications, J. Mol. Biol. 356 (2006) 497–509.

104. *Kundrot, C.E., Richards, F.M.* Crystal structure of hen egg-white lysozyme at a hydrostatic pressure of 1000 atmospheres, J. Mol. Biol. 193 (1987) 157–170.

105. *Katrusiak, A., McMillan, P.* High-Pressure Crystallography, NATO Science Series. II. Mathematics, Physics and Chemistry, Vol. 140 (2004) Kluwer, Dordrecht.

106. *Katrusiak, A., Dauter, Z.* Compressibility of lysozyme crystals by X-ray diffraction, Acta Crystallogr. D52 (1996) 607–608.

107. *Fourme, R., Kahn, R., Mezouar, M., Girard, E., Hoerentrup, C., Prange, T., Ascone, I.* High-pressure protein crystallography (HPPX): instrumentation, methodology and results on lysozyme crystals, J. Synchrotron Radiation 8 (2001) 1149–1156.

108. *Girard, E., Kahn, R., Mezouar, M., Dhaussy, A.-C., Lin, T., Johnson, J.E., Fourme, R.* The first crystal structure of a macromolecular assembly under high pressure: CpMV at 330 MPa, Biophys. J. 88 (2005) 3562–3571.

109. *Colloc'h, N. Girard, E., Dhaussy, A.-C., Kahn, R., Ascone, I., Mezouar, M., Fourme, R.* High-pressure macromolecular crystallography: The 140-MPa crystal structure at 2.3 Å resolution of urate oxidase, a 135-kDa tetrameric assembly, Biochim. Biophys. Acta. 1764 (2006) 391–397.

110. *Fourme, R., Girard, E., Kahn, R., Dhaussy, A.-C., Mezouar, M., Colloc'h, N., Ascone, I.* High-pressure macromolecular crystallography (HPMX): status and prospects, Biochim. Biophys. Acta. 1764 (2006) 384–390.

111. *Moreno A.J.D., Freire P.T.C., Melo F.E.A., Araujo Silva M.A., Guedes I., Mendes Filho J.* Pressure-induced phase transitions in monohydrated l-asparagine aminoacid crystals, Solid State Commun. 103(12) (1997) 655–658.

112. *Teixeira, A.M.R., Freire, P.T.C., Moreno, A.J.D., Sasaki, J.M., Ayala, A.P., Mendes Filho, J., Melo, F.E.A.* High-pressure Raman study of L-alanine crystal, Solid State Commun. 116(7) (2000) 405–409.

113. *Sasaki, J.M., Freire, P.T.C., Moreno, A.J.D., Melo, F.E.A., Guedes, I., Mendes-Filho, J., Shu, J., Hu, J., Mao, Ho-Kwang.* Single crystal X-ray diffraction in monohydrate L-asparagine under hydrostatic pressure. Science and Technology of High Pressure. Proceedings of AIRAPT-17, Ed. M.H. Manghnani, W.J. Nellis, and M.F. Nicol, University Press, Hyderabad, India (2000) 502–505.

114. *Boldyreva, E.V., Boldyrev, V.V.* Reactivity of Molecular Solids, John Wiley & Sons, Chichester, (1999).

115. *Boldyreva, E.V., Ahsbahs, H., Weber, H.-P.* A comparative study of pressure-induced lattice strain of α and γ polymorphs of glycine, Z. Kristallogr. 218 (2003) 231–236.

116. *Boldyreva, E.V.* High-pressure studies of the anisotropy of structural distortion of molecular crystals, J. Mol. Struct. 647 (2003) 159–179.

117. *Boldyreva, E.V.* Molecules in strained environment, in High-Pressure Crystallography, Ed. A. Katrusiak & P.F. McMillan, Kluwer, Dordrecht (2004) 495–512.

118. *Boldyreva, E.V.* High pressure and supramolecular systems, Russ. Chem. Bulletin. 7 (2004) 1315–1324.

119. *Boldyreva, E.V.* High-pressure induced structural changes in molecular crystals preserving the space group symmetry: anisotropic distortion / isosymmetric polymorphism, Cryst. Engineering 6(4) (2004) 235–254.

120. *Boldyreva, E.V.* High-pressure studies of the hydrogen bond networks in molecular crystals, J. Mol. Struct. 700(1–3) (2004) 151–155.

121. *Boldyreva, E.V., Drebushchak, T.N., Shakhtshneider, T.P., Sowa, H., Ahsbahs, H., Goryainov, S.V., Ivashevskaya, S.N, Kolesnik, E.N., Drebushchak, V.A., Burgina, E.B.* Variable-temperature and variable-pressure studies of small-molecule organic crystals, Arkivoc XII (2004) 128–155.

122. *Boldyreva, E.V., Ivashevskaya, S.N., Sowa, H., Ahsbahs, H., Weber, H.-P.* Effect of hydrostatic pressure on the γ-polymorph of glycine. 1. A polymorphic transition into a new δ-form, Z. Kristallogr. 220(1) (2005) 50–57.

123. *Goryainov, S.V., Kolesnik, E.N., Boldyreva, E.V.* A reversible pressure-induced phase transition in β-glycine at 0.76 GPa, Physica B Condensed Matter. 357(3–4) (2005) 340–347.

124. *Kolesnik, E.N., Goryainov, S.V., Boldyreva, E.V.* Different behavior of L- and DL-serine crystals at high pressures: phase transitions in L-serine and stability of the DL-serine structure, Doklady Chem. 404(2005) 61–64 (Rus.), or 169–172 (Engl.).

125. *Goryainov, S.V., Boldyreva, E.V., Kolesnik, E.N.* Raman observation a new (ζ) polymorph of glycine? Chem. Phys. Letters. 419(4–6) (2006) 496–500.

126. *Drebushchak, T.N., Sowa, H., Seryotkin, Yu.V., Boldyreva, E.V.* L-serine-III at 8.0 GPa, Acta Crystallogr. E62 (2006) o4052–o4054.

127. *Boldyreva, E.V., Sowa, H., Seryotkin, Yu.V., Drebushchak, T.N., Ahsbhas, H., Chernyshev, V.V., Dmitriev, V.P.* Pressure-induced phase transitions in crystalline L-serine studied by single-crystal and high-resolution powder X-ray diffraction, Chem. Phys. Letters. 429 (2006) 474–478.

128. *Murli, C., Sharma, S.M., Karmakar, S., Sikka, S.K.* α-Glycine under high pressures: a Raman scattering study, Physica B 339 (2003) 23–30.

129. *Dawson, A., Allan, D.R., Belmonte, S.A., Clark, S.J., David, W.I.F., McGregor, P.A., Parsons, S., Pulham, C.R., Sawyer, L.* Effect of high pressure on the crystal structures of polymorphs of glycine, Crystal Growth and Design. 5(4) (2005) 1415–1427.

130. *Moggach, S.A., Allan, D.R., Parsons, S., Sawyer, L., Warren, J.E.* The effect of pressure on the crystal structure of hexagonal L-cystine, J. Synchrotron Radiation. 12 (2005) 598–607.

131. *Moggach, S.A., Allan, D.R., Morrison, C.A., Parsons, S., Sawyer, L.* Effect of pressure on the crystal structure of L-serine-I and the crystal structure of L-serine II at 5.4 GPa. Acta Cryst. B61 (2005) 58–68.

132. *Moggach, S.A., Allan, D.R., Parsons, S., Sawyer, L.* Effect of pressure on the crystal structure of α-glycilglycine to 4.7 GPa; application of Hirshfeld surfaces to analyse contacts on increasing pressure, Acta Crystallogr. B62 (2006) 310–320.

133. *Moggach, S.A., Allan, D.R., Clark, S.J., Gutmann, M.J., Parsons, S., Pulham, C.R., Sawyer, L.* High-pressure polymorphism in L-cysteine: the crystal structures of L-cysteine-III and L-cysteine-IV, Acta Crystallogr. B62 (2006) 296–309.

134. *Moggach, S.A., Marshall, W.G., Parsons, S.* High-pressure neutron diffraction study of L-serine-I and L-serine-II, and the structure of L-serine-III at 8.1 GPa, Acta Crystallogr. B62 (2006) 815–825.

135. *Rieckhoff, K.E., Peticolas, W.L.* Optical second harmonic generation in crystalline amino acids, Science. 147(1965) 610–611.

136. *Misoguti, L., Bagnato, V.S., Zilio, S.C., Varela, A.T., Nunes, F.D., Melo, F.E.A., Filho, J. Mendes.* Optical properties of L-alanine organic crystals, Optical Materials 6(3) (1996) 147–152.

137. *Bhat, M.N., Dharmaprakash, S.M.* Effect of solvents on the growth morphology and physical characteristics of nonlinear optical γ-glycine crystals, *J. Crystal Growth* 242 (1–2) (2002) 245–252.

138. *Petrosyan, H.A., Karapetyan, H.A., Antipin, M.Yu., Petrosyan, A.M.* Non-linear optical crystals of L-histidine salts, J. Crystal Growth. 275 (2005) e1919–e1925.

139. *Petrosyan, A.M., Sukiasyan, R.P., Karapetyan, H.A., Antipin, M.Yu., Apreyan, R.A.* L-arginine oxalates, J. Crystal Growth. 275 (2005) e1927–e1933.

140. *Petrosyan, A.M., Karapetyan, H.A., Sukiasyan, R.P., Aghajanyan, A.E., Morgunov, V.G., Kravchenko, E.A., Bush, A.A.* Crystal structure and characterization of L-arginine chlorate and L-arginine bromate, J. Mol. Struct. (2005) 752, 144–152.

141. *Ramesh, K, Raj, S.G., Mohan, R., Jayavel, R.* Growth, structural and spectral analyses of nonlinear optical l-threonine single crystals, J Crystal Growth. 275(1–2) (2005) e1947–e1951.

142. *Lemanov, V.V., Popov, S.N., Pankova, G.A.* Piezoelectric properties of some crystalline amino acids and their complexes, Solid State Physics (Fiz. Tv. Tela). 44(10) (2002) 1840–1846.

143. *Bernal, J.D.* The crystal structure of the natural amino acids and related compounds. Z. Kristallogr. **78** (1931) 363–369.

144. *Boldyreva, E.V., Drebushchak, V.A., Drebushchak, T.N., Paukov, I.E., Kovalevskaya, Yu.A., Shutova, E.S.* Polymorphism of glycine. Thermodinamic aspects. Part I. Relative stability of the polymorphs, J. Therm. Analys. Calorim. 73 (2003) 409–418.

145. *Boldyreva, E.V., Drebushchak, V.A., Drebushchak, T.N., Paukov, I.E., Kovalevskaya, Yu.A., Shutova, E.S.* Polymorphism of glycine. Thermodinamic aspects. Part II. Polymorphic transitions, J. Therm. Analys. Calorim. 73 (2003) 419–428.

146. *Sakai, H., Hosogai, H., Kawakita, T.* Transformation of α-glycine to γ-glycine. J. Crystal Growth 116 (1992) 421–426.

147. *Pyne, A., Suryanarayanan, R.* Phase transitions of glycine in frozen aqueous solutions and during freeze-drying, Pharm. Res. 18 (2001) 1448–1454.

148. *Ferrari, E.S., Davey, R.J., Cross, W.I., Gillon, A.L., Towler, C.S.* Crystallization in polymorphic systems: the solvent-mediated transformation of β to α glycine, Crystal Growth Design. 3(1) (2003) 53–60.

149. *Harding, M.M., Long, H.A.* The crystal and molecular structure of L-cysteine, Acta Crystallogr. B24 (1968) 1096–1102.

150. *Khawas, B.* X-ray study of L-arginine HCl, L-cysteine, DL-lysine, and DL-phenylalanine, Acta Crystallogr. B27 (1971) 1517–1520.

151. *Görbitz, C.H., Dalhus, B.* L-cysteine. Monoclinic form. Redetermination at 120 K, Acta Crystallogr. C52 (1996) 1756–1759.

152. *Kerr, K.A., Ashmore, J.P., Koetzle, T.F.* A neutron diffraction study of L-cysteine, Acta Crystallogr. B31 (1975) 2022–2026.

153. *Kerr, K.A., Ashmore, J.P.* Structure and conformation of orthorhombic L-cysteine, Acta Crystallogr. B29 (1973) 2124–2127.

154. *Mathieson, A.M.* The crystal structures of the dimorphs of DL-methionine, Acta Crystallogr. 5 (1952) 332–341.

155. *Taniguchi, T., Takaki, Y., Sakurai, K.* The crystal structures of the α and β forms of DL-methionine, Bull. Chem. Soc. Jpn. 53 (1980) 803–804.

156. *Alagar, M., Krishnakumar, R.V., Mostad, A., Natarajan, S.* DL-Methionine at 105 K, Acta Crystallogr. E61 (2005) o1165–o1167.

157. *Ramachandran, E., Natarajan, S.* Gel-growth and characterization of β-DL-methionine, Cryst. Res. Technol. 41 (2006) 411–415.

158. *Hirokawa, S.* A new modification of L-glutamic acid and its structure, Acta Crystallogr. 8 (1955) 637–641.

159. *Marcoin, W., Duda, H., Kusz, J., Bzowski, B., Warcewski, J.* L-glutamic acid, Appl. Crystallogr. Conference 17th (1999) 40.

160. *Hirayama, N., Shirahata, K., Ohashi, Y., Sasada, Y.* L. Glutamic acid, Bull. Chem. Soc. Jpn. 53 (1980) 30–35.

161. *Lehmann, M.S., Nunes, A.C.* A short hydrogen bond between near identical carboxyl groups in the α-modification of L-glutamic acid, Acta Crystallogr. 36B (1980) 1621–1625.

162. *Lehmann, M.S., Koetzle, T.F., Hamilton, W.C.* L-Glutamic Acid, J. Cryst. Mol. Struct. 2 (1972) 225.

163. *Bernstein, J.* Polymorphism of L-glutamic acid: decoding the α-β phase relationship via graph-set analysis, Acta Crystallogr. B47 (1991) 1004–1010.

164. *Blagden, N., Davey, R.* Polymorph selection: challenges for the future, Crystal Growth & Design 3(6) (2003) 873–885.

165. *Cashell, C., Sutton, D., Corcoran, D., Hodnett, B.K.* Inclusion of the stable form of a polymorph within crystals of its metastable form, Crystal Growth and Design. 3(6) (2003) 869–872.

166. *Cashell, C., Corcoran, D., Hodnett, B.K.* Secondary nucleation of the β-polymorph of L-glutamic acid on the surface of α-form crystals, ChemComm. 374 (2003) 374–375.

167. *Ono, T., Horst, J.H., Jansens, P.J.* Quantitative measurement of the polymorphic transformation of L-glutamic acid using in-situ Raman spectroscopy, Crystal Growth and Design. 4(3) (2004) 465–469.

168. *Ono, T., Kramer, J.M., Horst, J.H., Jansens, P.J.* Process modelling of the polymorphic transformation of L-glutamic acid, Crystal Growth and Design. 4(6) (2004) 1161–1167.

169. *Frey, M.N., Lehmann, M.S., Koetzle, T.F., Hamilton, W.C.* Precision neutron diffraction structure determination of protein and nucleic acid components. XI. Molecular configuration and hydrogen bonding of serine in the crystalline amino acids L-serine monohydrate and DL-serine, Acta Crystallogr. B29 (1973) 876–884.

170. *Lahav, M., Weissbuch, I., Shavit, E., Reiner, C., Nicholson, G.J., Schurig, V.* Parity violating energetic difference and enantiomorphous crystals – caveats; reinvestigation of tyrosine crystallization, Origins of Life and Evolution of the Biosphere. (2006) 36, 151–170.

171. *Igarashi, K., Sasaki, Y., Azuma, M., Noda, H., Ooshima, H.* Control of polymorphs on the crystallization of glycine using a WWDJ batch crystallizer, Eng. Life Sci. 3 (2003) 159–163.

172. *Yu, Lian, Ng, Kingman.* Glycine crystallization during spray drying: the pH effect on salt and polymorphic forms, J. Pharm. Sci. 91(11) (2002) 2367–2375.

173. *Akers, M.J., Milton, N., Byrn, S.R., Nail, S.L.* Glycine crystallization during freezing: The effects of salt form, pH, and ionic strength, Pharm. Res. 12 (1995) 1457–1461.

174. *He, G., Bhamidi, V., Wilson, S.R., Tan, R.B.H., Kenis, P.J.A., Zukoski, C.F.* Direct growth of γ-glycine from neutral aqueous solutions by slow, evaporation-driven crystallization, Crystal Growth and Design. 6(8) (2006) 1746–1749.

175. *Allen, K., Davey, R.J., Ferrari, E., Towler, C., Tiddy, G.J.* The crystallization of glycine polymorphs from emulsions, microemulsions, and lamellar phases, Crystal Growth and Design. 2(6) (2002) 523–527.

176. *Lee, A.Y., Lee, I.S., Dette, S.S., Boerner, J., Myerson, A.S.* Crystallization on confined engineering surfaces: a method to control crystal size and generate different polymorphs, J. Amer. Chem. Soc. 127 (2005) 14982–14983.

177. *Sun, X., Garetz, B.A., Myerson, A.S.* Supersaturation and polarization dependence of polymorph control in the nontopochemical laser-induced nucleation (NPLIN) of aqueous glycine solutions, Crystal Growth and Design. 1 (2001) 5–8.

178. *Garetz, B.A., Matic, J., Myerson, A.S.* Polarization switching of crystal structure in the nonphotochemical light-induced nucleation of supersaturated aqueous glycine solutions, Phys. Rev. Letters. 89(17) (2002) 175501–4.

179. *Zaccaro, J., Matic, J., Myerson, A.S., Garetz, B.A.* Nontopochemical, laser-induced nucleation of supersaturated aqueous glycine produces unexpected γ-polymorph, Crystal Growth and Design. 1(1) (2001) 5–8.

180. *Aber, J.E., Arnold, S., Garetz, B.A., Myerson, A.S.* Strong dc electric field applied to supersaturated aqueous glycine solution induces nucleation of the γ-polymorph, Phys. Rev. Lett. 94 (2005) 145503–5.

181. *Boldyreva, E.V., Drebushchak, V.A., Drebushchak, T.N., Shutova, E.S.* Synthesis and calorimetric investigation of unstable β-glycine, J. Cryst. Growth. 241 (2002) 266–268.

182. *Torbeev, V.Yu., Shavit, E., Weissbuch, I., Leiserowitz, L., Lahav, M.* Control of crystal polymorphism by tuning the structure of auxiliary molecules as nucleation inhibitors. The β-polymorph of glycine grown in aqueous solutions, Crystal Growth & Design. 5(6) (2005) 2190–2196.

183. *Weissbuch, I., Torbeev, V.Yu., Leiserowitz, L., Lahav, M.* Solvent effect on crystal polymorphism: why addition of methanol or ethanol to aqueous solutions induces the precipitation of the least stable β form of glycine, Angew. Chem. Int. Ed. 11(10) (2005) 3039–3048.

184. *Weissbuch, I., Popovitz-Biro, R., Lahav, M., Leiserowitz, L.* Understanding and control of nucleation, growth, habit, dissolution and structure of two- and three-dimensional crystals using "tailor-made" auxiliaries, Acta Crystallogr. B51 (1995) 115–148.

185. *Kitamura, M., Ishizu, T.* Kinetic effect of L-phenylalanine on growth process of L-glutamic acid polymorph, J. Cryst. Growth. 192(1) (1998) 225–235.

186. *Cashell, C., Corcoran, D., Hodnett, B.K.* Effect of amino acid additives on the crystallization of L-glutamic acid, Crystal Growth & Design. 5 (2005) 593–597.

187. *Chattopadhyay, S., Erdemir, D., Evans, J.M.B., Ilavsky, J., Amenitsch, H., Segre, C.U., Myerson, A.S.* SAXS study of the nucleation of glycine crystals from a supersaturated solution, Crystal Growth & Design. 5(2) (2005) 523–527.

188. *Gidalevitz, D., Feidenhans'l, R., Matlis, S., Smilgies, D.-M., Christensen, M.J., Leiserowitz, L.* Monitoring in situ growth and dissolution of molecular crystals: towards determination of the growth units, Angew. Chem. Int. Ed. Engl. 36 (1997) 955–959.

189. *Gavezzotti, A.* Molecular Aggregation: Structure Analysis and Molecular Simulation of Crystals and Liquids, Oxford University Press, 2006.

190. *Nuraje, N., Su, K., Samson, J., Haboosheh, A., McCuspie, R.I., Matsui, H.* Self-assembly of Au nanoparticle-containing peptide nano-rings on surfaces, Supramolecular Chemistry 18(5) (2006) 429–434.

191. *Joshi, K.B., Verma, S.* Ordered self-assembly of a glycine-rich linear and cyclic hexapeptide: contrasting ultrastructural morphologies of fiber growth, Supramolecular Chemistry. 18(5) (2006) 405–414.

192. *Gekko, K., Hasegawa, Y.* Compressibility-structure relationship of globular proteins, Biochemistry. 25(21) (1986) 6563–6571.

193. *Gekko, K.* Compressibility gives new insight into protein dynamics and enzyme function, Biochim. Biophys. Acta. 1595 (2002) 382–386.

194. *Boldyreva, E.V., Tumanov, N.A., Ahsbahs, H., Dmitriev, V.P.* High-resolution X-ray diffraction study of the effect of pressure on β-glycine, in preparation.

195. *Boldyreva, E.V., Seryotkin, Yu. V., Ahsbahs, H., Dmitriev, V.P.* High-resolution X-ray diffraction study of the effect of pressure on L- and DL-alanine, in preparation.

196. *Boldyreva, E.V., Goryainov, S.V., Seryotkin, Yu.V., Ahsbhas, H., Dmitriev, V.P.* Pressure-induced phase transitions in the crystals of β-alanine, Vestnik NGU, Ser. Fizika (Proceed. NSU, Series Physics), 2(2) (2007) 30–35.

197. *Nye, J.F.* Physical properties of crystals. Their representation by tensors and matrices, Clarendon Press, Oxford (1957).

198. *Hazen, R., Finger, L.* Comparative Crystal Chemistry. Temperature, Pressure, Composition and Variation of the Crystal Structure, Wiley, NewYork USA (1982).

199. *Katrusiak, A.* High-pressure X-ray diffraction studies on organic crystals, Cryst. Res. Techn. 26 (1991) 523–531.

200. *Katrusiak, A.* General description of hydrogen-bonded solids at varied pressures and temperatures. In: Katrusiak, A., McMillan, P. (Eds). High-Pressure Crystallography, NATO Science Series. II. Mathematics, Physics and Chemistry, Vol. 140 (2004) Kluwer, Dordrecht, Netherlands 513–520.

201. *Su, C.-C., Chang, H.-C., Jiang, J.-C., Wei, P.-Y., Lu, L.-C., Lin, S.H.* Evidence of charge-enhanced C—H—O interactions in aqueous protonated imidazole probed by high pressure infrared spectroscopy, J. Chem. Phys. 119(20) (2003) 10753–10758.

202. *Chang, H.-C., Lee, K. M., Jiang, J.-C., Lin, M.-S., Chen, J.-S., Lin, I.J.B., Lin, S.H.* Charge-enhanced C—H—O interactions of a self-assembled triple helical spine probed by high-pressure, J. Chem. Phys. 117(4) (2002) 1723–1728.

203. *Goryainov, S.V., Boldyreva, E., Smirnov, M.B., Madyukov, I.A., Kolesnik, E.N.* Raman spectroscopy study of the effect of pressure on α-glycilglycine, Physica B (2007), submitted.

204. *Chernoby, G.B., Chesalov, Yu.A., Burgina, E.B., Drebushchak, T.N., Boldyreva, E.V.* Variable-temperature IR-spectroscopy study of the crystalline amino acids, dipeptides and polyaminoacids. I. Glycine, Russ. J. Struct. Chem. 48(2) (2007) 339–347.

205. *Pain, R.H.* Mechanisms of Protein Folding, Frontiers in Molecular Biology, Ser. Ed. B.D. Hames and D.M. Glover, Oxford University Press, Oxford UK (2000).

206. *Chatani, E., Kato, M., Kawai, T., Naiki, H. Goto, Y.* Main-chain dominated amyloid structures demonstrated by the effect of high pressure, J. Mol. Biol. 352 (2005) 941–951.

207. *Isenberg, H., Kjaer, K., Rapaport, H.* Elasticity of crystalline β-sheet monolayers, J. Amer. Chem. Soc. 128 (2006) 12468–12472.

CHAPTER 8

UNRAVELLING THE CHEMICAL BASIS OF THE BATHOCHROMIC SHIFT OF THE LOBSTER CARAPACE CAROTENOPROTEIN CRUSTACYANIN

JOHN R. HELLIWELL AND MADELEINE HELLIWELL

Abstract: This chapter will review the atomic structural details of the lobster carotenoid-protein complex, crustacyanin, discovered in the last 5 years, and also present, in that context, an overview description of several new carotenoid crystal structures, protein free, in various new crystal packing arrangements. The interrelationship of structure, dynamics and function and the tuning of the colours are well illustrated by these molecular structures. This field of structural chemistry research contains much that is seemingly magical, namely the colour shift and the mystery to us of how this has evolved in Nature

INTRODUCTION

The power of crystallography to "see atoms" (paraphrasing W L Bragg (1968)) in β-crustacyanin (β-CR) (Cianci et al 2002) indicated the candidate molecular parameters that are responsible for the bathochromic shift, which is most dramatically demonstrated by the colour change of lobsters on cooking, turning from blue/black to orange/red.

There have then followed studies, described in the second half of this chapter, seeking both to obtain more precise carotenoid structures and also to mimic the spectral shift in Nature's lobster but in a "simpler" way. Perhaps the *magic* of Nature cannot be copied?

The reason why lobsters have evolved this elaborate colouration mechanism may be to develop a predominant colour which allows this species to avoid ocean predators such as octopus. Although evidence that the octopus is colour blind has been presented by Messenger (1977), that work also showed that octopus vision is very sensitive to the hue and brightness of colours. The Australian Western Rock lobster interestingly has a lengthy "white" colour period before turning red (not black). Might camouflage issues be so different there? Or could it be that the colours of

193

J.C.A. Boeyens and J.F. Ogilvie (eds.), Models, Mysteries and Magic of Molecules, pp. 193–208.
© 2008 *Springer*.

a particular species of lobster are for different reasons, such as to absorb light of a particular wavelength (North, 1993)? Most likely this aspect of lobster colouration, and its evolution will, for the time being, remain a *mystery* to us.

PUBLIC INTEREST IN THE RESULTS

There has been considerable public interest in our crystallographic work on crustacyanin (see summary in Chayen et al 2003) and the related question:- "Why do lobsters change colour on cooking?" stimulated by our research. The effect of cooking lobster can be reasonably guessed at being due to denaturation of the crustacyanin carotenoprotein to the extent that the bound colouring agent, the carotenoid astaxanthin (ASX), becomes permanently liberated into a free, or freer form, coloured orange at this level of dilution (nb concentrated ASX in solid crystalline form is a deep purple colour).

We published our article in PNAS (Cianci et al 2002), a learned society publication with extremely wide readership, and accompanied this with a press release, which was quickly picked up by the media. The Liverpool Daily Post, for example, reported on our work with "Boffins solve lobster shell colouration mystery"! The TV news reported "a new method of delivering the antioxidant, possibly cancer preventing, ASX molecule, which is hydrophobic unless bound to a protein (like beta -CR)". Other newspapers referred to the chance to use protein engineering and site specific mutagenesis to alter the molecular recognition site of ASX and thus manipulate the colour, which might find applications in food colourants or in the design of genetically modified flowering plants with new colours. High school students formed a second wave of interest with emailed requests to us to help them with the details of our PNAS research article so as to complete their chemistry homework assignments; fortunately SWISSProt (Gerritsen, 2002) provided an excellent resume for us to send at that time. Later, more detailed science articles were presented in for example, Physics Today (Day, 2002), which included a description of the molecular basis of the narrowing of the energy gap between the highest occupied and the lowest unoccupied molecular orbitals ("HOMO LUMO gap"). Our technical breakthrough of using softer X-ray crystallography and xenon with sulphur anomalous scattering, reported in Acta Cryst D (Cianci et al 2001), along with the first new protein structure solved that way, was also highlighted (Day, 2002).

The publicity led to new directions of research too. A USA radio listener drew our attention to the Marine Research Center in Maine, USA, which has a collection of unusually coloured lobsters. Suitably inspired as amateur biologists we went on the hunt for these and found success whilst holidaying in Scotland where the Harbour Master at the charming village of the Isle of Whithorn showed us a colour spread of lobsters in his water tank. Intriguingly his fishermen reported a link between colour of lobsters and how deep the sea-bed lay. So, black lobsters, he said, came from shallow sandy waters and the lighter blue ones from deep rocky sea-beds, an observation consistent with the depth distribution of colour in Crustacea (Herring, 1985). This example illustrates the connection between whole biology and molecular biology and the balance of molecular complexity with the apparent simplicity of the colour of lobster shell.

PRACTICALITIES OF THE RESEARCH AND ITS HISTORICAL DEVELOPMENT

In terms of the practical details of the work, all the biochemical separations were done by Dr Peter Zagalsky of Royal Holloway college, University of London. The interest in the blue colouration of the shell of lobster *Homarus gammarus* (Figure 8-1) was initiated by Newbigin (1897,1898), who found that one of the group of pigments, now known as carotenoproteins, could be extracted with ammonium chloride; she suggested that these were composed of complex organic bases with lipochromes. Later, Verne (1921) proposed that the pigment was a combination of a carotene-like compound and a protein. The carotenoid astaxanthin (ASX) ($3,3'$-dihydroxy-β, β-carotene-$4,4'$-dione), was subsequently identified as the prosthetic group (Kuhn and Sorensen, 1938). Wald et al (1948) extracted the predominant complex from among other carotenoproteins in the carapace, established its protein nature and named it crustacyanin (CR).

Partly from Peter Zagalsky's work (see Zagalsky 1985, 2003 and references therein spanning more than 30 years) and from other major contributions (see the classical papers of Kuhn and Kuhn (1967), Buchwald and Jencks (1968) and

Figure 8-1. (a) The natural blue coloration of the shell of lobster *Homarus gammarus* (b) the colour of cooked lobster. From Dr P Zagalsky with permission (see color plate section)

Britton et al 1997), we know that the native carotenoprotein, α-CR, has a molecular weight of about 320 kDa and that it is a multi-macromolecular 16-mer complex of protein subunits with 16 bound ASX molecules (Figure 8-2). These subunits are electrophoretically distinct components, each of approximate molecular weight of 20 kDa, classified by amino acid composition, size estimation, and peptide mapping into two groups: Gene type I [CRTC]: protein forms A1, C1 (predominant) and C2 and Gene type II [CRTA]: protein forms A2 (predominant) and A3.

Pure β-CR (Figure 8-3b) was given to Prof Naomi Chayen of Imperial College London by Dr Peter Zagalsky. She grew the beautiful blue crystals of the protein (Figure 8-4a) over a 4 months period using a technique of crystal growth under

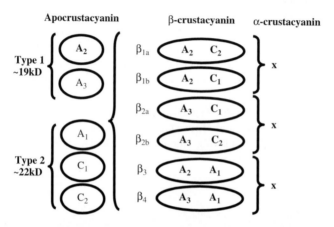

Figure 8-2. The crustacyanin subunits are electrophoretically distinct components, each of approximate molecular weight of 20 kDa, classified by amino acid composition, size estimation, and peptide mapping into two groups: Gene type I [CRTC]: protein forms A1, C1 and C2 and Gene type II [CRTA]: protein forms A2 and A3. From Boggon (1998) with permission

Figure 8-3. (a) ASX in hexane at a dilution similar to that in lobster crustacyanin (left) (b) beta-crustacyanin (right). From Dr P Zagalsky with permission (see color plate section)

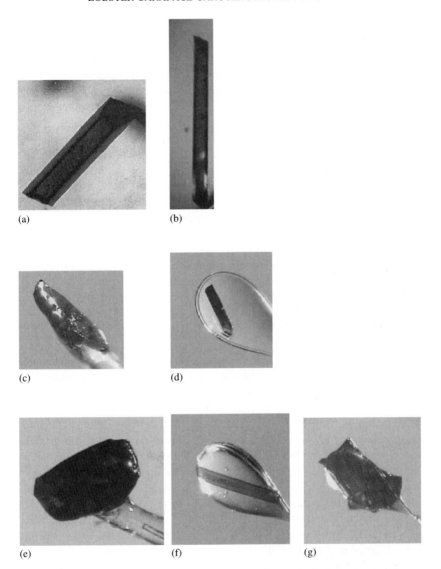

Figure 8-4. (a) Blue crystals of the protein β-CR (grown by Prof Naomi Chayen of Imperial College London; figure reproduced from Chayen (1998) with permission of the author and of IUCr Journals) and for comparison those of several unbound carotenoids (Bartalucci et al 2007) (b) chloroform solvate of ASX (c) pyridine solvate of ASX (d) unsolvated form of ASX (e) canthaxathin (f) zeaxanthin (g) β,β-carotene (see color plate section)

oil known as "microbatch", since the classic methods of crystallization repeatedly failed. She then presented the crystals, along with those of apocrustacyanin A1, C1 and C2 to us. Our Laboratory of Structural Chemistry in Manchester elucidated the structure of the β-CR at 3.2 Å (Figure 8-5) using X-ray crystallographic techniques; it was necessary first to establish the structure of apocrustacyanin A1 at 1.4 Å resolution (Cianci et al 2001) by innovative use of softer X-rays on the Synchrotron Radiation Source at Daresbury Laboratory (Helliwell, 2004). This employed a xenon heavy atom derivative prepared by placing a crystal in a high pressure of xenon gas. Four individual inert xenon gas atoms each occupied four separate pockets within the protein. The hand of the apocrustacyanin crystal structure was established using sulphur anomalous X-ray scattering also using softer X-rays. The apocrustacyanin A1 model at 1.4 Å served as a molecular replacement motif to solve the β-CR structure in detail, which is comprised of an identical A1 protein subunit and a sufficiently closely related apocrustacyanin A3 protein, which has 40% amino acid sequence identity to the A1 protein. Studying a protein at 3.2 Å is capable of yielding atomic positions and atomic displacement parameters but it is not optimum for some of the structure interpretation and in particular, for elucidating the details of the bound water. Our work was assisted by some other information however; firstly the apocrustacyanin A1 model at 1.4 Å diffraction resolution provided a level of detail where individual atoms are nearly separately resolved (which actually occurs at 1.2 Å or better diffraction resolution). Secondly, the β-CR, being composed of the two related A1 and A3 proteins, allowed structural chemistry cross-checks to be made. The amino acid sequence identity in the ASX binding regions was better than the 40% value overall, referred to above, so that common structural details were available for examination. Thirdly, for the ASX itself, a model was derived from the atomic resolution structure of the closely related carotenoid, canthaxanthin (β,β-carotene-4,4'-dione, Bart and MacGillavry, 1968). Thus the changes to the unbound ASX conformation could be elucidated rather than having to start from no structural information at all. Finally, our discovery of a bound water molecule at one end of each ASX at equivalent locations turned out to be of particular interest. The evidence for bound waters at such a diffraction resolution is the most testing aspect of a 3.2 Å diffraction analysis. Evaluation was made by comparing two types of difference electron density map namely 2Fo-Fc and Fo-Fc, which is "standard good practice". In addition the correspondence of equivalent chemical environments of each of these waters was encouraging. Finally, the cross checking for identically positioned bound waters in the superior diffraction 1.4 Å resolution apocrustacyanin A1 structure that we had determined previously, which were indeed present within 1 Å, was then overall compelling evidence. The fact that dehydration produces a colour change, which is reversible unlike cooking, now opens up a possible new protein crystallography experiment involving deliberate stages of dehydration of the β-CR crystal combined with simple colour monitoring.

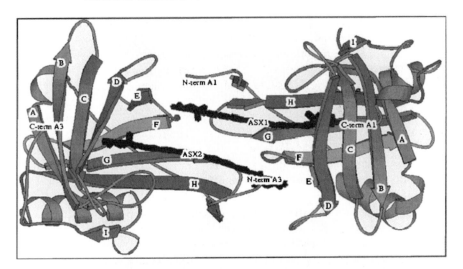

Figure 8-5. The structure of the β-CR at 3.2Å elucidated using X-ray crystallography. From Cianci et al 2002, with the permission of PNAS (see color plate section)

NEXT STEPS IN THE STRUCTURAL BIOLOGICAL CHEMISTRY RESEARCH

We are working to obtain crystals of the α-CR that diffract to higher resolution as well as undertake determination of the ultrastructure using other biophysical imaging techniques, such as solution X-ray scattering and cryo-electron microscopy (Chayen et al 2003). We also envisage exciting new projects working with the blue carotenoprotein, velallacyanin, of the Velella velella (Zagalsky, 1985), see Figure 8-6, and with the Australian Western Rock lobster which has a white phase (Figure 8-7); Wade and colleagues (Wade et al 2005, Wade, 2005) established that this was due to an esterification process of the intrinsic carotenoid.

Figure 8-6. A Velella velella (approx 1cm across) whose velallacyanin confers the distinctive blue colour on this sea creature (Zagalsky, 1985) (see color plate section)

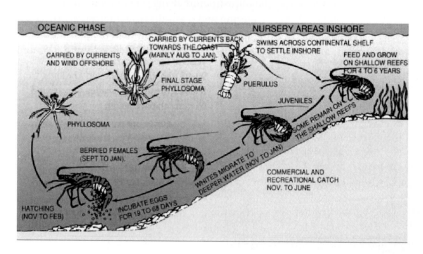

Figure 8-7. The Australian Western Rock lobster has a red phase (top) and a white phase (middle); reprinted from Comparative Biochemistry and Physiology, Part B 141 authors Wade et al (2005) entitled "Esterified astaxanthin levels in lobster epithelia correlate with shell colour intensity: Potential role in crustacean shell colour formation" pp. 307–313 Copyright (2005), with permission from Elsevier and the authors. The bottom schematic shows the life cycle; Figure from Wade 2005 with the permission of Dr. N. Wade and originally from the Western Australian Fisheries website http://www.fish.wa.gov.au/ (see color plate section)

DETAILS OF THE KEY MOLECULE "ASX" AND ITS INTERACTIONS IN β-CR

ASX, whose isomers are shown schematically in Figures 8-8a,b,c, is a carotenoid found in shrimps, crab, and lobsters, as well as in many invertebrate animals. The lobster β-CR study (Cianci et al 2002) has shown that the interactions with the protein causes the ASX molecule to change from an orange/red colour in the unbound form, to a deep slate-blue/black colour. At one end of each ASX, the 4-keto group is linked to a bound water molecule by a hydrogen bond. The other end of each ASX is connected by a hydrogen bond to the neighbouring protein subunit via its C4-keto group to the nitrogen atom of a His residue. A substantial conformational change of the ASX molecule was found to occur on binding to the protein, with the molecules both in the 6-s-trans conformation, the end-ring approximately coplanar with the polyene chains, and a significant bowing of the polyene chain compared with that found in the structure of canthaxanthin, which was used as a model for unbound ASX, in the absence of a crystal structure of unbound ASX at the time of the Cianci et al (2002) study. Theories that have been put forward as to the cause the bathochromic shift, based on this structural study, include the following:

- the coplanarisation of the β-end rings with the polyene chain increases the conjugation in the planar chromophoric system;
- an electronic polarisation effect stemming from interaction of the keto oxygens of the bound ASX molecules by H-bonding with histidine, and/or water;
- an exciton interaction due to the close proximity of the two bound ASX molecules;
- the bowing of the ASX molecule to relieve the strain in the all 6-s-trans conformation;
- A generally hydrophobic environment, where it is also known that dissolution in different solvents provokes different, albeit small, colour changes (Britton, 1995).

It is now generally accepted that the coplanarisation of the end rings accounts for up to a third of the overall shift of the absorption maximum from 472 nm in unbound ASX dissolved in hexane, to 580 nm in the β-crustacyanin molecule (Durbeej & Eriksson, 2003, 2004; van Wijk et al 2005; Ilagan et al 2005). In the unbound molecule, there is already some partial conjugation of the polyene chain into the end rings (Durbeej and Eriksson 2004), and the coplanarisation has the effect of extending this to 13 conjugated double bonds. Even in the recent literature, there is still disagreement as to where the additional bathochromic shift arises; Weesie et al 1999 proposed that protonation of the C-4 keto groups is responsible for polarisation of the chromophore, thus reducing the energy difference between the HOMO and LUMO orbitals; more recently Durbeej & Eriksson, 2003, 2004, put forward the theory that the polarisation effect is caused by H-bonding of the C-4 keto group particularly to a His residue, which they suggest must be protonated; van Wijk et al 2005 dismiss the polarisation theory, citing evidence from ^{13}C NMR spectroscopy studies and instead propose that the approach of the chromophores to within 7 Å of one another at 120° inclination gives rise to exciton interaction, leading to the observed spectral shift. The further shift in the absorption in the

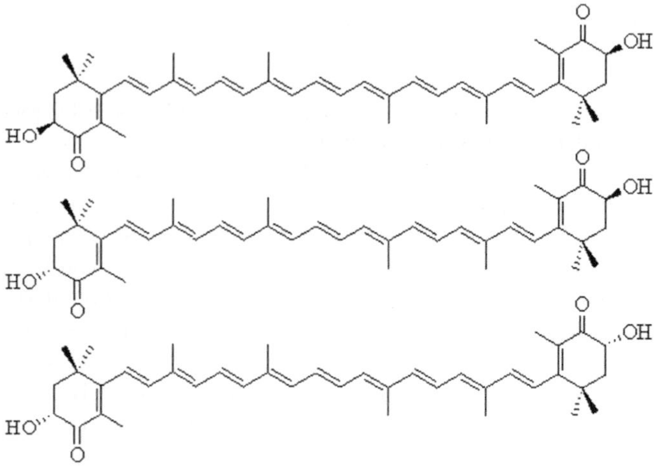

Figure 8-8. CHEMDRAW plots of (a) (3S,3S')-ASX (b) (3R,3S',meso)-ASX (c) (3R,3R')-ASX

β-CR to that seen in the α-CR arises, they argue, from increased exciton coupling caused by additional aggregation of ASX molecules. The effect of the bowing of the ASX molecule in the bound form is proposed by Durbeej and Eriksson 2004 to give a negligible effect on the colour shift.

CHEMICAL CRYSTALLOGRAPHY OF CAROTENOIDS

In order to investigate these colour tuning parameters further via as wide an ensemble of crystal structures as possible, we set about crystallising various carotenoids from a number of different solvents. Also we aimed to obtain much more precise structural information regarding the interactions of the end rings and different packing arrangements of the molecules than the β-CR at 3.2 Å resolution could provide. Thus we have successfully crystallised and determined the crystal structures of three crystal forms of ASX, a chloroform solvate, ASX-Cl a pyridine solvate, ASX-Py and an unsolvated crystal form ASX-un. In addition, we crystallised the closely related carotenoids, canthaxanthin, which differs from ASX in that the hydroxyl groups at the 3 and 3'- positions are absent and (3R,3'S,meso)-zeaxanthin, which differs from ASX in that the keto oxygen atoms at the 4 and 4'-positions are absent. The unsolvated crystal form of ASX which we have obtained (Bartalucci et al 2007) has been described previously by Hashimoto et al 2002, but the crystal data, which was not as precise as for our new structure, were not deposited in the Cambridge Crystallographic Data Centre (CCDC). The crystal structure of canthaxanthin has also been reported in Bart and MacGillavry, 1968, but again our new crystal structure (Bartalucci et al 2007) is more precise due to the use of modern crystallographic methods, as well as using sample cryocooling.

All these crystal structures reveal that the free carotenoid molecules adopt the 6-s-cis form, which is consistent with calculations and NMR spectroscopy

(Hashimoto et al 2002); the main difference between the conformations of these closely related carotenoids was found to be the angle by which the end ring is twisted out of the plane of the polyene chain; for the three ASX crystal structures and canthaxanthin, the conformations are very alike, with the C5-C6-C7-C8 torsion angles varying from $-43°$ to $-51°$. These results agree with calculations performed by Hashimoto et al 2002, which gave angles of $-47.7°$ and an average of $-46.5°$ for canthaxanthin and ASX respectively. However, this torsion angle for zeaxanthin derived from the crystal structure is $-74.8(3)°$, which is quite different to the calculated value of $-48.7°$. The effect of the torsion of $-74.8(3)°$ for zeaxanthin, which is much closer to $-90°$ than for the other carotenoids, is expected to be a reduction of the conjugation from the polyene chain, into the end rings, which should in turn affect the colour of the crystals (see below).

Comparison of the ASX conformations with that of ASX in β-CR shows firstly that ASX in the protein bound molecule is in the 6-s-trans conformation rather than the 6-s-cis conformation of the free molecules, which is clearly illustrated in Figure 8-9. Moreover, the end rings lie approximately in the plane of the polyene chain, and are not rotated out of the plane as seen in the free ASX crystal structures.

We can also compare the packing arrangements of the free carotenoids with the protein bound ASX in β-CR. The chloroform and pyridine solvates of ASX both form chains of molecules by pairwise end-to-end hydrogen bonding of the hydroxyl and keto oxygen atoms with hydrogen bonding distances of 2.790(3) and 2.829(6) respectively (Figure 8-10a,b). In addition there is a particularly strong interaction

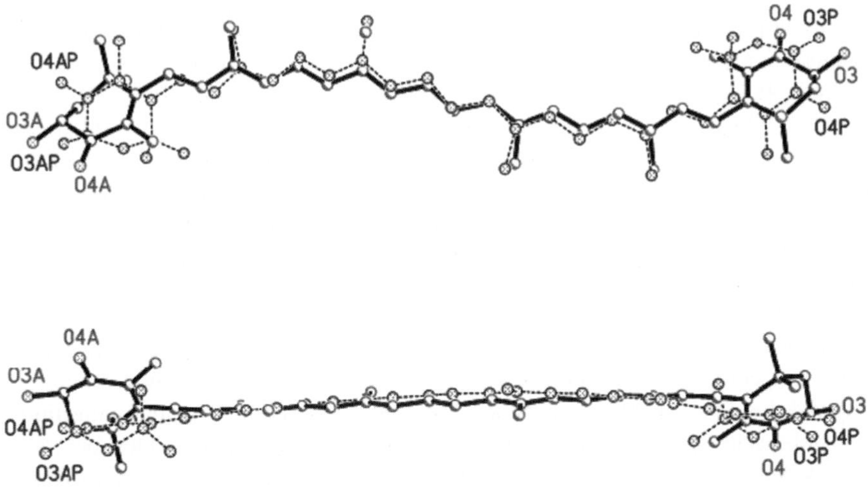

Figure 8-9. Conformation of unbound ASX-Cl (red) best overlay against protein bound ASX (blue); top is the view perpendicular to the plane of the polyene chain and bottom is the view edge on to the polyene chain. From Bartalucci et al (2007) with permission of IUCr Journals (see color plate section)

of the chloroform hydrogen atom with a hydroxyl oxygen atom as well as much weaker C—H hydrogen bond interactions of the methyl and pyridine H atoms with the hydroxyl or keto oxygen atoms of the ASX molecules Figures 8-10a,b). For the unsolvated ASX structure, the hydroxyl and keto oxygens form an intramolecular hydrogen bond of 2.656(4) Å and the molecules are linked into chains by much weaker C—H hydrogen bonds (Figures 8-10c) similar to those found in the structure of canthaxanthin. Finally for zeaxanthin, the most important interaction is hydrogen bonding of the hydroxyl oxygens with a distance of 2.655(4) Å which link the molecules into chains. In addition, for each carotenoid structure there are further π-π stacking interactions bringing the molecules into close proximity of between 3.61 and 3.79 Å apart, with the molecules one above the other (Figure 8-10d). All these arrangements are quite different to that found in the β-CR protein crystal structure, where the ASX molecules are organized in pairs with the keto oxygen at one end of each molecule H bonded to a protein bound water molecule; the hydrogen bonding distances to these water oxygens of 2.60 and 2.73 Å suggest that these are strong hydrogen bonds, but these distances are not precisely determined due to the low resolution limit of the data of 3.2 Å; the other end of each ASX molecule is hydrogen bonded to a protein histidine nitrogen atom. The ASX molecules are at an angle of about 120° to one another and their closest approach at the centre of the molecules is about 7 Å, much further apart than in the crystals of the free carotenoid molecules.

It is known that the nature of the solvent can alter the colour of a carotenoid solution (Britton, 1995), and we want to investigate the colours of the carotenoid samples not only in solution but also in the crystalline state to see whether the conformations and packing arrangements of the molecules in the crystal cause a change of colour versus those found in the solution state. Therefore the UV/Vis spectra have been measured of solutions of the carotenoids in chloroform. ASX and canthaxanthin have similar spectra in the visible region with just a single peak at 490 and 482 nm respectively. The solution state spectra of zeaxanthin and also β-carotene, which has the same back-bone as zeaxanthin, but with the hydroxyl groups removed, have almost identical solution state spectra to one another with a central peak and a shoulder on either side; the peak maxima of the central peak are at 461 and 462 nm respectively. Although the conformations of these molecules in solution are unknown, the fact that the solution state spectra of ASX and canthaxanthin are similar to one another suggests that the conjugation along the polyene back-bone and into the end rings, and therefore the conformation, is similar for each molecule, and likewise, for zeaxanthin and β-carotene; the shift to shorter wavelength for the latter two compounds reflects the fact that the length of the conjugated chain is reduced due to the absence of the keto oxygen atoms at the 4- and 4'- positions. The solid state spectra are much broader than the solution state spectra, but work is underway to precisely measure UV/Vis spectra of the carotenoid crystals. However, Figure 8-4 compares the crystals of the unbound carotenoids with the β-CR crystal, showing the big difference in colour between the free carotenoids and the protein bound ASX in β-CR. Qualitatively, the colours of the three crystal forms

(a)

(b)

(c)

(d)

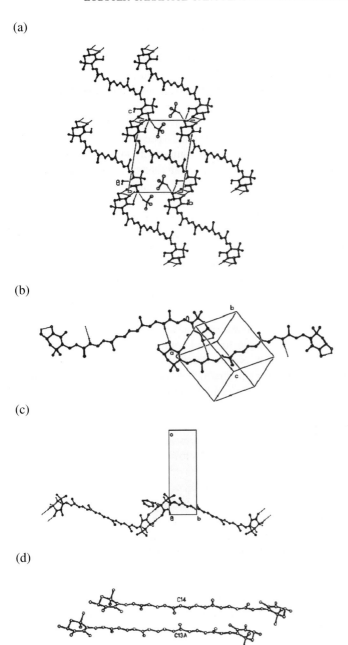

Figure 8-10. Packing arrangements of the carotenoid crystal structures; H atoms not involved in H bonding have been omitted for clarity. From Bartalucci et al (2007) with permission of IUCr Journals. (a) Packing of ASX-Cl viewed down the a-axis. (b) Packing arrangement of the ASX-py viewed down the a-axis. (c) Packing arrangement of ASX-un. (d) Plot showing the π-π stacking interaction of ASX-Cl

of unbound ASX crystals and those of canthaxanthin and even β,β-carotene are all very similar.

However, the crystal of (3R,3'S,meso)- zeaxanthin appears distinctly more yellow than those of the other unbound carotenoids suggesting that it is absorbing at shorter wavelength. This would be in keeping with the much reduced conjugation of the polyene chain into the end ring arising from the $-74.8(3)°$ C5-C6-C7-C8 torsion angle versus the angles of between -43 and $-51°$ for the other carotenoids leading to a greater degree of conjugation in these crystals. The torsion angle for these carotenoids in solution is unknown, but the calculations of Hashimoto et al 2002, suggest that all these molecules should have similar torsion angles between -40 and $-50°$, including (3R,3'S,meso)- zeaxanthin. The similarity of the solution state spectra of (3R,3'S,meso)- zeaxanthin and β-carotene suggest that the conformations of these two molecules in solution are probably very similar, but that the packing forces in the crystal have led to a quite different torsion angle for (3R,3'S,meso)-zeaxanthin in the solid state.

Overall, the effect of the additional intermolecular hydrogen bonding and π-π stacking interactions in the crystals of the free carotenoids seem to have very little effect on the colours of the crystals. This is in contrast to the coordination of the two ASX molecules in β-CR (see Figure 8-5), with hydrogen bonding of the keto oxygen atoms to bound water and histidine residues and their pairing with minimum distance of 7 Å and 120° tilt-angle to one another leading to the very large bathochromic shift that is observed in Nature. Further experiments will be carried out to vary these molecular tuning parameters in free carotenoid model compounds to see if the bathochromic shift in Nature can be replicated.

AN INSPIRING REMARK

From J D Bernal's book "The Origin of Life" (1967) Weidenfield and Nicholson Natural History page xv:-

- *Life is a partial, continuous, progressive, multiform, and conditionally interactive self-realization of the potentialities of atomic electron states . . . .*
- *This provisional definition of life . . . should be sufficient to show that all life has some material structures in common, in the form of enzymes, nucleic acids and essentially reproducing organs or molecules. . . . biology is the science of these substances and of their necessary interactions.*

The bathochromic shift in biology, as exemplified by the lobster crustacyanin case, is an amazing example of interactions of various molecules and of the relevance of atomic electron states as manipulated by those interactions.

POTENTIAL APPLICATIONS OF NEW COLOURATION STRUCTURAL CHEMISTRY

Sensor materials responding to chemical changes with colour changes obviously could prove useful in new devices. Also new food additives as colourants are of continual interest to the food industry. Furthermore the strong antioxidant properties of ASX have significant medical therapeutic value (see www.astaxanthin.org). Taking these potential and actual applications together with the curiosity driven research described above provides us with a compelling interest in the "magic" and "mystery" of these molecules and their 3D crystal structures.

ACKNOWLEDGMENTS

We thank our numerous co-authors in this programme of research for very fruitful collaborations notably Dr Peter Zagalsky, Royal Holloway and Bedford New College, London, the father of this work for us and also especially Prof Naomi Chayen, Imperial College, London and Dr Michele Cianci (now at EMBL Hamburg "PETRA – 3 Project") and Dr Pierre Rizkallah, CCLRC Daresbury Laboratory and also, for the unbound carotenoids research work, notably Prof Synnove Liaaen-Jensen (University of Trondheim, Norway). We thank SRS Daresbury Laboratory and the Joint Biology Programme of the UK Research Councils for synchrotron radiation beamtime used for the apocrustacyanin A1 and β-CR data collections on SRS Stations 7.2, 9.6, 9.5 and 14 respectively. JRH is grateful for the award of two research grants from The LeverhulmeTrust (with Prof Chayen and Dr Zagalsky). JRH thanks EPSRC and the University of Manchester for a PhD studentship for Dr M Cianci). We thank the EC for a Marie Curie Training Centre award which supported Giuditta Bartalucci and the Nuffield Foundation for a summer vacation studentship award, which supported Mr S Fisher). The Wellcome Trust and the BBSRC funded our University of Manchester Structural Chemistry Laboratory computing and graphics suite. We are very grateful to Dr George Britton (University of Liverpool), Prof John Sutherland and Prof Jonathan Clayden (The University of Manchester) for helpful discussions.

REFERENCES

J.C.J. Bart and C.H. MacGillavry Acta Cryst B24, (1968) 1587–1606.

Bernal, J D (1967) Weidenfield and Nicholson Publishers, London.

G. Bartalucci, J. Coppin, S. Fisher, G. Hall, J.R. Helliwell, M. Helliwelland S. Liaaen-Jensen Acta Cryst B (2006) 328–337.

T.J. Boggon "Biophysical and Structural Chemistry Studies of Lysozyme and Astaxanthin Binding Proteins" PhD thesis University of Manchester (1998).

W.L. Bragg X-ray crystallography Scientific American 219(1), (1968) 58–70.

G. Britton in "Carotenoids", Volume 1B: Spectroscopy, Edited by G. Britton, S. Liaaen-Jensen and H. Pfander, Birkhauser Verlag-Chapter2:UV/Visible Spectroscopy, E: Effects of Molecular Environment, 1. Effect of different solvents, (1995) 43–44.

G. Britton, R.J. Weesie, D. Askin, J.D. Warburton, L. Gallardo-Guerrero, F.J.H.M. Jansen H.J.M de Groot, J. Lugtenburg and J.P. Cornard J.C. Merlin Pure Appl. Chem. 69, (1997) 2075–2085.

M. Buchwald and W.P. Jencks Biochemistry 7, (1968) 844–859.

N.E. Chayen Acta Cryst D54, (1998) 8–15.

N.E. Chayen, M. Cianci, J.G. Grossmann, J. Habash, J.R. Helliwell, G.A. Nneji, J. Raftery, P.J. Rizkallah and P.F. Zagalsky Acta Cryst D59, (2003) 2072–2082.

M. Cianci, P.J. Rizkallah, A. Olczak, J. Raftery, N.E. Chayen, P.F. Zagalsky and J.R. Helliwell Acta Cryst. D57, (2001) 1219–1229.

M. Cianci, P.J. Rizkallah, A. Olczak, J. Raftery, N.E. Chayen, P.F. Zagalsky and J.R. Helliwell PNAS USA 99, (2002) 9795–9800.

C. Day Physics Today (November 2002) 16–17.

B. Durbeej and L.A. Eriksson Chem Phys Letters 375, (2003) 30–38.

B. Durbeej and L.A. Eriksson, *PCCP*, 6, (2004) 4190.

V.B. Gerritsen SwissProt Protien Spotlight Issue (2002) 2 (ISSN 1424–4721)

H. Hashimoto, T. Yoda, T. Kobayashi and A.J. Young J. Molec. Struct. 604, (2002) 125–146.

J.R. Helliwell J Synchrotron Rad 11, (2004) 1–3.

R.P. Ilagan, R.L. Christensen, T.W. Chapp, G.N. Gibson, T. Pascher, T. Polivka, and H.A. Frank J. Phys. Chem A, 109, (2005) 3120–3127.

R. Kuhn and N.A. Sorensen Z. Angew. Chem. 51, (1938) 465–466.

R. Kuhn and H. Kuhn Eur. J. Biochem. 2, (1967) 349–360.

J.B. Messenger J Exp Biol 70, (1977) 49–55.

M.I. Newbigin J.Physiol. 21, (1897) 237–257.

M.I. Newbigin Colour in Nature: a study in biology. (1898) John Murray: London.

A.C.T. North Biol. Sci. Rev. 5, (1993) 31–35.

Peter J. Herring Journal of Crustacean Biology, 5, (1985), 557–573.

A.A.C. van Wijk, A. Spaans, N. Uzunbajakava, C. Otto, H.J.M. de Groot, J. Lugtenburg and F. Buda JACS 127, (2005) 1438–1445.

J. Verne Bull.Soc.Zool.Fr. 46, (1921) 61–65.

N. Wade "Crustacean Shell Colour Formation and the White Phase of the Western Rock Lobster, Panulirus Cygnus". PhD thesis University of Queensland (2005).

N. Wade, K.C. Goulter, K.J. Wilson, M.R. Hall and B.M. Degnan Comparative Biochemistry and Physiology, Part B 141 (2005) 307–313.

G. Wald, N. Nathanson, W.P. Jencks and E. Tarr Biol. Bull. Mar. biol., Woods Hole, 95, (1948) 249–250.

R.J. Weesie, J.C. Merlin, J. Lugtenburg, G. Britton, F.J.H.M. Jansen and J.P. Cornard Biospectroscopy, 5, (1999) 19–33.

P.F. Zagalsky Methods Enzymol. 111, (1985) 216—247.

P.F. Zagalsky Acta Cryst D59, (2003) 1529–1531.

CHAPTER 9

TINY STRUCTURAL FEATURES AND THEIR GIANT CONSEQUENCES FOR PROPERTIES OF SOLIDS

ANDRZEJ KATRUSIAK

Abstract: The origin on an atomic scale of huge macroscopic effects of most materials can be subtle. No matter whether a property involves the formation of entire atomic and ionic assembles, supramolecular clusters or inter-phases, their understanding is invariably refined to certain minimal regions. Materials and properties applied and produced by mankind have an immense variety, and a formulation of general rules describing substances of specific types is interesting and timely. Apart from well known thermodynamic, statistical and physical laws, chemical "signposts" toward substances in new groups exhibiting required properties are also needed. The structure-property relations described here focus on hydrogen-bonded structures and their dielectric properties; they pertain to the smallest of atoms and the biological functions of macromolecules and dielectric properties of matter of current interest for technological applications. We show that the spontaneous polarisation in hydrogen-bonded crystals is related to tiny atomic displacements rather than to the largest molecular dipoles present in the structure. New dimensions of the structure-property relations are described for NH^{+}- - -N bonded ferroelectrics and relaxors. The structure-property relations described for hydrogen-bonded crystals are applicable also to substances without hydrogen bonds

INTRODUCTION

Until recently – in terms of human history, the scope of properties determining the conscious utility of materials was limited. A few centuries ago it mattered only that a material was flexible, hard, inflammable, mechanically and chemically resistant etc. (Tsoucaris and Lipkowski, 2003). With cultural and technological development, additional properties gradually become increasingly important, and to list all material properties now in demand is difficult. For example, the inception of electricity was followed by a demand for electric conductors, and specific dielectric properties are required for the developing electronic devices and their miniaturisation. An understanding of the living functions of biological tissue is fundamental for progress in agricultural and medical sciences, and for the preservation of our natural environment.

J.C.A. Boeyens and J.F. Ogilvie (eds.), Models, Mysteries and Magic of Molecules, pp. 209–218.
© 2008 *Springer*.

Symmetry belongs to the most general concepts governing non-scalar properties, such as thermal expansion, compressibility or spontaneous polarisation. Symmetry thus enables one to predict directly the type of tensor relations between the crystal directions and faces, the type of a phase transition, or whether a crystal can exhibit ferroelectricity. The symmetry itself fails, however, to predict the magnitude of thermal expansion or of spontaneous polarisation, and two substances with the same symmetry might exhibit, for example, one positive and the other negative thermal expansion under the same thermodynamic conditions. To evaluate the magnitudes of given properties, one must refer to most subtle features of structures at the atomic level, but one can show that symmetry relations at the atomic level can affect the magnitude of thermal expansion of crystals near phase transitions (Katrusiak, 1993).

Below, the interdependence of macroscopic properties on these microscopic features is described briefly for OH- - -O hydrogen-bonded crystals. The bistable homonuclear hydrogen bonds are convenient examples to explain the structure-property relations, because the H atom is readily locatable in the structure; in this way the mechanism of microscopic transformations can be followed. At the same time, H-bonded systems are important for understanding transformations of molecular and biological systems, or for prospective practical applications. At present there appear to exist no practical applications of electronic devices based on hydrogen-bonded materials.

Although properties of many materials depend on hydrogen bonds of various types, it is convenient to concentrate on strong hydrogen bonds, which are clearly distinguishable from other cohesive forces. It is also characteristic of strong hydrogen bonds that their polarization is typically reversed easily when the H transfers to the opposite atom: OH- - -O transforming into O- - -HO. In these strong hydrogen bonds the barrier of potential energy separating the two minima is small, and the crystal-lattice vibrations can activate the H-atom transfers. These H bonds are not the only interactions in the crystal, and other cohesive forces and molecular and ionic conformations can play important roles for specific properties of crystals. The contribution of strong hydrogen bonds to the properties of crystals should evidently be more suitable for observation than the contribution of weak hydrogen bonds.

The contribution of hydrogen bonds to the properties of crystals is not limited to H-bond dimensions, such as the H-donor · · · acceptor distance or the distances and angles involving the H atom only, but also requires that the arrangement of hydrogen-bonded molecules be taken into account. We show below that the interactions of a hydrogen bond with its surroundings and their role for the arrangement of molecules or ions are essential for their transformations and for possible triggering of the H-atom dynamics.

COUPLING OF THE H-ATOM SITE AND MOLECULAR ORIENTATION

The molecular or ionic orientation within a hydrogen-bonded aggregate formed by a bistable hydrogen bond can favour one H site; this interdependence is illustrated

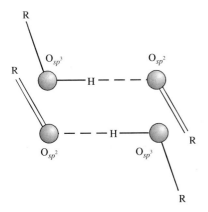

Figure 9-1. The dependence of the orientation of —O- - -HO═ bonded molecules (moieties) on the H-atom site in the hydrogen bond. Two hydrogen bonds with the H atom located at their opposite sites are shown schematically, one below the other, in this drawing to visualise the magnitude of molecular reorientation due to the H transfer. The hydrogen bonds are linear – the O-H- - -O angle is 180° – and the R-O-H and H- - -O-R angles assume ideal openings for the sp^3 and sp^2 hybridisations of the oxygen atoms, of 109° and 120°, respectively; R denotes the bulk of the molecule or ion, which is covalently bonded to the oxygen atoms

in Figure 9-1. One might argue that this coupling also operates in the reverse direction – the H site modifies the arrangement of H-bonded molecules or ions. The interdependence between the molecular orientation and the H site can be applied to determine which H site is occupied, when the experimental measurements are not sufficiently precise to specify the H location (Katrusiak, 1998). It is symptomatic that commonly H atoms are scarcely visible in x-ray diffraction measurements. H atoms are also lighter than other elements constituting the compounds. This small element of the crystal's structure introduces a polarisation of hydrogen bonds and molecules, and is coupled with molecular orientations in aggregates. Apart from the electronic structure of the molecules or ions, the H-atom sites are thereby correlated in the crystal structure.

H-DISORDERING AND STRUCTURAL TRANSFORMATIONS

The dynamical disordering of H atoms in hydrogen bonds affects the orientation of the hydrogen-bonded molecules. The two sites of the H atom generally become related by symmetry, and the angles on both sides of the hydrogen bonds are consequently identical. When the H atom jumps from one side of a hydrogen bond to the other, the electronic structure of the oxygen atoms immediately adjusts to the H-atom positions. Depending on the hybridisation of the oxygen atoms, the O—H direction alters with respect to the R—O bonds on both sides of the hydrogen bond, and for both sites of the H atoms, as shown in Figure 9-2. The O—H- - -O must

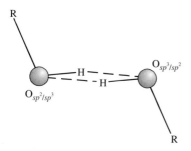

Figure 9-2. A schematic drawing of a time-averaged structure of a dynamically disordered hydrogen bond, with two half-occupied sites of the H atom indicated. Each H site satisfies the boundary conditions required by the coexistence of disparate R-O-H and H---O=R angles, 109° and 120° for the sp^3 and sp^2 hybridisations respectively, of oxygen atoms repeatedly being switched following the H hopping. The two hydrogen-bonded groups are related by symmetry, and the O-H---O angle(s) must deviate from the ideal 180° to satisfy the angular requirements for each H site; cf. Figure 9-1

deviate from 180° to satisfy the angular requirements of each site of the disordered H atom.

Due to the large mass of the oxygen and other atoms in the molecule relative to that of the H atom, these heavy atoms can not adjust their positions following the H-atom hopping at a large frequency, ca. 10^{10} Hz, and assume average positions.

The alteration of the O—H---O angle is indeed observed in structures in which the H atom becomes disordered (Katrusiak, 1992; 1993). The O—H and H---O distances do not alter significantly when the H atom becomes disordered, and therefore – because of the altered O—H---O angle – the O$\cdots$O distance decreases. This contraction of the O$\cdots$O distance contributes to the thermal expansion of the crystal in such a way that a region of anomalous thermal expansion is observed in the direction of hydrogen bonds about the critical temperature (T_c) when the H atoms become disordered (Kobayashi, Uesu, Mizutani & Enomoto, 1970; Katrusiak, 1993; Horvath, 1983; Horvath & Kucharczyk, 1981).

The process of H-atom disordering in hydrogen bonds is generally associated with phase transitions of hydrogen-bonded crystals. A crystal at a critical temperature, T_c, thus exhibits anomalous behaviour and alters its properties. Substances in a group exhibiting such behaviour are OH---O bonded ferroelectrics, which, according to the prototypical crystal of potassium dihydrogen phosphate, KH_2PO_4, are described as ferroelectrics of KDP type (Slater, 1941). The KDP crystal undergoes a paraelectric-ferroelectric phase transition at $T_c = 122$ K. Above T_c the crystal is paraelectric in space group I-$4d2$; below T_c the symmetry of these crystals decreases to $Fdd2$ – the crystals become ferroelectric and exhibit spontaneous polarisation, whereas above T_c the spontaneous polarisation disappears and the crystal enters the paraelectric phase. If the temperature is decreased below T_c, the crystal re-enters the ferroelectric phase, but the sense of the spontaneous polarisation might alter. In many ferroelectrics the direction of spontaneous polarisation does not alter: only its sense changes. There are also ferroelectric crystals in which both direction and sense of spontaneous polarisation alter – multidimensional ferroelectrics. In any case, ferroelectric crystals are intensively investigated because of their possible electronic applications, e.g. as binary memory devices or sensors.

H-BONDING AND THE DIRECTION OF SPONTANEOUS POLARISATION IN KDP

Although the off-centre H-atom site in an OH- - -O hydrogen bond clearly introduces a polarisation along the $O \cdots O$ direction, from the H-acceptor to the H-donor, it is generally a local feature compensated by the oppositely directed hydrogen bonds in the crystal structure. This feature is illustrated in the KDP structure (Nelmes, 1987); in this prototypical KH_2PO_4 crystal the OH- - -O hydrogen bonds are nearly parallel to the (001) plane, as shown in Figure 9-3a. Each PO_4 group is involved in four hydrogen bonds, one pair of them parallel and perpendicular to the other pair of parallel H-bonds. For each pair of the collinear hydrogen bonds the polarisations introduced by the H sites are opposite, and compensate along this direction (Figure 9-3a). There is thus no resultant component of spontaneous polarisation along the hydrogen bonds in the crystal (001) plane.

The spontaneous polarisation of the KDP crystal occurs along the [z] axis, approximately perpendicular to the hydrogen bonds. It is characteristic for most ferroelectrics of KDP type that the direction of spontaneous polarisation is perpendicular to the direction of hydrogen bonds. It is also observed for molecular crystals built of polar molecules that the molecules typically arrange in the crystal in such a way that their dipole moments compensate. The molecules or ions arrange in the crystal structure in a way that minimizes the energy of their interactions. Such molecules typically arrange head to tail into chains, and the chains in turn arrange antiparallel. The compounds of strongly polar molecules preferably form consequently centrosymmetric (e.g. $P2_1/c$, $P-1$, $Pbca$) or non-polar (e.g. $P2_12_12_1$) crystals. These symmetries compensate the molecular dipoles of molecules in three dimensions. When the substance 'chooses' to crystallize in a polar group, such as $P2_1$, $C2$ or $Fdd2$, it is typically the smallest component of the molecular dipole that compromises with the requirements of molecular packing, in particular electrostatic forces favouring antiparallel arrangement of the dipoles.

COUPLING BETWEEN H SITES AND LATTICE VIBRATIONS

The H-atom disordering in OH- - -O hydrogen bonds is a common feature of many crystals. Whereas a dynamical disorder of the H atom is commonly observed, the origin of this disorder and its relation to the crystal structure and lattice vibrations are poorly understood. As is well known, the H hopping becomes coupled with lattice vibrations in particular modes. When the temperature is decreased to T_c, these modes soften and their frequencies become zero at and below T_c. In structural terms, the vibrations propagating in the crystal lattice involve motion of types that

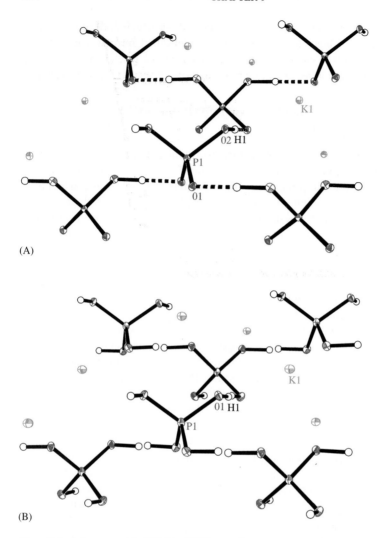

Figure 9-3. A fragment of the KH_2PO_4 (KDP) crystal structure viewed along the (001) plane (horizontal in these drawings), and perpendicular to the [z] axis (vertical): (A) in the ferroelectric phase at 102 K (i.e. 20 K below the phase transition at T_c=122 K); and (B) in the paraelectric phase at 127 K (i.e. 5 K above T_c). The non-H atoms are shown with thermal ellipsoids at the 50% probability level, and the sites of H atoms are represented as small circles of arbitrary size, half occupied in the paraelectric phase (B). The H bonds are indicated as dashed lines in drawing A

facilitate the H transfers. There might thus be vibrations of longitudinal or transverse types that squeeze and stretch the $O\cdots O$ distance in hydrogen bonds, and in this way decrease and increase the potential barrier separating the two H-atom sites. Decreasing the barrier itself does not, however, push the H atom toward the other site; this impulse can be produced by rotations of the H-bonded groups, as illustrated in Figure 9-4. Because of the electronic structure of the oxygen atoms, the R—O—H angle is a few degrees smaller than the H---O—R angle, but small bulky groups are prone to orientational vibrations that can easily reverse this angular dependence.

The "snapshot" in Figure 9-4A shows an instant at which the PO_4 groups are rotated by 3° in a direction favouring the left H-atom site in the hydrogen bond, and an instant later the PO_4 groups shown in Figure 9-4B are rotated in the opposite direction about the central phosphorus atom by 3°. At this second instant the right H site is energetically more favoured for the H atom than the left one. The rotations of small ions or molecules belong to vibrations of the types most easily activated at

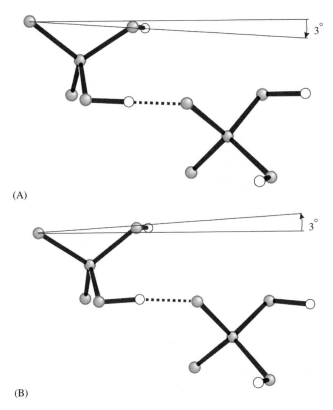

(A)

(B)

Figure 9-4. A fragment of the KH_2PO_4 (KDP) crystal structure drawing presented in Figure 9-3B modified in such a way that they show two snapshots of two H_2PO_4 groups tilted in the structure in the angular conditions to be (A) favourable for the left occupied position of the H-atom in the hydrogen bond; and (B) unfavourable for this H-site when the H_2PO_4 groups are tilted in the other direction. In Figure 9-4A the upper-left H_2PO_4 group is rotated clockwise by 3° and the bottom-right H_2PO_4 group is rotated anti-clockwise by 3°; the H_2PO_4 groups in Figure 9-4B are rotated in the opposite directions

a given temperature, and their amplitudes are larger than those of lattice vibrations of other modes. Subtle features of the molecular conformation can thus be related to the macroscopic properties of crystals. In Figure 9-3B the atomic displacements of the H atom perpendicular to the hydrogen bond (i.e. in the direction of spontaneous polarisation in the ferroelectric phase) is observed to be small. The ionic rotations and alterations of the hydrogen-bond geometry, which are coupled to the H transfers, are also subtle, and diffraction data at the highest resolution are required to observe them, but these subtle structural features are essential for an understanding of the structural origin of macroscopic properties of ferroelectric crystals.

OTHER DIMENSIONS OF STRUCTURE-PROPERTY RELATIONS

The structure-property relations present in the OH- - -O hydrogen bonds in ferroelectrics of KDP type are applicable also for hydrogen bonds of other types and for transformations of other types. In particular, it might be unimportant in which way the H atom is moved to the other site in the hydrogen bond. For example, in OH- - -O hydrogen bonds between hydroxyl groups of alcohol molecules the OH groups rotate; in this way the polarisation of the individual hydrogen bonds varies (Katrusiak, 1998). A similar process can be considered for hydrogen bonds in H_2O ices (Tajima, Matsuo & Suga, 1984; Katrusiak, 1996c). We found that yet another type of transformation of NH- - -N bonds can be responsible for huge macroscopic properties. The NH- - -N bonded molecules and ions are observed to form ferroelectric crystals with a large spontaneous polarisation (Szafrański, Katrusiak & McIntyre, 2002; Szafrański & Katrusiak, 2000; Katrusiak & Szafrański, 1999). The properties of NH- - -N hydrogen bonds differ considerably from those of OH- - -O bonds, but the H site in the bistable NH- - -N hydrogen bonds is coupled to the orientation of the hydrogen-bonded molecules or ions and to the positions of the counter ions (Katrusiak, 1999).

The structure-property relations discussed here have been observed in NH- - -N bonded crystals, and are connected with the transformations of these hydrogen bonds (Szafrański & Katrusiak, 2007; Katrusiak & Szafrański, 2006). These transformations have significant consequences for the dielectric properties of the crystals.

DABCO or pyrazine monosalts have been found to exhibit a strong dielectric response because H transfers alter the polarisation of the NH- - -N hydrogen bonds and monocations (Katrusiak & Szafrański, 2006; Szafrański & Katrusiak, 2004; 2007). This property occurs for centrosymmetric crystals, and can be rationalized by the formation of polar nano-regions. Because of the H transfers – i.e. NH^+- - -N transforming to N- - -H^+N, fragments of hydrogen-bonded linear chains change their polarisation, and small polar nanoregions are formed, as illustrated in Figure 9-5. Only a few H^+ transfers are required for a nanoregion to form, but they considerably alter the macroscopic properties and the dielectric response of these substances, in a process analogous to ferroelectric relaxors (Szafrański & Katrusiak, 2007).

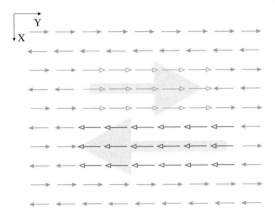

Figure 9-5. A fragment of a centrosymmetric crystal structure with the molecules (represented as arrows parallel to their molecular dipoles) NH$^+$---N hydrogen-bonded into antiparallel chains along [y] (the anions are neglected for clarity). The ideal crystal structure with antiparallel molecules in neighbouring chains is marked in green (full arrowheads). Due to defects in the fourth and seventh chains, in which 5 and 6 molecules have reversed orientation, respectively, two polar nanoregions are formed. The red and blue colours and open arrowheads mark these nanoregions, the polarisation of which is indicated with large grey arrows

CONCLUSIONS

Properties of materials originate from their microscopic structure, but the structural features resulting in the macroscopic effects are typically small. These structure-property relations have been exemplified for the spontaneous polarisation in OH---O bonded crystals of KDP type and NH---N hydrogen-bonded complexes. The described transformations of hydrogen bonds are applicable to explain spontaneous polarisation of ferroelectrics of KDP type, anomalous thermal expansion at T_c in ferroelectrics and in H_2O ice (Katrusiak, 1996c), the existence of the deuteration effect on T_c (Ichikawa, 1978; Katrusiak, 1995), the occurrence of the tricritical point (Bastie, Vallade, Vettur, Zeyen & Meister, 1981; Landau, 1937; Landau & Lifschitz, 1976; Schmidt, Western & Baker, 1976; Katrusiak, 1996b), the dependence of T_c on atomic displacements in crystal structure and hydrogen-bond dimensions (Katrusiak, 1992; 1993; 1995; 1998) and to understand the mechanism of phase transitions and polymorphism of hydrogen-bonded crystals (Katrusiak, 1990; 1991; 1992). The hydrogen bonds are directional and contribute in a simple manner to the structural transformations of crystals; they are therefore applicable for an understanding of the structural background of properties of materials of other types.

REFERENCES

1. Bastie, P., Vallade, M., Vettur, C., Zeyen, C.M.E. & Meister, H. (1981). Neutron diffractometry investigation of the tricritical point of KH_2PO_4. *J. Phys. Paris* **42**, 445–458.
2. Horvath, J. (1983). Lattice parameter measurements of $PbHPO_4$ single crystals by the ration method. *J. Appl. Cryst.* **16**, 623–628.

3. Horvath, J. & Kucharczyk, D. (1981). Temperature dependence of lattice parameters of PbHPO$_4$ and PbDPO$_4$ single crystals. *Phys. stat. solidi A* **63**, 687–692.

4. Ichikawa, M. (1978). The O-H *vs.* O $\cdots$ O distance correlation, the geometric isotope effect in OHO bonds, and its application to symmetric bonds. *Acta Cryst. B* **34**, 2074–2080.

5. Katrusiak, A. (1990). High-pressure X-ray diffraction study on the structure and phase transition of 1,3-cyclohexanedione crystals. *Acta Cryst. B* **46**, 246–256.

6. Katrusiak, A. (1991). Structure and phase transition of 1,3-cyclohexanedione crystals as a function of temperature. *Acta Cryst. B* **47**, 398–404.

7. Katrusiak, A. (1992). Stereochemistry and transformation of—OH- - -O$=$ hydrogen bonds. Part I. Polymorphism and phase transition of 1,3-cyclohexanedione crystals. *J. Mol. Struct.* **269**, 329–354.

8. Katrusiak, A. (1993). Geometric effects of H-atom disordering in hydrogen-bonded ferroelectrics. *Phys. Rev. B* **48**, 2992–3002.

9. Katrusiak, A. (1995). Coupling of displacive and order-disorder transformations in hydrogen-bonded ferroelectrics. *Phys. Rev. B* **51**, 589–592.

10. Katrusiak, A. (1996a) Stereochemistry and transformation of - -OH- - -O$=$ hydrogen bonds. Part II. Evaluation of T$_c$ in hydrogen-bonded ferroelectrics from structural data. *J. Mol. Struct.* **374**, 177–189.

11. Katrusiak, A. (1996b). Structural Origin of Tricritical Point in KDP-Type Ferroelectrics. *Ferroelectrics* **188**, 5–10.

12. Katrusiak, A. (1996c). Rigid H$_2$O molecule model of anomalous thermal expansion of ices. *Phys. Rev. Lett.* **77**, 4366-4369.

13. Katrusiak, A. (1998). Modelling hydrogen-bonded crystal structures beyond resolution of diffraction methods. *Pol. J. Chem.* **72**, 449-459.

15. Katrusiak, A. (1999) Stereochemistry and transformations of NH- - -N hydrogen bonds. Part I. Structural preferences for the H-site. *J. Mol. Struct.* **474**, 125–133.

15. Katrusiak, A. & Szafrański, M. (1999) Ferroelectricity in NH- - -N Hydrogen-Bonded Crystals. *Phys. Rev. Lett.* **82**, 576-579.

16. Katrusiak, A. & Szafrański, M. (2006). Disproportionation of pyrazine in NH$^+$ $\cdots$ N hydrogen-bonded complexes: new materials of exceptional dielectric response. *J. Am. Chem. Soc.* **128**, 15775–15785.

17. Kobayashi, J., Uesu, Y., Mizutani, I. & Enomoto, Y. (1970). X-Ray study on the thermal expansion of ferroelectric KH$_2$PO$_4$. *Phys. stat. solidi (a)* **3**, 63–69.

18. Landau, L.D. (1937). On the theory of phase transitions I. *Zh. Eksp. Teoret. Fiz.* **7**, 19–32 [in Russian]; *Sov. Phys. JETP* **26**.

19. Landau, L.D. & Lifschitz, E.M. (1976). *Statisticheskaia Fizika*, Izdatielstvo Nauka, Moscow. p. 536.

20. Nelmes, R.J. (1987). Structural studies of KDP and KDP-type transitions by neutron and X-ray diffraction: 1970–1985. *Ferroelectrics* **71**, 87–123.

21. Schmidt, V.H., Western, A.B. & Baker, A.G. (1976). Tricritical point in KH$_2$PO$_4$. *Phys. Rev. Lett.* **37**, 839–842.

22. Slater, J.C. (1941). Theory of the transition in KH$_2$PO$_4$. *J. Chem. Phys.* **9**, 16–33.

23. Szafrański, M., & Katrusiak, A. (2000). Thermodynamic behaviour of bistable NH$^+$- - -N hydrogen bonds in monosalts of 1,4-diazabicyclo[2.2.2]octane. *Chem. Phys. Lett.* **318**, 427–432.

24. Szafrański, M., Katrusiak, A. & McIntyre, G.J. (2002). Ferroelectric order of parallel bistable hydrogen bonds. *Phys. Rev. Lett.* **89**, 215507–1–4.

25. Szafrański, M., & Katrusiak, A. (2004). Short-range ferroelectric order induced by proton transfer-mediated ionicity. *J. Phys. Chem.* **108**, 15709–15713.

26. Szafrański, M., & Katrusiak, A. (2007) – being submitted.

27. Tajima, Y., Matsuo, T. & Suga, H. (1984). Calorimetric study of phase transition in hexagonal ice doped with alkali hydroxides. *J. Phys. Chem. Solids* **45**, 1135–1144.

28. Tsoucaris, G. and Lipkowski, J. (2003). Molecular and Structural Archaeology: Cosmetic and Therapeutic Chemicals. Kluwer, Dordrecht Netherlands.

CHAPTER 10

POLYMORPHISM IN LONG-CHAIN *N*-ALKYLAMMONIUM HALIDES

GERT J. KRUGER, DAVE G. BILLING AND MELANIE RADEMEYER

Abstract: The structures and properties of long-chain organic molecules provide a fascinating glimpse into polymorphism and the packing forces that operate between molecules in the solid state. These molecules are roughly cylindrical in shape and normally stack in layers consisting of densely packed molecules, all in parallel, much like pencils in a box. Two factors influence their molecular packing in a crystal lattice – the forces between the end groups in a layer and the van der Waals interactions between the alkyl chains in a layer. Both forces contribute to the lattice energy, and their relative strengths determine the crystal structure. Most long-chain compounds exhibit polymorphism because of their conformational flexibility and the weak intermolecular forces directing the packing.

A study of the structural properties of primary *n*-alkylammonium halides enabled us to evaluate the relative importance of intermolecular forces of the two types as the polar ends of the layers form strong hydrogen-bonded networks that dominate the packing on crystallization. Here we illustrate trends in the crystal chemistry of the *n*-alkylammonium halides, and the investigative methods, using as examples *n*-octadecylammonium chloride, $C_{18}H_{37}NH_3^+Cl^-$, and their related bromides and iodides

INTRODUCTION

A typical long-chain molecule contains an extended hydrocarbon chain, also called a polymethylene chain, that makes up a substantial part of the molecule. The structural formula thus includes a section, $-(CH_2)_n-$, in which n is between one and about fifty. Functional groups might be present, typically attached to one or both ends of the molecule, but commonly attached to more than one alkyl-chain segment. The chains might also be saturated or unsaturated. Compounds of these types include the commercially important normal alkanes, fatty acids, fatty alcohols, long-chain esters, aldehydes, soaps, *n*-alkylammonium halides and the even more complicated triacylglycerols and phospholipids. These compounds are found in diverse products that include pharmaceuticals, detergents, surfactants, waxes, cosmetics, emulsifiers, stabilizers and bactericides. Extensive reviews regarding the properties and

J.C.A. Boeyens and J.F. Ogilvie (eds.), Models, Mysteries and Magic of Molecules, pp. 219–231.

industrial applications of these compounds are found in several reference works and books, including "Crystallization and Polymorphism of Fats and FattyAcids" (Garti and Sato, 1988), "The Physical Chemistry of Lipids" (Small, 1986), "The Chemistry and Technology of Waxes" (Warth, 1947) and "Waxes: Chemistry, Molecular biology and functions" (Hamilton, 1995), "Crystallography of the Polymethylene Chain" (Dorset, 2005).

Primary and tertiary n-alkylammonium halides are widely used as surfactants. The amphiphilic nature of the molecules produces micelles in water, resulting in superior detergent properties. Some have also been shown to exhibit bactericidal properties. Despite their industrial importance, little fundamental information on the polymorphism of the primary n-alkylammonium halides was available when we began our structural investigation. Our focus here is thus on the structural properties of primary n-alkylammonium halides – compounds with straight, saturated hydrocarbon chains that contain one terminal functional group.

SHORTHAND NOTATION USED IN THIS PAPER

Throughout this paper we represent all alkylammonium halides according to a notation CnX, in which n is the number of carbon atoms in the long chain and X is the halide anion. If water be present in the crystal lattice, the notation CnX.H_2O refers to the monohydrate forms. For example, C14Cl indicates $C_{14}H_{29}NH_3^+Cl^-$ and C16Br.H_2O indicates $C_{16}H_{33}NH_3^+Br^-.H_2O$.

POLYMORPHISM

Polymorphism is the ability of a substance to exist in multiple crystal forms. This condition might be due firstly to the variations in packing of fairly "rigid" molecules in the unit cell, secondly to varied conformations of molecules (known as conformational polymorphism), and thirdly to the incorporation of solvent molecules into the crystal lattice (forming solvates, also known as pseudo-polymorphism). Most long-chain compounds exhibit polymorphism. This tendency is attributed to the weak intermolecular forces directing the packing of the molecules and the conformational flexibility of the long chain. Parameters such as temperature, solvent system, impurities, and the rate of crystallization and saturation might influence which polymorphic form is obtained. Samples grown from a melt typically exhibit more complicated crystal structures than samples crystallized from solution because of the formation of metastable polymorphs.

MOLECULAR CONFORMATIONS AND FORCES DIRECTING THE MOLECULAR PACKING

In the most stable conformation of the alkyl chain all carbon atoms are coplanar; this plane is called the carbon zigzag plane. The ideal C—C bond distance is 1.54Å, and the distance between alternating carbon atoms is 2.54Å. The ideal C—C bond angle is 112°. The extended conformation is also called the *trans* conformation. This all-*trans* conformation of the chain might be distorted by forces between the functional groups, inducing slight deviations resulting in the curvature of the molecule, or in significant deviations of the torsional angles from the ideal value 180°, sometimes leading to *gauche* conformations of the bonds in the chain.

If no functional group be present in the long-chain molecule (i.e. a *n*-alkane), the only interaction between neighbouring molecules involves the weak van der Waals forces with minimal directing ability. When functional groups are present at one end of the long-chain molecule, forces other than the van der Waals interactions between the chains direct the packing. These forces include ionic forces and hydrogen-bonding interactions, and might be the major forces directing the packing. The packing in the lattice is influenced by two factors – the force between the end groups in a layer and the van der Waals interactions between the alkyl chains in a layer. Both forces contribute to the lattice energy; their relative strengths determine the crystal structure.

Kitaigorodskii (1961) presented a geometric analysis of all possible structures of *n*-alkanes based on close packing, but his treatment is inadequate when describing the packing of long-chain molecules with functional groups that form hydrogen bonds.

MOLECULAR PACKING

In all known crystal structures of long-chain compounds with a simple functional group at a terminal position, molecules pack in layers, also called lamella, with the terminal groups forming the surfaces of the layers Figure 10-1. The molecules in the layers might be perpendicular or tilted relative to the end-functional group plane, and the layers might be stacked in various ways. Layered packing (with

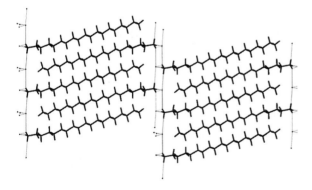

Figure 10-1. Interdigitated and tilted packing in *n*-octadecylammonium iodide crystals

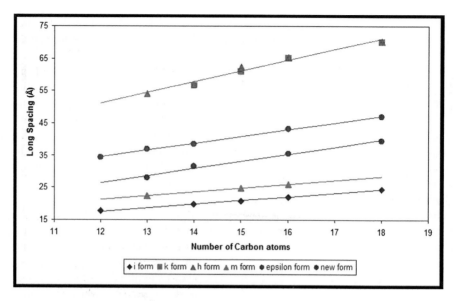

Figure 10-2. Long spacings of *n*-alkylammonium chlorides in all five polymorphic forms observed at room temperature

interdigitated or non-interdigitated molecules) of two types is commonly observed for end-functionalised long-chain molecules.

HOMOLOGOUS SERIES

Long-chain molecules with the same functional group but varied chain length form compounds in a homologous series. When all compounds belonging to this homologous series crystallize as the same polymorph, this series is called isostructural. If the distance between layers (or long spacing) be linearly proportional to the number of carbon atoms in the long chain, this condition indicates isostructural equivalence that is verifiable with X-ray powder diffraction (Figure 10-2).

THE ODD-EVEN EFFECT

The odd-even effect refers to the variation of properties of *n*-alkanes and *n*-alkane derivatives with carbon atoms of odd or even number in the alkyl chain. This odd-even effect reflects the varied packing of odd and even compounds regardless of the functional group present. The odd compounds form an isostructural series, and likewise the even compounds. This condition results in a distinct structure and physical properties for compounds with carbon atoms in the chain of odd and even number (Figure 10-3).

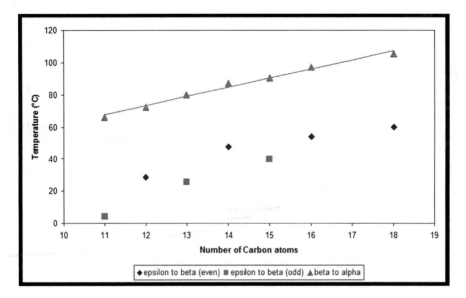

Figure 10-3. Temperatures of phase transition of the epsilon form of *n*-alkylammonium bromides, illustrating the odd-even effect

HIGH-TEMPERATURE PHASES

On increasing temperature, some long-chain molecules might undergoseveral phase transitions to intermediate phases before melting. Busico and Vacatello (1983) called this phenomenon thermotropic polymorphism. The consecutive phases vary in their degree of disorder in the aliphatic chain, namely deviation from the extended, all-*trans* arrangement of chains, and might be due to rotation, longitudinal displacements and dynamic *gauche* bonds. With increasing temperature the chain interactions weaken and atoms in the chain might reorient in the lattice. The packing in high-temperature phases is hexagonal, the best close packing for rod-like molecules with almost circular cross section. The high-temperature sequence of phase transitions observed for a specific homologous series depends on the end functional group or groups. No pre-melting transitions are observed for long-chain fatty acids, but long-chain *n*-alkanes, *n*-alcohols and simple esters undergo a transition to a "rotator phase" with much disorder just below the melting point. More complicated sequences of phase transitions are observed for *n*-alkylammonium halides (Terreros et al. 2000).

EXPERIMENTAL METHODS

Range of n-Alkylammonium Halides Investigated

Much information on the structures and thermal behaviour of alkylammonium halides is available in the crystallographic literature. As compounds with short chains crystallize readily as large single crystals, suitable for diffraction measurements, their structures (chain lengths less than twelve carbons) dominate the literature. We investigated compounds with 11–18 carbon atoms in the saturated hydrocarbon chain and halide anions chloride, bromide and iodide, i.e. C_nH_{2n+1} NH_3^+ X^- with $n = 11-18$ and X = Cl, Br and I. A chain length in this range was chosen for reasons that include gaps in the structural information already available, availability of starting materials and ease of crystallization. A full review of structural information obtained from preceeding and our own work (Rademeyer, 2003) will be published soon. The results presented here include information from both sources.

Synthesis

n-Alkylammonium halide salts are prepared from commercially available long-chain n-alkylamines and the corresponding halide acids according to this reaction scheme:

$$C_nH_{2n+1}NH_2 + HX \rightarrow C_nH_{2n+1}NH_3{}^+X^-$$

The primary amine is dissolved in lukewarm trichloromethane and the acid added dropwise. The resulting alkylammonium salts precipitate during the addition of the acid, in slight excess to ensure the complete formation of the salt. The precipitation of the halide is further encouraged in cooling with ice. For purification, recrystallization should be repeated at least three times. The final product is then crystallized under various conditions required to form single crystals of the various polymorphs of satisfactory quality. Environmental factors that might influence the formation of a specific polymorph include choice of solvent, temperature of crystallisation, and rate and method of crystallization. Despite considerable effort, as with most long-chain compounds, single crystals of satisfactory quality are rarely obtained. The long-chain n-alkylammonium halides crystallize as thin plates that are fragile and difficult to handle. For example, of twenty-one compounds that we investigated, single crystals of sufficient quality were obtained for only nine compounds, six of them hydrates.

Investigation of Structures

Two major experimental techniques available for the investigation of the polymorphism of n-alkylammonium halides are X-ray diffraction and thermal analysis. X-ray diffraction is the most informative technique because it provides detailed information on the molecular conformation and the packing of polymorphs.

Single crystal X-ray diffraction enables the determination of atomic positions in the unit cell, allowing a three-dimensional visualization of the structure. Because the exact packing of molecules in the polymorphic forms is established, the crystal structures of polymorphs are directly comparable.

In powder X-ray diffraction, the three-dimensional information of a single-crystal diffraction-data set is compressed into one dimension, immensely complicating the structure determination, but it remains the most important tool for structure and identification. A powder pattern provides a unique fingerprint of a crystalline substance, and therefore also of the polymorphs of a compound. The simplest application is to reveal or to confirm the existence of distinct polymorphs. In some cases it is possible to determine crystal-structure models of a specific polymorph from powder patterns, which can then be refined with Rietveld's technique.

The long spacings in a crystal are calculable from the powder pattern on multiplying the d values of the small-angle lamellar reflections by the non-zero index of the reflection – *00l, 0k0* or *h00* reflections depending on cell choice. The average of these values is the experimentally determined long spacing. This value indicates the repeat distance of the lamella in the structure, and is unique to a specific polymorph of a compound (Figure 10-4).

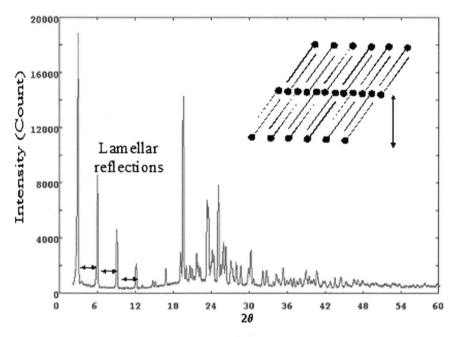

Figure 10-4. X-ray powder diffraction pattern of a *n*-alkylammonium halide. The arrows indicate how the difference between the d values is related to the lamellar thickness

Thermal Analysis

Through differential scanning calorimetry (DSC), thermogravimetric analysis (TGA) and thermal microscopy (TM), thermal analysis (TA) is useful to investigate the behaviour of a compound as a function of temperature (Figure 10-5). Whereas hot-stage microscopy is a qualitative and visual technique, DSC and TGA provide quantitative measurements that enable one to calculate the enthalpies and stoichiometry of thermal events. An advantage of such thermal analysis is that only a small sample is required. This condition is especially useful in regard to polymorphism because only a small amount of a specific polymorph is typically available. These techniques are thus ideal for the study of sequences of unique thermotropic phase transitions.

In a DSC scan, the difference of energy input (heat flow) into a sample and into a reference material is plotted as a function of temperature. In a plot of heat flow vs. temperature, endothermic minima correspond to desolvation of solvates or phase transitions of polymorphs, and exothermic maxima correspond to crystallization or decomposition. Integration over the area of a transition feature yields the associated transition enthalpy.

The most important temperature in the analysis of a DSC feature is the onset temperature because this temperature is not influenced by the sample size. The

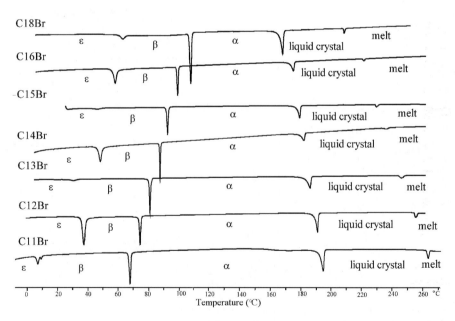

Figure 10-5. Thermal analysis by DSC of phase transitions of the ε form of melt-crystallized *n*-alkylammonium bromides. Note the trends with chain length

temperature at the extremum, called the peak temperature, is influenced by the size of the sample, and might vary with the sample size of the same polymorph. When transition temperatures are compared, the onset temperature should thus be used.

In thermogravimetry the variation of the mass of a sample with temperature is monitored in a thermobalance comprising a microbalance, a furnace and a temperature controller. This apparatus allows the calculation of the fractional loss of mass and thus the stoichiometry of a solvate by TGA (Caira, 1998) (Figure 10-6).

In hot-stage microscopy a sample is examined with a microscope while being heated with variation of temperature at a convenient rate. The instrument consists of a thermo-microscope with a hot stage, a lamp and recording devices, commonly a camera or video camera. The rate of heating is programmable and readily adjustable. With these techniques both structural and morphological alterations occurring in a sample on heating become observable. This information complements that from DSC.

On enclosing a sample in a heating stage or oven, variable-temperature X-ray powder diffraction gives the technique a thermal dimension. A diffraction pattern is then collected at various temperatures; the specific polymorph present at a given temperature thus becomes identifiable (Figure 10-7).

These experimental techniques are complementary to one another. Single crystal X-ray diffraction allows only the study of polymorphs of *n*-alkylammonium halides at specific temperatures, commonly room temperature. In contrast thermal analysis yields no information about the structural nature of compounds, but allows the determination of temperatures and sequences of phase transitions (Figure 10-8). In some instances powder diffraction fails to distinguish two polymorphic forms, but DSC scans indicate their distinction.

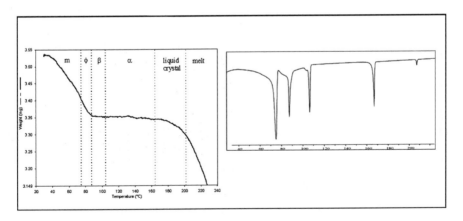

Figure 10-6. TGA and DSC scans of crystals of C18Br.H$_2$O, one *n*-alkylammonium bromide monohydrate. Temperatures of phase transitions as observed by DSC are indicated with dotted lines

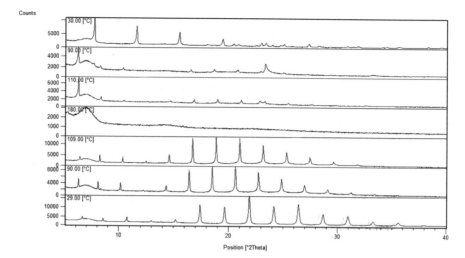

Figure 10-7. Variable-temperature X-ray powder diffraction patterns of solution-crystallized *n*-octadecylammonium iodide. The thermal sequence begins at the top and shows the polymorphs forming after phase transitions indicated by DSC scans

SEQUENCES OF PHASE TRANSITIONS

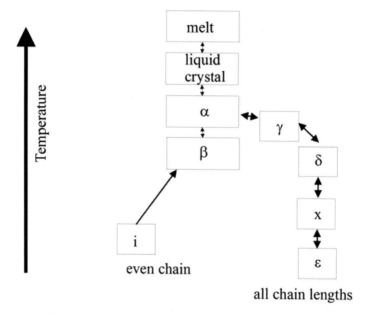

Figure 10-8. Thermotropic phase transitions observed for *n*-alkylammonium iodides

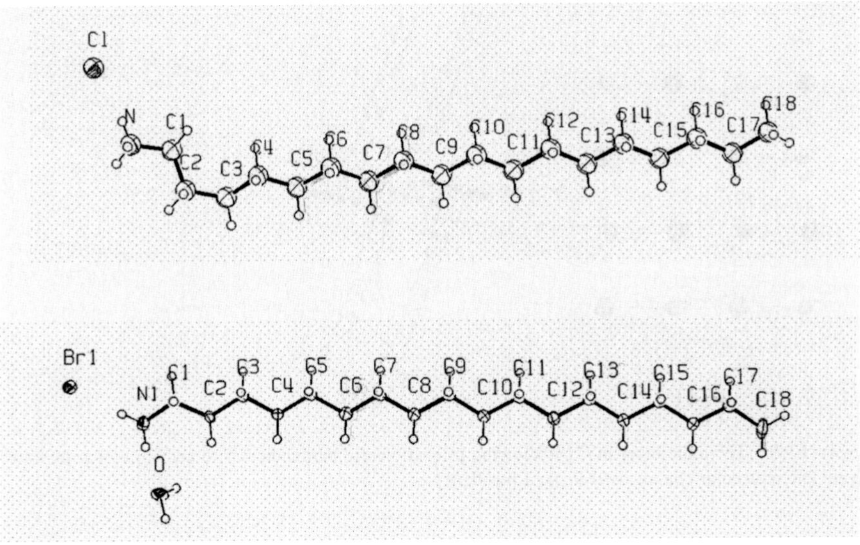

Figure 10-9. The *gauche* bond between C2 and C3 in the *k* form of C18Cl and the extended, all-*trans* conformation in C18Br.H$_2$O

MOLECULAR CONFORMATIONS

Alkane molecules in crystals that are stable at room temperature normally exist in an extended all-*trans* conformation. For all known crystal structures of the *n*-alkylammonium halides, the long-chain molecule deviates from the ideal all-*trans* conformation because of hydrogen-bonding interactions. In some structures this deviation is severe, and results in *gauche* bonds in the molecular chain (Figures 10-9 and 10-11). The long-chain molecule remains mostly in the *trans* conformation for all known structures.

THE IONIC LAYER

The bonds between anions and cations in the ionic layer invariably involve hydrogen atoms. The hydrogen-bonding networks that form reveal that hydrogen-bonding interactions of the maximum possible number are present and that all hydrogen atoms on the ammonium groups and water molecules, when present, participate (Figure 10-10).

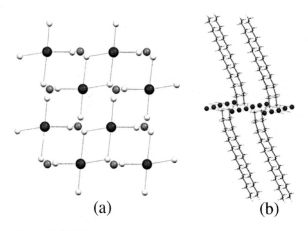

(a)   (b)

Figure 10-10. The ionic layer in C18I showing projections (a) perpendicular to and (b) along the ionic layer. The molecular chains are truncated for clarity

PATTERNS OF CRYSTAL PACKING

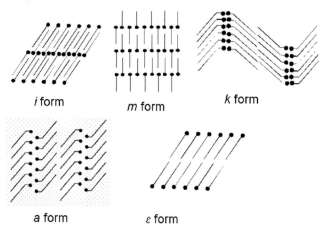

i form   *m* form   *k* form

a form   *ε* form

Figure 10-11. A cartoon of the packing patterns observed in the *n*-alkylammonium halides

ACKNOWLEDGEMENT

We thank Professor Erie Reynhardt and Ms. Rosalie Scholtz for introducing us to the alkylammonium halides.

REFERENCES

1. Busico, V., Vacatello, M. (1983). *Mol. Cryst. Liq. Cryst.* **97**, 195–207.
2. Caira, M.R. (1998). *Topics in current chemistry.* **198**, 163–280.

3. Dorset, D.L. (2005). *Crystallography of the Polymethylene Chain.* IUCr Monocraphs on Crystallography 17, Oxford Science Publictions. Oxford UK.
4. Garti, N., Sato, K. (1988). *Crystallization and polymorphism of fats and fatty acids.* Surfactant Science Series, Marcel Dekker. New York USA.
5. Hamilton, R.J. (1995). *Waxes: Chemistry, Molecular Biology and Functions.* The Oily Press, Dundee. Scotland.
6. Kitaigorodskii, A.I. (1961). *Organic Chemical Crystallography.* Chapter 4. Consultants Bureau, New York USA.
7. Rademeyer, M. (2003). PhD thesis, RAU University.
8. Small, D.M. (1986). *The Physical Chemistry of Lipids.* Plenum Press. New York.
9. Terreros, A., Galera-Gomez, P.A., Lopez-Cabarcos, E. (2000). *J. Therm. Anal. Cal.* **61**, 341–350.
10. Warth, A.H. (1947). *The chemistry and Technology of Waxes.* Reinhold Publishing Corporation, New York USA.

CHAPTER 11

MYSTERIOUS CRYSTALLOGRAPHY

From Snow Flake to Virus

ALOYSIO JANNER

Abstract: Despite the absence of translational symmetries, snow flakes and biomacromolecules share properties of crystals and/or quasicrystals, such as an underlying lattice structure and crystallographic scaling leaving the lattice invariant.

At the morphological level, one observes in axially symmetric proteins linear and planar crystallographic scalings, whereas 3-dimensional scaling occurs in icosahedral viruses.

In all the cases considered so far, the lattices involved are integral. This property implies the existence of a metric tensor with integral entries, up to one real lattice parameter, as in the cubic case. One finds, in particular, isometric hexagonal lattices (with $c = a$). Similar and additional properties allow one to speak of strongly correlated biomacromolecular structures

INTRODUCTION

In the years after the discovery of aperiodic crystals (incommensurate modulated, intergrowth and quasicrystals) crystallography was for me a very rich and open field of research, but not mysterious. Even the surprising combination in snow crystals of sixfold circular rotations with hyperbolic rotations [1], leading to hexagrammal scaling symmetry, fitted into the whole because the atomic positions in ice are invariant with respect to both types of crystallographic rotations [2].

During the 1994 meeting of the American Crystallographic Association (ACA) held in Atlanta, the observation of a pentagrammal scaling in cyclophilin represented the turning point. The structure of cyclophilin A in complex with cyclosporin A had shortly before been determined [3]. The scaling relation is not primarily between atomic positions, as in pentagonal and decagonal quasicrystals, but is a morphological property which relates the external boundary with the central hole. The pentagrammal scaling relation can be recognized in the ribbon diagram of the cyclophilin-cyclosporin complex appearing in the logo of the ACA meeting.

233

J.C.A. Boeyens and J.F. Ogilvie (eds.), Models, Mysteries and Magic of Molecules, pp. 233–254.
© 2008 *Springer*.

Since then, many other biomacromolecules showed analogous polygrammal scaling relations.

Molecular forms bring concepts back developed during the nineteenth century in which crystal growth forms were systematically investigated. At that time, atoms were not considered to be real but only as a way of expressing chemical laws [4], and lattice periodicity an hypothesis compatible with the empirical law of rational indices with no consequence for the physical nature of crystals [5]. The present molecular situation is reversed: one knows that there are atoms and where the atoms are. A molecular lattice allows an interpretation of the molecular morphology, also expressible in terms of rational indices, but without a theoretical basis, even if one can speak of *molecular crystallography* [6].

Indeed, molecular crystallography seems to be a self-contradicting concept. Crystallography is by definition the science of crystals and a molecule is not a crystal. An analogous contradiction arises while speaking of *aperiodic crystals*, because lattice periodicity is the fundamental property characteristic for crystals [7]. Without a lattice structure there is no crystallography, but, as one has learned from incommensurate crystals, the lattice is not necessarily three-dimensional, and one sees from snow flakes and from biomacromolecules that a lattice structure need not imply translational symmetry. In aperiodic crystals the lattice periodicity is a higher dimensional one, and the lattice occurring in various biomacromolecules is determined by properties other than lattice translations [8]. At present, the evidence given is purely geometrical with no physical or chemical foundation. According to this point of view, snow flakes and biomacromolecules are mysterious, not crystallography.

There is, however, more. The lattices underlying the molecular forms have the remarkable property to be integral. *Integral lattices* are characterized by metric tensors with integral entries, up to a real lattice parameter, as in the cubic lattice case [9]. Being aware of this, one discovers that many crystals have (in the ideal case) integral lattices, as revealed by sharp peaks in the frequency distribution of crystal lattices as a function of ratios of lattice parameters [10, 11]. This property is not explained with the known crystallographic laws. Again and again, crystals and biomacromolecules appear as weft and warp of the same mysterious crystallography.

What is presented here summarizes fifteen years of research devoted to the relation between crystals and molecules, with results scattered in several articles published mainly in Acta Crystallographica, from which most figures of this paper have been taken in a version more or less modified and with the permission of the International Union for Crystallography (IUCr). Here the attention is focused on molecules through specific examples illustrating the basic ideas; more information is available in papers published elsewhere and quoted further on.

SNOW CRYSTALS

Everybody agrees that snow flakes are normal crystals of which the macroscopic forms are based on lattice periodicity at an atomic level. Like any other crystal growth form, a snow flake is non-periodic. No problem arises thereby because its symmetry is explained by crystallographic laws applied at the macroscopic level. These laws, however, fail to explain the lattice structure observed in many snow crystals [12]: snow crystals are a model of a kind for the molecular case.

Consider as a first example, among many possibilities, the facet-like snow flake BH53.1, taken from the book of Bentley and Humphreys [13] with figures labeled BHp.q; here p denotes the page and q the successive figure on that page. One recognizes several hexagonal patterns of increasing size (Figure 11-1). These patterns are a record of the growth and indicate (in my view) an alternation of slow and rapid rates of growth, with correspondingly greater or less concentration of impurity.

To associate the vertices of the central hexagon to points of an hexagonal lattice generated by the corresponding position vectors is trivial. Not trivial, however, is that the vertices of the following hexagons belong to the same lattice. In the present case the external boundary is scaled by a factor four with respect to the central hexagon. The growth process is apparently temporally periodic, producing a macroscopic lattice structure distinct from the periodic lattice of the microscopic atomic positions. This interpretation does not explain why other points of the growth lattice match special points of the complex patterns of the snow flake, which can also be labeled by the integral components (the *indices*) of the corresponding lattice points. Their interpretation requires the crystallographic hyperbolic rotations mentioned above [1, 2]. In the case of BH53.1 these indices range between −4 and +4. In particular, the mid-edge positions of the second and fourth hexagons produce a hexagrammal relation between the vertices of the star hexagon {6/3}. The Schäfli symbol {n/m} denotes the star polygon obtained on joining the *m*th successive vertices of a regular *n* polygon.

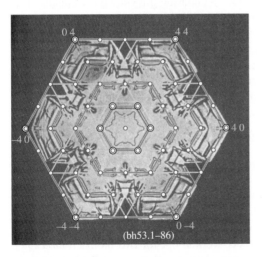

Figure 11-1. Facet-like snow flake from Bentley & Humphreys (1931, 1962) *Snow Crystals* (courtesy Dover) with an indexed hexagonal growth lattice superimposed (from [15], courtesy IUCr)

Figure 11-2. Dendritic snow flake from Bentley & Humphreys (courtesy Dover) with hexagonal growth lattice points. The transition from a facet-like growth of the central region to a dendritic growth corresponds to a triplication of the lattice parameter (from [12], courtesy IUCr)

The second example is the dendritic snow crystal BH167.8 (Figure 11-2), also taken from the book of Bentley and Humphreys [13]. The central part is similar to the facet-like BH53.1 and involves three successive hexagons (instead of four). The largest hexagon marks the transition to a dendritic growth with branching sites at points of a growth lattice three times as large as in the central region. The six extremal points of this snow flake have accordingly indices $\pm[9\ 0]$, $\pm[9\ 9]$, $\pm[0\ 9]$ with respect to the central lattice and indices $\pm[3\ 0]$, $\pm[3\ 3]$, $\pm[0\ 3]$ with respect to the dendritic growth lattice, which is an index 3 sublattice of the preceeding one. A hexagrammal pattern marks the transition between the two regimes of growth, with the hexagram as the star hexagon $\{6/3\}$.

AXIAL-SYMMETRIC BIOMACROMOLECULES

Molecular crystallography at a morphological level is based on enclosing forms with vertices at points of a non-periodic *form lattice*. The enclosing forms of a given biomacromolecule are mutually related according to scaling transformations, leaving the lattice invariant or transforming it to a sublattice or a superlattice. As already mentioned, a remarkable property of the form lattices observed so far is their being integral. These basic ideas are worked out for two biomacromolecules that involve hexagonal and pentagonal form lattices: the proteins R-phycoerythrin and the cyclophilin A, respectively.

Isometric Hexagonal Form Lattice of R-phycoerythrin

In the axial projection, the hexamer R-phycoerythrin has an hexagonal central hole and an hexagonal envelope, despite the fact that the rotational symmetry of the hexamer is only trigonal [14]. Non-trivial is the empirical observation that the six vertices of the projected molecular envelope lie at points of the two-dimensional hexagonal lattice defined by the vertices of the central hole. Envelope and hole are related according to a planar crystallographic scaling with scaling factor four, but not by lattice translations [15] (Figure 11-3).

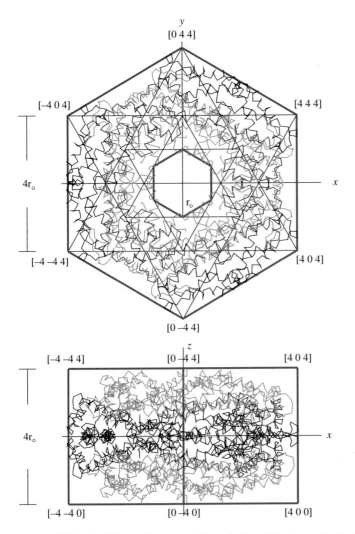

Figure 11-3. Indexed isometric hexagonal form lattice of the hexameric R-phycoerythrin with hexagrammal scaling relations between central hole and external boundary (adapted from [15], courtesy IUCr)

With r_0 denoting the hexagonal radius of the central hole, the radius of the envelope is $4r_0$ and equal to the height of the hexamer. It follows that the three-dimensional form lattice is hexagonal with lattice parameters $a = c = r_0$ and, therefore, isometric hexagonal because the axial ratio is $\gamma = c/a = 1$. This lattice, denoted $1\text{-}\Lambda_{hex}$, is integral. Indeed the metric tensor of the lattice basis $b = \{a_1, a_2, a_3\}$ with Cartesian components $a_1 = r_0(1, 0, 0)$, $a_2 = r_0(-1/2, \sqrt{3}/2, 0)$, $a_3 = r_0(0, 0, 1)$ is given by the Gram matrix,

$$(1) \qquad g(b) = \frac{r_0^2}{2} \begin{pmatrix} 2 & -1 & 0 \\ -1 & 2 & 0 \\ 0 & 0 & 2 \end{pmatrix},$$

which has integral entries, up to the factor $r_0^2/2$.

The envelope is radially scaled with a factor four from the chosen lattice unit cell and, as already pointed out, in a planar scaling relation with the central hole according to the same scaling factor. The planar scaling is the same as already observed in the facet-like snow crystal of Figure 11-1, and can be expressed in terms of two successive mid-edge hexagrammal scalings. All vertices of these various molecular forms have integral indices.

Isometric Pentagonal Form Lattice of Cyclophilin A

Not all axial-symmetric form lattices are isometric, but isometric lattices play a privileged role even in crystals [10, 11]. A pentagonal isometric lattice occurs in the pentameric and decameric configurations of cyclophilin A [3]. As in the previous example, the planar scaling relation between envelope and central hole is expressible in terms of a star polygon. Typical in this respect is the cyclophilin pentamer [16]. Hole and envelope have vertices belonging to the same pentagonal lattice, and are related with a planar pentagrammal scaling $S_{\{5/2\}}$. The scaling factor between the two regular pentagons defined by the star pentagon $\{5/2\}$ is $-1/\tau^2$, in which τ is the golden ratio $(\sqrt{5}+1)/2$ and the minus sign indicates that the pentagons are in opposite orientation.

In the decameric configuration the dyadically related cyclophilin pentamers take a peculiar orientation with respect to the twofold axes, not required by the point group 52 of the whole. Pairs of corresponding vertices of the pentagonal enclosing form of the two pentamers, at the projected positions of Glu15, subdivide the edges of the pentagonal envelope of the decamer in the pentagrammal ratios $\tau : 1 : \tau$. This property is expressible in terms of the linear scaling Y_{1/τ^3}. A scaling transformation indicated with Y_λ transforms the Cartesian coordinates of a point according to $Y_\lambda(x, y, z) = (x, \lambda y, z)$. In Figure 11-4, the y-coordinates of the points labeled with P and Q are scaled by a factor $1/\tau^3$ from the y coordinate of the corresponding pentagonal points C and D of the envelope: $P = Y_{1/\tau^3} C$ and $Q = Y_{1/\tau^3} D$. The scalings $S_{\{5/2\}}$ and Y_{1/τ^3} are crystallographic and occur also among atomic positions of the decagonal quasicrystal $Al_{78}Mn_{22}$ [17].

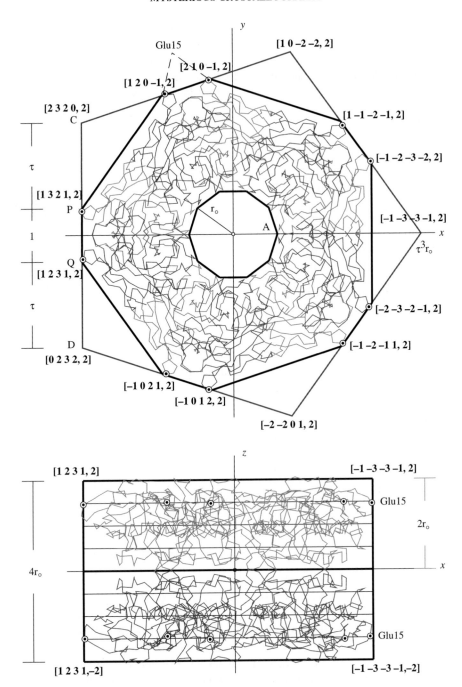

Figure 11-4. The indexed form lattice of the decameric cyclophilin A is isometric pentagonal, as explained in the text (adapted from [16], courtesy IUCr)

Crystallographic signifies that the transformations leave the pentagonal lattice invariant.

The concept of a pentagonal lattice (and in general that of a *polygonal lattice*) has been introduced by Yamamoto in the context of axially symmetric aperiodic crystals [18]. As this term might be unfamiliar to many people, some explanations are required before showing that the form lattice of cyclophilin is isometric pentagonal.

Consider the decagonal hole of the decamer, as represented in axial projection in Figure 11-4, and choose, in the x, y plane, vectors a_1, a_2, a_3, a_4 pointing from the center to four successive but non-adjacent vertices of the regular decagon. Together with a fifth vector a_5 taken along the rotation axis z, they form a basis of rank 5 and dimension 3 of the pentagonal lattice. Hence these vectors span a three-dimensional space and are linearly independent on the rationals $\mathbb{Q}$, and thus on the integers $\mathbb{Z}$. The position vector r_P of a point in the three-dimensional space expressed with integral linear combinations of the vectors of the basis $b = \{a_1, a_2, a_3, a_4, a_5\}$ is accordingly uniquely given by the components, called *indices*, of r_P in the basis b.

$$(2) \qquad r_P = [n_1 \ n_2 \ n_3 \ n_4, n_5] = \sum_{i=1}^{i=5} n_i a_i \in \Lambda_{pent}, \quad n_i \in \mathbb{Z},$$

in which n_1, n_2, n_3, n_4 are planar indices, separated for practical reasons by a comma from the axial index n_5. The set of all positions with integer indices defines the pentagonal three-dimensional lattice $\Lambda_{pent}(a, c) = \Lambda_{pent}(a, \gamma)$, in which $a = |a_i|, i = 1, \ldots 4$ and $c = |a_5|$ are lattice parameters and $\gamma = c/a$ is the axial ratio. The points of the lattice Λ_{pent} are dense in space and, therefore, not a lattice from the Euclidean point of view, but a $\mathbb{Z}$-module, a projection in space of a five-dimensional Euclidean lattice. Only points with small integral indices have a structural meaning. This case is applicable for the vertices of the molecular forms, which are discrete, so that one can safely admit Λ_{pent} as a form lattice. In the pentagonal basis b the pentagrammal scaling $S_{\{5/2\}}$ and the linear scaling Y_{1/τ^3} are expressed with the invertible integer matrices:

$$(3) \qquad S_{\{5/2\}}(b) = \begin{pmatrix} \bar{2} & 1 & 0 & \bar{1} & 0 \\ 0 & \bar{1} & 1 & \bar{1} & 0 \\ \bar{1} & 1 & \bar{1} & 0 & 0 \\ \bar{1} & 0 & 1 & \bar{2} & 0 \\ 0 & 0 & 0 & 0 & 1 \end{pmatrix}, \quad Y_{1/\tau^3}(b) = \begin{pmatrix} 0 & 1 & \bar{1} & 1 & 0 \\ 1 & \bar{1} & 2 & \bar{1} & 0 \\ \bar{1} & 2 & \bar{1} & 1 & 0 \\ 1 & \bar{1} & 1 & 0 & 0 \\ 0 & 0 & 0 & 0 & 1 \end{pmatrix}.$$

These transformations leave, therefore, the pentagonal lattice invariant.

One can here return to cyclophilin as represented in Figure 11-4. Denoting by r_0 the radius of the decagonal central hole, one finds that the radius of the pentagon enclosing in projection the decamer is $\tau^3 r_0$. The planar scaling S_{τ^3} in the basis b

is integral invertible, as well as the fivefold rotation R_5. Both leave the pentagonal lattice invariant.

$$
(4) \qquad R_5(b) = \begin{pmatrix} 0 & 0 & 0 & \bar{1} & 0 \\ 1 & 0 & 0 & \bar{1} & 0 \\ 0 & 1 & 0 & \bar{1} & 0 \\ 0 & 0 & 1 & \bar{1} & 0 \\ 0 & 0 & 0 & 0 & 1 \end{pmatrix}, \quad S_{T^3}(b) = \begin{pmatrix} 1 & 2 & 0 & \bar{2} & 0 \\ 0 & 3 & 2 & \bar{2} & 0 \\ \bar{2} & 2 & 3 & 0 & 0 \\ \bar{2} & 0 & 2 & 1 & 0 \\ 0 & 0 & 0 & 0 & 1 \end{pmatrix}.
$$

The height of the pentamer is $2r_0$ and that of the decamer $4r_0$, as indicated in Figure 11-4. The form lattice of cyclophilin in both conformations is therefore the isometric pentagonal lattice $1-\Lambda_{pent}$ with lattice parameters $a = c = r_0$ and axial ratio $\gamma = 1$. According to this choice, all vertices of the enclosing forms and of the central hole have integral indices, as readily computed using the scaling matrices given.

STRONGLY CORRELATED BIOMACROMOLECULAR SYSTEMS

The proteins considered in the preceeding section are three-dimensional, unlike the planar model adopted for the snow flakes. The three-dimensional lattice of hole and envelope is axial and has, therefore, two independent lattice parameters: the parameter a in the rotational plane and the parameter c along the axis. In all axial proteins investigated so far, a and c are structurally related. For example, for both R-phycoerythrin and cyclophilin one has $c = a$, generating so-called *isometric lattices*. Other simple ratios (such as $\sqrt{3}$) occur also, a property not expressible in terms of the known crystallographic laws. As a single parameter suffices to connect geometry and metrical structure, the corresponding molecular system is described as *strongly correlated* [15, 16, 19]. The non-accidental character of these intriguing crystallographic relations becomes evident in compound systems and in complexes when the same laws apply to the components and to the whole.

Cyclophilin A is a simple example: pentamer and decamer share the same isometric form lattice. Moreover, the mutual orientation of the pentamers in the decameric configuration is fixed through a linear scaling transformation leaving the form lattice invariant. The geometry of the entire biomacromolecular system and of its components is expressible in term of the single parameter r_0, which is the shortest distance of the monomeric chain from the fivefold axis.

A further striking example is given by the Pyrococcus abyssi Sm core (PA Sm). In the free state and in complex with RNA, the PA Sm has point symmetry 72 and consists of a sandwich of two heptameric rings in the same orientation and dyadically related [20]. In the two states, the folding of the monomers differs only slightly, so that the corresponding molecular forms of PA Sm heptamer (central hole and envelope) are the same.

The heptagonal lattice requires seven basis vectors: six $a_1, \ldots, a_6$, perpendicular to the sevenfold axis, pointing from the center to vertices of a regular heptagon and one a_7 along the rotation axis. The lattice parameters are $a = |a_1| = \ldots = |a_6|$ and $c = |a_7|$. In the basis $b = \{a_1, \ldots, a_7\}$, a lattice point P has seven integer indices,

$$(5) \qquad r_P = \sum_{i=1}^{i=7} n_i a_i = [n_1 \ n_2 \ n_3 \ n_4 \ n_5 \ n_6, n_7] \in \Lambda_{hept}(a, c),$$

with again a comma separating the planar indices from the axial index. In this basis, the planar scaling that relates the external envelope of the PA Sm heptamer to the central hole with radii r_e and r_0, respectively, has scaling factor $\mu = 2 - 2\cos(2\pi/7) + 2\cos(4\pi/7) = 0.3079\ldots$, is integral invertible [21], and leaves the heptagonal lattice invariant:

$$(6) \qquad S_\mu(b) = \begin{pmatrix} 3 & \bar{2} & 1 & 0 & \bar{1} & 2 & 0 \\ 0 & 1 & \bar{1} & 1 & \bar{1} & 1 & 0 \\ 2 & \bar{2} & 2 & \bar{1} & 0 & 1 & 0 \\ 1 & 0 & \bar{1} & 2 & \bar{2} & 2 & 0 \\ 1 & \bar{1} & 1 & \bar{1} & 1 & 0 & 0 \\ 2 & \bar{1} & 0 & 1 & \bar{2} & 3 & 0 \\ 0 & 0 & 0 & 0 & 0 & 0 & 1 \end{pmatrix}, \qquad \mu = 0.3079\ldots$$

In the free case the heptagonal radius r_e of the envelope is $8u$, with u the distance of each heptamer from the twofold axis. The height h of the heptamer, which is also the height of the monomer, is $h = 7u$ and the total height H of the double ring is $H = 16u$ (see Figure 11-5). The form lattice of the single and the double heptamer is therefore isometric heptagonal $1 - \Lambda_{hept}(u)$, with lattice parameter $u = a = c$ [16].

In the complex configuration the RNA binds at two sites. One site, at which one finds the radial outer RNA, is between the heptamers in a space enlarged with respect to the free state [20]. Surprisingly enough, the form lattice remains the same [19]. The only alteration of parameters is the distance d between the two rings, which is $4u$ in the complex instead of $2u$ in the free state, implying an increase of the total height H from $16u$ to $18u$ (Figure 11-6). Despite the changes in chemical bindings, the biomacromolecular system remains apparently in the same strongly correlated state. This condition appears applicable also for the RNA subsystem, as discussed elsewhere [19].

The geometry of all the various enclosing forms of the PA Sm core depends on only the same single lattice parameter u of the heptagonal isometric lattice $1 - \Lambda_{hept}(u)$.

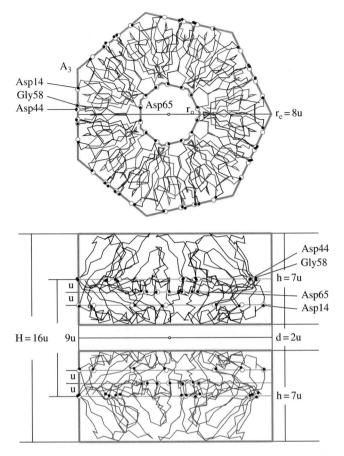

Figure 11-5. The form lattice of the double heptameric ring of the pyrococcus abyssi Sm core in the free state conformation is isometric heptagonal. The lattice parameter is given by the distance *u* of each heptamer from a twofold axis (from [16], courtesy IUCr)

ICOSAHEDRAL VIRUSES

Many viruses have icosahedral symmetry. Their nucleic-acid component (DNA or RNA) is contained in the hole of a capsid formed by the package of coat proteins. A first question then arises, whether in these viruses the external envelope of the capsid is related according to a crystallographic scaling to the viral hole in a way similar to that in axially symmetric biomacromolecules.

An icosahedral capsid has 12 fivefold, 20 threefold and 30 twofold symmetry axes. This situations yields a second question, whether clusters of coat proteins having one of these axial symmetries share a form lattice of the same type as in the axially symmetric biomolecules discussed above.

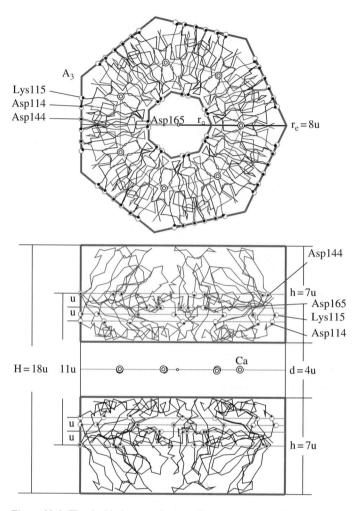

Figure 11-6. The double heptameric ring of pyrococcus abyssi Sm core in complex with RNA (not shown) has the same form lattice as in the free state. Only the distance d between the heptamers is doubled: $d = 4u$ instead of $2u$ (from [19], courtesy IUCr)

To provide an answer to these questions, one virus strain has been examined in detail: the Rhinovirus of various serotype [22–26]. The result is amazing. Both questions receive an affirmative answer for the virus as a whole and for the subsystems of each coat protein VP1, VP2, VP3 and VP4, respectively. The morphological properties observed are, moreover, independent of the serotype 16, 14, 3, 2 and 1A, respectively [27].

The form lattices for the threefold and the fivefold cases have already been presented. The form lattices for the twofold case are orthorhombic and possibly tetragonal or cubic and not simply monoclinic, because of the three mutually orthogonal dyadic axes implied by the icosahedral symmetry. It remains to introduce the icosahedral lattice.

Icosahedral Lattice

The form lattice of a viral capsid with icosahedral symmetry requires six basis vectors, which can be chosen pointing from the center to the non-aligned vertices of an icosahedron. Indicating within parentheses and brackets the components of a vector r with respect to the orthonormal basis $e = \{e_1, e_2, e_3\}$ and to the symmetry-adapted icosahedral basis $b = \{a_1, a_2, \ldots, a_6\}$, respectively, one can choose the vectors a_i according to

$$
\begin{aligned}
a_1 &= a_0(1, 0, \tau) &&= [100000], & a_2 &= a_0(\tau, 1, 0) &&= [010000], \\
a_3 &= a_0(0, \tau, 1) &&= [001000], & a_4 &= a_0(-1, 0, \tau) &&= [000100], \\
a_5 &= a_0(0, -\tau, 1) &&= [000010], & a_6 &= a_0(\tau, -1, 0) &&= [000001].
\end{aligned}
$$

These vectors are linearly independent on the rationals. The set of all their integral linear combinations defines a $\mathbb{Z}$-module of dimension 3 and rank 6 called an *icosahedral lattice* Λ_{ico}, in a way similar to the axially symmetric case. A lattice point P has position vector r_P with integer indices n_i

$$
(7) \qquad r_P = \sum_{i=1}^{i=6} n_i a_i = [n_1 n_2 n_3 n_4 n_5 n_6] \in \Lambda_{ico}, \quad n_i \in \mathbb{Z}.
$$

The icosahedral group 235 is defined with

$$
(8) \qquad 235 = \{\alpha, \beta \mid \alpha^5 = \beta^3 = (\beta\alpha)^2 = 1\}.
$$

In the orthonormal basis e the two generators α and β are represented with the rotation matrices $R_5(e)$ and $R_3(e)$, of orders 5 and 3, respectively:

$$
(9) \qquad R_5(e) = \frac{1}{2}\begin{pmatrix} 1 & -\tau & \tau-1 \\ \tau & \tau-1 & -1 \\ \tau-1 & 1 & \tau \end{pmatrix}, \quad R_3(e) = \frac{1}{2}\begin{pmatrix} \tau & 1-\tau & 1 \\ \tau-1 & -1 & -\tau \\ 1 & \tau & 1-\tau \end{pmatrix}.
$$

In the basis b the same rotations are given by integral matrices:

$$
(10) \qquad R_5(b) = \begin{pmatrix} 1 & 0 & 0 & 0 & 0 & 0 \\ 0 & 0 & 0 & 0 & 0 & 1 \\ 0 & 1 & 0 & 0 & 0 & 0 \\ 0 & 0 & 1 & 0 & 0 & 0 \\ 0 & 0 & 0 & 1 & 0 & 0 \\ 0 & 0 & 0 & 0 & 1 & 0 \end{pmatrix}, \quad R_3(b) = \begin{pmatrix} 0 & 0 & 1 & 0 & 0 & 0 \\ 1 & 0 & 0 & 0 & 0 & 0 \\ 0 & 1 & 0 & 0 & 0 & 0 \\ 0 & 0 & 0 & 0 & \bar{1} & 0 \\ 0 & 0 & 0 & 0 & 0 & \bar{1} \\ 0 & 0 & 0 & 1 & 0 & 0 \end{pmatrix}.
$$

Indexed Icosahedral Polyhedra

Applying the icosahedral group to the position [100000] one obtains the vertices of an icosahedron with indices given by the permutations of $[\pm 100000]$. One states that this indexed icosahedron has the point [100000] as generator. In a similar way, the point [111000] generates the 20 vertices of a dodecahedron with indices given by

(11) $\tau^2(1, 1, 1) = [111000], \quad \tau(0, 1, \tau^2) = [101100], \quad \dots, \quad \tau(-\tau^2, 0, 1) = [0\bar{1}010\bar{1}]$.

Scaling this dodecahedron by a factor $1/\tau^2$ one obtains a rescaled dodecahedron with vertices

(12) $(1, 1, 1) = \dfrac{1}{2}[111\bar{1}1\bar{1}], \quad (0, \dfrac{1}{\tau}, \tau) = \dfrac{1}{2}[1\bar{1}11\bar{1}1], \quad \dots, \quad (-\tau, 0, \dfrac{1}{\tau}) = \dfrac{1}{2}[1\bar{1}\bar{1}1\bar{1}\bar{1}]$.

The combination of these rescaled dodecahedral points with the vertices of the icosahedron indicated above yields another indexed polyhedron with icosahedral symmetry: the *triacontahedron* discovered by Kepler in 1611. The triacontahedron has 32 vertices, 12 icosahedral and 20 dodecahedral ones and 30 rhombic faces, or 60 triangular ones; in the latter case it is then denoted *ico-dodecahedron*. The triacontahedron is the projection in space of a six-dimensional hypercube and has two points [200000] and $[111\bar{1}1\bar{1}]$ as generators with integral indices (Figure 11-7). Further polyhedra with icosahedral symmetry and vertices at icosahedral lattice points are obtainable from one or more generators with integer indices [28].

Rhinovirus Capsid

The capsid. The capsid of the rhinovirus is encapsulated between two triacontahedra: one enclosing the external surface of the capsid and one delimiting the central hole. The two polyhedra are related through a radial scaling S_τ with scaling factor τ, as shown in Figure 11-8.

This scaling transformation is crystallographic.

(13) $S_\tau(e) = \begin{pmatrix} \tau & 0 & 0 \\ 0 & \tau & 0 \\ 0 & 0 & \tau \end{pmatrix}, \quad S_\tau(b) = \dfrac{1}{2} \begin{pmatrix} 1 & 1 & 1 & 1 & 1 & 1 \\ 1 & 1 & 1 & -1 & -1 & 1 \\ 1 & 1 & 1 & 1 & -1 & -1 \\ 1 & -1 & 1 & 1 & 1 & -1 \\ 1 & -1 & -1 & 1 & 1 & 1 \\ 1 & 1 & -1 & -1 & 1 & 1 \end{pmatrix}.$

S_τ is integral when expressed in a body-centred icosahedral basis. On applying $S_{1/\tau} = S_\tau^{-1}$ to the indices of the 12 icosahedral vertices generated by [200000] of the triacontahedron, one obtains the corresponding ones of the $1/\tau$-scaled triacontahedron:

(14) $[\bar{1}11111], \quad [1\bar{1}1\bar{1}\bar{1}1], \quad \dots, \quad [\bar{1}\bar{1}11\bar{1}1]$.

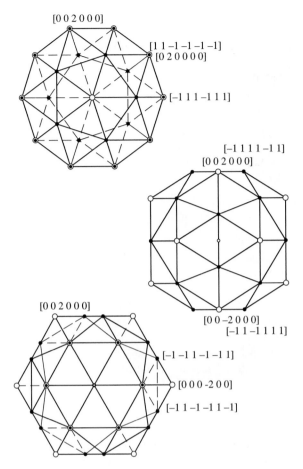

[0 0 2 0 0 0]

[1 1 –1 –1 –1 –1]
[0 2 0 0 0 0]

[–1 1 1 –1 1 1]

[–1 1 1 1 –1 1]
[0 0 2 0 0 0]

[0 0 –2 0 0 0]
[–1 1 –1 1 1 1]

[0 0 2 0 0 0]

[–1 –1 1 –1 –1 1]

[0 0 0 -2 0 0]

[–1 1 –1 –1 1 –1]

Figure 11-7. Indexed ico-dodecahedron (triacontahedron) viewed along the fivefold, twofold and three-fold axes, respectively. It has 12 icosahedral vertices (empty circles) and 20 dodecahedral vertices (filled circles) all belonging to the same icosahedral lattice (adapted from [27], courtesy IUCr)

In a similar way, from the 20 dodecahedral vertices generated with $[111\bar{1}1\bar{1}]$, one finds the indices of the dodecahedral vertices of the $1/\tau$-scaled triacontahedron:

(15) $[00022\bar{2}]$, $[02002\bar{2}]$, … , $[\bar{2}02020]$.

Accordingly, all vertices of the molecular form encapsulating the capsid of the rhinovirus are at points of the same icosahedral lattice, proving that the icosahedral lattice is indeed the form lattice for the viral capsid, which has its envelope and hole related by a three-dimensional crystallographic scaling. This property provides an answer to the first question.

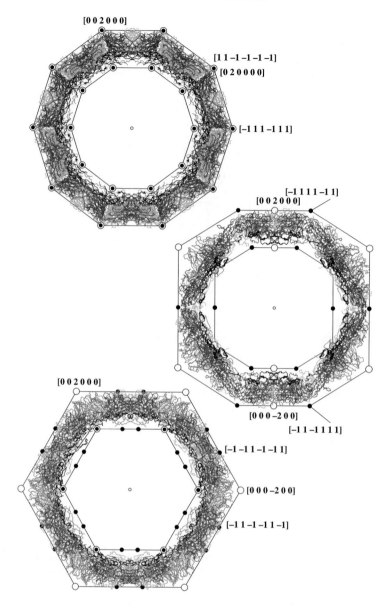

Figure 11-8. The capsid of the human rhinovirus is encapsulated between two ico-dodecahedra, one external and one internal scaled with a factor $1/\tau$, with τ the golden ratio. All vertices belong to the same icosahedral lattice. Only the vertices and the monomeric chains of the four coat proteins VP1, VP2, VP3 and VP4 in the various equatorial regions are plotted in projected views along the fivefold, the twofold and the threefold axes, respectively (adapted from [27], courtesy IUCr)

Numerous molecular clusters with axial symmetry are obtainable from the coat proteins of the rhinovirus, by applying to one or more of these proteins the axial subgroups of the icosahedral group, each producing its own form lattice. Here the general situation is illustrated by few representative clusters of one of the four coat proteins taken from one serotype. Further information is available elsewhere [27]. **Orthorhombic clusters.** In general, the form lattice of clusters with symmetry 222 is orthorhombic with lattice parameters a, b, c. These parameters are in integral relation with the half-edge $a_c = \tau a_0$ of the cube circumscribing the triacontahedron of the capsid and a_0 the icosahedral lattice parameter:

$$(16) \qquad a_c = \tau a_0 = z_1 a = z_2 b = z_3 c, \quad \text{integers} \quad z_1, z_2, z_3.$$

The triple of integers z_1, z_2, z_3 can serve to indicate the corresponding lattice, which is integral because the metric tensor of the basis vectors is proportional to one with integral entries. Figure 11-9 shows the enclosing form of four PV3 dimers related according to point group 222, subgroup of the icosahedral group. In this case, the form lattice is tetragonal with lattice parameters $a = a_c/9, b = c = a_c/8$. Indicated (in projection) are rectangular prismatic enclosing forms and refined ones, all with vertices at lattice points. Inside each form one finds the C_α-backbone chains of two PV3 proteins. For each serotype of the rhinovirus, of 60 possible tetramers with 222 symmetry of the four coat proteins, 28 have cubic, 14 tetragonal and 18 orthorhombic form lattices, with z_i values ranging from 6 to 11.

Hexagonal clusters. For each coat protein there are 10 hexamers and 20 trimers with axial point symmetries 32 and 3, respectively. The lattice parameters a, c of the hexagonal form lattices observed are in integral relation with the hexagonal radius r_h of the triacontahedron of the capsid projected along the threefold axis and of the height H of the hexamer, respectively,

$$(17) \qquad na = r_h = \frac{2\tau}{\sqrt{3}} a_0, \quad mc = H = \frac{m}{n} \gamma r_h, \quad \text{integers} \quad m, n,$$

in which a_0 is the icosahedral lattice parameter and $\gamma = c/a$ is the axial ratio. One finds a similar relation in terms of the height h of the trimer, which is also the height of a monomer.

In the particular case of the enclosing forms of the PV1 hexamer shown in Figure 11-10, one finds $n = m = 4$ and $H = r_h$; the corresponding form lattice becomes isometric hexagonal. The relation between this lattice and the external boundary is the same as for the R-phycoerythrin of Figure 11-3 and in two dimensions for the facet-like snow crystal of Figure 11-1.

For all hexagonal clusters the form lattices are integral. Of 40 hexameric clusters, 27 form lattices are rationally equivalent with the isometric lattice $1\text{-}\Lambda_{hex}$, seven with the lattice $\sqrt{3}\text{-}\Lambda_{hex}$ and six with the lattice $\sqrt{2}\text{-}\Lambda_{hex}$.

Pentagonal clusters. For each coat protein one has six decamers with symmetry 52 and 12 pentamers with symmetry 5 about the fivefold axis. The enclosing

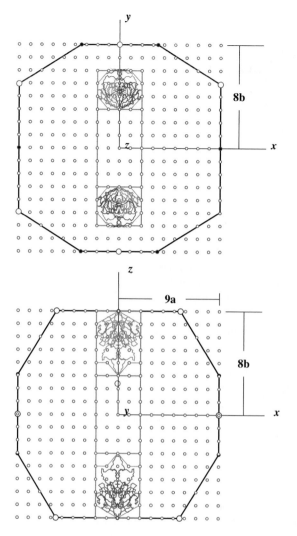

Figure 11-9. The cluster with 222 symmetry of four VP1 coat proteins taken from the human rhinovirus is enclosed in molecular forms with vertices at points of an integral tetragonal lattice with lattice parameters $a = \tau a_0/9$, $b = c = \tau a_0/8$, in which a_0 is the icosahedral lattice parameter and τ the golden ratio (adapted from [27], courtesy IUCr)

forms of the clusters with pentagonal symmetry can be characterized in a way similar to that of the hexagonal case. One example should suffice to convince that these clusters have decagrammal planar symmetry and integral pentagonal form lattices $\Lambda_{pent}(a, c)$ obeying morphological rules similar to those in the example of cyclophilin. Figure 11-11 shows the indexed forms of a VP3 decamer.

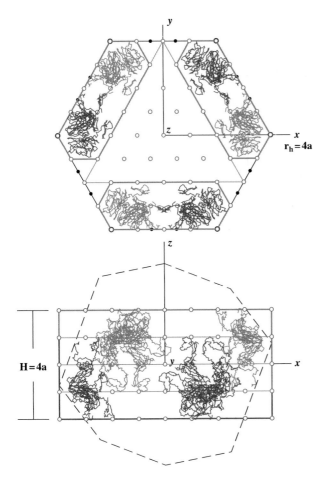

Figure 11-10. The cluster with symmetry 32 of a hexamer of the PV1 coat protein in the human rhinovirus has molecular forms with vertices at points of an isometric hexagonal lattice with parameters $a = c = r_h/4$ at which r_h is the hexagonal radius of the capsid projected along the threefold axis. The boundary of the capsid projected along the y-axis is indicated with dashed lines

The axial lattice parameter c is given by the height h of the monomer. The planar lattice parameter a is equal to $\tau^2 r_d$ with r_d the decagonal radius of the capsid. The height H of the decamer is $8c$ and equal to $a/\sqrt{2}$. The decagrammal scaling $S_{\{10/3\}}$ with scaling factor τ leaves the pentagonal lattice invariant; the lattice parameter $a' = r_d$ becomes equivalent to a, and $\Lambda_{pent}(a, c)$ is an integral lattice, rationally equivalent to $\sqrt{2}$-Λ_{pent}. Of the 24 decameric clusters with 52 symmetry, 22 have a form lattice rationally equivalent to the isometric pentagonal lattice 1–Λ_{pent}, one is equivalent to $\sqrt{2}$-Λ_{pent} and one to $\sqrt{3}$–Λ_{pent}.

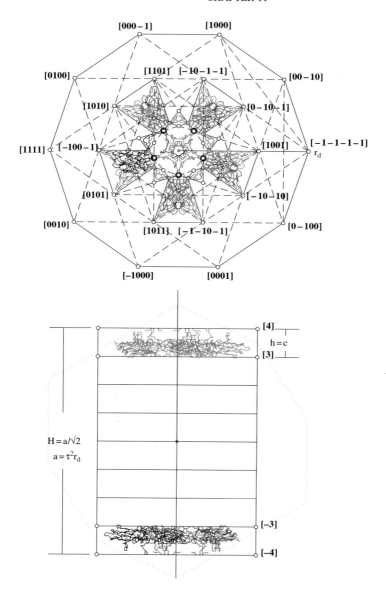

Figure 11-11. The VP3 decamer selected from the human rhinovirus has molecular forms with vertices at the pentagonal integer lattice with axial ratio $\sqrt{2}$. In the upper view along the fivefold axis only the planar indices are indicated, together with decagrammal scaling relations (dashed lines), in which r_d is the decagonal radius of the capsid projected along the fivefold axis. In the lower part, axial indices are given. In the case presented, the height H of the decamer is eight times the height h of the monomer (from [27], courtesy IUCr)

CONCLUSION

Condensed matter is not necessarily crystalline. In an analogous way one might not expect all biomacromolecules to have crystallographic properties, but there are many that possess a crystallographic structure, at least at the morphological level considered. In several cases it has been shown that these properties extend to a selected number of atomic positions [8, 16, 19, 29, 30]. In the molecular case the role of the missing lattice translations is taken over by lattice-invariant scalings. The crystallographic scale-rotation groups for molecules, analogous to space groups for crystals, have not yet been derived; this condition is a severe handicap to progress in this field.

The present molecular investigation has revealed unknown crystal properties that occur in nature (such as the isometric hexagonal crystals) but that are not yet taken into account in crystallography. To mention mysterious crystallography is a way to express that there are missing theoretical elements in the present understanding of the laws for aggregates of atoms and of molecules. Only considering the whole (periodic and aperiodic crystals together with molecules), one can hope to arrive at comprehensive crystallographic laws shifting further forward the frontiers of mystery.

REFERENCES

1. A. Janner, Phys. Rev. B, 43 (1991) 13206–13214.
2. A. Janner, Acta Cryst. A, 53 (1997) 615–631.
3. H. Ke, D. Mayrose, P.J. Belshaw, D.G. Alberg, S.L. Schreiber, Z.Y. Chang, F.A. Etzkorn, S. Ho and Ch.T. Walsh, Structure, 2 (1994) 33–44.
4. F. Klein, Vorlesungen über die Entwicklung der Mathematik im 19. Jahrhundert, Chelsea, New York, 1956, p. 344.
5. G. Friedel, Leçons de cristallographie, Hermann, Paris, 1911, p. 4 and p. 23.
6. A. Janner, Acta Cryst. A, 57 (2001) 378–388.
7. T. Janssen, Physics Reports, 168 (1988) 55–113.
8. A. Janner, Crystal Engineering, 4 (2001) 119–129.
9. A. Janner, Acta Cryst. A, 60 (2004) 198–200.
10. R. de Gelder and A. Janner, Acta Cryst. B, 61 (2005) 287–295.
11. R. de Gelder and A. Janner, Acta Cryst. B, 61 (2005) 296–303.
12. A. Janner, Acta Cryst. A, 58 (2002) 334–345.
13. W.A. Bentley and W.J. Humphreys, Snow Crystals, McGraw-Hill, 1931, reprinted by Dover, New York, 1962.
14. W. Chang, T. Jiang, Z. Wan, J. Zhang, Z. Yang and D. Liang, J. Mol. Biol., 262 (1996) 721–731.
15. A. Janner, Acta Cryst. D, 61 (2005) 247–255.
16. A. Janner, Acta Cryst. D, 61 (2005) 256–268.
17. A. Janner, Acta Cryst. A, 48 (1992) 884–901.
18. A. Yamamoto, Acta Cryst. A, 52 (1996) 509–560.
19. A. Janner, Acta Cryst. D, 61 (2005) 269–277.
20. S. Thore, C. Mayer, C. Sauter, S. Weeks and D. Suck, J. Biol. Chem., 278 (2003) 1239–1247.
21. A. Janner, Acta Cryst. D, 59 (2003) 783–794.
22. A.T. Hadfield, W. Lee, R. Zhao, M.A. Oliveira, I. Minor, R.R. Rueckert and M.G. Rossmann, Structure, 5 (1997) 427–441.
23. E. Arnold and M.G. Rossmann, J. Mol. Biol., 211 (1990) 763–801.

24. R. Zhao, D.C. Pevear, M.J. Kremer, V.L. Giranda, J.A. Kofron, R.J. Kuhn and M.G. Rossmann, Structure, 4 (1996) 1205–1220.
25. N. Verdaguer, D. Blaas and I. Fita, J. Mol. Biol., 300 (2000) 1179–1194.
26. S. Kim, T.J. Smith, M.S. Chapman, M.G. Rossmann, D.C. Pevear, F.J. Dutko, P.J. Felock, G.D. Diana and M.A. McKinlay, J. Mol. Biol., 210 (1989) 91–111.
27. A. Janner, Acta Cryst. A, 62 (2006) 270–286.
28. A. Janner, Acta Cryst. A, 62 (2006) 319–330.
29. A. Janner, Z. Kristallogr., 217 (2002) 408–414.
30. A. Janner, Acta Cryst. D, 59 (2003) 795–808.

CHAPTER 12

CLUSTERS IN F-PHASE ICOSAHEDRAL QUASICRYSTALS

ZORKA PAPADOPOLOS, OLIVER GRÖNING AND ROLAND WIDMER

INTRODUCTION

All known stable quasicrystals are either icosahedral or decagonal. The experimentally best studied quasicrystal is the large monograin of icosahedral Al-Pd-Mn, ($Al_{70}Pd_{21}Mn_9$) [1], followed by icosahedral Al-Cu-Fe ($Al_{62}Cu_{25.5}Fe_{12.5}$) [2]. Both quasicrystals belong to the icosahedral F-phase.

We studied the structure of the bulk terminations according to a model $\mathcal{M}(\mathcal{T}^{*(2F)})$ of icosahedral quasicrystals of an F-phase (see Refs. [3–5]). The model in a physical space is based on an icosahedral tiling [6] $\mathcal{T}^{*(2F)}$, projected from the D_6, six-dimensional root lattice [7]. See the next two sections. The bulk terminations of quasicrystals we define [8–10] in the fourth section by generalising the Bravais rule [11], obeyed by almost all crystals [12], to quasicrystals as well.

A surface orthogonal to a symmetry axis of a quasicrystal is obtained by a cut of a monograin orthogonal to the axis. The *clean surface* appears as a result of polishing the surface with a diamond paste, bombarding it in an ultrahigh vacuum (UHV) with argon ions (sputtering), and finally by heating it several times to nearly the melting point (annealing). The most stable clean surfaces of icosahedral Al-Pd-Mn (i-AlPdMn) and of icosahedral Al-Cu-Fe (i-AlCuFe), which are the fivefold followed by the twofold surfaces [13], are *terrace-stepped*. Under the assumptions that the clean surfaces are not reconstructed, a sequence of terraces offers insight into the single crystal on various levels along the symmetry axis, presented in the fourth section. In the fifth and final section we compare the highly resolved images, from a scanning tunnelling microscope (STM), of the stable surfaces to their simulations [14] on the *bulk terminations* defined as in Ref. [10].

255

J.C.A. Boeyens and J.F. Ogilvie (eds.), Models, Mysteries and Magic of Molecules, pp. 255–281.
© 2008 *Springer*.

Among other results, we conclude that the Bergman [15] and Mackay [16] polyhedra, which perform a decoration on the icosahedral tiling $\mathcal{T}^{*(2F)}$ in the model $\mathcal{M}(\mathcal{T}^{*(2F)})$ can not be energetically stable constituents, i.e. the *clusters*, because these are cut by both the fivefold and the twofold stable surfaces.

In the i-AlPdMn compound, 70% of the atomic positions are occupied by Al atoms. The i-AlCuFe [2], with 62% Al atoms is described with the same model of atomic positions (see Refs. [17], [4], [3]), and probably some other F-phase quasicrystals. There also exist STM images of the fivefold surfaces of i-AlCuFe. Both the large scale STM images [18] and the highly resolved STM images [19] that show the local atomic configurations on the surfaces are similar to the corresponding images of i-AlPdMn, presented in "Surfaces represented by the bulk terminations, Bravais' rule" and "Clusters on bulk terminations". The simulations of the STM images of i-AlPdMn that we present are made under the assumptions that all atoms in a termination show the same topographic contrast regardless of their chemical nature; these are hence comparable to the STM images of i-AlCuFe. The striking similarity of the STM simulations to the real images is due to the geometry of the model alone. The information provides us even with some hints important for the chemistry of the model.

ICOSAHEDRAL TILING $\mathcal{T}^{*(2\mathbf{F})}$ PROJECTED FROM THE $\mathbf{D_6}$ ROOT-LATTICE

Both quasicrystals icosahedral Al-Pd-Mn (i-AlPdMn) and Al-Cu-Fe (i-AlCuFe) that belong to the F-phase are best modelled through a single model of the atomic positions that we introduce in this section.

The first step is to define an icosahedral F-phase tiling. A quasiperiodic tiling and related quasilattice of the vertex points is defined with a *module* and a *window*. A synonym for the *window* is an *atomic surface*. When we refer to an F-phase, we mean that the module M_F is obtained as the icosahedrally projected D_6 root lattice into $\mathbb{E}_3$. The diagram of D_6 presented in Figure 12-1 (*left*) is the extended Dynkin diagram. Each simple root (a basis vector) is marked with a nod. The two vectors related to nods connected with a single line subtend between them an angle $(2\pi)/3$. Each pair of unconnected nods subtend between them an angle $\pi/2$. With each simple root is associated a Weyl reflection in a hyperplane, passing the origin, perpendicular to the root. The six simple roots generate the Weyl group W of the lattice. The extending root $e_1 + e_2$, labelled with a solid (dark) circle, generates an affine reflection [7] in a hyperplane orthogonal to the root, passing the point $(e_1 + e_2)/2$. These seven hyperplanes bound a fundamental simplex S. A Voronoi domain $V(q = 0)$ is obtained with the Weyl group W acting on the fundamental simplex [6] S. A tiling of the space $\mathbb{E}_6$ with Voronoi cells centred at the lattice points D_6 labelled by q (or q_{D_6}) is generated either with translations (integer combinations of the simple roots) or with action of the introduced affine reflection on $V(q = 0)$.

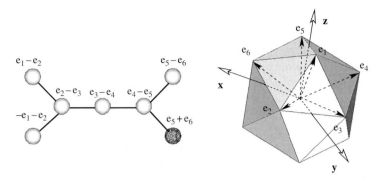

Figure 12-1. (*left*) Extended Dynkin diagram of the D_6 root lattice. With e_i, $i = 1, 2 \ldots 6$ we denote the standard $\mathbb{Z}_6$ basis $\{e_i \mid (e_i, e_j) = \delta_{i,j}, i, j = 1, \ldots 6\}$. (*right*) Embedding of the icosahedral group into the Weyl group of the D_6 lattice: an icosahedrally projected standard $\mathbb{Z}_6$ basis onto the observable (tiling) space, $\mathbb{E}_\parallel$; x-, y- and z-axes are the twofold symmetry axes of the icosahedron. Here (in the right picture), by e_i, $i = 1, 2 \ldots 6$ we mean the icosahedrally projected basis vectors $\{(e_i)_\parallel\}$, as is clear from the geometry of the image

We embed the icosahedral group into the Weyl group W, such that the icosahedral group acts as a signed permutation of the six fivefold axes, in both the three-dimensional tiling space $\mathbb{E}_\parallel$ and the three-dimensional coding space $\mathbb{E}_\perp$. The standard $\mathbb{Z}_6$ basis is projected into $\mathbb{E}_\parallel$ onto the six vectors along the fivefold symmetry axes (as in Figure 12-1 (*right*)), and also into $\mathbb{E}_\perp$. The coordinate axes (x, y and z) are chosen along the three twofold symmetry axes of an icosahedron

$$(1) \qquad \left(e_{1\parallel} \quad \cdots \quad e_{6\parallel} \right) = \frac{1}{\sqrt{2(\tau+2)}} \begin{pmatrix} 0 & 1 & -1 & -\tau & 0 & \tau \\ 1 & \tau & \tau & 0 & -1 & 0 \\ \tau & 0 & 0 & 1 & \tau & 1 \end{pmatrix}$$

$$(2) \qquad \left(e_{1\perp} \quad \cdots \quad e_{6\perp} \right) = \frac{1}{\sqrt{2(\tau+2)}} \begin{pmatrix} 0 & \tau & -\tau & 1 & 0 & -1 \\ \tau & -1 & -1 & 0 & -\tau & 0 \\ -1 & 0 & 0 & \tau & -1 & \tau \end{pmatrix}.$$

We choose a Voronoi cell V as a fundamental domain of D_6. When one hits a three-dimensional boundary of a Voronoi cell, by cutting $\mathbb{E}_6$ parallel to the observable space, one projects into the observable space $\mathbb{E}_\parallel$ the dual (three-dimensional) boundary. The result appears as a tile in $\mathbb{E}_\parallel$. The dual boundary is defined as a convex hull of all lattice points of which Voronoi domains contain the hit boundary. Instead of cutting the six-dimensional Voronoi cells, one can define a procedure on a *single* projected Voronoi cell $V(0)$ with a hierarchy of all its lower-dimensional boundaries into the orthogonal space $\mathbb{E}_\perp$, $V_\perp(0) \equiv W$, the *window* [6]. In case of D_6, the window W has an outer shape of a triacontahedron; see Figure 12-2 (*top*).

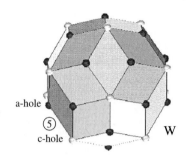

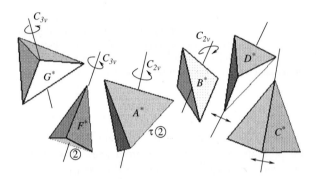

Figure 12-2. (*top*) The *window* of the tiling $\mathcal{T}^{*(2F)}$ is a triacontahedron. (*bottom*) The six tetrahedra are the tiles (Ref. [6]). The symbols ⑤ and ③ are the standard lengths: $③/\sqrt{3} = ⑤/\sqrt{\tau+2} = ②/2$ ($=1/\sqrt{2(\tau+2)}$), in which $\tau = (1+\sqrt{5})/2$. The standard distances are used in both the observable space $\mathbb{E}_{\parallel}$ and the coding space $\mathbb{E}_{\perp}$ parallel to the directions of the symmetry axes

The tiles of the observable space $\mathbb{E}_{\parallel}$ are the six golden tetrahedra on Figure 12-2 (*bottom*). This procedure is called *cut-and-project* (see Ref. [6] and references therein), and the resulting tiling is denoted by $\mathcal{T}^{*(2F)}$.

A *window* marks in six-dimensional space the points of the D_6 lattice that after the icosahedral projection become the quasilattice points, i.e., the window chooses from the dense module M_F in three-dimensional space a discrete set of points.

A MODEL OF ICOSAHEDRAL QUASICRYSTALS BASED ON THE TILING $\mathcal{T}^{*(2F)}$

The model of Katz and Gratias (i-AlCuFe) [2] and the model of Boudard (i-AlPdMn) [1] can be pictured as special realizations of a single model of the atomic positions [17].

Beginning with the quasilattice points of the tiling $\mathcal{T}^{*(2F)}$, which are all in the class $q \equiv q_{D_6}$, we attain the model $\mathcal{M}(\mathcal{T}^{*(2F)})$ that contains three quasilattices of types q, b and a (see Table 12-1).

Table 12-1. Points of D_6^ω lattice, reciprocal to D_6, split into four classes with respect to the D_6 translations. The symbol e denotes an even integer and o an odd one. N.B.: we use the "five-fold coordinate system", see Figure 12-1 (*right*). The authors in Refs. [2], [1], [17] use the "threefold coordinate system", equivalent to the basis $\{e_1, e_2, e_3, e_4, -e_5, e_6\}$. All these basis vectors, as in Figure 12-1 (*right*) are projected in $\mathbb{E}_\parallel$, $(e_i)_\parallel$

class-criterion	class	Ref.[2]	Refs.[1, 17]
$\frac{1}{2}(e_1, \ldots, e_6)$; $\frac{1}{2}\sum_i e_i = e$	q_{D_6}	n'	n_0
$\frac{1}{2}(e_1, \ldots, e_6)$; $\frac{1}{2}\sum_i e_i = o$	b	n	n_1
$\frac{1}{2}(o_1, \ldots, o_6)$; $\frac{1}{2}\sum_i o_i = o$	a	bc	bc_1
$\frac{1}{2}(o_1, \ldots, o_6)$; $\frac{1}{2}\sum_i o_i = e$	c		

To specify the scale in the model, we use the standard distances, denoted ⑤, ②
and ③ along the fivefold, twofold and threefold axes respectively, which are related
by $③/\sqrt{3} = ⑤/\sqrt{\tau+2} = ②/2$ $(=1/\sqrt{2(\tau+2)})$, in which $\tau = (1+\sqrt{5})/2$. The standard distances are used in both the observable space $\mathbb{E}_\parallel$ and the coding space $\mathbb{E}_\perp$.
The lines and intervals along the fivefold directions, if marked in colour, are in all
figures red, along twofold blue, and along threefold yellow. The standard distance
⑤$(=1/\sqrt{2})$ in $\mathbb{E}_\parallel$ is set to be 4.561 Å for i-AlPdMn and 4.465 Å for i-AlCuFe.

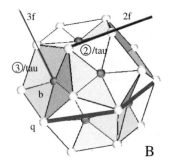
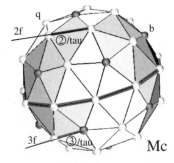

Figure 12-3. (*left*) A Bergman (B) and (*right*) a Mackay (Mc) polyhedron as an atomic decoration of
the icosahedral tiling $\mathcal{T}^{*(2F)}$ in the model $\mathcal{M}(\mathcal{T}^{*(2F)})$ (see Ref.[3]). The standard length along the twofold
directions is ②, $\tau^{-1}② = 2.96$ Å, and along the threefold direction is ③, $\tau^{-1}③ = 2.57$ Å, $\tau = (\sqrt{5}+1)/2$.
By convention, twofold directions are presented in blue, threefold in yellow. A cut of B with a fivefold
plane from Figure 12-18 is a pentagon, of edge length ②$=4.77$ Å and a cut of Mc with the same plane
is a decagon, of edge length $\tau^{-1}② = 2.96$ Å. Both polygons are marked with dark blue thick lines (see
color plate section)

The model is obtained on decorating every second vertex (of type a) in the tiling $\tau\mathcal{T}^{(P)}$ (along the edges) by a Bergman polyhedron from Figure 12-3 (*left*), as shown in Figure 12-4. The Mackay polyhedra, Figure 12-3 (*right*), appear centred on almost all other vertex positions of type q of the tiling $\tau\mathcal{T}^{(P)}$; i.e. the model resembles a chess board of Bergman and Mackay polyhedra.

We know that the primitive tiling [20], $\mathcal{T}^{(P)}$ of $\mathbb{E}_\parallel$, by the acute and obtuse rhombohedra can be locally derived [6] from the tiling $\mathcal{T}^{*(2F)}$, as indicated in Figure 12-5.

This global rule for decoration of the tilings $\mathcal{T}^{(P)}$ and $\mathcal{T}^{*(2F)}$ by Bergman and Mackay polyhedra becomes reduced to the representative tiles of the coexisting tilings with either acute and obtuse rhombohedra ($\tau\mathcal{T}^{(P)}$) or the inscribed tetrahedra G* and F* ($\tau\mathcal{T}^{*(2F)}$), as in Figures 12-6 and 12-7.

For each representative position on the tiles in $\mathbb{E}_\parallel$ one finds a representative polyhedron in the orthogonal $\mathbb{E}_\perp$ coding space (labelled with the same numbers, respectively); in this way one systematically generates the three windows for the three translational classes of the points with respect to D_6. These three classes are labelled with q, b and a; see Table 12-1. The result for two windows W_q and W_b is presented in Figures 12-6 and 12-7 respectively. See also Ref. [4].

The third window W_a, for the a positions, which are the centres of the Bergman polyhedra, is a triacontahedron of edge length τ^{-1}⑤, see Figure 12-9. Both presented W_q and W_b windows in Figures 12-6 and 12-7 have some fixed, deterministic points and some probabilistic positions [4]. To obtain a completely deterministic model, we require that the minimal atomic distances along the fivefold, twofold and threefold directions exist and have lengths τ^{-1}⑤, τ^{-1}② and τ^{-1}③, respectively. The second requirement is that the positions are in accordance with arrows of the inflation rules [21] on the tiles of the tiling $\tau\mathcal{T}^{*(2F)}$, as pointed out in Figure 12-8. See also [5].

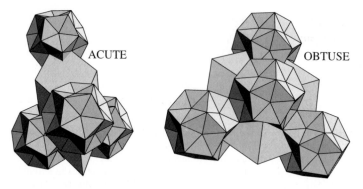

Figure 12-4. One centres the Bergman polyhedra at the a-class of points contained into the quasilattice of the primitive tiling [20] $\tau\mathcal{T}^{(P)}$

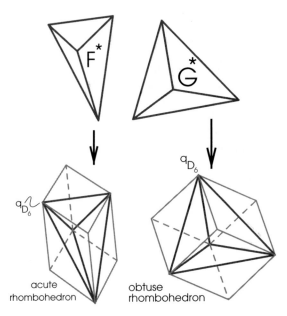

Figure 12-5. The tiling $\mathcal{T}^{(P)}$ can be derived from the F-phase tiling $\mathcal{T}^{*(2F)}$ upon replacing the two tetrahedra F* and G* by the acute and obtuse rhombohedra [6] respectively. By convention, fivefold directions are in red, twofold in blue (see color plate section)

The final shape of the three windows for the model with deterministic positions W_q, W_a and W_b is as in Figure 12-9.

SURFACES REPRESENTED BY THE BULK TERMINATIONS, BRAVAIS' RULE

Atomically well defined surfaces for the experimental investigations on i-AlPdMn with a scanning tunneling microscope (STM) were prepared from large ($8 \times 8 \times 2$ mm^3) monograin crystals oriented and cut orthogonal to the corresponding high symmetry axis. The surface is polished to a mirror-like finish with diamond paste down to a grain size 0.1 micrometre. A surface is prepared in an ultrahigh vacuum with repeated cycles of argon-ion sputtering and annealing to temperatures near the melting point of the crystal. After the preparation one images the *clean* surface with the STM. A consensus has emerged that the *clean surface* is itself quasicrystalline, i.e. as in the bulk model. Before imaging, each surface is tested with low-energy electron diffraction (LEED).

The model is a superposition of three icosahedral quasilattices, q, a and b, of atomic positions in the physical space $\mathbb{E}_{\parallel}$, as explained in "A model of icosahedral quasicrystals based on the tiling $\mathcal{T}^{*(2F)}$". These are defined in the Table 12-1.

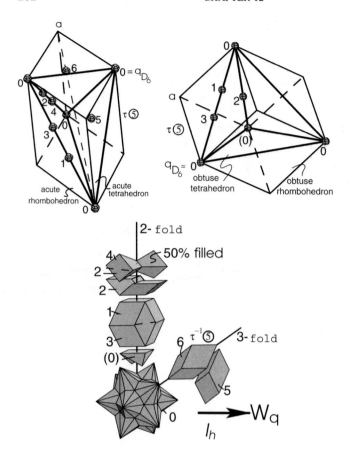

Figure 12-6. (*top-left*) Decoration in $\mathbb{E}_{\parallel}$ of the acute tetrahedron F^* (inscribed into the acute rhombohedron) by representative atomic positions in class q_{D_6}. (*top-right*) Decoration in $\mathbb{E}_{\parallel}$ of the obtuse tetrahedron G^* (inscribed into the obtuse rhombohedron) by representative atomic positions in class q_{D_6}. (*bottom*) The corresponding representative coding polyhedra in $\mathbb{E}_{\perp}$ for atomic positions in class q_{D_6}. After the icosahedral group I_h is applied, one attains a *window* W_q with some positions being probabilistic

Each plane orthogonal to a threefold or fivefold axis contains exclusively the points of a single quasilattice, q, a or b, but each plane orthogonal to a twofold axis might contain points of all three quasilattices, as shown in Table 12-2.

As described in Ref. [8] and references cited therein, there is a *coding space*, $\mathbb{E}_{\perp}$, containing three *windows*, W_q, W_a and W_b (shown in Figure 12-9), and a *∗-map* (effected by changing τ to $-1/\tau$ everywhere) [22] that takes each point of one quasilattice into a point of the module M_F in the corresponding *window*. Conversely (also under the *∗-map*), the module points in the *windows* produce all atomic positions and *define* the model $\mathcal{M}(\mathcal{T}^{*(2F)})$. The *∗-map* is not continuous: it

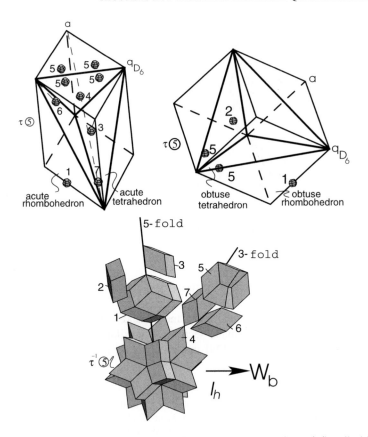

Figure 12-7. (*top-left*) Decoration in $\mathbb{E}_{\parallel}$ of the acute tetrahedron F^* (inscribed into the acute rhombohedron) by representative atomic positions in class *b*. (*top-right*) Decoration in $\mathbb{E}_{\parallel}$ of the obtuse tetrahedron G^* (inscribed into the obtuse rhombohedron) by representative atomic positions in class *b*. (*bottom*) The corresponding representative coding polyhedra in $\mathbb{E}_{\perp}$ for atomic positions in class *b*. After the icosahedral group I_h is applied, one attains a *window* W_b with some positions being probabilistic. The positions labelled with numbers 5, 6 and 7 are probabilistic [4]

maps a discrete unbounded quasilattice to a dense point set bounded with a *window*. It does, however, map lines and planes in physical space $\mathbb{E}_{\parallel}$ to lines and planes in coding space $\mathbb{E}_{\perp}$ and preserves orthogonality. It is also reversible from $\mathbb{E}_{\perp}$ to $\mathbb{E}_{\parallel}$. The atomic positions on a given plane P, orthogonal to a chosen axis $z_{\parallel}$ (five-, three- or twofold along $z_{\parallel}$), that belong to a given class h ($= q$, a or b) arise as the inverse images under the $*$-map of the points of M_F in the intersection of the window W_h with the image plane P^*, which is orthogonal to $z_{\perp}$ in $\mathbb{E}_{\perp}$ (see Ref. [23] and, in particular, Figure 12 in Ref. [8]). Hence the atomic density function $\rho(z_{\parallel})$ of planes orthogonal to a given axis $z_{\parallel}$, which is an erratic discrete function on the physical space axis $z_{\parallel}$, is a continuous function (which we also designate $\rho(z_{\perp})$) on the coding space axis $z_{\perp}$, and can be graphed as in Figures 12-11, 12-13.

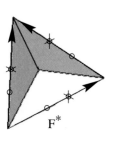

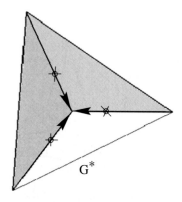

Figure 12-8. The choice of atomic positions respecting the directions of the inflation arrows [21] attached to the edges of the tiles $F_\parallel^*$ and $G_\parallel^*$ lead to the final form of the window $W_{q_{D_6}}$

The atomic densities of terminations are thus most conveniently calculated and visualized in $\mathbb{E}_\perp$.

For ordinary lattices the density of points in a plane depends on only the orientation of the plane, but the density of a plane section of quasilattices q, a, b is a product of two factors: the *module factor* [24] that depends on the orientation of

Table 12-2. The atomic positions $x = \frac{1}{2}(n_1, \ldots, n_6)$ in a single fivefold or threefold plane belong to a single class, but twofold planes might contain atomic positions of all classes. A unit vector normal to an ifold plane ($i = 5$, 3 or 2) is denoted with $n_\parallel^i$. The symbol e denotes an even integer and o an odd one. The scalar products are given in the units $\kappa_3 = ③/3$, $\kappa_5 = ⑤/\sqrt{5}$, $\kappa_2 = ②/4$, in which $⑤/\sqrt{\tau+2} = ③/\sqrt{3} = ②/2 = 1/\sqrt{2(\tau+2)}$. We use a fivefold coordinate system of six unit orthogonal basis vectors projected icosahedrally, see Figure 12-1 (*right*). (Some authors working with similar models use a threefold coordinate system [1, 2, 17].)

Class criterion	Class	$n_\parallel^3 \cdot x_\parallel [\kappa_3]$	$n_\parallel^5 \cdot x_\parallel [\kappa_5]$	$n_\parallel^2 \cdot x_\parallel [\kappa_2]$
$\frac{1}{2}(e_1, \ldots, e_6)$; $\quad \frac{1}{2}\sum_i e_i = e$	q_{D_6}	$e + e\tau$	$e + e\tau$	$e + e\tau$
$\frac{1}{2}(e_1, \ldots, e_6)$; $\quad \frac{1}{2}\sum_i e_i = o$	b	$o + o\tau$	$o + e\tau$	$e + e\tau$
$\frac{1}{2}(o_1, \ldots, o_6)$; $\quad \frac{1}{2}\sum_i o_i = o$	a	$o + e\tau$	$e + o\tau$	$e + e\tau$
$\frac{1}{2}(o_1, \ldots, o_6)$; $\quad \frac{1}{2}\sum_i o_i = e$	c	$e + o\tau$	$o + o\tau$	$e + e\tau$

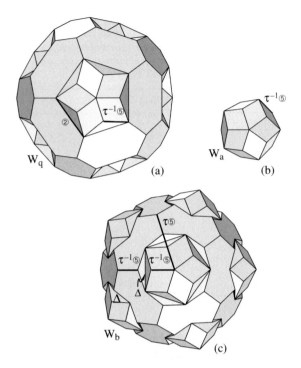

Figure 12-9. The final windows W_q, W_a, W_b in the coding space $\mathbb{E}_\perp$; they define the deterministic geometric model $\mathcal{M}(\mathcal{T}^{*(2F)})$ of atomic positions based on the icosahedral D_6 module M_F. The model $\mathcal{M}(\mathcal{T}^{*(2F)})$ describes both i-AlPdMn and i-AlCuFe. (a) W_q with edge lengths $\tau^{-1}\text{⑤}$ and $\text{②} = 2\text{⑤}/\sqrt{\tau+2}$. (b) W_a is a triacontahedron of edge length $\tau^{-1}\text{⑤}$. (c) W_b is obtained on taking the marked tetrahedra ($\blacktriangle$) away from the triacontahedron of edge length $\tau\text{⑤}$. The tetrahedron $\blacktriangle$ has two mirror symmetry planes and edges of lengths $\tau^{-1}\text{⑤}$, $\tau^{-2}\text{⑤}$ and ②. The windows fulfil the closeness condition: i.e. there are no forbidden (short) distances in the model. The volumes of the windows are in a proportion $Vol(W_b) : Vol(W_q) : Vol(W_a) = (6\tau+8) : (8\tau+2) : 1$, which defines the relative frequencies of the b, q and a sites in $\mathcal{M}(\mathcal{T}^{*(2F)})$

the plane, (see Row 8 in Table 12-3) and the *window factor* that is the area of the section of the window by the $*$-mapped plane in coding space.

Row 9 of Table 12-3 presents the maximum density of planes in the main symmetry directions. As already noted in Ref. [13], there are twofold planes denser than the densest fivefold or threefold planes even though experimental evidence indicates that the fivefold sputtered and annealed surfaces are the most stable. In the light of this condition we propose a modification to the Bravais rule to take into account the *layers of planes* orthogonal to the main symmetry directions.

Fivefold Terminations

On a large scale (>100 nm), the fivefold surfaces (orthogonal to the fivefold symmetry axis) present sequences of flat terraces with characteristic terrace step heights

Table 12-3. $\mathcal{M}(\mathcal{T}^{*(2F)})$ data of i-AlPdMn. Row 1: shortest interatomic distances parallel to the axis $(③/\sqrt{3} = ⑤/\sqrt{\tau+2} = ②/2$; the standard distance ⑤ for i-AlPdMn is 4.561 Å). Rows 2–4: three shortest interplanar separations orthogonal to the axis. Rows 5–7: other interplanar separations of neighboring planes. Row 8: D_6 module factor. Row 9: maximum absolute atomic density of planes. The corresponding data for i-AlCuFe are obtained on setting the standard length ⑤ to 4.465 Å

	5fold	2fold	3fold
shortest interatomic distances	$\tau^{-1}⑤$	$\tau^{-1}②$	$\tau^{-1}③$
	2.82 Å	2.96 Å	2.57 Å
interplanar distances: s	$\tau^{-3}⑤/(\tau+2)$	$\tau^{-3}②/2$	$\tau^{-4}③/3$
	0.30 Å	0.57 Å	0.20 Å
interplanar distances: $m = \tau s$	0.48 Å	0.92 Å	0.33 Å
interplanar distances: $l = \tau m$	0.78 Å	1.48 Å	0.53 Å
interplanar distance: $2m$			0.65 Å
interplanar distance: τl			0.86 Å
interplanar distance: $2l$	1.56 Å		
D_6 module factor	$1/\sqrt{5}$	$1/4$	$1/3$
Densest planes (absolute)	0.086 Å^{-2}	0.101 Å^{-2}	0.066 Å^{-2}

$m = 4.08$ Å and $l = 6.60$ Å (see Ref. [25], Ref. [8], and Refs. quoted therein). See Figure 12-10. The corresponding sequences were observed on fivefold surfaces of i-AlCuFe [18], as already discussed in Ref. [8].

Apart from the fivefold surfaces, the twofold surfaces were imaged. These surfaces also exhibit terraces, visible in Figures 12-14, 12-15, but the fivefold surfaces are the most stable, followed by the twofold surfaces, whereas the threefold surfaces become faceted, i.e. are breaking into islands that tend to orientate themselves differently [13].

Visible in Figure 12-10, the typical heights of terrace steps on the fivefold surface are 4.08 Å and 6.60 Å. From Figures 12-14 and 12-15 of the twofold surfaces the common heights are circa 6.3 Å and 10.2 Å. The typical heights on surfaces of both kinds are considerably larger than the minimum interplane distance in the corresponding directions, as listed in Table 12-3. Two questions arise. What are the structures in terms of atomic planes of the terraces? Which atomic plane terminates a fivefold or a twofold surface?

In Ref. [23] we sought to determine a criterion for an appearance of a fivefold surface. At that time it was generally expected [17] that the smaller, Bergman polyhedra in the model $\mathcal{M}(\mathcal{T}^{*(2F)})$ are energetically stable clusters. In our investigations we managed to implant a sequence of the terraces according to Schaub et al [25] such that the most dense fivefold layers of Bergman clusters were directly below the terraces, but we stated that some less dense layers of the Bergmans must be cut by the stable fivefold surfaces. Later, in [8, 9] we tried to formulate a variant of the Bravais rule for quasicrystals. According to the Bravais rule, which is generally valid for crystals, the most stable surfaces are the densest atomic planes in

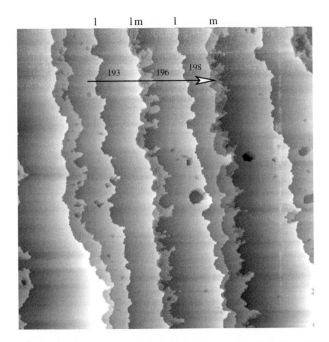

l   l m   l   m

193   196   198

Figure 12-10. An STM image of a fivefold surface, size $1750 \times 1750\,\text{nm}^2$, of i-AlPdMn. This image is taken from Ref. [9] (rotated by 180°), produced by J. Ledieu. In Ref. [26] a Fibonacci sequence of the step heights (circa) $m = 4.08\,\text{Å}$ and $l = 6.60\,\text{Å}$ was measured on the surface. We fix a marked subsequence of steps l, l, m, l, m from left–right, downwards, corresponding to a subsequence of the $qbbq$-terminations in $\mathcal{M}(\mathcal{T}^{*(2F)})$ from Figure 12-12 (*bottom*). Numbers 193, 196 and 198 label the tiling fivefold planes in the model

the bulk [11]. In [10] we observed that several Å thick atomic layers of (equal) maximum density appear in correct sequences as terrace-like sequences of the surfaces. This density is a *planar* density, averaged in a layer. We repeat here this consideration for the most stable fivefold surfaces.

We introduce a bundle of highly dense fivefold planes, or thin, plane-like layers (buckled planes). A fivefold termination can be considered to be a layer consisting of a (q, b) layer and a (b, q) layer, each containing two planes with a spacing $0.48\,\text{Å}$; see the right side of Figure 12-11 (*bottom*). The (q, b) and the (b, q) layers are *thin layers* or *buckled planes*. The *thick* layer that contains four planes with spacings q_1-plane, $0.48\,\text{Å}$, b_1-plane, $1.56\,\text{Å}$, b_2-plane, $0.48\,\text{Å}$, q_2-plane is a candidate for a fivefold termination. For a bundle we define an effective (averaged) density of internally contained thin layers[1]/planes

$$(3) \qquad \rho_{5f}(z_\perp) = (\rho_{q_1}(z_\perp) + \rho_{b_1}(z_\perp))/2 + (\rho_{q_2}(z_\perp) + \rho_{b_2}(z_\perp))/2.$$

---

[1]Under a *thin layer* we consider a layer of 2–3 planes of stacked atoms, on a distance significantly smaller than $0.86\,\text{Å}$.

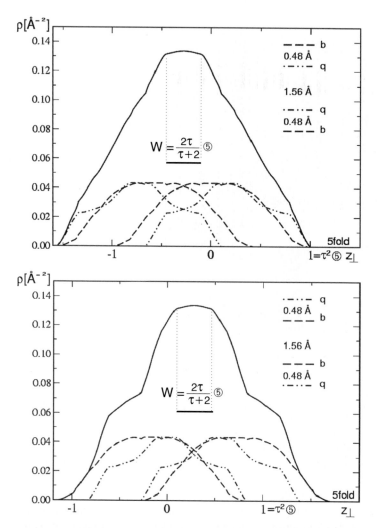

Figure 12-11. (*top*) Density graph $\rho_{5f}(z_\perp)$ of the fivefold (b, q, q, b) layers, with spacings as in the image. The symbol ⑤ is the standard distance along a fivefold axis. (*bottom*) Density graph $\rho_{5f}(z_\perp)$ of the fivefold (q, b, b, q) layers, with spacings as in the image. The supports of the plateaus in cases (*top*) and (*bottom*) are equally broad W, and encode the Fibonacci sequence with step heights $l = 6.60\,\text{Å}$ and $L = m + l = 10.68\,\text{Å}$. The plateaus are equally high, i.e. the terminations are equally dense, but the density graph (*top*) is slightly steeper in the region $\tau W/W$ than in case (*bottom*)

As stated already, whereas the density of the bundle independent of its position $(z_\parallel)$ in the bulk along the symmetry axis in $\mathbb{E}_\parallel$ is *not a function* (because, from a plane to a plane in a sequence, the density $\rho(z_\parallel)$ increases discontinuously in its value), in the coding space $\mathbb{E}_\perp$ it *is a function* [9] of $z_\perp$, the smooth density

bqqb-terminations

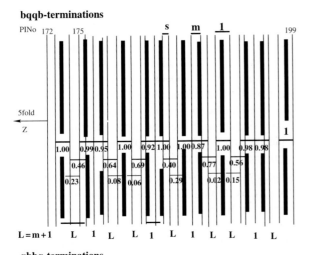

qbbq-terminations

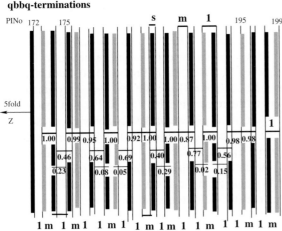

Figure 12-12. In the finite segment of $\mathcal{M}(\mathcal{T}^{*(2F)})$ thin lines mark the fivefold tiling $\mathcal{T}^{*(2F)}$ planes, labelled from 172 to 199. Plane No 175 is at the position $z = 50.0000$ Å in [3], No 199 on $z = -47.0741$ Å. The distances between the tiling planes are $s = 2.52$ Å, $m = \tau s$ and $l = \tau m$, $\tau = (1 + \sqrt{5})/2$. The fivefold layers of 6.60 Å broad Bergmans are marked by the l-intervals. Below each is its relative density. (*top*) The Fibonacci sequence $\{l, L = \tau l = l + m\}$ of the (b,q,q,b)-terminations presented relative to the sequence of the fivefold planes of the tiling. (*bottom*) The Fibonacci sequence $\{m, l\}$ of the (q,b,b,q)-terminations presented relative to the sequence of the fivefold planes of the tiling. Possible terminations marked in gray (pink) are less dense; hence we expect them to appear as smaller terraces on a surface (see Figure 12-10)

graph $\rho(z_\perp)$. In Figure 12-11 (*bottom*) is presented the density graph of the averaged planar density $\rho_{5f}(z_\perp)$ in the terminating layer of planes as in expression (3). The area below the plateau along $z_\perp$, the support of the plateau, marks the layers with equal maximum densities, the **terminations**. Each module point in the support along the fivefold symmetry axes $z_\perp$ in $\mathbb{E}_\perp$ corresponds one to one to

$z_\parallel$, a position of the penetration of a fivefold symmetry axis (in $\mathbb{E}_\parallel$) into a single terminating layer.

One can also bundle the dense (b, q) and (q, b) plane-like layers into a bundle (b, q, q, b), see Figure 12-11 (*top*). Such a layer contains five planes with spacings b_1-plane, 0.48 Å, q_1-plane, 0.78 Å, a-plane, 0.78 Å, q_2-plane, 0.48 Å, b_2-plane. The a-plane, in the middle, has a density (<0.013 Å^{-2}), negligible compared to the densities of the surrounding plane-like layers in the $(b, q, (a), q, b)$ layer. Neglecting the a-plane we hence define an effective density as in the case of the (q, b, b, q) layers.

The height and width of the plateaus on the density graphs have the same size, the width is $\frac{2\tau}{\tau+2}$ ⑤ (compare (*top*) and (*bottom*) of Figure 12-11) and encode the Fibonacci sequence of terrace heights $l = \frac{2\tau^2}{\tau+2}$ ⑤ $= 6.60$ Å and $L = \tau l = 10.68$ Å. But the density graph of the layer (b, q, q, b) (Figure 12-11 (*top*)) is slightly steeper in the region which is the complement of W in τW ($\tau W/W$) than the graph of the layer (q, b, b, q) (Figure 12-11 (*bottom*)) and causes that the appearance of the terrace height 4.08 Å is less probable to appear. To a terrace height 4.08 Å is related (either above or below) a (q,b,b,q) layer of large densities $\rho_{5f}(z_\perp)$, but *smaller* than the terminating one, 0.134 Å. The 4.08 Å terrace height was observed at even the highest annealing temperature 1070 K (see Ref.[25]), and is clearly detected in Figure 12-10, although to the 4.08 Å steps are related the smaller terraces [26], as proven for a short Fibonacci sequence $\{l, l, m, l, m\}$ in Figure 12-10. Terminations marked in gray (pink) in Figure 12-12 (*bottom*) are the less dense terminations, connected (from above or below) to the terrace heights 4.08 Å. For that reason we declared in Ref. [10] that the support of the plateau of the density graph in the case of the (q, b, b, q) atomic layers defines the sequence of fivefold bulk terminations. Another solution is not completely excluded. In Ref. [27], through analysis by x-ray photoelectron diffraction, the authors *confirm* the distances from the surface plane to the next to be $d_{12} = 0.48$ Å and from the first to the next buckled layer to be $d_{24} = 1.56$ Å, but it confirms equally well both solutions of the fivefold thick layer terminations, if the chemistry of the model were not given. We return to this problem in "Clusters on bulk terminations".

In Figure 12-12 we show the positions of the atomic layers 2.52 Å thick as candidates for the fivefold bulk terminations, relative to the positions of the layers of the 6.60 Å broad Bergman polyhedra in the model. Thin lines mark the tiling $\mathcal{T}^{*(2F)}$ planes orthogonal to the fivefold z-axis. On the finite segment of the image, the tiling planes are labelled from 172 to 199. The tiling planes contain only the points in the class q (see Tables 12-1 and 12-2). The fivefold planes of $\mathcal{T}^{*(2F)}$ are on the neighbouring distances $s = 2.52$ Å, $m = \tau s$ and $l = \tau m$, $\tau = (1 + \sqrt{5})/2$. The tiling $\mathcal{T}^{*(2F)}$ is decorated by the Bergman polyhedra. The fivefold layers of Bergmans, 6.60 Å $= l$ broad, are marked in the figure with the l-intervals. Below each l-interval is written the relative density of the corresponding layer of Bergmans in the model. The most dense layers have densities that equal unity.

According to Figure 12-12, the most stable fivefold terminations (independent of which candidate, (q, b, b, q) or (b, q, q, b), is the correct fivefold termination) in the model pass through the dense layers of the Bergman polyhedra; hence these can not be energetically stable clusters. The Mackay polyhedra can not be the stable clusters either, because of the "chess-board" nature of the decoration of the tiling $\mathcal{T}^{*(2F)}$ by Bergman and alternating Mackay polyhedra, and because the Mackay polyhedra are larger than the Bergmans: compare their size in Figure 12-3. Hence, the fivefold layers of the Mackay polyhedra are also cut by the fivefold terminations. This result, that the clusters in a stable quasicrystal are themselves unstable is in agreement with the general phenomenon observed in molecular crystallography, that stable molecules are located on an unstable lattice, and unstable (metastable) molecules (clusters) on a stable lattice. This condition was proven in particular in the case of 1,2,3,5-tetra-O-acetyl-beta-D-ribofuranose, see Ref. [28] and Refs. quoted therein, and similar examples are discussed elsewhere: see the recent article by Nangia et al [29].

The height of the plateau of the graph ρ_{5f} in Figure 12-11 (*top*)/(*bottom*) defines the densities of the terminations to be $0.134\,\text{Å}^{-2}$ (Table 12-4). Both solutions for the sequences of the fivefold terminations are presented in Figure 12-12. Whereas the volume density of the model (for i-AlPdMn) is $0.065\,\text{Å}^{-3}$ (averaged on the patch of size $100 \times 100 \times 100\,\text{Å}^3$), the *volume* density in the fivefold terminating layer, which is exactly $2.52\,\text{Å}$ broad, is $0.106\,\text{Å}^{-3}$.

Similar results are applicable for any icosahedral quasicrystal described with the $\mathcal{M}(\mathcal{T}^{*(2F)})$ model, in particular for i-AlCuFe [2].

Twofold Terminations

In the case of twofold terminations, we consider a layer of four atomic planes with spacings $(abq)_1$-plane, $1.48\,\text{Å}$, $(bq)_1$-plane, $0.92\,\text{Å}$, $(bq)_2$-plane, $1.48\,\text{Å}$, $(abq)_2$-plane. (abq)-plane implies that the twofold plane contains atomic positions of all classes, a, b and q, see Tables 12-1 and 12-2. (bq)-plane implies that the plane contains atomic positions of only classes a and b. For a bundle we define an effective (averaged) density of planes $\rho_{2f}(z_\perp)$

$$(4) \qquad \rho_{2f}(z_\perp) = (1/4)[\rho_{(abq)_1}(z_\perp) + \rho_{(bq)_1}(z_\perp) + \rho_{(bq)_2}(z_\perp) + \rho_{(abq)_2}(z_\perp)].$$

For this *thick*, $3.88\,\text{Å}$ broad twofold layer the maximum of $\rho_{2f}(z_\perp)$ is a perfectly flat plateau (Figure 12-13). The height of the plateau defines the effective density of terminations to be $0.086\,\text{Å}^{-2}$ (Table 12-4). The support of the width of the plateau equals $W = (1/2)②$ (see Figure 12-13) and encodes the Fibonacci sequence of twofold terminations with terrace heights $S = \frac{\tau^2}{2}② = 6.3\,\text{Å}$ and $L = \tau S = 10.2\,\text{Å}$.

The heights of the larger twofold terraces were measured to be $S = 6.2\,\text{Å}$ and $L = 9.5\,\text{Å}$ (see Figure 12-14(b)) in agreement with the predicted values. The twofold layers intertwining the terminations have densities not greater than $0.079\,\text{Å}^{-2}(<0.086\,\text{Å}^{-2})$. The pits within the big terraces (see Figure 12-14(b)) are

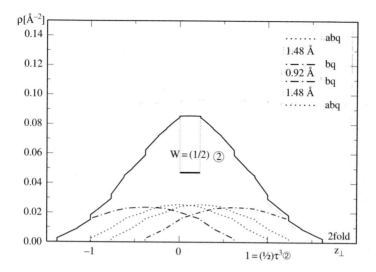

Figure 12-13. Density graph $\rho_{2f}(z_\perp)$ of the "thick" twofold layers, with spacings *abq*-plane, 1.48 Å, *bq*-plane, 0.92 Å, *bq*-plane, 1.48 Å, *abq*-plane. The plateau of the graph defines the terminations. The width of the support of the plateau equals exactly ②/2 and defines the Fibonacci sequence of twofold terminations with step heights $S = \frac{\tau^2}{2}② = 6.3\,\text{Å}$ and $L = \tau S = 10.2\,\text{Å}$

explained by the large distances between the atomic planes inside of the "thick" terminating layer, 1.5 Å, 2.4 Å and 3.9 Å (see Figure 12-13). These values satisfactorily reproduce the measured values 2.4 Å and 3.6 Å (see Figure 12-14(b)).

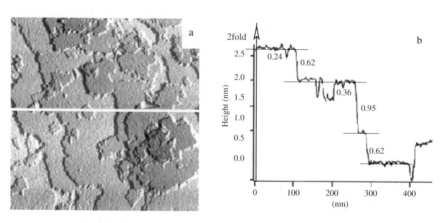

Figure 12-14. Image formed from V. Fournée [9, 30]: (a) STM image of a twofold terrace-stepped surface of i-AlPdMn, size $500 \times 500\,\text{nm}^2$. (b) Correlated measurements, height profile along the white line in (a) with step heights. The large terraces have heights 0.62 nm and 0.95 nm on which are superimposed smaller terraces (pits) with heights 0.24 nm and 0.36 nm

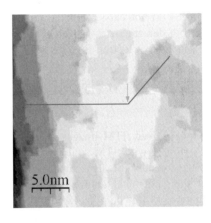
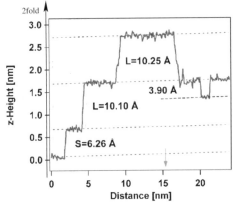

Figure 12-15. (*left*) 25 × 25 nm² STM image of a twofold terrace-stepped surface of i-AlPdMn. (*right*) Height profile measurement along the solid red line (N. B. the red line in this figure is NOT related to a fivefold direction) in the (*left*) image with the corresponding step heights. A single pit of 3.9 Å is detected (see color plate section)

In our STM image of a clean, terrace-like twofold surface, we measure the steps between terraces of the heights 6.26 Å, 10.10 Å and 10.25 Å (see Figure 12-15), in agreement with values predicted with the density graph. Our well prepared clean surface is not rich in pits, although the measured one in Figure 12-15 is 3.9 Å deep, exactly as predicted by the configuration of the *thick* twofold layer termination; see the right side of Figure 12-13.

The **threefold terminations** might be modelled as the *thick* layers of atomic planes in $\mathcal{M}(\mathcal{T}^{*(2F)})$, see Table 12-4, but, inspecting the intertwining threefold layers, between the "terminating" ones, we calculate that these have densities *comparable* to those of the "terminating" layers! We know also that the threefold surfaces facet readily [13], and some correlated STM measurements, (as those in

Table 12-4. Relative and absolute densities of planes and layer terminations orthogonal to five-, two- and threefold symmetry axes in $\mathcal{M}(\mathcal{T}^{*(2F)})$ of i-AlPdMn. The corresponding data for i-AlCuFe are similar

	5fold	2fold	3fold
Densest planes (abs.)	0.086 Å^{-2}	0.101 Å^{-2}	0.066 Å^{-2}
Densest thick layers (abs.)	0.134 Å^{-2}	0.086 Å^{-2}	0.058 Å^{-2}
Densest thick layers (rel.)	1	0.64	0.44

Figures 12-14 and 12-15) for the threefold surfaces do not exist. That the three-fold planes in i-AlPdMn show no strong variation of densities was proven also by M. Erbudak (ETH Zürich) and coworkers with secondary-electron imaging (SEI), see Figure 8(a) in Ref. [10].

CLUSTERS ON BULK TERMINATIONS

In what follows, we present our results on highly resolved STM images of clean surfaces of icosahedral $Al_{70}Pd_{21}Mn_9$, which we compare with STM simulations on the model $\mathcal{M}(\mathcal{T}^{*(2F)})$. In the STM simulations we consider the same topographic contrast for all atomic positions regardless of their chemical nature. Hence, if slightly scaled, these images present STM simulations of *any* compound described with the model of atomic positions $\mathcal{M}(\mathcal{T}^{*(2F)})$.

Scanning tunnelling microscopy is in principle a method extremely sensitive to a surface and probes the structure of the uppermost atomic layers of a sample. In the case of Al(111) the atomic plane density of the surface corresponds to 0.14 Atoms/$Å^2$ with a plane distance 2.25 Å in the (111) direction. In the case of the fivefold i-AlPdMn the situation is otherwise: we have typically a smaller plane density in the range 0.05 to 0.08 Atoms/$Å^2$; we neglect the planes of small density also present but irrelevant for our discussion. The reduced atomic-plane density results in a correspondingly reduced interplane spacing with typical values between 0.5 Å and 1.5 Å (see Table 12-3). In this respect a *compact* atomic plane is obtained only in the case of the fivefold i-AlPdMn surface if we consider two consecutive planes or more[2]. Compact here is understood to imply an atomic plane density comparable to the dense Al(111). Bearing this in mind the STM image of the fivefold i-AlPdMn surface is clearly given by two or more atomic planes. This condition can also be understood from the corrugation[3] about 1.5 Å in the experimental STM image of this surface. In comparison Al(111) shows a corrugation less than 0.5 Å depending on the tunneling conditions. In our simulations of the STM images we consider four to five atomic planes, which result in a total atomic plane density of order 0.3 Atoms/$Å^2$, which is roughly twice the value of the dense Al(111) surface. Considering more planes has no effect on the simulation of the STM image as they are masked by the top layers.

The STM simulations are based on a simplified atomic charge model [31], in which a spherical shape of the valence charge density is assumed for the atoms. These spherical shells of fixed radius (solid sphere) are superimposed in three dimensions for all atomic positions considered in the simulation. From this super-position the z(x,y) (z parallel to the plane normal) contour of the uppermost shells is

---

[2] A *compact plane*, a *thin layer* and a *buckled plane* are synonymous
[3] Corrugation means that the atoms at the surface can be seen in an interval along the direction vertical to the surface. The interval is centred at the 0-level of the surface.

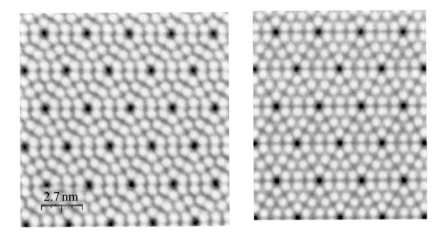

Figure 12-16. (left) Experimental STM image of Si(111) 7×7 reconstruction with U =1.6 V sample bias. *(right)* Solid sphere STM simulation using the structure model of Si 7×7 reconstruction by Tong et al [32]

computed. This z(x,y) contour is then convoluted with a two-dimensional gaussian function to take into account the finite STM tip resolution.

The validity of our STM simulation can be estimated on a standard material, such as the periodic Si(111) 7×7 surface structure, see Figure 12-16.

Clusters on Fivefold Surfaces

We performed the STM simulations of a fivefold surface on candidates for the bulk-terminating layers, (b,q,(a)q,b)-layer at $z = 47.96$ Å (see Ref.[3]) in $\mathcal{M}(\mathcal{T}^{*(2F)})$, and on a (q,b,b,q)-layer at $z = 50.00$ Å (see Ref.[3]). We used the vertical range (between black and white) 1.5 Å, in both simulations and the STM images, indicated by the corrugation 1.5 Å on the fivefold surfaces. On the STM image one notices two characteristic fivefold symmetric configurations, a "white flower" (wF) and a "dark star" (dS), see Ref. [8]. The wF and the dS were observed also on STM images of the fivefold surface of i-AlCuFe; see for example Figure 4(a) in Ref. [19]. The "white flowers", and in particular the "dark stars", are evidently better reproduced on the (q,b,b,q)-layers; compare *(top-left)* and *(top-right)* images to the *(bottom)* image in Figure 12-17. Whereas the fivefold "dark star" on a (b,q,(a)q,b)-layer, Figure 12-17 *(top-left)* is the large dark pentagon, on the (q,b,b,q)-layer, Figure 12-17 *(top-right)*, the dark pentagonal star has a similar shape and size as on the STM image; see Figure 12-17 *(bottom)* and Figure 12-18. The accented fivefold shape of the wF and some parts of the dS is determined by the shape of the W_q window, i.e. the shape of its intersection by the coding fivefold plane in $\mathbb{E}_\perp$, ("Surfaces represented by the bulk terminations, Bravis' Rule"). Such an intersection

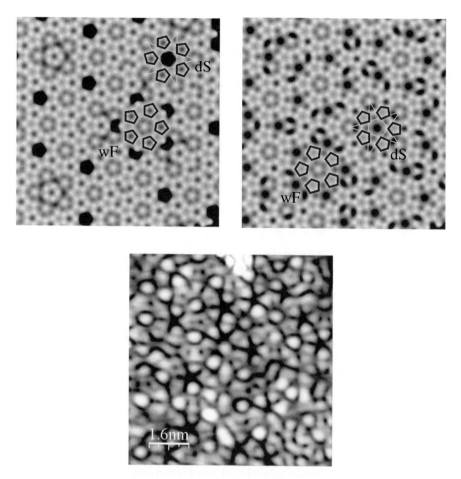

Figure 12-17. STM simulations of a fivefold surface on candidates for the bulk-terminating layers (*top-left*) on a (b,q,(a),q,b)-layer at z = 47.96 Å in the patch of $\mathcal{M}(\mathcal{T}^{*(2F)})$ (see Ref.[3]), (*top-right*) on a (q,b,b,q)-layer at z = 50.00 Å in the patch of $\mathcal{M}(\mathcal{T}^{*(2F)})$ (see Ref.[3]). Both images are 80 × 80 Å²; the vertical range is 1.5 Å. (*bottom*) An STM image of the real fivefold clean surface of i-AlPdMn, 80 × 80 Å²; the vertical range is 1.5 Å, I = 100pA, U = 300meV. The candidates for the observed fivefold symmetric local configurations, "white flower" (wF) and the "dark star" (dS) are marked on simulations of layers of both kinds. By convention, twofold directions are marked in blue. The wF and the dS are not marked on the STM image in this Figure, (*bottom*); these are marked in Figure 12-18 (*right*). The same local configurations were observed on the STM images of the fivefold surface of i-AlCuFe; see Figure 4(a) in Ref. [19] (see color plate section)

is present in Figure 12 of Ref. [8]. The reason that the dS resembles a pentagonal dark star in (*top-right*) of Figure 12-17, and not a dark pentagon (*top-left*) of Figure 12-17 involves the particular five atoms in the b-plane, 0.48 Å below the surface, which are coded in the model by the polyhedra 1, 2 and 3 in the window W_b; see Figure 12-7. Hence, the maximum dense (q,b,b,q)-layers are the fivefold

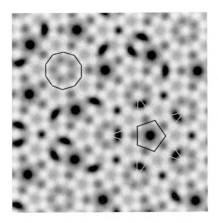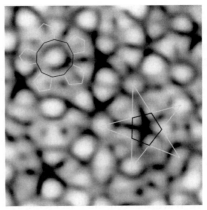

Figure 12-18. (*left*) An STM simulation of the fivefold termination 50×50 Å^2 large, taken from Figure 12-17 (*top-right*); (*right*) An STM image of the real surface 50×50 Å^2 large, taken from Figure 12-17 (*bottom*). By convention, twofold directions are in blue. The intersection of a Bergman polyhedron by the fivefold surface (*right*) and by the termination (*left*) are framed with a pentagon of edge length ② $= 4.77$ Å, marked in dark blue. The intersection of a Mackay polyhedron is a decagon of edge length τ^{-1}② $= 2.96$ Å ($\tau = (\sqrt{5}+1)/2$), marked in dark blue. Compare the shape and the scale of the intersection to Figure 12-3 (see color plate section)

bulk terminations in the model $\mathcal{M}(\mathcal{T}^{*(2F)})$, due to the almost spherical shape of the window W_b and the strongly concave shape in the fivefold directions of the window W_q.

The (q,b,b,q)-layers are not favoured as fivefold bulk terminations by the dynamical LEED simulations, neither in i-AlPdMn [33] nor in i-AlCuFe [18], likely because the top q-planes in the terminations are less rich in Al atoms than the corresponding top b-planes in the (b,q,(a)q,b)-layers. Dynamical LEED confirms the (b,q,(a)q,b)-layers as terminations on which the top surface plane is an Al-plane (93% Al by i-AlPdMn [33] and 90% Al by i-AlCuFe [18]); the next plane, 0.48 Å below, contains 49% Al in i-AlPdMn [33] and 45% Al in i-AlCuFe [18]. The dynamical LEED is a model-dependent method. The model-independent method of low-energy ion scattering (about LEIS see Ref. [34] and Refs. quoted therein) confirms that the surface of i-AlPdMn is rich in Al: it counts 88% Al in the top plane, and 69% Al in the plane 0.38 Å below (corresponding to a distance 0.48 Å in the bulk). We hence conclude that the present positions of Al atoms in the $\mathcal{M}(\mathcal{T}^{*(2F)})$ model, in which the chemistry was mainly adopted from the Boudard model [1], are probably incorrect. If (q,b,b,q)-layers of the maximum density are the terminations, the q-sites in $\mathcal{M}(\mathcal{T}^{*(2F)})$, labelled 1, 3, 2(alone), 5 and 6 in Figure 12-6, should be predominantly occupied by Al atoms, what is only partly fulfilled in the present model [35], [3]. If the maximum dense (b,q,(a)q,b)-layers were terminations, the surface b-planes in the terminations would be rich in Al; hence the b-sites labelled 1, 2 and 3 in Figure 12-7 should be predominantly occupied by Al atoms, like the present $\mathcal{M}(\mathcal{T}^{*(2F)})$ (see Refs.[35] and [3]) with chemistry adopted from Ref. [1].

The white flower (wF) on the termination (q,b,b,q) is defined by the five Bergman polyhedra *below* the surface in a special fivefold symmetric arrangement. From each Bergman below the surface, a shining pentagon, of edge length $\tau^{-1}② = 2.96$ Å, is seen, marked in dark blue in Figure 12-17 (*top-right*). Such a configuration of Bergmans enforces a Mackay polyhedron in the middle [36], [8] drawn in Figure 12-18 (*left*). When we compare it to Figure 12-18 (*right*) we conclude that the fivefold surface cuts the Mackay polyhedra, and the intersection is a *decagon* of edge length $\tau^{-1}② = 2.96$ Å. The decagon is marked in thick blue in Figure 12-3 (*right*). The dark star (dS) on the termination (q,b,b,q) is defined also by five Bergman polyhedra *below* the surface, but in another fivefold symmetric arrangement; see Figure 12-17 (*top-right*). Such a fivefold configuration of Bergmans enforces the sixth Bergman polyhedron in the middle [8], drawn in Figure 12-18 (*left*). When we compare it to Figure 12-18 (*right*) we conclude that the fivefold surface cuts this sixth Bergman polyhedron in the center of each fivefold configuration, and the intersection is a *pentagon* of edge length $② = 4.77$ Å. The pentagon is marked in thick blue in Figure 12-3 (*left*).

Clusters on Twofold Terminations

The STM simulation of twofold surfaces on the twofold terminations introduced in "Twofold terminations" is presented in Figure 12-20, with the STM image of the twofold surface of i-AlPdMn. From a comparison of the STM image (*top-left*) with its simulation (*top-right*), only the large feature characteristic, the dark lines along some twofold directions in the twofold plane, appear in a sequence, and are on equal mutual distances on both simulations and STM images. The dark lines are additionally marked with light blue dotted lines, on both the STM image and its simulation; see Figure 12-20 (*top-left*) and (*top-right*). The local configurations that we relate to the cut Bergman and the cut Mackay clusters by the twofold termination

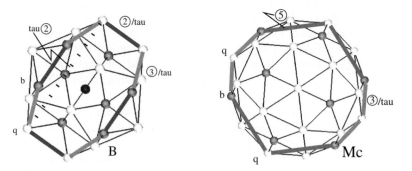

Figure 12-19. Bergman (B) and Mackay (Mc) polyhedra. On each an equator, orthogonal to a twofold direction, is marked. $\tau^{-1}② = 2.96$ Å, $② = 4.80$ Å, $\tau② = 7.76$ Å, $\tau^{-1}③ = 2.57$ Å, $⑤ = 4.56$ Å. By convention, if marked in color, fivefold directions are in red, twofold in blue, threefold in yellow (see color plate section)

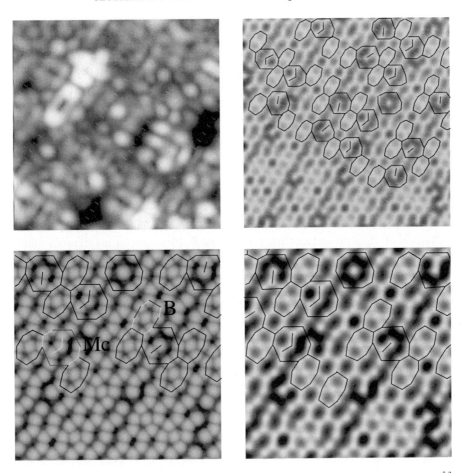

Figure 12-20. (*top-left*) Experimental STM image of the twofold surface of i-AlPdMn, $80 \times 80 \,\text{Å}^2$ large, with 2.5 Å vertical range (the corrugation on the surface has a range 2.5 Å) and $I = 100$ pA and $U = 300$ meV. (*top-right*) STM simulation with gaussian convolution filter of the "sphere model", $80 \times 80 \,\text{Å}^2$ large on a twofold termination in the model $\mathcal{M}(\mathcal{T}^{*(2F)})$ at $x = 49.3818 \,\text{Å}$ (see Ref.[3]) with vertical range 2.5 Å. (*bottom-left*) The "sphere model" STM simulation of the terminating layer at $x = 49.3818 \,\text{Å}$ (see Ref.[3]) $50 \times 50 \,\text{Å}^2$. On it we mark in colour a Bergman (B) and a Mackay (Mc) cluster, cut by the termination on the equator (compare the shape and the scale to Figure 12-19). By convention, if marked in color, fivefold directions are in red, twofold in blue, threefold in yellow. (*bottom right*) STM simulation with gaussian convolution filter of the "sphere model" on a twofold termination in the model $\mathcal{M}(\mathcal{T}^{*(2F)})$ at $x = 49.3818 \,\text{Å}$, $50 \times 50 \,\text{Å}^2$ large with vertical range 1.5 Å (see color plate section)

(see Figure 12-19) and mark on the STM simulations (see (*top-right*), (*bottom-left*) and (*bottom-right*)) images in Figure 12-20, are unrecognizable on the STM image (see (*top-left*)) in Figure 12-20, likely because of the 2.5 Å large corrugation on the twofold surfaces.

CONCLUSIONS

We have here reviewed our work on the surfaces of icosahedral quasicrystals since 1997. We have discovered a confirmation that the maximum dense fivefold (q,b,b,q)-layers are the fivefold bulk terminations in the model $\mathcal{M}(\mathcal{T}^{*(2F)})$. This conclusion contradicts the chemistry of the model $\mathcal{M}(\mathcal{T}^{*(2F)})$, adopted from the Boudard model. Not only have we shown that the atomic decorations in a shape of Bergman and Mackay polyhedra can not be stable clusters but also we have drawn the appearance of the cut polyhedra on the surfaces that matches the image of the real material on the fivefold surface.

ACKNOWLEDGEMENTS

Z. Papadopolos thanks G. Kasner for his programming that was crucial for the realisation of ideas on surfaces since 1997, and P. Pleasants for his rigorous mind through which the formulation of the Bravais rule by quasicrystals became accepted, V. Kecman for sorting the data from $\mathcal{M}(\mathcal{T}^{*(2F)})$ for the twofold surfaces, V. Fournée for the STM image presented in Figure 12-14, J. Ledieu for the image presented in Figure 12-10, and finally P. A. Thiel and C. Yenks for many stimulating discussions and for the decision to put the model $\mathcal{M}(\mathcal{T}^{*(2F)})$ on the Website of the Laboratory in Ames.

O. Gröning and R. Widmer thank M. Feuerbacher for providing the icosahedral Al-Pd-Mn monograin used in the STM work presented in Figures 12-15, 12-17, 12-18 and 12-20, and acknowledge financial support from the Swiss National Science Foundation (SNF) under contract 200021-112333/1.

REFERENCES

1. M. de Boissieu, P. Stephens, M. Boudard, C. Janot, D. L. Chapman and M. Audier, *J. Phys.: Condens. Matter* **6**, (1994) 10725.

2. A. Katz and D. Gratias, in *Proc. of the 5th Int. Conf. on Quasicrystals*, eds. C. Janot and R. Mosseri (World Scientific, Singapore, 1995) p. 164.

3. http://www.quasi.iastate.edu/Structure%20Dbase%20Info.html.

4. Z. Papadopolos, P. Kramer and W. Liebermeister, in *Proc. of the Int. Conf. on Aperiodic Crystals, Aperiodic 1997*, edited by M. de Boissieu, J.-L. Verger-Gaugry and R. Currant (World Scientific, Singapore, 1998), p. 173.

5. Z. Papadopolos, G. Kasner, P. Kramer and D. E. Bürgler, in *Mat. Res. Soc. Symp. Proc.* Vol. 553, *"Quasicrystals"*, Boston 30.11.–02.12.1998, edited by Jean–Marie Dubois, Patricia A. Thiel, A.-P. Tsai and K. Urban, p. 231.

6. P. Kramer, Z. Papadopolos and D. Zeidler, in *Symmetries in Science V: Algebraic structures, their representations, realizations and physical applications*, edited by B. Gruber and L. C. Biedenharn (Plenum Press, 1991), p. 395.

7. J. H. Conway and N. J. A. Sloane, *Sphere Packings, Lattices and Groups* (Springer, New York 1988).

8. Z. Papadopolos, G. Kasner, J. Ledieu, E. J. Cox, N. V. Richardson, Q. Chen, R. D. Diehl, T. A. Lograsso, A. R. Ross and R. McGrath, *Phys. Rev. B* **66**, (2002) 184207.

9. Z. Papadopolos, P. Pleasants, G. Kasner, V. Fournée, C. J. Jenks, J. Ledieu and R. McGrath, *Phys. Rev. B* **69**, (2004) 224201.

10. Z. Papadopolos and G. Kasner, *Phys. Rev. B* **72**, (2005) 094206.
11. A. Bravais, Etudes crystallographiques, première partie: du cristal considéré comme un simple assemblage de points, (presented 26.02.1849) J. Ecole. Polytech., XXXIV Cahier, pp. 13–25 in *Études crystallographiques*, Gauthier-Villars, Paris 1866.
12. J. D. H. Donnay and D. Harker, *Am. Mineral.*, **22**, (1937) 446.
13. Z. Shen, W. Raberg, M. Heinzig, C. J. Jenks, V. Fournée, M. A. Van Hove, T. A. Lograsso, D. Delaney, T. Cai, P. C. Canfield, I. R. Fisher, A. I. Goldman, M. J. Kramer and P. A. Thiel, *Surf. Sci.* **450**, (2000) 1.
14. O. Groening, R. Widmer, P. Ruffieux and P. Groening, *Philos. Mag.* **86**, (2006) 773.
15. Gunnar Bergman, John L. T. Waugh and Linus Pauling, *Acta Cryst.* **10** (1957) 254.
16. A. L. Mackay, *Acta Cryst.* **15** (1962) 916.
17. V. Elser, *Philos. Mag. B* **73**, (1996) 641.
18. T. Cai, F. Shi, Z. Shen, M. Gierer, A. I. Goldman, M. J. Kramer, C. J. Jenks, T. A. Lograsso, D. W. Delaney, P. A. Thiel, M. A. Van Hove, *Surf. Sci.* **495**, (2001) 19.
19. T. Cai, J. Ledieu, R. McGrath, V. Fournée, T. Lograsso, A. Ross, P. Thiel, *Surf. Sci.* **526**, 2003) 115.
20. P. Kramer, Z. *Naturf.* **41a**, (1986) 897.
21. Z. Papadopolos, C. Hohneker and P. Kramer, *Discrete Math.* **221**, (2000) 101.
22. R. V. Moody, in *The Mathematics of Long-Range Aperiodic Order*, edited by R. V. Moody, (Kluwer, 1997) p. 403.
23. G. Kasner, Z. Papadopolos, P. Kramer and D. E. Bürgler, *Phys. Rev. B* **60**, (1999) 3899.
24. P. Pleasants, in *Covering of Discrete Quasiperiodic Sets: Theory and Applications to Quasicrystals*, Springer Tracts in Modern Physics Vol. 180, edited by P. Kramer and Z. Papadopolos (Springer, Berlin, 2002), p. 185.
25. T. M. Schaub, D. E. Bürgler, H.-J. Güntherodt and J. B. Suck, *Phys. Rev. Lett.* **73**, (1994) 1255.
26. J. Ledieu, E. J. Cox, R. McGrath, N. V. Richardson, Q. Chen, V. Fournée, T. A. Lograsso, A. R. Ross, K. J. Caspersen, B. Unal, J. W. Evans and P. A. Thiel, *Surf. Sci.* **583**, (2005) 4.
27. J.-C. Zheng, C. H. A. Huan, A. T. S. Wee, M. A. Van Hove, C. S. Fadley, F. J. Shi, E. Rotenberg, S. R. Barman, J. J. Paggel, K. Horn, Ph. Ebert and K. Urban, *Phys. Rev. B* **69**, (2004) 134107.
28. P. Bombicz, M. Czugler, R. Tellgren, A. Kálmán, *Angew. Chem. Int. Ed.* **42** (2003) 1957.
29. S. Roy, R. Banerjee, A. Nangia and G. L. Kruger, *Chem. Eur. J.* **12** (2006) 3777.
30. V. Fournée, A. R. Ross, T. A. Lograsso, J. W. Evans, P. A. Thiel, *Surf. Sci.* **537**, (2003) 5.
31. J. Tersoff and D.R. Haman, *Phys. Rev. B* **31**, (1985) 805.
32. S. Y. Tong, U. Huang, C. M. Wei, W. E. Packard, F. K. Men, G. Glander, M. B. Webb, *Journal of Vacuum Science and Technology A6* **3**, (1988) 615.
33. M. Gierer, M. A. Van Hove, A. I. Goldman, Z. Shen, S.-L. Chang, C. J. Jenks, C.-M. Zhang and P. A. Thiel, *Phys. Rev. Lett.* **78**, 467 (1997); M. Gierer, M. A. Van Hove, A. I. Goldman, Z. Shen, S.-L. Chang, P. J. Pinhero, C. J. Jenks, J. W. Anderegg, C.-M. Zhang and P. A. Thiel, *Phys. Rev. B* **57**, (1998) 7628.
34. R. Bastasz, J. A. Whaley, T. A. Lograsso and C. J. Jenks in *Philos. Mag.* **86**, (2006) 855.
35. G. Kasner, Z. Papadopolos in *Mat.Res.Soc.Symp.Proc.* Vol. 643 (2000), *Quasicrystals: Preparation, Properties and Applications*, edited by E. Belin-Ferré, P. A. Thiel, A.-P. Tsai and K. Urban, p. K9.7.1
36. G. Kasner, Z. Papadopolos, *Philos. Mag.* **86**, (2006) 813.

CHAPTER 13

PROTEIN-PROTEIN INTERACTIONS IN THE CYANOBACTERIAL KaiABC CIRCADIAN CLOCK

MARTIN EGLI*, REKHA PATTANAYEK AND SABUJ PATTANAYEK

Abstract: The discovery that the central oscillator of the cyanobacterial KaiABC circadian clock can be reconstituted in vitro by the protein components KaiA, KaiB and KaiC renders this biological timer a unique target for biochemical and structural studies. The oscillator can be monitored through changes in the KaiC phosphorylation status that is modulated by KaiA and KaiB. As the 24-h period of the recombinant clock remains unaltered as a result of modest variation of temperature, interactions between the three Kai proteins not only form the basis for rhythmic control of levels of KaiC phosphorylation but also provide temperature compensation. A profound understanding of how this biological timer works requires a dissection of the functions of, and interactions between, the three proteins. Three-dimensional structures of the individual Kai proteins have been determined, and the KaiA-KaiC complex has been studied using hybrid structural methods. This chapter provides an overview of progress in the characterization of the cyanobacterial circadian clock with an emphasis on structural aspects of individual Kai proteins and the binary KaiA-KaiC complex

INTRODUCTION

Circadian clocks are endogenous biological timers that rhythmically regulate numerous processes with a period of roughly 24 h and exhibit temperature compensation (Dunlap et al 2004). Circadian clocks exist in various eukaryotic systems including mammals, plants, fungi and insects, and have been found also in cyanobacteria (Johnson, 2004; Iwasaki and Kondo, 2004); the latter are the simplest organisms known to possess a clock. In the model organism *Synechococcus elongatus* PCC 7942, the *kaiA*, *kaiB* and *kaiC* genes that form a cluster on the chromosome

---

* Correspondence: martin.egli@vanderbilt.edu

J.C.A. Boeyens and J.F. Ogilvie (eds.), Models, Mysteries and Magic of Molecules, pp. 283–299.
© 2008 *Springer*.

were shown to be essential for proper circadian function (Ishiura et al 1998). The following basic properties of this biological timer have emerged: (i) circadian rhythm is lost when KaiC protein is overexpressed continuously due to shutdown of *kaiBC* expression, whereas transient increases of KaiC serve to set the phase of the rhythm (Ishiura et al 1998; Xu et al 2000); (ii) in continuous light conditions the proportions of *kaiBC* mRNA and KaiC protein oscillate in a circadian fashion and exhibit a phase shift (Xu et al 2000); (iii) KaiA and KaiC are positive and negative regulators, respectively, of *kaiBC* transcription (Ishiura et al 1998); (iv) because practically all promoter activities in cyanobacteria underlie circadian rhythm, the Kai clock system might appear not to work in a clock-gene specific fashion, but to control a process that governs genome-wide expression the mechanism of which is unknown (Liu et al 1995; Xu et al 2003; Johnson, 2004; Nakahira et al 2004); (v) the proteins encoded by the *kai* genes – KaiA, KaiB and KaiC – interact with each other in vitro and in vivo (Iwasaki et al 1999; Taniguchi et al 2001), and KaiC constitutes the central component of the protein complex (Kageyama et al 2003); (vi) KaiC is an auto-kinase and an auto-phosphatase in vitro and in vivo (Nishiwaki et al 2000; Iwasaki et al 2002; Xu et al 2003), and the clock speed is correlated with the level of phosphorylation (Xu et al 2003), and (vii) both in vitro and in vivo, KaiA enhances phosphorylation of KaiC, and KaiB antagonizes the action of KaiA (Iwasaki et al 2002; Williams et al 2002; Kitayama et al 2003; Xu et al 2003 Kageyama et al 2006) (Figure 13-1). The observation that Kai proteins (KaiA and KaiC) can positively and negatively regulate *kaiBC* transcription (Ishiura et al 1998) rendered the cyanobacterial clock consistent with an oscillatory (TTO) feedback model involving transcription and translation, believed to be at the core of all self-sustaining biological timers (Dunlap et al 2004).

Recent observations have provided clear evidence that in *S. elongatus* a TTO feedback model is not valid. One advance occurred when the behaviour of the cyanobacterial KaiABC clock was scrutinized under constant dark conditions. Originally such an experiment had disclosed that the phase of rhythm in *S. elongatus* was not affected significantly when bacteria were switched back to conditions of continuous light following a period of constant dark (Xu et al 2000). In the dark, the metabolism of *S. elongatus* including RNA and protein syntheses is normally suppressed, but Kondo and coworkers reported a robust circadian rhythm under a constant dark condition in the presence of transcription inhibitors in excess proportions that almost quantitatively block the synthesis of RNA and protein (Tomita et al 2005). Despite the absence of rhythmic accumulation of Kai proteins and the lack of kaiA and kaiBC mRNA, KaiC phosphorylation exhibited a robust circadian rhythm for more than two days. The cyanobacterial circadian clock is therefore able to function without synthesis *de novo* of clock gene mRNA and the proteins encoded by them, and the period is accurately determined without transcriptional feedback.

These findings define a minimal timing loop in vivo that functions without transcription and translation and is temperature-compensated (Figure 13-1). The three Kai proteins accordingly comprise the minimal components of the circadian

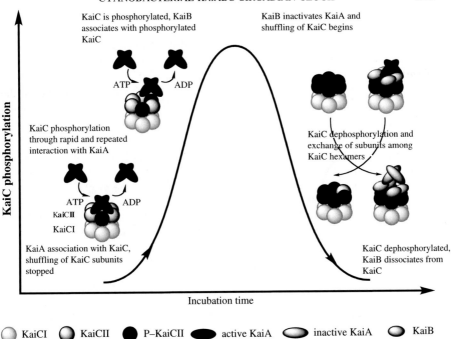

KaiC is phosphorylated, KaiB associates with phosphorylated KaiC

KaiB inactivates KaiA and shuffling of KaiC begins

KaiC phosphorylation through rapid and repeated interaction with KaiA

ATP ADP

KaiC dephosphorylation and exchange of subunits among KaiC hexamers

ATP ADP
KaiCII
KaiCI

KaiA association with KaiC, shuffling of KaiC subunits stopped

KaiC dephosphorylated, KaiB dissociates from KaiC

KaiC phosphorylation

Incubation time

⬤ KaiCI ⬤ KaiCII ⬤ P–KaiCII ⬬ active KaiA ⬭ inactive KaiA ⬤ KaiB

Figure 13-1. Model of the KaiC phosphorylation cycle. Schematic diagrams illustrate enhancement of KaiC phosphorylation (or inhibition of dephosphorylation) by KaiA dimer (left) and inactivation of KaiA by KaiB (right; adapted from Kageyama et al 2006). Only the KaiCII domains harbor phosphorylation sites (Xu et al 2004; see text)

oscillator and provide the output for the regulation of the general mechanism of transcription (Tomita et al 2005), perhaps using two associated histdine kinases – SasA and CikA (Schmitz et al 2000) – as signal mediators possibly to affect DNA superhelicality (Johnson, 2004). These observations raised also the spectre that KaiA, KaiB and KaiC might form a robust oscillator in vitro that exhibits rhythmic phosphorylation and dephosphorylation of KaiC and compensates for temperature changes (Figure 13-2). This condition was indeed demonstrated (Nakajima et al 2005), making the KaiABC system a unique target for a biochemical and structural dissection of the inner workings of a molecular timer.

STRUCTURAL STUDIES OF KAI PROTEINS

Three-dimensional structures based on crystallographic data and NMR data from solutions are available for all three Kai proteins from various cyanobacterial systems (Johnson & Egli, 2004; Golden , 2004) (Table 13-1). With regard to a structural characterization, the components of the cyanobacterial clock are the best studied, such that far more is known about them than the cogs of the eukaryotic circadian clocks

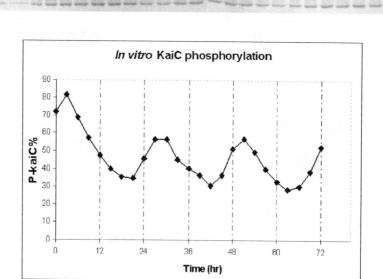

Figure 13-2. KaiC phosphorylation rhythm in vitro monitored over 72 h. Gel image courtesy of Ximing Qin and Tetsuya Mori (Johnson laboratory, Vanderbilt University)

for which only one partial structure has been reported (Yildiz et al 2005). Following the initial NMR determination of the structure of the N-terminal pseudo-receiver domain of KaiA from *S. elongatus* (Williams et al 2002) and EM investigations focusing on KaiC (Mori et al 2002; Hayashi et al 2003), high-resolution structural information for all Kai proteins emerged in 2004. The crystal structure of full-length KaiA was published for *S. elongatus* and revealed a domain-swapped arrangement with three dimer interfaces, one of which connects the N-terminal receiver domain with the C-terminal KaiC-interacting domain (Ye et al 2004) (Figure 13-3). The structures of the C-terminal dimerization and KaiC-interacting domain of KaiA from *Thermosynechococcus elongatus* BP-1 were solved separately by X-ray crystallography (Uzumaki et al 2004) and NMR (Vakonakis et al 2004a). The crystal structure and mutational data implicated grooves above the dimerization interface on opposite faces of the dimer as potential sites for interaction with KaiC.

A further crystal structure of the C-terminal domain of KaiA and a structure of full-length KaiB from the cyanobacterium *Anabaena* PPC7120 revealed a thioredoxin-like fold for the latter (Garces et al 2004) (Figure 13-4). This work also identified similarities in the dimensions and electrostatic potentials of particular regions in the KaiA and KaiB dimers as well as similar spacings between conserved arginine pairs on the surfaces of the respective Kai proteins. A crystal structure of

Table 13-1. Structures of cyanobacterial circadian clock proteins

Kai protein	Organism	Technique	Reference	PDB code[a]
KaiA N-terminal domain	PCC7942 *Synechococcus elongatus* (*S. elongatus*)	NMR	Williams et al 2002	1m2e
KaiA full-length	*S. elongatus*	X-ray	Ye et al 2004	1r8j
KaiA full-length	PCC7120 *Anabaena* (*Anabaena*)	X-ray	Garces et al 2004	1r5q
KaiA C-terminal domain	*Thermosynechococcus elongatus* BP-1 (*T. elongatus*)	X-ray	Uzumaki et al 2004	1v2z
KaiA C-terminal domain	*T. elongatus*	NMR	Vakonakis et al 2004a	1q6a
KaiB full-length	*Anabaena*	X-ray	Garces et al 2004	1r5p
KaiB full-length	PCC6803 *Synechocystis*	X-ray	Hitomi et al 2005	1wwj
KaiB full-length (T64C mutant)	*T. elongatus*	X-ray	Iwase et al 2005	1vgl
KaiB full-length (wild type)	*T. elongatus*	X-ray	Pattanayek et al unpubl. data	—
KaiC full-length	*S. elongatus*	X-ray	Pattanayek et al 2004	1tf7[b]
			Xu et al 2004	1u9i
KaiA - KaiC peptide complex	*T. elongatus*	NMR	Vakonakis & LiWang, et al 2004	1suy
KaiA - KaiC complex	*T. elongatus* / *S. elongatus*	X-ray/ EM	Pattanayek et al 2006	2gbl
SasA N-terminal domain	*S. elongatus*	NMR	Vakonakis et al 2004b	1t4y

[a] http://www.rcsb.org (Berman et al 2000).
[b] The 1tf7 and 1u9i entries are based on the same crystallographic data, but in 1u9i phosphate groups were added to T432 and S431 in six and four subunits, respectively.

KaiB from *Synechocystis* PCC6803 revealed formation of a tetramer with a positively charged perimeter, a negatively charged center and a zipper of aromatic rings important for oligomerization (Hitomi et al 2005). Additional evidence was based on mutational data that appeared to demonstrate the importance of the tetrameric state of KaiB for proper clock function. In the crystal structure of a *T. elongatus* mutant KaiB protein, a similar tetramer motif was found (Iwase et al 2005). The relevance of the tetrameric state of KaiB for its role in the control of the KaiC phosphorylation state has, however, been doubted as the protein appears to bind consistently to KaiC as a dimer, as judged from experiments using gel filtration chromatography (Kageyama et al 2006).

We determined the crystal structure of the full-length KaiC protein from *S. elongatus* (Pattanayek et al 2004). The structure of the central and largest protein from the cyanobacterial clock revealed the formation of a homo-hexamer in the

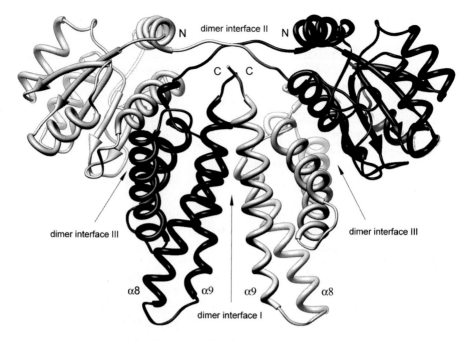

Figure 13-3. Crystal structure of the domain-swapped KaiA dimer from *S. elongatus* (Ye et al 2004).
Figures 13-3 – 13-7 were produced with Chimera (Huang et al 1996)

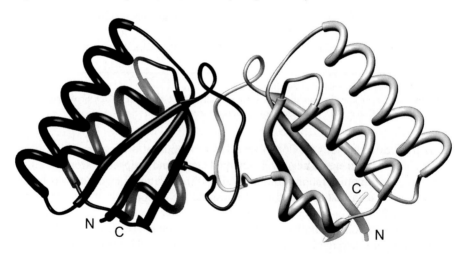

Figure 13-4. Crystal structure of the KaiB dimer from *Anabaena* (Garces et al 2004)

shape of a double torus with a central pore and 12 ATP molecules bound between
the interfaces of monomers (Figure 13-5). The C-terminal 21 residues of KaiC
monomers were partly disordered in the original crystal structure, indicating great
conformational flexibility in this region for the unbound state of KaiC. Vakonakis

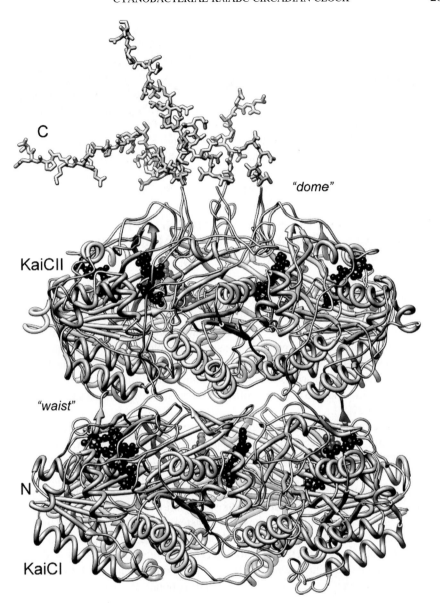

Figure 13-5. Crystal structure of the KaiC hexamer from *S. elongatus* (Pattanayek et al 2004). The model for full-length KaiC (519 amino acids) in the C-terminal region is complete for only two subunits (Pattanayek et al 2006). Atoms of the twelve ATP molecules bound between the KaiCI and KaiCII domains of individual subunits are shown as black spheres

and LiWang reported the NMR structure of a complex in solution between the dimeric C-terminal KaiA domain and 30mer peptides derived form the C-terminus of KaiC for the cyanobacterium *T. elongatus* BP-1 (Vakonakis & LiWang, et al 2004). Subsequent efforts to trace the C-terminal region of KaiC molecules in maps of electron density yielded a complete model for full-length KaiC from *S. elongatus* in the case of two subunits (Pattanayek et al 2006). The NMR structure of the monomeric N-terminal sensory domain of the SasA histidine kinase in solution has also been described (Vakonakis et al 2004b). Although KaiB shares with SasA and the thioredoxin family the initial beta-alpha-beta folding topology, the remaining structures and sequences diverge considerably (Hitomi et al 2005; Vakonakis et al 2004b).

DETERMINATION OF PHOSPHORYLATION SITES IN KaiC AND CONSEQUENCES OF THEIR MUTATION TO ALANINE FOR FUNCTION IN VITRO AND IN VIVO

The structure determined for *S. elongatus* KaiC was based on crystals grown from a mixture of proteins exhibiting various levels of phosphorylation as the protein had been purified as a hexamer and in the presence of Mg^{2+} and ATP (Pattanayek et al 2004). Following completion of the crystallographic model of the KaiC hexamer, inspection of difference electron-density maps allowed the identification of three sites, T432, S431 and T426 (Figure 13-6), of phosphorylation in the KaiCII domain; the KaiCI domain seems to contain no phosphorylation site (Xu et al 2004). Two residues, T432 and S431, were confirmed independently by mass spectrometry (Nishiwaki et al 2004).

The three serine and threonine residues, when mutated to alanine individually, render the clock arhythmic in vivo (Xu et al 2004). Individual T426A, S431A or T432A mutations as well as double mutations to alanine alter the phosphorylation patterns, and the triple mutant (T426/S431/T432→A) is no longer phosphorylatable. Mutation of Ser and Thr residues does not affect hexamerization. All phosphorylation sites are located in the KaiCII half; phosphorylation proceeds across subunits, and the presence of phosphate groups is consistent with a more stable subunit interface (Xu et al 2004). Binding of ATP or ADP between the KaiCII domains of adjacent subunits is expected also to affect the stability of the complex. Lys and/or Arg residues can thus interact with the γ-phosphate group of ATP across the interface; such interactions are absent when ADP is bound (Hayashi et al 2006).

A STRUCTURAL MODEL OF THE COMPLEX BETWEEN KaiA AND KaiC

An intriguing feature of the cyanobacterial KaiABC circadian clock is that analysis of the structure and function of the central timer requires no concern with input and output. Beyond an understanding of how three proteins are able both to sustain a stable oscillation with a period of 24 h and to do

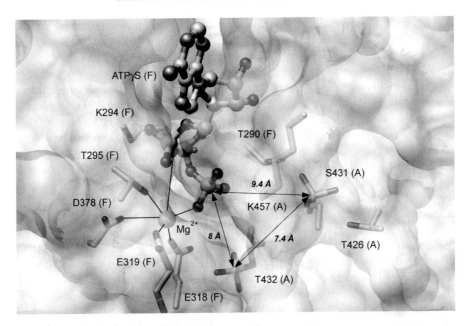

Figure 13-6. Location of phosphorylation sites in the KaiCII domain (T432, S431 and T426) at the interface between subunits A and F in the KaiC hexamer from *S. elongatus* (Xu et al 2004) The phosphoryl transfer occurs across subunits; selected distances in Å between the γ-phosphate and phosphorylated residues are shown

so in a temperature-compensated fashion, it is also important to acquire insight into how photoreceptors, and perhaps other sensors, are coupled with the clock (Schmitz et al 2000; Zhang et al 2006; Ivleva et al 2006). Similarly, how the ATP-dependent phosphorylation cycle driven by interactions between the three Kai proteins relates to global rhythmic control of gene expression (Nakahira et al 2004) remains to be worked out, although some players involved in output signaling have been identified (Katayama et al 1999; Iwasaki et al 2000; Ditty et al 2003; Takai et al 2006). In terms of an analysis of the output mechanism, the Kondo group has reported the identification of a protein, SasR, that interacts with SasA and has a leucine zipper DNA-binding motif (Kondo, 2005).

Based onsequence alignments, KaiC was shown to be a member of the RecA/DnaB superfamily of proteins (Leipe et al 2000), but, unlike classical helicases, KaiC is the result of a gene duplication and is composed of two hexameric rings (Figure 13-5). A 3D-structural alignment between the KaiCI or KaiCII hexameric rings and helicases revealed clear deviations in diameter, channel size and ATP position (Pattanayek et al 2004). Such alignments exhibited a fit that was somewhat inferior to superimpositions of the monomeric proteins. The best correspondence was found to exist between hexameric rings of KaiC halves and the F1-ATPase (Abrahams et al 1994), and was unanticipated from an alignment of the primary sequences. In light of these observations, KaiC is unlikely to act as a helicase, consistent with the results of gel shift experiments that demonstrate the need for KaiC at picomole concentrations to cause a shift with poly-dT or forked oligodeoxynucleotides at femtomole concentrations (Mori et al 2002). As there is currently no experimental evidence that proves KaiC to be a helicase, it appears unlikely that clock-controlled regulation of genes involves a direct interaction between KaiC and DNA. The similarities at the structural level between F1-ATPase and the hexamers formed by the KaiCI and KaiCII halves are also unlikely to extend to the functional level. The molecular machine that produces ATP is anchored in the membrane, and features a central stalk that rotates inside the channel formed by the trimer of $\alpha\beta$-heterodimers. Neither the KaiA nor the KaiB dimer exhibit a conformation that indicates the possibility of insertion into the KaiC channel (Figures 13-3–13-5), and they have been shown to exert their functions as dimers, not monomers (Kageyama et al 2006).

Based on yeast two-hybrid screens, early attempts to map the binding sites of KaiA on KaiC resulted in the identification of two candidate regions in KaiC involving the C-terminal 60 and 100 amino acids of the KaiCI and KaiCII domains, respectively (Taniguchi et al 2001). In a model of the hexamer that had the KaiCI and KaiCII domains arranged tail to tail, the two regions were expected to lie close together. However, the arrangement head to tail of the two KaiC halves observed in the crystal structure places the putative KaiA-interacting sites at a significant distance from each other (Pattanayek et al 2004). One encompasses the dome-shaped surface formed by C-terminal regions of KaiCII domains, and the other is located in the constricted waist region between KaiCI and CII and includes the 15-amino acid peptide linking the two (Figure 13-5). Both deviating topologies of these sites – a concave surface in the waist and a convex dome surface on KaiCII – and the fact that KaiCI appears devoid of phosphorylation sites raise doubts about the need for an interaction between KaiA and KaiCI. The presumed function of KaiA is either to enhance phosphorylation of KaiC or to inhibit dephosphorylation (Figure 13-1), but the absence of phosphorylation sites, and hence kinase and phosphatase activity by KaiCI, renders unnecessary such a regulation.

Vakonakis and LiWang observed specific binding between a KaiCII C-terminal peptide and the C-terminal domains of the KaiA dimer from *T. elongatus* (Vakonakis & LiWang, et al 2004); the corresponding peptide at the C-terminus of KaiCI showed no binding. This observation is consistent with regulation by KaiA of the level of KaiC phosphorylation affecting only KaiCII. This finding prompted us to reexamine the electron density above the C-terminal dome in the KaiC hexamer crystal structure from *S. elongatus*, leading to complete models for full-length KaiC in two subunits and an addition of several residues to the remaining four (Figure 13-5) (Pattanayek et al 2006). Deletion of the C-terminal 25 residues in KaiC abolishes complex formation with KaiA in vitro and clock rhythmicity in vivo; the deletion does not affect hexamerization (Pattanayek et al 2006). Binding between a C-terminal peptide from a KaiC subunit and the KaiA dimer sheds no light on the mechanism according to which the latter enhances KaiC phosphorylation. A study focusing on the proteins from *T. elongatus* demonstrated that a single KaiA dimer is capable of upregulating KaiC phosphorylation to a virtually saturated level (Hayashi et al 2004a). The interaction between KaiA and KaiC is apparently dynamic in nature, involving rapid and repeated binding of KaiA to C-terminal peptides from KaiC subunits (Figure 13-1; Kageyama et al 2006).

Using a combination of X-ray crystallography, electron microscopy and assays in vitro and in vivo with native and mutant proteins from *S. elongatus* and *T. elongatus*, we have developed a model for the KaiA-KaiC 1:1 complex. This model leaves intact the binding interface between the KaiCII C-terminal peptide and the KaiA dimer worked out with solution NMR (Vakonakis & LiWang, et al 2004). The conformation of the peptide in the NMR structure and that of the C-terminal portion of one KaiC subunit in the crystal structure of full-length KaiC are similar (Pattanayek et al 2006). This discovery made possible replacement of that C-terminal peptide (from *S. elongatus*) by the NMR peptide with the C-terminal domains of KaiA dimer bound (from *T. elongatus*). With account taken of the EM-based envelope of the KaiA-KaiC 1:1 complex, the KaiA dimer based on the crystal structure of the full-length protein from *S. elongatus* (Ye et al 2004) was superimposed on the model of the KaiA dimer (C-terminal domains only) - KaiC complex. The resulting model of the complex has the α8-loop-α9 portion of the C-terminal domain of a KaiA monomer (Figure 13-3) in close proximity to the nucleobase portion of ATP bound between two KaiC subunits (Figure 13-7). The model discloses no detail of the interactions between KaiA and KaiC at this site, but main-chain atoms of residues in the apical KaiA helix-loop-helix region, of which mutation critically affects the period of the clock, lie as close as 12 Å from ATP.

There exists potentially a second binding site between KaiA and KaiC. The first involves the KaiA dimer and the flexible C-terminal peptide of a KaiC subunit, and the second a seemingly more transient interaction between a helix-loop-helix region of a KaiA monomer and the ATP-binding cleft formed between the KaiCII domains from two subunits. There are several scenarios for how this second interaction might affect the extent of phosphorylation at residues T432, S431 and T426. For example, sealing the cleft that harbors ATP might increase the residence period of the latter. Alternatively, the contact with KaiA might result in a conformational change of residues and facilitate the transfer of the ATP γ-phosphate group. In the crystal structure, the T432residues in all six subunits and S431 residues in four subunits are phosphorylated (Xu et al 2004). The side-chain oxygen atoms of T432 and S431 are more than 8 Å away from the γ-phosphate group of ATP, and the conformations of subunitinterfaces observed in the crystal are unconducive to phosphoryl transfer. A structure of non-phosphorylated KaiC hexamer with bound ATP is lacking, and no experimental data provide insight into the conformational changes that the subunit interface undergoes as a result of one or more of the above residues becoming phosphorylated. What appears clear is that this second interaction is not tight, consistent with a rapid and repeated association and dissociation of potentially just a single KaiA dimer on the dome-shaped surface of KaiCII (Figure 13-1). One is tempted to draw an analogy between this mode of interaction

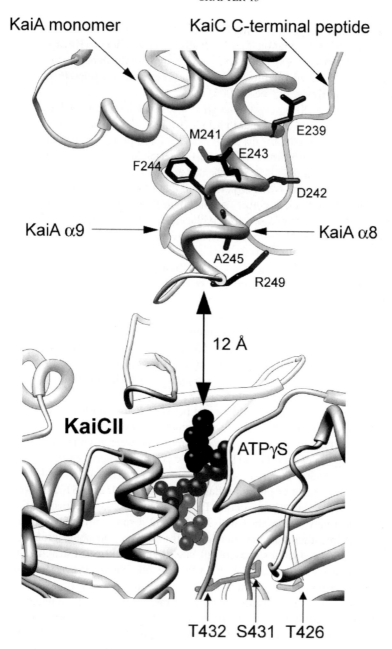

KaiA monomer   KaiC C-terminal peptide

M241   E239

E243

F244

D242

KaiA α9 ————▶   ◀———— KaiA α8

A245

R249

12 Å

KaiCII   ATPγS

T432 S431 T426

Figure 13-7. EM-based model of the 1:1 KaiA-KaiC complex from *S. elongatus* (Pattanayek et al 2006). Phosphorylation sites for a single KaiC subunit and selected residues in the α8-loop-α9 region of KaiA are highlighted

Figure 13-8. Artistic rendering of the interaction between KaiA dimer (*Parthenos sylvia subsp. lilacinus* – clipper butterfly) and KaiC hexamer (Tiger Lily)

for KaiA and KaiC and that of a butterfly drinking plant nectar and pollinating a flower. The butterfly (KaiA) hovers near a stamen (KaiC peptide) and eventually touches two petals (subunits; Figure 13-8, left), before moving to the next stamen or petals (Figure 13-8, right), thus pollinating the flower (phosphorylating KaiC).

DIVERGENT FUNCTIONS OF THE KAICI AND KAICII DOMAINS

There is mounting evidence for distinct roles of the two hexameric KaiC rings that comprise the central cog of the KaiABC clock in sustaining the phosphorylation rhythm. The crystal structure revealed formation of hydrogen bonds between P-loop residues and the nucleobase moiety of ATP molecules bound between KaiCI domains from adjacent subunits. Conversely, these hydrogen bonds are absent in the ATP binding pockets between subunits in the KaiCII ring. There is instead a tighter grip around the β- and γ-phosphates there (Pattanayek et al 2004, 2006). The structural data are consistent with distinct affinities for ATP by the KaiCI and KaiCII halves. The affinity for ATP in the CI half is accordingly greater than in the CII half (Hayashi et al 2004b). Work with proteins from *T. elongatus* demonstrated that the KaiCI domain expressed separately forms stable rings in the presence of ATP, but no hexamer formation was seen with KaiCII domains (Hayashi et al 2006). Beyond these differences in the recognition of and binding affinity for ATP, the two domains exhibit also topological (the C-terminal peptide tentacles protrude only from the KaiCII domains) and electrostatic differences (the N-terminal dome is negatively and the C-terminal dome is positively polarized) (Pattanayek et al 2004). Most importantly, only KaiCII contains Thr and Ser residues that become phosphorylated, and KaiA seems to interact with only the KaiCII half. These observations together support a conclusion that the KaiCI hexamer serves as a structural platform whereas the KaiCII hexamer constitutes the business end of the homo-hexameric complex. Conformational changes as a result of KaiA-mediated phosphorylation might affect mostly the KaiCII half. Although no model of the interaction between KaiB and KaiC has been proposed, KaiB likely performs its role as a KaiA-antagonist also at the KaiCII end.

SUMMARY AND OUTLOOK

A dissection of the structure and function of the cyanobacterial KaiABC circadian clock offers the prospect of understanding a molecular timer – a nanoclock – in unprecedented detail. Whether key features of this clock, namely maintenance of a stable oscillation and temperature compensation decoupled from transcription and translation, are unique or will be established for other clocks in higher organisms remains to be seen. Significant progress has been made over the past two years in the analysis of the KaiABC clock. The availability of 3D-structures for proteins KaiA, KaiB and KaiC has enabled an examination of the interactions between them. X-ray, NMR and EM data with the results of assays in vitro and in vivo were thus compiled into a model of the 1 : 1 KaiA-KaiC complex. The model features two binding sites between the proteins that are both located on the outer surface of KaiC. There is no evidence for the central KaiC channel being used by either KaiA or KaiB for regulation of the level of phosphorylation of KaiC. Only the KaiCII hexameric ring that harbors all phosphorylation sites is likely contacted by KaiA and KaiB. The KaiCI and KaiCII domains that are the result of a gene duplication have divergent functions: the CI hexamer serves as a structural platform and is conformationally more rigid, whereas the CII hexamer is the functional center, and conformational changes in KaiCII domains triggered by phosphorylation and dephosphorylation are key to the generation of the rhythm with a ca. 24 h period. Application of hybrid structural methods will likely provide insight into the conformational properties of the binary KaiB-KaiC and the ternary KaiABC complexes, but only X-ray crystallography in combination with modeling of the dynamic processes underlying the interactions between the three clock components will disclose the atomic details required to understand the mechanism of this molecular timer. A central problem that remains to be solved is the origin of the temperature compensation – the independence of the clock period of temperature within a limited range – seen with the KaiABC clock reconstituted in vitro. Isolation of mutant proteins that lack temperature compensation and insight into potentially altered interactions between Kai proteins as a consequence of specific mutations might yield an improved understanding of this fascinating property exhibited by a complex of three proteins with bound ATP.

Acknowledgements

The US National Institutes of Health (grant R01 GM073845) supports our work on the cyanobacterial circadian clock. We thank our collaborators at Vanderbilt University, Drs. Carl H. Johnson and Phoebe L. Stewart for sharing their insight.

REFERENCES

1. Abrahams, J.P., Leslie, A.G.W., Lutter, R., and Walker, J.E. (1994). Structure at 2.8 Å resolution of F1 ATPase from bovine heart mitochondria. Nature *370*, 621–628.
2. Berman, H.M., Westbrook, J., Feng, Z., Gilliland, G., Bhat, T.N., Weissig, H., Shindyalov, I.N. and Bourne, P.E. (2000) The Protein Data Bank. Nucleic Acids Res. *28*, 235–242.
3. Ditty, J.L., Williams, S.B., and Golden, S.S. (2003). A cyanobacterial circadian timing mechanism. Annu. Rev. Genet. *37*, 513–543.
4. Dunlap, J.C., Loros, J.J., and DeCoursey, P.J. (2004). Chronobiology: Biological Timekeeping (Sunderland, MA: Sinauer).
5. Garces, R.G., Wu, N., Gillon, W., and Pai, E.F. (2004). Anabaena circadian clock proteins KaiA and KaiB reveal potential common binding site to their partner KaiC. EMBO J. *23*, 1688–1698.
6. Golden, S.S. (2004). Meshing the gears of the cyanobacterial circadian clock. Proc. Natl. Acad. Sci. USA *101*, 13697–13698.
7. Hayashi, F., Suzuki, H., Iwase, R., Uzumaki, T., Miyake, A., Shen, J.-R., Imada, K., Furukawa, Y., Yonekura, K., Namba, K., and Ishiura, M. (2003). ATP-induced hexameric ring structure of the cyanobacterial circadian clock protein KaiC. Genes to Cells *8*, 287–296.
8. Hayashi, F., Ito, H., Fujita, M., Iwase, R., Uzumaki, T., and Ishiura, M. (2004a). Stoichiometric interactions between cyanobacterial clock proteins KaiA and KaiC. Biochem. Biophys. Res. Comm. *316*, 195–202.
9. Hayashi, F., Itoh, N., Uzumaki, T., Iwase, R., Tsuchiya, Y., Yamakawa, H., Morishita, M., Itoh, S., and Ishiura, M. (2004b). Roles of two ATPase-motif-containing domains in cyanobacterial circadian clock protein KaiC. J. Biol. Chem. *279*, 52331–52337.
10. Hayashi, F., Iwase, R., Uzumaki, T., and Ishiura, M. (2006). Hexamerization by the N-terminal domain and intersubunit phosphorylation by the C-terminal domain of cyanobacterial circadian clock protein KaiC. Biochem. Biophys. Res. Comm. *348*, 864–872.
11. Hitomi, K., Oyama, T., Han, S., Arvai, A.S., and Getzoff, E.D. (2005). Tetrameric architecture of the circadian clock protein KaiB: a novel interface for intermolecular interactions and its impact on the circadian rhythm. J. Biol. Chem. *280*, 19125–19137.
12. Huang, C.C., Couch, G.S., Pettersen, E.F., and Ferrin, T.E. (1996). Chimera: an extensible molecular modeling application constructed using standard components. Pacific Symposium on Biocomputing *1*, 724.
13. Ishiura, M., Kutsuna, S., Aoki, S., Iwasaki, H., Andersson, C.R., Tanabe, A., Golden, S.S., Johnson C.H., and Kondo, T. (1998). Expression of a gene cluster *kaiABC* as a circadian feedback process in cyanobacteria. Science *281*, 1519–1523.
14. Ivleva, N.B., Gao T., LiWang A.C., and Golden, S.S. (2006). Quinone sensing by the circadian input kinase of the cyanobacterial circadian clock. Proc. Natl. Acad. Sci U.S.A., *103*, 17468–17473.
15. Iwasaki, H., Taniguchi, Y., Kondo, T., and Ishiura, M. (1999). Physical interactions among circadian clock proteins, KaiA, KaiB and KaiC, in cyanobacteria. EMBO J. *18*, 1137–1145.
16. Iwasaki, H., Williams, S.B., Kitayama, Y., Ishiura, M., Golden, S.S., and Kondo, T. (2000). A KaiC-interacting sensory histidine kinase, SasA, necessary to sustain robust circadian oscillation in cyanobacteria. Cell *101*, 223–233.
17. Iwasaki, H., Nishiwaki, T., Kitayama, Y., Nakajima, M., and Kondo, T. (2002). KaiA-stimulated KaiC phosphorylation in circadian timing loops in cyanobacteria. Proc. Natl. Acad. Sci. USA *99*, 15788–15793.

18. Iwasaki, H., and Kondo, T. (2004). Circadian timing mechanism in the prokaryotic clock system of cyanobacteria. J. Biol. Rhythms *19*, 436–444.

19. Iwase, R., Imada, K., Hayashi, F., Uzumaki, T., Morishita, M., Onai, K., Furukawa, Y., Namba, K., and Ishiura, M. (2005) Functionally important substructures of circadian clock protein KaiB in a unique tetramer complex. J. Biol. Chem. *280*, 43141–43149.

21. Johnson, C.H. (2004). Precise circadian clocks in prokaryotic cyanobacteria. Curr. Issues Molec. Biol. *6*, 103–110.

21. Johnson, C.H., and Egli, M. (2004). Visualizing a biological clockwork's cogs. Nature Struct. Mol. Biol *11*, 584–585.

22. Kageyama, H., Kondo, T., and Iwasaki, H. (2003). Circadian formation of clock protein complexes by KaiA, KaiB, KaiC, and SasA in cyanobacteria. J. Biol. Chem. *278*, 2388–2395.

23. Kageyama, H., Nishiwaki, T., Nakajima, M., Iwasaki, H., Oyama, T., and Kondo, T. (2006). Cyanobacterial circadian pacemaker: Kai protein complex dynamics in the KaiC phosphorylation cycle in vitro. Mol. Cell *23*, 161–171.

24. Katayama, M., Tsinoremas, N.F., Kondo, T., and Golden, S.S. (1999). *cpmA*, a gene involved in an output pathway of the cyanobacterial circadian system. J. Bacteriol., *181*, 3516–3524.

25. Kitayama, Y., Iwasaki, H., Nishiwaki, T., and Kondo, T. (2003). KaiB functions as an attenuator of KaiC phosphorylation in the cyanobacterial circadian clock system. EMBO J. *22*, 1–8.

26. Kondo, T. (2005). Unpublished results reported at meetings in 2005.

27. Leipe, D.D., Aravind, L., Grishin, N.V., and Koonin, E.V. (2000). The bacterial replicative helicase DnaB evolved from a RecA duplication. Genome Res. *10*, 5–16.

28. Liu, Y., Tsinoremas, N.F., Johnson, C.H., Lebedeva, N.V., Golden, S.S., Ishiura, M., and Kondo, T. (1995). Circadian orchestration of gene expression in cyanobacteria. Genes Dev. *9*, 1469–1478.

29. Mori, T., Saveliev, S.V., Xu, Y., Stafford, W.F., Cox, M.M., Inman, R.B., and Johnson, C.H. (2002). Circadian clock protein KaiC forms ATP-dependent hexameric rings and binds DNA. Proc. Natl. Acad. Sci. USA *99*, 17203–17208.

30. Nakajima, M., Imai, K., Ito, H., Nishiwaki, T., Murayama, Y., Iwasaki, H., Oyama, T., and Kondo, T. (2005). Reconstitution of circadian oscillation of cyanobacterial KaiC phosphorylation in vitro. Science *308*, 414–415.

31. Nakahira, Y., Katayama, M., Miyashita, H., Kutsuna, S., Iwasaki, H., Oyama, T., and Kondo, T. (2004). Global gene repression by KaiC as a master process of prokaryotic circadian system. Proc. Natl. Acad. Sci. USA *101*, 881–885.

32. Nishiwaki, T., Iwasaki, H., Ishiura, M., and Kondo, T. (2000). Nucleotide binding and autophosphorylation of the clock protein KaiC as a circadian timing process of cyanobacteria. Proc. Natl. Acad. Sci. USA *97*, 495–499.

33. Nishiwaki, T., Satomi, Y., Nakajima, M., Lee, C., Kiyohara, R., Kageyama, H., Kitayama, Y., Temamoto, M., Yamaguchi, A., Hijikata, A., Go, M., Iwasaki, H., Takao, T., and Kondo, T. (2004). Role of KaiC phosphorylation in the circadian clock system of *Synechococcus elongatus* PCC 7942. Proc. Natl. Acad. Sci. USA *101*, 13927–13932.

34. Pattanayek, R., Wang, J., Mori, T., Xu, Y., Johnson, C.H., and Egli, M. (2004). Visualizing a circadian clock protein: crystal structure of KaiC and functional insights. Mol. Cell *15*, 375–388.

35. Pattanayek, R., Williams, D.R., Pattanayek, S., Xu, Y., Mori, T., Johnson, C.H., Stewart, P.L., and Egli, M. (2006). Analysis of KaiA–KaiC protein interactions in the cyano-bacterial circadian clock using hybrid structural methods. EMBO J. *25*, 2017–2028.

36. Schmitz, O., Katayama, M., Williams, S. B., Kondo, T., and Golden, S.S. (2000). CikA, a bacteriophytochrome that resets the cyanobacterial circadian clock. Science *289*, 765–768.

37. Takai, N., Nakajima, M., Oyama, T., Kito, R., Sugita, C., Sugita, M., Kondo, T, and Iwasaki, H. (2006). A KaiC-associating SasA–RpaA two-component regulatory system as a major circadian timing mediator in cyanobacteria. Proc. Natl. Acad. Sci. USA *103*, 12109–12114.

38. Taniguchi, Y., Yamaguchi, A., Hijikata, A, Iwasaki, H., Kamagata, K., Ishiura, M., Go, M., and Kondo, T. (2001). Two KaiA-binding domains of cyanobacterial circadian clock protein KaiC. FEBS Lett. *496*, 86–90.

39. Tomita, J., Nakajima, M., Kondo, T., and Iwasaki, H. (2005). Circadian rhythm of KaiC phosphorylation without transcription-translation feedback. Science *307*, 251–254.

40. Uzumaki, T., Fujita, M., Nakatsu, T., Hayashi, F., Shibata, H., Itoh, N., Kato, H., and Ishiura, M. (2004). Role of KaiA functional domains in circadian rhythms of cyanobacteria revealed by crystal structure. Nature Struct. Mol. Biol. *11*, 623–631.

43. Vakonakis, I., Sun, J., Wu, T., Holzenburg, A., Golden, S.S., and LiWang, A.C. (2004a). NMR structure of the KaiC-interacting C-terminal domain of KaiA, a circadian clock protein: Implications for the KaiA-KaiC Interaction. Proc. Natl. Acad. Sci. USA *101*, 1479–1484.

43. Vakonakis, I., Klewer, D.A., Williams, S.B., Golden, S.S., and LiWang, A.C. (2004b). Structure of the N-terminal domain of the circadian clock-associated histidine kinase SasA. J. Mol. Biol. *342*, 9–17.

43. Vakonakis, I., and LiWang A.C. (2004). Structure of the C-terminal domain of the clock protein KaiA in complex with a KaiC-derived peptide: implications for KaiC regulation. Proc. Natl. Acad. Sci. U.S.A. 101, 10925–10930.

44. Williams, S.B., Vakonakis, I., Golden, S.S., and LiWang, A.C. (2002). Structure and function from the circadian clock protein KaiA of *Synechococcus elongatus*: a potential clock input mechanism. Proc. Natl. Acad. Sci. USA *99*, 15357–15362.

45. Xu, Y., Mori, T., and Johnson, C.H. (2000). Circadian clock-protein expression in cyanobacteria: rhythms and phase setting EMBO J. *19*, 3349–3357.

46. Xu, Y., Mori, T., and Johnson, C.H. (2003). Cyanobacterial circadian clockwork: roles of KaiA, KaiB, and the *kaiBC* promoter in regulating KaiC. EMBO J. *22*, 2117–2126.

47. Xu, Y., Mori, T., Pattanayek, R., Pattanayek, S., Egli, M., and Johnson, C.H. (2004) Identification of key phosphorylation sites in the circadian clock protein KaiC by crystallographic and mutagenetic analyses. Proc. Natl. Acad. Sci. U.S.A. *101*, 13933–13938.

48. Ye, S., Vakonakis, I., Ioerger, T.R., LiWang, A.C., and Sacchettini, J.C. (2004). Crystal structure of circadian clock protein KaiA from *Synechococcus elongatus*. J. Biol. Chem. *279*, 20511–20518.

49. Yildiz, Ö, Doi, M., Yujnovsky, I., Cardone, L., Berndt, A., Henning, S., Schultze, S, Urbanke, C., Sassone-Corsi, P., and Wolf, E. (2005). Crystal structure and interactions of the PAS repeat region of the *Drosophila* clock protein PERIOD. Mol. Cell. *17*, 69–82.

50. Zhang, X., Dong, G., and Golden S.S. (2006). The pseudo-receiver domain of CikA regulates the cyanobacterial circadian input pathway. Mol. Microbiol. *60*, 658–668.

CHAPTER 14

PROTEIN-PROTEIN DOCKING USING THREE-DIMENSIONAL REDUCED REPRESENTATIONS AND BASED ON A GENETIC ALGORITHM

ANDY BECUE, NATHALIE MEURICE[†], LAURENCE LEHERTE
AND DANIEL P. VERCAUTEREN[*]

Abstract: An original scoring function dedicated to the docking of biological macromolecules is implemented in complementarity research within an automated algorithm. As these systems involve complicated atomic structures, we use for each partner reduced representations obtained by topological analysis of electron density maps at medium resolution, and develop specific terms for the characterization of the intermolecular interactions including a geometric fit based on the knowledge in a statistical survey, an electronic interaction potential using an expression of modified Coulomb type, and a penalty score based on detection of steric clashes. To validate the strategy, we performed automated docking runs, based on genetic algorithms (GA) for various protein-protein complexes including enzyme-inhibitor and antibody-antigen. For most complexes, the GA-proposed fit solutions have *rmsd* values below 3 Å relative to the native structures

Key words: Macromolecular Complementarity, Docking, Electron Density, Crystallographic Medium Resolution, Critical Points, Genetic Algorithms

INTRODUCTION

Theoretical models of bio-macromolecules are concerned with the development of efficient drugs, the mechanisms of signal transduction, and gene regulation by transcription factors. All these phenomena have in common the capital role played by intermolecular interactions between the partners and their ability to recognize patterns. The current developments in structural genomics are, however, continuously

---

[*] E-mail: daniel.vercauteren@fundp.ac.be
[†] F.N.R.S. Scientific Research Worker

J.C.A. Boeyens and J.F. Ogilvie (eds.), Models, Mysteries and Magic of Molecules, pp. 301–323.
© 2008 *Springer.*

proposing many protein structures, experimentally determined or theoretically predicted, from which still only little functional information is available. To develop
strategies to predict how two or more molecules can interact – the *docking* problem – is thus essential. This term refers to all methods that have as their objectives
to find the correct orientations and schemes of interaction between several partners,
beginning with their three-dimensional (3D) coordinates.

Docking between biomolecular partners is viewed at three levels of complexity according to the nature of the partners of interest – drugs, proteins, or DNA
oligomers. With regard to protein-ligand (P-L) recognition, efficient methods of
drug design should clearly be able to test probe molecules in large collections and
to propose putative active drugs, which would diminish the experimental costs of
extensive and combinatorial tests [1]. In this sense, strategies to discover drugs have
yielded numerous powerful tools that predict the binding modes of small molecules
(ligands) at active sites of proteins [1–4]. For such a purpose, algorithms are based
on flexible atomic representations of potent drugs [5, 6] and on locally or fully
flexible receptors [7].

Protein-protein (P-P) docking is obviously more complicated because both partners are huge and have numerous degrees of freedom [8, 9]. Even if the flexibility
of small molecules can be efficiently modeled, it is still difficult to contemplate
the full flexibility of a protein, which makes complementarity studies between
such macromolecules a complicated computational task [10]. To avoid protracted
calculations, one must consider several simplifications of the problem, including
simplified models. Predictions of protein-DNA (P-DNA) association are still at
a pioneer theoretical stage. The lack of efficient methods is due mainly to the
highly flexible DNA strings, which undergo deformations during the binding of the
proteins, and the lack of a clearly comprehended recognition code [11].

In this context, we developed a complete strategy for docking, beginning with an
in-house building of reduced representations of macromolecules based on the calculation of electron density maps of proteins at medium crystallographic resolution,
followed by their topological analysis, as described by Becue et al [12]. To quantify
the interactions and affinities between the molecular partners in a given configuration, we developed a complete scoring function to suit these representations. To
perform the complementarity research, we then built an entirely automated docking
method, based on a genetic-algorithm (GA) core.

Before describing the results that we obtained and to ensure that we introduce
thoroughly the numerous aspects of the docking problem, we treat three major
aspects that one must consider when developing a method for macromolecular
docking [10]: (i) the molecular representation that is best suited to the docking
strategy; (ii) the exploration algorithm that covers the potential-energy landscape
and (iii) the scoring function that associates a quality score to each proposed solution. These three elements can be combined independently, but are somehow linked
as they all influence the efficiency of the method. In Materials and Methods, we
expound our strategy, with a description of the macromolecular sets with which we
worked. In Results, we describe the construction of the fitness expression and its

application to the docking of P-P complexes. We discuss finally the efficiency of the scoring function and the abilities of our docking algorithm.

Representation of Interacting Molecular Systems

The major difficulty in modeling the behavior of proteins is the great computational cost; the use of fully flexible atomic representations is consequently generally prohibitive. A current way to accelerate the docking process is to consider rigid partners; this solution has the advantage that it allows to decrease greatly the number of degrees of freedom [13, 14]. The docking can then be compared to a *lock-and-key* problem with only six degrees of freedom, three rotational and three translational, for the moving molecules. This approach yields, however, optimal results only if the molecular structures vary little between the associated and non-associated states [15]. If many conformational alterations occur, principally from the interfacial residues or even from the main chain, the performance of the docking problem becomes rapidly degraded. Moreover, the ultimate aim of all docking methods is to be able to handle *unbound* proteins, of which structures have been solved independently of any complex. An additional difficulty induced by the conformational alterations between *bound* and *unbound* partners therefore limits strongly the rigid-body approach [16]. In general, that approach is thus combined with some blurring of the surface representation, e.g. by tolerating some interpenetration [17, 18].

Another solution involves reduced representations of the macromolecules. For this purpose, many simplified models of proteins were developed first for folding processes [19–22] and subsequently for docking problems [17, 23]. The various proposed models differ in their degree of simplification. A residue may be described by only one point per side chain [21, 24, 25] or more [20, 22, 26]. Other researchers prefer to use discrete positions with a grid representation of the molecules [4, 27]. A small interval enables obtaining a nearly atomic representation [18]; a larger one yields a sharply reduced model [17, 23]. Many other techniques have been proposed to simplify the systems, e.g. spherical harmonics [28, 29], Connolly surfaces [8, 30, 31] etc.

We used reduced representations of the macromolecules obtained by topological analysis of distributions of electron density calculated at medium crystallographic resolution, i.e. 2.85 Å, as explained by Becue et al [12]. According to this method, each molecule is represented with a set of critical points, which are the points at which the gradient of the density vanishes, *i.e.*, peaks, passes, pales, and pits. Among critical points of these four kinds, we considered only the peaks as they correspond to local maxima of electron density, that is, atomic groups or chemical functions. Amino acids are thereby represented by one peak for the backbone and zero to two peaks for the side chain, according to their size and chemical composition (Figure 14-1). An advantage of the proposed method, relative to others that encompass only geometrical parameters such as centres of mass, is that each peak is associated with numerical descriptors – position, electron density, eigenvalues, and eigenvectors, which confer on the peaks an anisotropic shape.

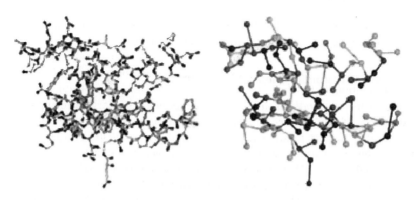

Figure 14-1. Atomic structure and reduced representation in terms of a graph of peaks of a protein chain (PDB file: 1B69). The peaks are colour-coded per amino-acid type. Visualizations of the atomic and reduced representations were obtained on combining the Swiss PDB Viewer (http://www.expasy.org/spdbv) and the ray-tracing POV-Ray (http://www.povray.org) programs (see color plate section)

This approach has been successfully applied to various systems in our laboratory, e.g. complementarity between cyclodextrins and ligands [32, 33] or similarity between small molecules [34, 35].

Exploration Algorithm

Within a complementarity study, the goal of the exploration algorithm is to locate, rapidly and precisely, the most favorable relative position of the two partners, which is generally represented with a global minimum in the energy landscape. Most docking problems are, however, characterized with a complicated topology having many local minima dispersed over the surface; the global minimum is typically hidden among them. An exhaustive exploration of the conformational space is difficult as it would require great computational resources, and even greater if the flexibility of both partners must be considered. Much effort is thus devoted to develop intelligent algorithms for exploration. We cite, non-exhaustively, Monte-Carlo (MC) optimization [36–38], which randomly explores a potential-energy surface and retains the best solutions; molecular dynamics (MD) [7, 39], based on the resolution of Newton's equations of motion, and which typically requires extensive computations, and is thus applied mainly when the molecular conformations or configurations are near the final solution. According to the temperature, both MC and MD techniques might fail to overcome large energy barriers and thus become retained in local minima. Fourier correlations have also been proposed to perform rapidly an exhaustive search by superimposing rigid structures digitized on grids [40]; the principal inconvenience of these methods is that a new set of grids must be recalculated after each molecular rotation [28]. Finally, we cite evolutionary algorithms [2], and in particular genetic algorithms (GA) [8, 31, 41], which belong to *simulated evolution methods* and transpose numerically the *Laws of Evolution*, as established by Darwin; they can be viewed as guided procedures for random search over the energy landscape [8, 9, 42, 43], which, with the concept of populations of candidates, make them well adapted to multiple solutions.

We developed a GA strategy specifically dedicated to our reduced representations. Initially developed by Holland [44], GA are considered powerful and versatile techniques of optimization [8, 9, 43, 45]; their robustness and the great freedom for the user, in terms of parameterization, searching strategies etc., constitute their principal advantages. They have thus become widely used to study molecular conformation [46, 47], molecular similarity [48], protein folding [46, 49–51], and protein docking [8, 31, 42, 52]. Comparing flexible docking programs, Bursulaya et al [53] showed that tools based on GA, i.e. GOLD [42] and AutoDock [54], performed reliably with about half of the structures correctly predicted. For comparison, reconstruction algorithms attained 30% with DOCK [55], 35% with FlexX [56], and 76% with ICM [57] that uses a Monte-Carlo approach. In our laboratory, we successfully applied GA based on peak representations of small molecules to perform similarity studies [34, 58, 59]. We cite the work of Goldberg [41] and Cartwright [60] for detailed descriptions of the various numerical operators that must be used to govern a GA.

Scoring Function

A crucial point in the development of a docking method resides in the design of an efficient scoring function. It must be sufficiently complete to be able to associate the best scores with the solutions that are observed, but it must be also sufficiently simple to allow a rapid evaluation of millions or billions of possible conformations. The effective definition and parameterization of the fitting expression has thus an impact on the efficiency of the docking algorithm – in terms of speed of completion and quality of the proposed solutions, signifying an evaluation of the quantity of false positives towards correct configurations. We distinguish scoring functions of three main kinds: (i) force fields, which allow the simulation of (non-)binding interactions like those used mainly in molecular mechanics or molecular dynamics, e.g. with Amber [2, 36] and CHARMM [61, 62]; (ii) empirical scores, which are composed of an ensemble of experimentally determined physico-chemical terms and geometric parameters contributing to the entropy, the desolvation, or the detection of hydrogen bonds [56, 63] and (iii) scores based on knowledge, which are constructed statistically based on the assumption that frequently encountered situations are energetically favorable [64]. Many parameters are generally considered in the various fitting equations available in the literature, e.g. shape complementarity, steric clashes, hydrogen bonds, surfaces of contact between the partners, paired-residue potential, electrostatic interactions etc. Below we present the terms that we retained in our scoring function.

Materials and Methods

To develop and to parameterize our scoring function, we based our strategy on the observation of an ensemble of macromolecular complexes. For this purpose, we considered a set composed exclusively of protein-protein (P-P) complexes. We chose COMBASE, established by Sali et al [ftp://salilab.org/pub/ilya/], which provides a non-redundant list of 475 P-P complexes with at least 1000 Å² of interfacial contacts; the 3D coordinates of all referenced structures were retrieved from the Protein Data Bank (PDB) [65, 66]. We then parameterized and tested our GA-based docking program on considering 15 protein-protein complexes, which were not included in the first large set of macromolecular complexes (Table 14-1).

Using a procedure as explained by Becue et al [67], the PDB files were collected and read with our program (FORTRAN 90) that identifies all partners encoded in the files, isolates the protein chains from each other, and creates a specific coordinate file for each partner. In a second step, we generated the reduced representations of the macromolecules in terms of critical-point graphs, using the method described by Becue et al [12].

RESULTS

Here we first explain the development of the scoring function with the description and the parameterization of each term composing the function. We then present the building and optimization of the genetic algorithm, and the results of the docking experiments.

Scoring Function

Using our simplified protein models, we defined descriptors in a set particularly well adapted to the peaks. For this, we combined the peak properties – electron density, anisotropic volume etc. – with the physico-chemical characteristics of the molecular subunits associated with each peak [12]. In this scoring function:

$$(1) \qquad Fitness \quad (\%) = \frac{\alpha PP + \beta Cb}{\alpha + \beta} - \gamma St_{clashes}$$

the purpose of the PP term is to evaluate the quality of the contact interface between two protein chains. Of two parts, a paired-residue term scores the amino-acid (aa) recognition, and a value of solvent-accessible surface area (SASA) quantifies the geometric matching. The purpose of the Cb term is to bring to the fore the effective complementarity of electronic charges by evaluating the electrostatic interactions at the interface. The $St_{clashes}$ term is a penalty score in the form of detection of steric clashes that substantially decrease the score if too great an interpenetration between the partners is detected. The global strategy of our scoring function is based on the normalization of each term to 100%. For an effective combination between all terms, they can all be weighted by varying the parameters α, β, and γ. A derivation of those scoring terms is briefly described below.

Table 14-1. Description of the protein-protein complexes used in the parameterization and test sets, in terms of the numbers of atoms and PKs, interface area, and global shape

PDB ID	Name	# atoms (chain_IDs)	# peaks (chain_IDs)	Interface[†] [Å^2]	Shape of the interface
• Parameterization set					
1aks	Alpha Trypsin	1153 (A), 879 (B)	231 (A), 189 (B)	5648.6	tangled up
1cgj	Serine Protease/Inhibitor	1799 (E), 436 (I)	455 (E), 106 (I)	1989.1	pocket shape
1dok	Monocyte Chemoattractant Protein 1	597 (A), 604 (B)	141 (A), 138 (B)	2099.1	planar and narrow
1ecm	Chorismate Mutase Domain	741 (A), 774 (B)	179 (A), 190 (B)	4857.8	α-helices dimerization
1ubs	Tryptophan Synthase	1945 (A), 2963 (B)	466 (A), 718 (B)	2794.2	planar and circular
• Test set					
1brs	Barnase (G Specific Endonuclease)/Barstar	864 (A), 695 (D), 878 (B), 665 (E), 839 (C), 699 (F)	209 (A), 165 (D), 221 (B), 160 (E), 199 (C), 173 (F)	1670 (A–D), 1753 (B–E), 1594 (C–F)	
1gla	Glycerol Kinase/Glucose-Specific Factor III	1201 (F), 3778 (G)	299 (F), 918 (G)	1312	
1igc	IGG1 Fab Fragment/Domain III of Protein G	3326 (L+H), 446 (A)	832 (L+H), 113 (A)	1422	

Table 14-1. (Continued)

PDB ID	Name	# atoms (chain_IDs)	# peaks (chain_IDs)	Interface[†] [Å^2]	Shape of the interface
1nmb	Fab NC10/Neuraminidase (N)	1798 (L+H), 3061 (N)	442 (L+H), 761 (N)	1496	
1spb	Subtilisin Prosegment/ Subtilisin	557 (P), 1860 (S)	149 (P), 476 (S)	2315	
1stf	Papain/Inhibitor Stefin B mutant	1655 (E), 789 (I)	394 (E), 189 (I)	1843	
1tgs	Trypsinogen/Pancreatic Trypsin Inhibitor	1646 (Z), 416 (I)	413 (Z), 106 (I)	1788	
1udi	Uracil-DNA Glycosylase/ Inhibitor	1818 (E), 654 (I)	444 (E), 161 (I)	2028	
3hfl	IGG1 Fab Fragment/ Lysozyme (Y)	3250 (L+H), 1001 (Y)	806 (L+H), 242 (Y)	1847	
4htc	Alpha-Thrombin/ Recombinant Hirudin	253 (L), 2069 (H), 447 (I)	61 (L), 510 (H), 110 (I)	2246 (L–H) 3657 (H–I)	

[†] The interfaces were calculated from the graphs of PKs, following the procedure explained in Becue et al [2004].

Paired-residue potential (PP)

After having generated the graphs of peaks for all protein chains individually, we restored them together in their crystalline conformations. We then performed a statistical study by collecting all inter-distances between the amino acids (aa), labeled as *backbone* (BB) or *side-chain* (SD) peak. As each aa can be paired with 20 others, this data collection yielded the construction of 20 tables of inter-distance distributions per aa (ALA-ALA, ALA-ARG etc.), with the additional distinction between SD-SD, SD-BB, and BB-BB peaks. For each aa$_i$-aa$_j$ pair, the values were normalized to have a maximum value of 1.0 in each table. As an illustration, the histograms obtained for CYS-CYS and ARG-GLU appear in Figure 14-2. For the

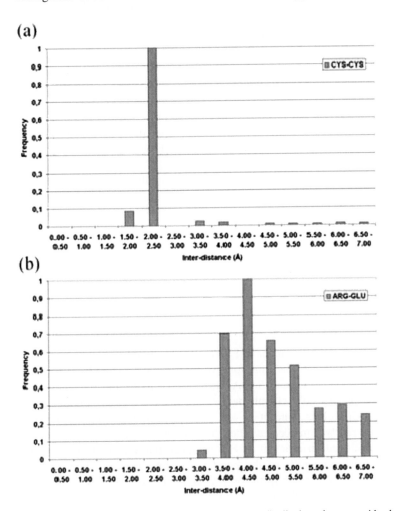

Figure 14-2. Histogram of normalized inter-distance distributions between side-chain peaks for CYS-CYS and ARG-GLU amino-acid pairs

CYS-CYS pair, the inter-distance distribution is extremely narrow and centred on an interval 2.00–2.50 Å; these two aa are hence likely to be found with such a distance (typically a disulfide bond). For the ARG-GLU pair, the inter-distance distribution is broader and centred on an interval 4.00–4.50 Å. These two aa present chains longer than CYS and are thus more flexible. We thus obtained a distribution of the possibilities according to which they can combine with each other at a small distance. Completing this procedure, we constructed a set of reference tables for each aa_i-aa_j pair (named PP_{ij}).

As much work has already been devoted to the properties of P-P interfaces, we decided to combine the PP_{ij} statistical term, describing the inter-distance distributions, with a score for aa pairing likeliness, as given by Glaser et al [64]; observing 621 P-P interfaces, they computed the residue-residue contacts between the 20 aa. They normalized each number by the residue volume, to obtain a criterion of propensity of residue-residue contact $G_{ij}(v)$:

$$(2) \qquad G_{ij}(v) = A.\log\left(\frac{Q_{ij}(v)}{w_i w_j}\right)$$

in which A is a parameter arbitrarily set to 10, $Q_{ij}(v)$ represents the number of contacts between the residues normalized by the residue volumes, and w_i and w_j denote the normalized frequencies of residues i and j.

By combining our statistical results, which encompass only distance distributions between the aa_i-aa_j pairs, with the Glaser parameters, which represent the likeliness of formation of these pairs, we constructed a large knowledge-based potential dedicated to the protein dimerization of the form:

$$(3) \qquad PP' = \frac{\displaystyle\sum_i^{N_{PK}^i}\sum_j^{N_{PK}^j} PP_{ij}(d)G_{ij(v)}}{\displaystyle\sum_i^{N_{PK}^i}\sum_j^{N_{PK}^j} G_{ij(v)}}$$

in which N_{PK}^i and N_{PK}^j are the number of PK for proteins i and j.

As a second contribution to the final form of the PP term, we added a part for the solvent-accessible surface area (SASA). Beyond the molecular recognition in terms of preferred distances, the complementarity of shape constitutes an important descriptor for docking. This concept is well defined on considering the surface that is buried between two molecules, after their complexation. For this purpose, we previously developed an efficient way to calculate the SASA of our reduced models [67], and, consequently, of the interface of dimerization. Many articles are concerned with P-P dimers [68]; to form stable complexes, the contact surface between two proteins is generally considered to exceed 1200 Å^2 [69], but additional analysis of the P-P interface proved that the binding interface is not necessarily the one that shows the greatest buried surface [16]. Correct normalization of the contribution of the interface between two partners in contact is, moreover, difficult, as we observed no relationship between the interface values and other parameters related to the structure or composition of the partners. For instance, large complexes can dimerize with a small interface value, e.g. the aldehyde ferredoxin oxydoreductase (PDB code: 1AOR), whereas small complexes might present a large contact area, e.g. the coat protein of GA Bacteriophage (PDB code: 1UNA). Investigating 75 protein-nucleic acid complexes (DNA and RNA), Nadassy et al [70] concluded that the interface areas vary between 1120 and 5800 Å^2, with an average value $3000 \pm 1200 \, \text{Å}^2$. They found also that the minimum interface value for stable association is similar to the value observed for P-P complexes, that is, $1200 \, \text{Å}^2$.

Considering these observations, we decided not to implicate directly the interface value in the scoring functions. On the contrary, we preferred to use this parameter in the form of a penalty-like score, when the interface value is less than $1200 \, \text{Å}$. When completing our knowledge-based recognition score with the interface contribution, we thus obtained the final expression of the *PP* geometric score:

$$(4) \qquad PP = \frac{\displaystyle\sum_{i}^{N_{PK}^i} \sum_{j}^{N_{PK}^j} PP_{ij}(d) G_{ij(v)}}{\displaystyle\sum_{i}^{N_{PK}^i} \sum_{j}^{N_{PK}^j} G_{ij(v)}} \cdot \frac{Interface}{1200} 100(\%)$$

When the interface value exceeds $1200 \, \text{Å}^2$, the ratio (*Interface*/1200) is set to 1.0, as its only role is to penalize the solutions presenting a small interface value during docking. We then applied the *PP* recognition scores to the set of P-P complexes. The results are similar between the various complexes in their crystalline configuration, with an average value of $65.3 \pm 6.6\%$ for the *PP* score. This result allows one to assume, in a first approximation, that the docking of unknown complexes produces solutions of a similar range.

Beyond their complementarity studies, Norel et al [16] affirmed that native complexes do not necessarily show the largest proportion of non-polar surface that would be adopted by the same partners in another configuration, nor the most hydrogen bonds. For all these reasons, we decided not to introduce any polarity or hydrogen-bond contribution into our scoring function.

Coulomb pseudo-potential (Cb)

As several aa might be charged and might lie in close contact during the docking experiment, electrostatic interactions might play an important role in the association between two proteins. We chose a simplified potential expression, specifically the classic Coulomb interaction, as it is well adapted to reduced representations [8, 71, 72]. For this purpose, we associated a positive unit charge with the Lys, Arg, and His SD, and a negative unit charge with the Glu and Asp SD.

Moreover, the evaluation of the Coulomb potential, for the full range of inter-peak distances, might become protracted, because of the many pairs to be considered. It is thus necessary to limit the extent of the calculation, particularly for macro-molecules. We modified the Coulomb expression with a force-switching function dedicated to the Coulomb interaction, as established by Field [73]:

$$
(5) \quad U_{Cb} = \begin{cases} \dfrac{Q_i Q_j}{R} + \dfrac{8 Q_i Q_j}{\gamma} \left[r_{on}^2 r_{off}^2 (r_{off} - r_{on}) - \dfrac{1}{5}(r_{off}^5 - r_{on}^5) \right] & \text{if } R \leq r_{on} \\[3mm] Q_i Q_j \Big[A\Big(\dfrac{1}{R} - \dfrac{1}{r_{off}}\Big) + B(r_{off} - R) + C(r_{off}^3 - R^3) \\[2mm] \qquad + D(r_{off}^5 - R^5) \Big] & \text{if } r_{on} < R \leq r_{off} \\[3mm] 0 & \text{if } R > r_{off} \end{cases}
$$

in which Q_i and Q_j are the electronic charges of elements i and j separated by distance R. The other parameters are:

$$
r_{on} = 8\text{Å}; \; r_{off} = 12\text{Å}; \; \gamma = (r_{off}^2 - r_{on}^2)^3
$$

$$
A = \frac{r_{off}^4 (r_{off}^2 - 3r_{on}^2)}{\gamma}; \; B = \frac{6 r_{on}^2 r_{off}^2}{\gamma}; \; C = \frac{-(r_{off}^2 + r_{on}^2)}{\gamma}; \; D = \frac{2}{5\gamma}
$$

According to Equation 5, the Coulomb potential is progressively set to zero while conserving a smooth $1/R$ evolution.Figure 14-3 illustrates the force-switching

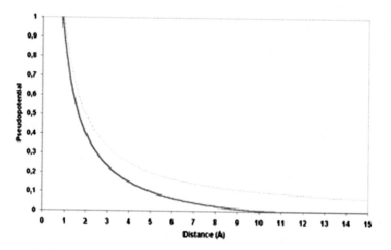

Figure 14-3. Illustration of the force-switching function [73] applied to the Coulomb potential calculated between two unit positive charges as a function of the distance. The dashed line depicts the straight Coulomb potential evolution

function when applied to the Coulomb potential calculated between two "+1" charges as a function of distance.

For the evaluation of each P-P complex, we thus considered a total electrostatic score based on the separation of favorable potentials (U_{favor}), i.e. due to opposite charges interacting, and unfavorable ones ($U_{unfavor}$), i.e. due to like charges interacting, both calculated using equation 5. We then evaluated the *Cb* score as:

$$(6) \qquad Cb = \frac{U_{favor}}{U_{favor} + U_{unfavor}} 100.0(\%)$$

Steric Clashes (St$_{clashes}$)

Most methods employ steric clashes as a filter to eliminate solutions presenting too great an interpenetration between partners [30, 31]. When using atomic representations, each atom can be viewed as a hard sphere of radius equal to its van der Waals value. However, when using reduced representations, one must define a set of radii to be compatible with each subunit. Furthermore, *blurring* is sometimes necessary when considering a rigid-body approach [18]. Peaks obtained by topological analysis of the electron density present a notably anisotropic shape; each peak is described with three eigenvalues and eigenvectors that provide information about the local behaviour of the electron density. In 3D, the peaks are thus represented with ellipsoids of revolution [74].

To compute the detection of steric clashes, we used a Lennard-Jones 6–12 pseudo-potential expression U_{ij}, defined in Equation 7. Such a potential, which was successfully applied to the complementarity of several host-guest systems [74], rapidly increases if two elements, e.g. atoms, spheres or peak, approach one another too closely, and interpenetrate. We performed the calculation of U_{ij} for each peak$_{ij}$ pair of all complexes:

$$(7) \qquad U_{ij} = -\frac{A_{ij}}{r_{ij}^6} + \frac{B_{ij}}{r_{ij}^{12}}$$

in which $A_{ij} = V_i.V_j$ and $B_{ij} = A_{ij}.(r_i+r_j)^6$, V_i and V_j are the peak volumes and r_i, r_j, and r_{ij} are the internal radii of the two ellipsoids and the distance between them, respectively. A steric clash was considered if the pseudo-potential value exceeded a limiting value, i.e. 5.0 in this case.

When applied to our macromolecular sets, we observed no interpenetration for most complexes kept in their crystalline configuration; only a few showed some steric clashes – 20 with one *bump*, nine with two *bumps* and one with three *bumps*. Some fuzziness was thus introduced through *tolerated* interpenetrations. If the number of clashes is less than the tolerance threshold, no *bump* is considered in the steric penalty score. We decided to set this number to three, according to the results above. To obtain a penalty score that is expressible as per cent, we limited the number of non-tolerated clashes to 30, and calculated the proportion of *bumps* relative to this value:

$$(8) \qquad St_{clashes} = \frac{(N_{clashes} - N_{threshold})}{N_{max}} 100.0(\%)$$

in which $N_{threshold}$ is set to 3 and N_{max} to 30. Beyond this limit, the penalty score equals 100%.

Weighting of the Scoring Terms (α, β, and γ)

To favor the knowledge-based *PP* score, which relies on a thorough statistical study, towards the *Cb* pseudo-potential, we optimized the weights to the following values: $\alpha = 3.5$, $\beta = 1.0$, and $\gamma = 0.9$. When applying the scoring function to the P-P complexes used in this work, we obtained a satisfactory regularity in the final scores with an average value of 63.7 $\pm$ 8.8%, as presented in Figure 14-4.

Genetic Algorithm

Presentation of the algorithm

To simulate the natural laws of evolution, GA are based on a population of numerical chromosomes, which contain the key variables of the problem and thus represent potential solutions. For a docking experiment, each chromosome should code for a possible orientation of one partner *versus* another. In our GA code (FORTRAN 90), we retained the solution generally applied to rigid-body complementarity studies,

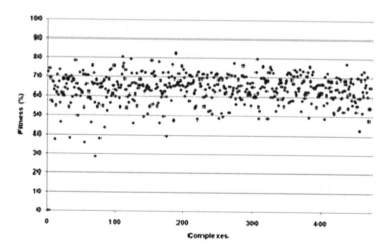

Figure 14-4. Final scores (in per cent) obtained when applying the fitness expression (eq. 1) to the set of 475 protein-protein complexes. The weights are $\alpha = 3.5$, $\beta = 1.0$, $\gamma = 0.9$

that is, six genes per movable partner, while the other was fixed. The first three variables code for the position of the centre of mass of the moving peak graph, and the last three for the Euler rotation angles. For each candidate in a population, the selection operator evaluates their quality (score) according to Equation 1 and constitutes a temporary population, which evolves by mutation and mating (crossover) to form a new one. Among various selection strategies [75], we selected a roulette scheme, as described by Cartwright [60].

Parameterization runs

GA are particularly sensitive to the various parameters, as their combination strongly influences the performance of the computation experiment. We focused on the optimization of the principal parameters that govern a GA experiment: N_{gener} – the number of cycles, N_{pop} – the size of the artificial population, and P_{mut} and P_{cross} – the rates of evolution (mutation and crossover, respectively). A satisfactory configuration of the GA is crucial, as it strongly influences the efficiency of the algorithm in terms of quality of the solutions and duration of computation. To evaluate the effect of each parameter, independent experiments were run on five P-P complexes of which the crystalline configuration is to be retrieved by the GA and of which the size and shape of dimerization interfaces differ (Table 14-2). By varying one parameter at a time, we observed how the GA behaved according to each parameter set. The initial values were set to those obtained by Meurice et al [34] in their similarity studies: $N_{gener} = 5000$, $N_{pop} = 100$, $P_{cross} = 60.0\%$, $P_{mut} = 0.1\%$ (the following notation occurs further in the paper: 5000 / 100 / 60.0 / 0.1). The ranges of tested values are presented in Table 14-2. After each complementarity study, the best 50 solutions proposed during the GA application were retained.

GA experiments are generally analysed on plotting the evolution of both the means (or the cumulative means) of the successive population fits to ensure that the GA proceeds to a global optimization and the means of the largest scores stored in the *best* list, which is continuously updated. If such charts are well adapted to problems of molecular superposition, in which the fitness can be set to the distance between the atoms to reflect directly the quality of the superposition, they do not always provide correct information about the performance of a docking algorithm. In complementarity studies, candidates with high scores might present large differences with the native structure, that is, false positives. By running docking

Table 14-2. Ranges of values that have been considered in the parameterization study, for each variable governing the GA based protein-protein docking

Parameter	Range of values	Step
Number of cycles (N_{gener})	3000 to 5000	500
Number of chromosomes (N_{pop})	100 to 500	100
Crossover rate (P_{cross})	50 to 70%	5
Mutation rate (P_{mut})	0.1 to 0.5%	0.1

experiments with complexes of known structure, we were able to calculate the root mean square deviations (*rmsd*) between the 3D structures of each complex obtained at the completion of the GA and the crystalline ones. The *rmsd* values were calculated on the all-atom structures, which were readily retrieved by the algorithm after the GA completion, as the original atomic positions are known and can be back superimposed on the graphs of peaks. To assess the efficiency of our GA during the parameterization, we determined whether the best solutions that were finally proposed were either optimal or false positives.

On analysis of the best combinations of GA parameters and the corresponding *rmsd* values, after executing 100 complementarity studies (five complexes times 20 configurations), it rapidly appeared that more than one optimal configuration of the GA scheme might yield satisfactory results (*rmsd* < 3 Å). To decide which combination of parameters appeared to be optimal, we analysed more profoundly the influence of each parameter. We concluded that three configurations of the GA could be conserved for further runs (Table 14-3). On analysing the results, we observed that 3000 cycles are sufficient to perform all the complementarity studies; no improvement was observed beyond this value as most of the best solutions are retrieved before 3000 iterations. When studying the effect of the population size on the quality of the proposed solution, we observed that $N_{pop} = 200$ represented a great improvement in terms of *rmsd* values, compared to $N_{pop} = 100$. A larger value of N_{pop} brings no additional improvement. The duration of computation, proportional to the size of the population, $N_{pop} = 200$, represents an effective compromise between the quality of the proposed solutions and the speed of completion. P_{mut} and P_{cross} have a less important role than N_{pop}. For instance, a mutation rate 0.4% is well adapted to $N_{pop} = 100$. It was however more difficult to find an optimal crossover rate (different from the original 60.0%) when considering $N_{pop} = 200$. Consequently, we retained $P_{cross} = 65.0\%$ and $P_{mut} = 0.4\%$ for $N_{pop} = 100$, and $P_{cross} = 60.0\%$ and $P_{mut} = 0.1$ or 0.5% for $N_{pop} = 200$. For illustration, Figure 14-5 represents the superposition of the crystalline structures of the five complexes that served for this parameterization with the least *rmsd* values, as proposed by the GA in its configuration 3000 / 200 / 60.0 / 0.1.

Validation runs

To validate and to test the three sets of parameters, we applied the GA docking to ten additional P-P complexes, producing 13 P-P docking experiments as some

Table 14-3. Description of the three GA parameter sets that emerged from the parameterization study

Parameter sets	N_{gener}	N_{pop}	P_{cross} [%]	P_{mut} [%]
#1	3000	100	60.0	0.4
#2	3000	200	60.0	0.1
#3	3000	200	60.0	0.5

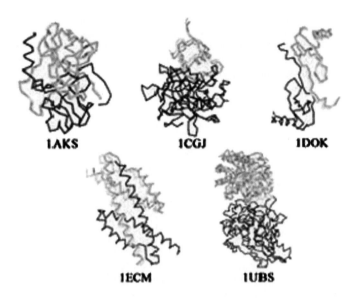

Figure 14-5. Cα trace superimpositions of the native solutions (blue and yellow) and the best candidates obtained with the GA in its 3000 / 200 / 60.0 / 0.1 configuration (green), for five complexes that composed our parameterization set – *1aks, 1cgj, 1dok, 1ecm,* and *1ubs.* Visualizations were obtained on combining the Swiss PDB Viewer (http://www.expasy.org/spdbv) and the ray-tracing POV-Ray (http://www.povray.org) programs (see color plate section)

complexes contained multiple protein chains (See Table 14-1 – Test set). We mixed complexes of varied types and sizes, that is, enzyme-inhibitor (PDB codes: *1brs, 1stf, 1tgs, 1udi, 4htc*), antibody-antigen (PDB codes: *1igc, 1nmb, 3hfl*), and others (*1gla, 1spb*). Details of the P-P structures and results, in terms of *rmsd* values for the most nearly optimal candidates, are presented in Table 14-4.

For each enzyme-inhibitor complex, and for *1gla* and *1spb*, the GA proposes at least one solution near that of the native structure, for one or more parameter sets. For example, for complex *1brs*, the three parameter sets yielded solutions with *rmsd* values 1.76 (32nd solution), 2.35 (11th one) and 2.86 Å (26th one), respectively. It was more difficult to bind correctly the antibody-antigen complexes, especially *1nmb* and *3hfl* for which *rmsd* values greater or near 3 Å were obtained for all three GA configurations (Table 14-4). Parameter set #1, with $N_{pop} = 100$, yielded the poorest results, with only five complexes on thirteen showing an *rmsd* value less than 3.0 Å. This observation confirms the conclusion made after the N_{pop} study: a population size of 200 is necessary to obtain satisfactory results. The two other parameter sets show superior performances, with seven and eight near-optimal structures found for ten complexes (not including the antibody-antigen ones), respectively. Set two provides superior solutions at lower ranks, among 50 proposed candidates. For illustration, some best proposed structures are presented in Figure 14-6, for complexes *1brs* (Enzyme/Inhibitor), *1udi* (Enzyme/Inhibitor),

Table 14-4. Candidates with the lowest *rmsd* values, with their rank between parentheses, retrieved among the solutions proposed by our GA for the 13 additional protein-protein complexes. The different parameter sets were those obtained at the end of the parameterization study (see Table 14-3 for details). The four characters in the first column refer to the complex PDB codes, and the letters between parentheses in the second column to the corresponding protein chains

Complexes	# atoms	# peaks	Lowest *rmsd* among the 50 solutions [Å] (rank)		
			Parameter set #1	Parameter set #2	Parameter set #3
• **Enzyme/Inhibitor**					
1brs	864 (A), 695 (D)	209/165	1.76 (32)	2.35 (11)	2.86 (26)
1brs	878 (B); 665 (E)	221/160	3.56 (36)	3.25 (37)	5.62 (43)
1brs	839 (C), 699 (F)	199/173	3.50 (35)	3.50 (35)	1.77 (14)
1stf	1655 (E), 789 (I)	394/189	4.38 (44)	1.79 (28)	3.36 (39)
1tgs	1646 (Z), 416 (I)	413/106	4.31 (13)	2.57 (42)	1.22 (21)
1udi	1818 (E), 654 (I)	444/161	2.15 (17)	2.15 (2)	1.85 (5)
4htc	253 (L), 2069 (H)	61/510	6.16 (39)	4.72 (2)	2.60 (22)
4htc	2069 (H), 447 (I)	510/110	1.46 (20)	1.42 (9)	1.52 (24)
• **Antibody/Antigen**					
1igc	3326 (L+H), 446 (A)	832/113	1.74 (8)	1.73 (14)	2.09 (16)
1nmb	1798 (L+H), 3061 (N)	442/761	9.02 (26)	4.91 (21)	9.16 (20)
3hfl	3250 (L+H), 1001 (Y)	806/242	3.77 (29)	4.28 (15)	3.36 (32)
• **Others**					
1gla	1201 (F), 3778 (G)	299/918	3.31 (32)	0.76 (5)	1.50 (22)
1spb	557 (P), 1860 (S)	149/476	1.54 (19)	1.58 (27)	0.96 (26)

1igc (Antibody/Antigen), and *1gla* (other) using this set. We thus concluded that the parameter combination 3000 / 200 / 60.0 / 0.1 is the most suitable for our protein-protein GA-driven docking application.

DISCUSSION

False Positives

For the parameter set we selected, 12 P-P complexes of 18 in total, i.e. five constituting the parameterization set and 13 constituting the test set, showed no or few structures with large *rmsd* values (>10 Å) among the top ten solutions. It is however not rare that the nearly optimal solutions are commonly lost among an ensemble of high-scored solutions with large *rmsd* values; these are called *false positives*. The ranks of the best or nearly optimal solutions thus increase, and those are sometimes found at the bottom of the *best* list. For our GA, the presence of false positives arose partially because we promoted the variability in terms of relative orientations in our *best* list, by avoiding two excessively similar solutions from being recorded together (redundancy is avoided). Candidates of two sorts could

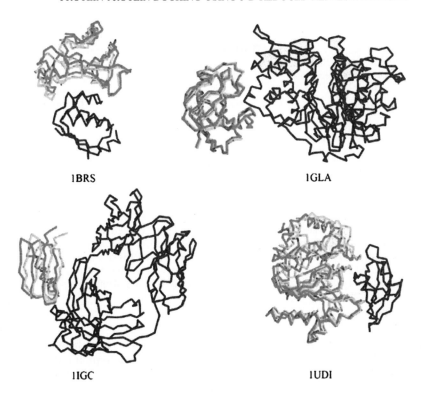

1BRS

1GLA

1IGC

1UDI

Figure 14-6. Cα trace superimpositions of native solutions (blue and yellow) and the best candidates obtained with the GA in its 3000 / 200 / 60.0 / 0.1 configuration (green), for four complexes that composed our validation set – *1brs (A and D chains), 1udi, 1igc, and 1gla.* Visualizations were obtained on combining the Swiss PDB Viewer (http://www.expasy.org/spdbv) and the ray-tracing POV-Ray (http://www.povray.org) programs (see color plate section)

indeed be obtained at the completion of a docking experiment – those combining small *rmsd* values and large scores, and those presenting large *rmsd* values with large fitness scores. The former constitute the nearly optimal solutions to be found by the GA, the latter are false positives. For instance, we encountered solutions wherein the P-P dimerization part was retrieved with only a small deviation from the native structure, but which produced great differences in the outer parts because rigid partners were considered (Figure 14-7). Consequently, the atoms that are far from the interface inevitably produce an increased *rmsd* value. A false positive of this kind might be strongly limited by limiting the *rmsd* evaluation to the Cα at the interface, i.e. to any atom that is within 10 Å from an atom of the other partner. This condition would allow to diminish the *rmsd* value by measuring the deviations at the dimerization part, without penalizing the nearly optimal solutions with large *rmsd* values. We encountered also solutions wherein the orientation of one partner had absolutely nothing in common with the native solution. These constituted the

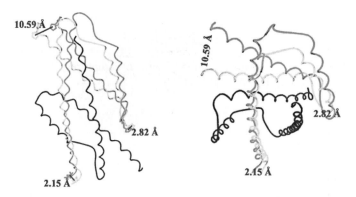

Figure 14-7. Ribbon representation (front and top views) of one proposed structure (*1ecm*) for which the interface with the partner (blue) is satisfactorily recognized by our GA (green), but which shows increasing divergence in the higher parts when compared with the crystalline structure (yellow). Distance calculations were performed on three amino acids, i.e. Asn5, His67, and His95, which present *rmsd* values 2.15, 2.82, and 10.59 Å, respectively. Visualizations were obtained on combining the Swiss PDB Viewer (http://www.expasy.org/spdbv) and the ray-tracing POV-Ray (http://www.povray.org) programs (see color plate section)

real false positives, as the fitness expression was unable to distinguish between the obtained complexes and the structure of the complex in its native form. It is difficult to determine whether a solution is a false positive, which should indicate that expert intervention remains necessary after such automated docking program. From the infinite possible orientations between the two partners, the GA has greatly restricted the searching domain to only a few. The proposed low-resolution models might then be retransformed into all-atom representations and be introduced into energy minimization or in complementary energetic filters, and consequently facilitating, or minimizing, the task of the researcher.

Comparison Between Our GA Strategy and Other Docking Methods

The literature contains many GA docking strategies, which differ in their molecular representations, fitness expressions, and exploration algorithms, which condition impedes their comparision. The approach developed by Gardiner et al [8] seems most similar to ours. In their work, also based on reduced representations of proteins, the molecular models were obtained on topological analysis of Connolly surfaces, which yielded critical point sets describing the local curvature of the surfaces of the two partners. The fitness function was based mainly on a surface complementarity score, with the combination of geometric matching, hydrogen-bonding potential and an interpenetration penalty score. Those authors used discrete positions on grid dots to place the partners. They performed complementarity studies on 34 complexes, mixing enzyme-inhibitors and antibody-antigen complexes, and for each one the GA was run 20 times to constitute a list of 100 candidates. The results obtained for the native complexes are similar to ours, with *rmsd* values <3.0 Å for nearly all complexes. The rank of the best nearly native structures is also generally located among the top 50 solutions. They also described the difficulties for the GA to predict correctly antibody-antigen complexes, with six on eight cases in which the GA found at least one success among the proposed candidates.

Li et al [62] investigated complementarity on three antibody-antigen complexes, i.e. camelid antibody-α-amylases. Their method was based on simplified representations of proteins, with one sphere per residue, combined with an energy-based scoring function, including electrostatic contributions and Gibbs energy of desolvation. Using the docking algorithm implemented in DOCK, Cherfils et al [76] found nearly native solutions in the top five for one complex, but they failed to predict two others, partially because of a poor definition of the exploration space. They also identified several differences between antibody-antigen and enzyme-inhibitor complexes, essentially in terms of interface compositions, hydrophobicity and electrostatic aspects.

About the interface sizes, Vakser et al [17] performed complementarity studies on low-resolution models of protein chains. By ignoring all atomic details, they sought to describe the role played by large-scale structural motifs. They tested their method on 475 cocrystallized protein-protein complexes and concluded that 52% of the complexes showed low-resolution recognition. A further analysis of their results showed the major role played by the interface of dimerization on the large-scale motif association, that is, 37% for 1000–2000 Å^2, 52% for 2000–4000 Å^2, and 76% for interfaces superior to 4000 Å^2. We link this observation partially with the difficulty of predicting the structure of antibody-antigen complexes: the interfaces of *1igc, 1nmb*, and *3hfl* are all less than 2000 Å^2, and present few interfacial contacts. These two observations explain why our algorithm, based on a rigid-body representation combined with a knowledge-based scoring function, has difficulty in correctly combining the reduced representations of such complexes.

CONCLUSION

Our objective was the development of a totally automated strategy dedicated to the docking of macromolecular systems, beginning with the construction of a scoring function to the parameterization and operation of the optimization algorithm for docking. For this purpose, we proposed a complete strategy based on an efficient technique of artificial intelligence, that is, the Genetic Algorithms, combined with reduced representations of the molecular partners. The results were highly encouraging, except for the antibody-antigen complexes. Our GA was able to retrieve the nearly optimal solution in each case. As future work, we plan to consider a statistical pair-wise potential of the same type, and electrostatic and geometric matching, to predict protein-DNA complexes and describe the role played by larger-scale recognition.

ACKNOWLEDGEMENTS

For the calculations performed at the Interuniversity Scientific Computing Facility (ISCF), installed at FUNDP, we acknowledge the financial support of FNRS-FRFC and the Loterie Nationale for the convention n° 2.4578.02. AB and NM thank the Fonds National de la Recherche Scientifique of Belgium for their appointments as Ph.D. Research Fellow and Scientific Research Worker, respectively.

REFERENCES

1. Harding SE, Chowdhry BZ (2001) Protein-Ligand Interactions: Structure and Spectroscopy Oxford University Press, New York
2. Yang J-M, Kao C-Y (2000) J Comput Chem 21:988
3. Taylor RD, Jewsbury PJ, Essex JW (2002) J Comput Aid Mol Des 16:151
4. Venkatachalam CM, Jiang X, Oldfield T, Waldman M (2003) J Mol Graph Model 21:289
5. Lorber DM, Shoichet BK (1998) Protein Sci 7:938
6. Daeyaert F, de Jonge M, Heeres J, Koymans L, Lewi P, Vinkers MH, Janssen PAJ (2004) PROTEINS 54:526
7. Mangoni M, Roccatano D, Di Nola A (1999) PROTEINS 35:153
8. Gardiner EJ, Willett P, Artymiuk PJ (2001) PROTEINS 44:44
9. Gardiner EJ, Willett P, Artymiuk PJ (2003) PROTEINS 52:10
10. Halperin I, Ma B, Wolfson H, Nussinov, R (2002) PROTEINS 47:409
11. Pabo CO, Nekludova L (2000) J Mol Biol 301:597
12. Becue A, Meurice N, Leherte L, Vercauteren DP (2003) Acta Crystallogr D 59:2150
13. Smith GR, Sternberg MJE (2002) Curr Opin Struct Biol 12:28
14. Comeau SR, Gatchell DW, Vajda S, Camacho CJ (2004) Bioinformatics 20:45
15. Najmanovich R, Kuttner J, Sobolev V, Edelman M (2000) PROTEINS 39:261
16. Norel R, Petrey D, Wolfson HJ, Nussinov R (1999) PROTEINS 36:307
17. Vakser IA, Matar OG, Lam CF (1999) P Natl Acad Sci USA 96:8477
18. Palma PN, Krippahl L, Wampler JE, Moura JJG (2000) PROTEINS 39:372
19. Levitt, M (1976) J Mol Biol 104:59
20. Wallqvist A, Ullner M (1994) PROTEINS 18:267
21. Kolinski A, Skolnick J (1998) PROTEINS 32:475
22. Betancourt MR (2003) PROTEINS 53:889
23. Tovchigrechko A, Wells CA, Vakser IA (2002) Protein Sci 11:1888
24. Levitt M, Warshel A (1975) Nature 253:694
25. Kolinski A, Skolnick J (2004) Polymer 45:511
26. Hassinen T, Peräkylä M (2001) J Comput Chem 22:1229
27. Ben-Zeev E, Eisenstein M (2003) PROTEINS 52:24
28. Ritchie DW, Kemp GJL (1999) J Comput Chem 20:383
29. Cai W, Shao X, Maigret B (2002) J Mol Graph Model 20:313
30. Ausiello G, Cesareni G, Helmer-Citterich M (1997) PROTEINS 28:556
31. Wang J, Hou T, Chen L, Xu X (1999) Chemometr Intell Lab 45:281
32. Leherte L, Latour T, Vercauteren DP (1995) Supramol Sci 2:209
33. Leherte L, Latour T, Vercauteren DP (1996) J Comput Aid Mol Des 10:55
34. Meurice N, Leherte L, Vercauteren DP (1998) SAR QSAR Environ Res 8:195
35. Leherte L (2001) J Math Chem 29:47
36. Trosset J-Y, Scheraga HA (1998) P Natl Acad Sci USA 95:8011
37. Trosset J-Y, Scheraga HA (1999) J Comput Chem 20:412
38. Fernández-Recio J, Totrov M, Abagyan R (2003) PROTEINS 52:113
39. Fitzjohn PW, Bates PA (2003) PROTEINS 52:28

40. Katchalski-Katzir E, Shariv I, Eisenstein M, Friesem AA, Aflalo C, Vakser IA (1992) P Natl Acad Sci USA 89:2195
41. Goldberg DE (1989) Genetic Algorithms in Search, Optimization, and Machine Learning, Addison-Wesley, Reading, USA
42. Jones G, Willett P, Glen RC, Leach AR, Taylor R (1997) J Mol Biol 267:727
43. Westhead DR, Clark DE, Murray CW (1997) J Comput Aid Mol Des 11:209
44. Holland JH (1975) Adaptation in Natural and Artificial Systems, MIT Press, Cambridge, USA
45. Lucasius CB, Kateman G (1993) Chemometr Intell Lab 19:1
46. Jin AY, Leung FY, Weaver DF (1999) J Comput Chem 20:1329
47. Mekenyan O, Dimitrov D, Nikolova N, Karabunarliev S (1999) J Chem Inf Comp Sci 39:997
48. Hou TJ, Wang JM, Liao N, Xu XJ (1999) J Chem Inf Comp Sci 39:775
49. Jones DT (1994) Protein Sci 3:567
50. Cui Y, Chen RS, Wong WH (1998) PROTEINS 31:247
51. Contreras-Moreira B, Fitzjohn PW, Offman M, Smith GR, Bates PA (2003) PROTEINS 53:424
52. Oshiro CM, Kuntz ID, Dixon JS (1995) J Comput Aid Mol Des 9:113
53. Bursulaya BD, Totrov M, Abagyan R, Brooks III CL (2003) J Comput Aid Mol Des 17:755
54. Morris GM, Goodsell DS, Halliday RS, Huey R, Hart WE, Belew RK, Olson AJ (1998) J Comput Chem 19:1639
55. Ewing TJA, Kuntz ID (1997) J Comput Chem 18:1175
56. Rarey M, Kramer B, Lengauer T, Klebe G (1996) J Mol Biol 261:470
57. Totrov M, Abagyan R (1997) PROTEINS 29 Suppl. 1:215
58. Meurice N, Leherte L, Vercauteren DP, Bourguignon J-J, Wermuth CG (1997) In: van de Water-beemd H, Testa B, Folkers G (eds) Computer-Assisted Lead Finding and Optimization. VHCA, Basel, p 497
59. Leherte L, Meurice N, Vercauteren DP (2000) J Chem Inf Comp Sci 40:816
60. Cartwright HM (1993) Applications of Artificial Intelligence in Chemistry, Oxford University Press, New York, USA
61. Vieth M, Hirst JD, Kolinski A, Brooks III CL (1998) J Comput Chem 19:1612
62. Li CH, Ma XH, Chen WZ, Wang CX (2003) PROTEINS 52:47
63. Verdonk ML, Cole JC, Hartshorn MJ, Murray CW, Taylor RD (2003) PROTEINS 52:609
64. Glaser F, Steinberg DM, Vakser IA, Ben-Tal N (2001) PROTEINS 43:89
65. Berman HM, Battistuz T, Bhat TN, Bluhm WF, Bourne PE, Burkhardt K, Feng Z, Gilliland GL, Iype L, Jain S, Fagan P, Marvin J, Padilla D, Ravichandran V, Schneider B, Thanki N, Weissig H, Westbrook JD, Zardecki C (2002) Acta Crystallogr D 58:899
66. Weissig H, Bourne PE (2002) Acta Crystallogr D 58:908
67. Becue A, Meurice N, Leherte L, Vercauteren DP (2004) J Comput Chem 25:1117
68. Jones S, Thornton JM (1995) Prog Biophys Mol Biol 63:31
69. Stites WE (1997) Chem Rev 97:1233
70. Nadassy K, Wodak SJ, Janin J (1999) Biochemistry 38:1999
71. Moont G, Gabb HA, Sternberg MJE (1999) PROTEINS 35:364
72. Camacho CJ, Gatchell DW, Kimura SR, Vajda S (2000) PROTEINS 40:525
73. Field MJ (1999) A Practical Introduction to the Simulation of Molecular Systems, Cambridge University Press, Cambridge, UK
74. Leherte L, Allen FH (1994) J Comput Aid Mol Des 8:257
75. Maeder M, Neuhold Y-M, Puxty G (2004) Chemometr Intell Lab 70:193
76. Cherfils J, Duquerroy S, Janin J (1991) PROTEINS 11:271

CHAPTER 15

TROPOLONE AS NEUTRAL COMPOUND AND LIGAND IN PALLADIUM COMPLEXES

GIDEON STEYL AND ANDREAS ROODT

Abstract: Focusing mainly on the solid-state effect of halogenation on compounds of tropolone type in the solid state, we report the character and functionality of these molecules. The behaviour of bromo-functionalised tropolone molecules is discussed relative to the tropolone parent compound. To investigate the large influence of bromination, we extracted structural data from the Cambridge Crystallographic Database covering Br...Br non-bonded interactions. Molecules of tropolone type π-stack in a fashion similar to benzene derivatives, with an interplanar distance *ca.* 3.72 Å. Palladium complexes with tropolonato ligands are presented as examples of the effect of coordination that ligands of these types induce on metal centres. These complexes form highly labile species in the presence of tertiary phosphines; the $[Pd(TropBr_3)Br(PCy_3)]$ complex is reported as an intermediate of these labile species, in which the initial $[Pd(TropBr_3)_2]$ complex acts as a bromo source in the presence of the PCy_3 ligand. Significant Br...Br interactions at distance *ca.* 3.6 Å stabilise the $[Pd(TropBr_3)Br(PCy_3)]$ complex in the solid state. Significant π-stacking is observed for palladium(II) tropolonato derivatives in the solid state

INTRODUCTION

In 1945, Dewar [1] predicted correctly the structure of tropolone (2-hydroxy-2,4,6-cycloheptatrienone, Figure 15-1, I.1), based on known natural products of colchicines and stipitatic acid [2]. In identifying the unique molecular structure of tropolone and its derivatives, i.e. a non-benzenoid "aromatic" compound, a new field in chemistry was launched. The mysteries involved in compounds of tropolone type remain incomplete, partly because of the duplicitous nature of tropolones: these can act as compounds of either aromatic or alkene type, with selective reactivity towards nucleo- or electrophiles.

The key to the mystery of tropolone and its derivatives is the seven-membered ring and the highly mobile tautomeric system (Figure 15-1, I.1). In addition to this

J.C.A. Boeyens and J.F. Ogilvie (eds.), Models, Mysteries and Magic of Molecules, pp. 325–340.
© 2008 *Springer.*

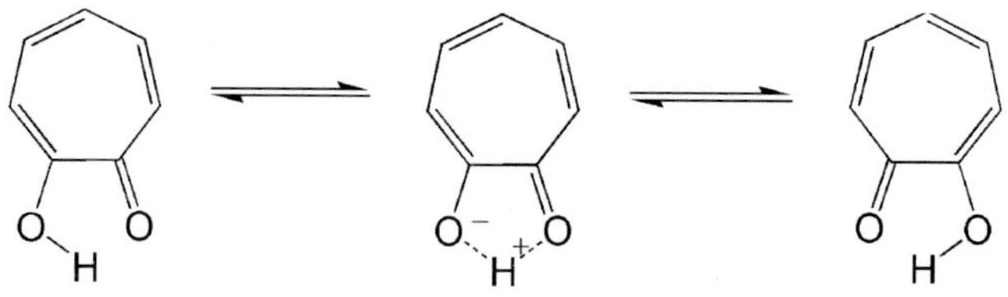

I.1

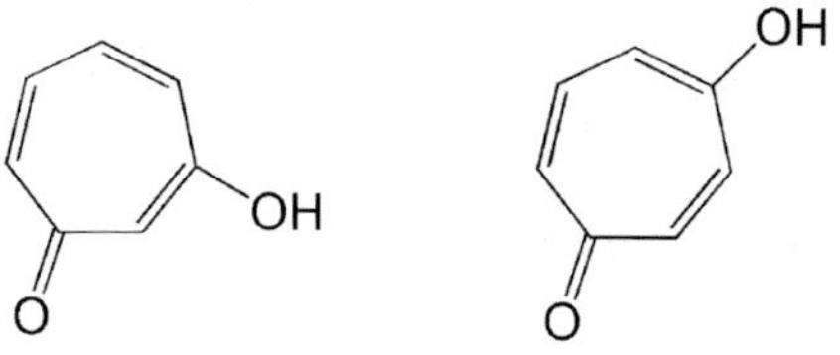

I.2   I.3

Figure 15-1. Tropolone tautomeric system and isomers [1, 3]

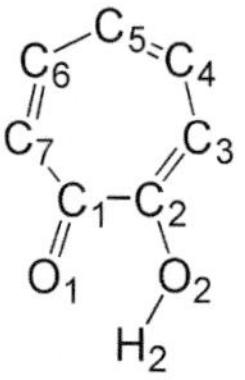

Figure 15-2. Numbering system of tropolone and its derivatives used throughout here

stabilization through the tautomeric forms, hydrogen bonding plays an important role in the stability of these tropolones. Tropolone (I.1) is more stable than the isomeric hydroxytropones (I.2) and (I.3), in which intramolecular hydrogen bonding is absent and a greater separation of charge is observed.

The numbering of compounds of tropolone type conforms to the standard IUPAC format (Figure 15-2). Tropolone that is the trivial name for 2-hydroxy-2,4,6-cycloheptatrienone is in general use, thus greatly simplifying the numbering of these compounds. In the literature reference to tropolone derivatives sometimes follows an alphanumeric method; i.e. hinokitiol can be either α- or β- or γ-isopropyl tropolone, indicating the three or four or five position, respectively, of the isopropyl moiety on the tropolone core, Figure 15-2.

Because tropolone is scarce in nature [4], occurring only in lower plants and fungi [5], limited information is available on these compounds. Tropolone derivatives exhibit remarkable pharmacological effects; for instance,

3,7-dihydroxytropolone [6] is known to act as a competitive inhibitor of inositol monophosphatase, which is a cause of manic depression. Even the mono-substituted 3-hydroxytropolone is recognized as an antibiotic [7]. The addition of morpholine and bromo moieties is reported to inhibit the Hepatitis C virus [8].

The use of bidentate ligands in organometallic chemistry is commonly illustrated by their successful application in homogeneous catalysis [9] and bio-medical [10] systems. In the field of radiopharmaceuticals, tropolonato metal complexes $[M^{III}(Trop)_3]$ ($M^{III} = Ga^{3+}$, In^{3+}) have been reported [11], with only the M^{III} oxidation state being important for gallium and indium under physiological conditions. The tropolonato ligand is notable as it forms neutral lipophilic complexes that might penetrate cell membranes [12, 13]. A catalytic application involving the use of tropolone derivatives is found with aminotroponimiato yttrium, and has been reported to serve as an active catalyst for alkyne hydroamination [14]. The highly stable complex [Rh(Trop)(diop)] has been used for the asymmetric hydrosilation of acetophenone [15].

That tropolone is such an effective chelating agent contributes to its wide application in chemistry. The tropolonate ion forms a five-membered chelating ring of which the "bite", that is, the oxygen-to-oxygen distance, is smaller than that in the β-diketonate, i.e. acetylacetonate. The "bite" is also well illustrated by the bite angle, O-M-O, in metal complexes.

The range of tropolonato complexes that have been investigated *via* crystallographic techniques is limited basically to transition-metal elements in the first row. In total, 52 tropolone complexes are known, of which 33 are first-row systems; most such complexes contain either copper (17) or cobalt (7). The main focus of these studies is typically the anti-microbial activity [16, 17] or magnetic properties [18]. In contrast to these limited crystallographic data, tropolone as ligand for transition-metal elements is long known to act as an effective chelating agent [19]. Tropolone complexes of actinides that have been investigated are found useful to extract [20] these metals.

Tropolone and its derivatives thus have clearly significant applications in medicinal chemistry and homogeneous catalysis. To illustrate the notable behaviour of tropolone, we discuss the solid-state behaviour of a selection of pure tropolone, its derivatives and organometallic complexes in the following sections.

EXPERIMENTS

In general, the abbreviations, e.g. HTrop, HTropBr$_3$, etc. denote the compound in its neutral form, whereas Trop$^-$, TropBr$_3^-$ etc. indicate the mono-anionic form of the ligands. We prepared the HTropBr$_3$ compound according to the literature [3]. For experimental procedure, i.e. synthesis, characterisation of compounds, underlying results described here we refer the reader to preceding work [3, 21]. The crystal structures that we report here (HTropBr$_3$ (**1**), [Pd(TropBr$_3$)Br(PCy$_3$)] (**2**), [Pd(Br)$_2$(PCy$_3$)$_2$] (**3**)) [22, 23] were solved with standard techniques [24, 25].

TROPOLONE COMPOUNDS

Compounds of tropolone type have a duplicitous nature in that the molecule can behave as either an aromatic or an allylic functionality. Tropolones can be brominated according to a method similar to that for alkenes, with tautomerisation restoring the cycloheptatriene backbone. In contrast to this simple halogenation with bromine, other halogenated products form less readily, requiring more advanced organic synthetic pathways.

Existing compounds of tropolone type are divisible into two general groups: one group comprises compounds in which the keto-enol functionalities are unaltered and the carbon backbone is modified, HTrop [26] (Figure 15-1 I.1), 3,5,7-HTropBr$_3$ [3, 27] (Figure 15-3), 5-HTropNO$_2$ [28], HHin (4-HTropisopropyl) [29], 3,7-HHinBr$_2$ [30] and 5-HTropCN [28]; in another group, the keto-enol moiety can be altered to include other hetero-atoms such as nitrogen, sulfur and selenium.

Tropolone and Tropolone Derivatives in the Solid State

The properties on compounds of tropolone type in the solid state, i.e. the geometrical parameters and interactions between molecules, can be most revealing in unlocking the mysteries of this molecule's behaviour. Hydrogen tunnelling occurs in the solid state [31, 32], and tropolone derivatives exhibit liquid-crystalline properties [33, 34]. In what follows we discuss first-generation derivatives of tropolone, with special reference to hydrogen bonding, π-π stacking and halogen-halogen interactions.

Figure 15-3 shows a polymorph of HTropBr$_3$ (**1**) [3] to illustrate a halogenated tropolone molecule; this structure (**1**) has a single molecule crystallised in the asymmetric unit. Table 15-1 presents the most important bond distances and angles.

Compared to the previously reported structure of HTropBr$_3$ [3] that contained two molecules in the asymmetric unit, both twinned, an overlay of these two polymorphs

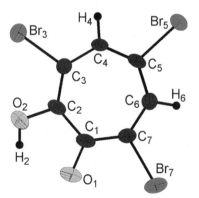

Figure 15-3. Diamond drawing of 3,5,7-tribromotropolone (HTropBr$_3$) (**1**) (50% probability ellipsoids), showing the numbering scheme

Table 15-1. Bond distances (Å) and angles (°) of HTropBr$_3$ (1)

C_1—O_1	1.221(11)	C_1—C_2	1.422(14)
C_2—O_2	1.243(14)	O_1—C_1—C_2	104.5(9)
C_3—Br_3	1.824(10)	O_2—C_2—C_1	122.3(9)
C_5—Br_5	1.840(10)	O_1—C_1—C_2—O_2	4.9(15)
C_7—Br_7	1.783(9)	C_7—C_1—C_2—C_3	8.5(18)

Table 15-2. Comparative data of tropolone derivatives

Compounds	$O_1\ldots O_2$	C_1=O_1	C_2—O_2	C_1—C_2	O_1=C_1—C_2—O_2	Ref.
	(Å)	(Å)	(Å)	(Å)	(°)	
HTrop	2.55	1.26	1.33	1.45	1.61	26
3,5,7-HTropBr$_3$	2.40	1.22	1.24	1.42	5.02	27
3,5,7-HTropBr$_3$	2.49	1.31	1.33	1.43	−2.00	3
3,5,7-HTropBr$_3$	2.54	1.32	1.34	1.45	1.00	3
3-Cl-HTrop	2.54	1.25	1.35	1.45	0.69	35
5-CN-HTrop	2.56	1.25	1.33	1.47	1.32	28
5-NO$_2$-HTrop	2.59	1.24	1.33	1.48	2.63	28
5-NO$_2$-HTrop	2.57	1.25	1.33	1.47	0.71	28
HHin	2.53	1.26	1.33	1.46	1.64	29
3,7-HHinBr$_2$	2.53	1.25	1.36	1.46	6.68	30

C=O and C—O indicates the "carbonyl" and "C—OH" moieties of the enol form, respectively. Average esd's reported for the structures are *ca.* 0.01–0.02 Å.

indicates that significant differences exist between them with an RMS overlay error 0.214 Å. A few halogenated or nitrogen-containing compounds exist that can be accessed directly from the tropolone parent compound (Table 15-2).

Tropolone derivatives in a range presented in Table 15-2 clearly show that the bond distances for the carbonyl groups (O_1=C_1/O_2—C_2) vary insignificantly upon ring functionalization, within the indicated standard deviations. The most notable change occurs in the compound HTropBr$_3$ (1), with a decreased length of the bond in the carbonyl (C_1=O_1) moiety. The hydroxyl carbonyl (C_2—O_2) moieties in the nitro and tribromotropolone derivatives have the shortest bonds. Secondary effects hence play a major role in determining geometrical parameters. One such effect reflects the "aromatic" properties of the tropolone ring system, thus compensating for electron-withdrawing or -donating moieties. Despite the torsional twist of the keto-enol functionality as indicated in Table 15-2, the system remains nearly planar, with the sterically hindered HHinBr$_2$ compound displaying the greatest distortion. The O...O distances of tropolone derivatives are near 2.5 Å, except both HTropBr$_3$ systems. We discuss further this effect with reference to hydrogen bonding and packing.

The solid-state behaviour of tropolones indicates that hydrogen-bridged dimeric units of tropolone compounds dominate the solid-state behaviour of these molecules.

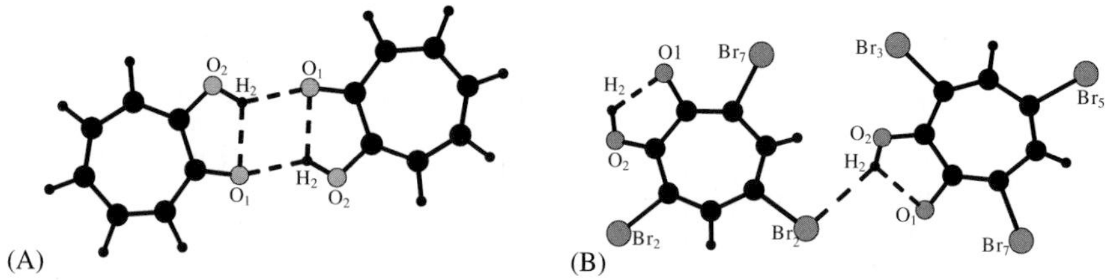

Figure 15-4. Hydrogen-bonding network in HTrop (A) and HTropBr$_3$ (B)

On adding other electronegative (X) centres to the tropolone core, a shift of hydrogen bonding was observed away from these dimeric units, towards a more pronounced X...H—O interaction; see Figure 15-4. A threshold seems to arise at which the change of hydrogen-bonding mode occurs in the solid state. If one or two electron-withdrawing moieties are added to tropolone, such as chloro [36], nitro [28] or dibromo [30], the dimeric unit stays intact; see Figure 15-4 (A). The addition of a further electron-withdrawing moiety tips the balance away from this dimeric unit, as observed from the HTropBr$_3$ structures. The Br...H—O interaction replaces the O...H—O interaction, causing a cessation of the dimeric units, Figure 15-4 (B).

The Br...H—O interactions observed in HTropBr$_3$ structures [27, 30] agree satisfactorily [O..Br 3.39(1), 3.49(1), 3.344(3) Å and O—H...Br 115(1), 118(2), 144(3)°] with values reported by Desiraju and Steiner [36] (3.47(4) Å and 134(4)°). The disordered structure of HTropBr$_3$ [3] has less effective directionality than the current polymorph structure (**1**) presented in Figure 15-3.

In our evaluation of the solid-state behaviour of brominated tropolones, our attention became focused on Br...Br interactions. In the HTropBr$_3$ (**1**) compound presented here, there exist two Br...Br interactions with distances of order 3.369(2) and 3.462(1) Å. As the van der Waals radius of bromine is 1.95 Å, the observed interaction in the solid state is significant. The same trend was observed for the HTropBr$_3$ [3] and HHinBr$_2$ [30] compounds, with observed distances 3.626(5) and 3.616(1) Å, respectively.

CSD Evaluation of Br...Br Interactions in the Solid State

To extend our investigation into interactions of these types between bromine atoms, we conducted a search of the Cambridge Crystallographic Database (CSD) [37] on the organic-molecule subset displaying C—Br...Br interactions. Br...Br non-bonded contact interactions were considered first in a range 2.4–4.6 Å. The probability of long-range non-bonded interactions more distant than 4.6 Å were investigated, but yielded no further information. Secondly and more specifically, the angle formed with C—Br and the non-bonded bromine atom was selected as a parameter to indicate a possible directionality of the evaluated interactions. The distance typical of Br$_2$ (Br—Br) in its covalent bond is of order 2.3 Å.

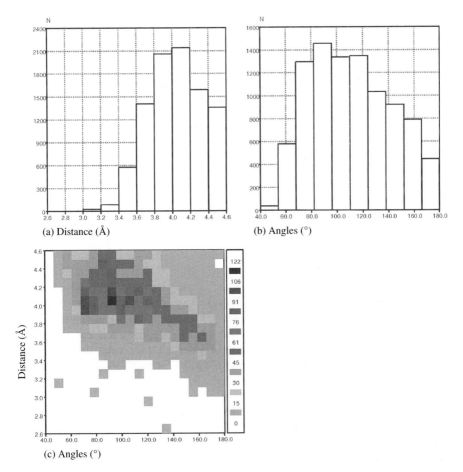

Figure 15-5. Histograms of non-bonded Br...Br interaction distances (a), C—Br...Br angle (b) and a scattergram of distance vs. angle (c)

A histogram of non-bonded Br...Br interactions indicates that successes in a significant number were obtained in a region 3.6–4.6 Å, see Figure 15-5 (a), with a mean value 4.04 Å. According to the data, the directionality of the C—Br...Br interaction ranges from 40–180°, with a maximum in the interval 80–120°. This maximum is deceptive because C—H...Br interactions are active also in this range. If a linear interaction exist, it would fall in a range 130–180°; see Figure 15-5(b). As the data ranges show a correlation, a scattergram plot of the distance *vs.* non-bonded angle was constructed as shown in Figure 15-5 (c). The large range of these results is attributed to inherent properties of the bromine atom. A covalently bound bromo moiety (C—Br) can both polarize the bond and donate to the carbon atom, generating a nucleophilic centre on the bromine atom. Br...Br interactions might thus alleviate this decrease of electron density in the solid state. Non-bonded interactions in two distinct groups are identified from the scattergram; see Figure 15-5 (c). An area of maximum overlap is observed from 3.8–4.6 Å and 70–120°, indicating a weaker perpendicular interaction over a larger distance.

This effect is attributed to the inherent nature of the bromo moiety – hyper-polarization of the C—Br bond and donation to the carbon atom. A stronger interaction, with a smaller population than the weaker interaction, is observed in a region 3.6–4.0 Å and 130–170°. According to the linear extent of this interaction and the small distances involved, an interaction of this type must play a significant role in stabilising the solid state. Similarly, the observed Br...Br interactions in the solid state for brominated tropolone compounds (HTropBr$_3$ and HHinBr$_2$) correlate well with structural data reported in the CSD (Figure 15-5).

Intramolecular π-π Stacking

A secondary aspect of the solid-state interactions in tropolone compounds is the ability to form π-stacking units. This effect is due largely to the delocalisation of the cycloheptatriene ring system over the total molecule. Typical π-π interactions in benzene molecules range from 3.5–4.0 Å (mean 3.81 Å) and with a nearly parallel orientation of 0–4° (mean 0.67°) between individual benzene molecules. In tropolone compounds the interaction is of order 3.72(3) Å with an inter-planar angle between tropolone molecules 0.7(8)°. We present in Table 15-3 a selection of tropolone compounds [in total 21 entries].

A single compound such as HTropBr$_3$ can either form part of a π-stacking unit (Table 15-3) or have no significant π-interaction in the solid state. A synergistic effect might exist between π-stacking and Br...Br interactions in the unit cell. The HTropBr$_3$ (**1**) molecule presented here exhibits the strongest Br...Br interaction in the solid state, with these interactions overriding any other and secondary weak effect. In the published structure of HTropBr$_3$ [3], secondary effects such

Table 15-3. Comparative π-stacking data for selected tropolone compounds

Compounds	Interplanar distance	Angle between planes
	(Å)	(°)
HTrop	3.594	6.14
3,5,7-HTropBr$_3$	–	–
3,5,7-HTropBr$_3$	3.570	0.00
3-Cl-HTrop	3.613	0.00
5-CN-HTrop	3.791	0.02
5-NO$_2$-HTrop	3.670	0.00
HHin	3.780	0.00
3,7-HHinBr$_2$	–	–

as π-stacking might play a large role in structuring the solid-state ordering of the molecule.

Conclusion

In tropolone compounds discussed above, three major effects contribute to the solid-state behaviour of the molecules: hydrogen bonding between electronegative groups such as the dimeric units of tropolone can significantly order the solid state; π-stacking also readily occurs, with an observed inter-planar distance of the same order as that for benzene molecules; if electronegative groups such as bromo moieties are present, the likelihood of Br...Br interactions might overshadow other packing effects observed in the solid state.

PALLADIUM TROPOLONATO COMPLEXES

Because palladium complexes are reactive in various solvents (Scheme 15.1), we present the following tropolonate complexes as a case study. The interpretation of NMR spectra, which typically contain multiple features, can be exhausting to clarify quantitatively, and the application of X-ray crystallography can thus partially circumvent this problem.

The initial complexes, [Pd(Trop)$_2$] [38] and [Pd(TropBr$_3$)$_2$], are well behaved systems that are stable in air and readily synthesised in high yield. Beginning with a bis-tropolonato palladium(II) complex, one expects the addition of one equivalent of a tertiary phosphine in solution to result in coordination to the palladium(II) centre; see Scheme 15.1, V.2 Step A. Initially, this species was perceived to be a single stable product, allowing the characterisation of the derivative [Pd(Trop)$_2$(PCy$_3$)] [39], but it soon became clear that a second possible route was available, Scheme 15.1, V.3–V.4. The reduction of Pd(II) to Pd(0) occurs readily in the presence of phosphine (PR$_3$) to form the well known Pd(0) complex, [Pd(PR$_3$)$_4$]. This product in turn can activate halogenated species in solution via oxidative addition to generate the products indicated in steps [C] and [D] of Scheme 15.1 [36, 43].

The effect of the solvent [40] should not be ignored in these systems, in which dichloroethane as solvent contributed in the formation of a [Pd(Cl)$_2$(PPh$_3$)$_2$].DCE complex; see Scheme 15.1, V.9. This product was obtained from the [Pd(TropBr$_3$)$_2$] complex, indicating that chloro ligands are more competitive in solution than bromo moieties. Because of the solvent system, the concentration of chloro species is present in a much greater ratio than that of the bromo species of which the ligand is the direct source. Chloro ligands can also be scavenged directly from solution via oxidative addition as reported for the [Pd(Trop)(Cl)(PPh$_3$)] [41] complex, upon addition of hydrochloric acid in a small proportion. This effect resulted in the loss of only a single tropolonato moiety, which can be replaced by the chloro ligand, resulting in a variant of Scheme 15.1, V.5.

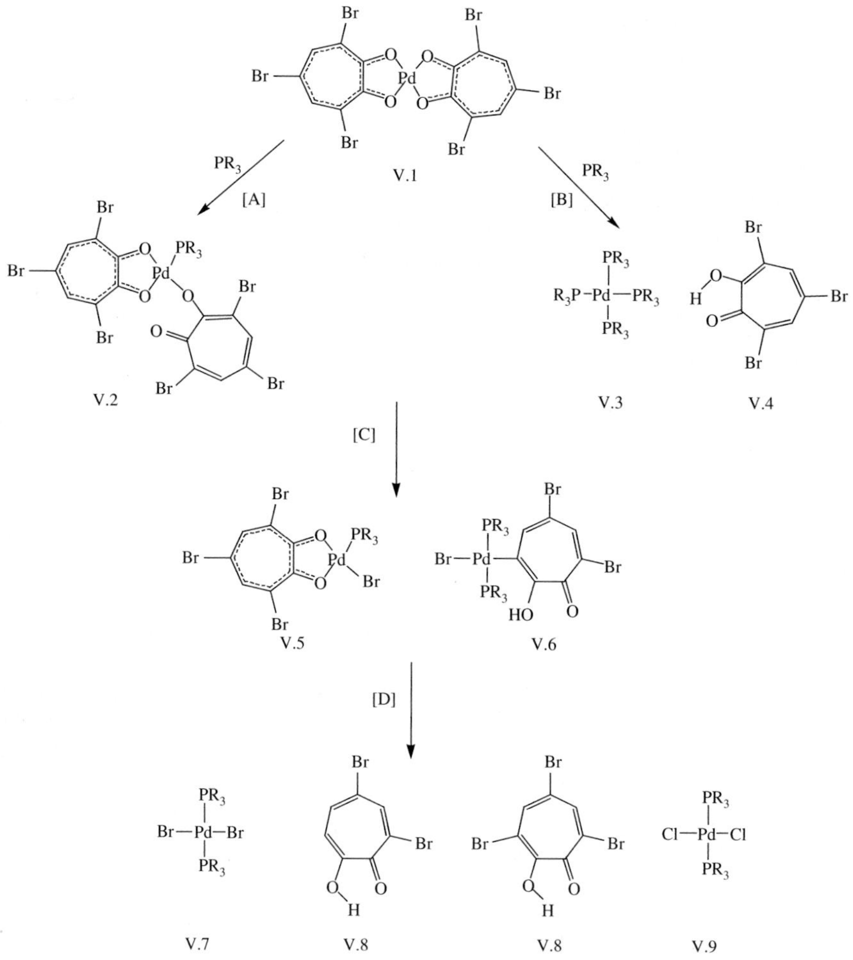

Scheme 15.1 Proposed products of $[Pd(TropBr_3)_2]$ and PR_3 in solution

We present two illustrative solid-state examples of crystallised products from the reaction of $[Pd(TropBr_3)_2]$ with PCy_3. To confirm the total reaction path, the $[Pd(Br)_2(PCy_3)_2]$ complex (Scheme 15.1, V.7) was isolated in yield *ca.* 80% from the reaction mixture after *ca.* two days. During the initial stages (after *ca.* 8 h) of the reaction, crystals in a small amount from an intermediate species were isolated from the reaction mixture, identified *via* X-ray diffraction as the complex $[Pd(TropBr_3)Br(PCy_3)]$ (**2**) (Scheme 15.1, V.5).

General crystallographic data of these compounds $[Pd(TropBr_3)Br(PCy_3)]$ (**2**) (Figure 15-6) and $[Pd(Br)_2(PCy_3)_2]$ (**3**) (Figure 15-9) [22] with the most important bond distances and angles are reported in Tables 15-4 and 15-5, respectively.

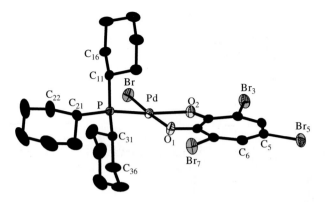

Figure 15-6. Diamond drawing of [Pd(TropBr$_3$)Br(PCy$_3$)] (**2**) (50% probability ellipsoids), showing the numbering scheme. For the phenyl groups, the first digit refer to the ring number and the second digit to the atom in the ring. Hydrogen atoms omitted for clarity

Table 15-4. Bond distances/Å and angles/deg of [Pd(TropBr$_3$)Br(PCy$_3$)]

Pd—O$_1$	2.022(2)	O$_1$—Pd—P	92.10(7)
Pd—O$_2$	2.109(2)	O$_2$—Pd—P	166.60(6)
Pd—P	2.2281(9)	Br—Pd—P	93.85(2)
Pd—Br	2.3945(4)	Br—Pd—O$_1$	174.04(6)
O$_1$—Pd—O$_2$	77.96(9)	O$_1$—C$_1$—C$_2$—O$_2$	−0.8(4)

Table 15-5. Selected geometrical parameters of [Pd(Br)$_2$(PCy$_3$)$_2$]

Pd—Br	2.4292(1)	P—Pd—P	180
Pd—P	2.3689(3)	Br—Pd—P—C$_{11}$	141.49(5)
Br—Pd—Br	180	Br—Pd—P—C$_{21}$	−101.58(5)
P—Pd—Br	91.253(9)	Br—Pd—P—C$_{31}$	14.50(5)

[Pd(TropBr$_3$)Br(PCy$_3$)] (**2**) crystallised in an asymmetric unit with a slightly distorted square-planar geometry about the palladium(II) centre. The palladium(II) atom is elevated 0.0808(2) Å above the plane defined by four atoms O$_1$, O$_2$, P and Br, forming a coordination polyhedron. The distortion is further evident from the non-linear bond angles O$_1$—Pd—Br and O$_2$—Pd—P (Table 15-4). The Pd—O bond distances are slightly larger than those reported for the complexes [Pd(Trop)$_2$(PCy$_3$)] [39] and [Pd(Trop)$_2$] [38]. The O—Pd—O bidentate bite angle (77.96(6)°) is significantly smaller than that reported for complexes [Pd(Trop)$_2$(PCy$_3$)] [39] (80.30(5)°) or [Pd(Trop)$_2$] [38] (81.8(1)°). The diminished bite angle is contrary to expectation, because the introduction of electron-withdrawing groups into the tropolonato ligand should have displayed the opposite effect; the Pd—O bond distances should decrease and the bite angle should thus

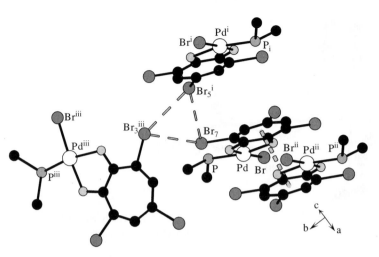

Figure 15-7. Partial unit cell of [Pd(TropBr$_3$)Br(PCy$_3$)] (2), illustrating π-π stacking and Br...Br interactions. Hydrogen atoms, cyclohexyl rings omitted for clarity [Symmetry codes: (i) –x, –y, –z; (ii) 1–x, –y, –z; (iii) 0.5–x, 0.5+y, z]

increase. As this effect does not occur, we assume that secondary packing effects play an enhanced role.

No classic hydrogen bonds are observed in the complex [Pd(TropBr$_3$)Br(PCy$_3$)], but intermolecular π-stacking between tropolonato units is observed with distance 3.476 Å. A further unique triangular intermolecular interaction is found between three separate bromo moieties of three palladium units; see Figure 15-7. The distinctive bromo...bromo interactions are nearly equidistant: Br$_7$...Br$_3^{iii}$, Br$_7$...Br$_5^i$ and Br$_3^{iii}$...Br$_5^i$ with intermolecular distances 3.616(5), 3.692(1) and 3.638(1) Å, respectively. Symmetry codes are reported in the caption of Figure 15-7. The addition of three bromo groups to the backbone of tropolone assisted in stabilising the solid-state ordering, acting as electron-donating functionalities.

Only a single triad is shown in Figure 15-7, but all bromo groups on the tropolonato ring in the structure are involved in this extended polymeric three-dimensional interaction (Figure 15-8). Thus, in the solid state, a single tribromotropolonato moiety is affected by both three triad interactions of bromo groups and π-stacking with a non-associated tropolonato ring system. The π-stacking component in turn interacts with bromo groups in the opposite direction, resulting in a stair conformation throughout the solid state.

About the final product in the reaction (Scheme 15.1), the complex [Pd(Br)$_2$(PCy$_3$)$_2$] was reported *via* a direct synthesis [42]. The complex crystallises on an inversion centre, with the tricyclohexylphosphine moieties in an eclipsed conformation; see Table 15-5. The Pd—P and Pd—Br distances are comparable to those reported in the literature, with a slightly increased bond precision due to the lower temperature study.

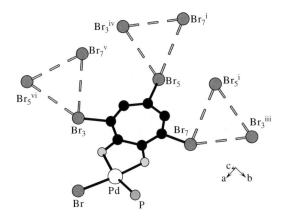

Figure 15-8. Molecular representation of the bromo triad units in [Pd(TropBr$_3$)Br(PCy$_3$)] (2). Hydrogen atoms, cyclohexyl rings omitted for clarity. [Symmetry codes: (i) –x, –y, –z; (ii) 1–x, –y, –z; (iii) 0.5–x, 0.5+y, z; (iv) –0.5+x, –0.5–y, –z; (v) 0.5–x, –0.5+y, z; (vi) 0.5+x, –0.5–y, –z]

A link might exist between the Pd—Br bond and the absence of intermolecular interactions (Pd—Br...H) observed for complexes [Pd(TropBr$_3$)Br(PCy$_3$)] and [Pd(Br)$_2$(PCy$_3$)$_2$]. The preference of palladium(II) to bond to halogen atoms in the presence of π-donors is well known, i.e. in the Suzuki coupling or Heck reactions [43]. This effect is beyond our scope here, although it begs the question why are no intermolecular interactions observed.

Because the reacting species defined in Scheme 15.1 comprise only the initial compounds, [Pd(Trop)$_2$] or [Pd(TropBr$_3$)$_2$], and a tertiary phosphine group, the range of products observed is astonishing, again underlining the reactivity and complexity of Pd(II) catalyst systems. In addition to the above, [Pd(TropBr$_3$)$_2$] can act as its own source of bromine to form the complex [Pd(Br)$_2$(PCy$_3$)$_2$] during the reaction cycle; see Figure 15-9. Because ligands of tropolonato type in palladium complexes have a unique character, the ligand can masquerade either as a stabilising, halogen-donating or -leaving moiety in the reaction mixture.

CONCLUDING REMARKS

The simplicity of the tropolone molecule belies its fascinating character in both solution and the solid state. The "aromatic" properties of these compounds make varied interactions observable in the solid state; including hydrogen tunnelling, π-π stacking and bromo...bromo interactions. The addition of a metal centre does not suppress this character, but can instead add further dimensions to the complexes formed. Palladium(II) complexes with tropolone can result in a myriad products, in which tropolone can act as a source of bromo moieties to assist in the reaction path. Although only a small fraction of our research team's work has been presented, the

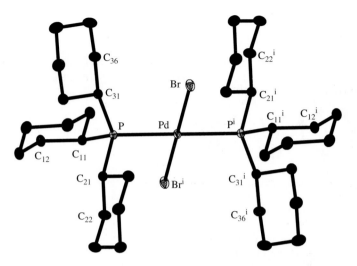

Figure 15-9. Diamond drawing of $[Pd(Br)_2(PCy_3)_2]$ (3) (50% probability ellipsoids), showing the numbering scheme. Hydrogen atoms are omitted for clarity. Cyclohexyl rings: the first digit refers to the ring number, the second digit to the atom number in the ring. [Symmetry code: (i) –x,–y, –z]

mystical character of tropolone is expected to deliver soon additional twists in the tale that will challenge our understanding of the chemistry of this compound.

ACKNOWLEDGEMENTS

SASOL, the South African National Research Foundation (SA-NRF/THRIP, grant number GUN 2068915) and University of the Free State provided financial assistance. We thank Dr. A. Muller, Mr. Leo Kirsten and Mr. J.M. Janse van Rensburg for data collection. Opinions, findings, conclusions or recommendations expressed in this material are those of the authors and do not necessarily reflect the views of the SA-NRF.

REFERENCES

1. M.J.S. Dewar, Nature, 155 (1945) 50.
2. M.J.S. Dewar, Nature, 155 (1945) 141.
3. G. Steyl, A. Roodt, S. Afr. J. Chem., 59 (2006) 21.
4. E. Ellington, J. Bastida, F. Viladomat, V. Simanek, C. Codina, Biochem. Syst. Ecol., 31 (2003) 715.
5. R.F. Angawi, D.C. Swenson, J.B. Gloer, D.T. Wicklow, Tet. Lett., 44 (2003) 7593.
6. S.R. Piettre, A. Ganzhorn, J. Hoflack, K. Islam, J.-M. Hornsperger, J. Am. Chem. Soc., 199 (1997) 3201.
7. H.A. Kierst, G.G. Markoni, F.T. Counter, P.W. Ensminger, N.D. Jones, M.O. Chaney, J.E. Toth, N.E. Allen, J. Antibiotics, 12 (1982) 1651.
8. A.M. Boguszewska-Chachulska, M. Krawczyk, A. Najda, K. Kopanska, A. Stankiewicz-Drogon, W. Zagorski-Ostoja, M. Bretner, Biochem. and Biophys. Res. Comm., 341 (2006) 641.
9. P.W.N.M. van Leeuwen, M.A. Zuideveld, B.H.G. Swennenhuis, Z. Freixa, P.C.J. Kramer, K. Goubitz, J. Fraanje, M. Lutz, A.L. Spek, J. Am. Chem. Soc., 125 (2003) 5523.

10. M. Yamato, J. Ando, K. Sakaki, K. Hashigaki, Y. Wataya, S. Tsukagoshi, T. Tashiro, T. Tsuruo, J. Med. Chem., 35 (1992) 267.

11. F. Nepveu, F. Jasanada, L. Walz, Inorg. Chim. Acta, 211 (1993) 141.

12. L. Hendershott, R. Gentilcore, F. Ordway, J. Fletcher, R. Donati, Eur. J. Nucl. Med., 7 (1982) 234.

13. H.F. Kotze, A. du P. Heyns, M.G. Lotter, H. Pieters, J. Roodt, M.A. Sweetlove, P.N. Badenhorst, J. Nucl. Med., 32 (1991) 62.

14. M.R. Burgstein, H. Berberich, P.W. Roesky, Organometallics, 12 (1998) 1452.

15. H. Brunner, A. Knott, R. Benn, A. Rufinska, J. Organomet. Chem., 295 (1985) 211.

16. K. Nomiya, A. Yoshizawa, K. Tsukagoshi, N.C. Kasuga, S. Hirakawa, J. Watanabe, J. Inorg. Biochem., 98 (2004) 46.

17. K. Nomiya, A. Yoshizawa, N.C. Kasuga, H. Yokoyama, S. Hirakawa, Inorg. Chim. Acta, 357 (2004) 1168.

18. A. Bencini, A. Beni, F. Constantino, A. Dei, D. Gatteschi, L. Sorace, Dalton Trans. (2006) 722.

19. P.L. Pauson, J. Chem. Soc. (1954), 11.

20. D. Dyrssen, Acta Chem. Scand., 9 (1955) 1567.

21. The palladium(II) derivatives were prepared from the initial complexes [Pd(Trop)$_2$] or [Pd(TropBr$_3$)$_2$] on adding tricyclohexylphosphine (PCy$_3$) in equimolar proportion to the solution. The [Pd(TropBr$_3$)Br(PCy$_3$)] complex was prepared in acetone and crystallised in the initial batch; a second crop of [Pd(Br)$_2$(PCy$_3$)$_2$] after decreasing the volume by half was collected ca. two days after the preparation.

22. General for 1, 2 and 3; Three-dimensional data were collected on a Bruker Apex II CCD diffractometer at 100(2) K; MoK$_\alpha$ radiation (0.71073 Å); reflections corrected for Lorentz and polarization effects and absorption corrections applied using SADABS; structures solved with direct methods and successive Fourier synthesis (SHELX-97); hydrogen atoms placed in calculated positions with fixed isotropoic thermal parameters; R = $[(\Sigma\Delta F)/(\Sigma F_o)]$; wR = $\Sigma[w(F_o^2-F_c^2)^2]/\Sigma[w(F_o^2)^2]^{1/2}$;

   1: Emp. Formula C$_7$H$_3$Br$_3$O$_2$; FW. 358.82; crystal system Monoclinic; space group P2$_1$/c; a 7.655(2), b 14.966(3), c 8.623(2) Å, α 90, β 112.88(3), γ 90°; V/ Å^3 910.1(3); Z 4; D$_c$/g.cm^{-3} 2.619; μ/mm^{-1} 13.25; Tmax/Tmin ; F(000) 664; crystal size/mm; θ limit/° 1.86 to 28.27; index $-9 \le$ h ≤ 9, $-19 \le$ k ≤ 8, $-9 \le$ l ≤ 9; collected refl. 3859; independent refl. 1659; R$_{int}$ 0.0471; Obs. Refl. [I $> 2(\sigma)$I] 1045; data/restr./param. 1659/0/111; goodness of fit 1.043; R/wR (I $> 4(\sigma)$I) 0.0591, 0.1696; R/wR (all data) 0.0956, 0.2175; extension coefficient 0.010(3); $\Delta\rho_{max}$, $\Delta\rho_{min}$./eÅ^{-3} 1.306, -1.049. CSD reference code: 614697.

   2: Emp. Formula C$_{20}$H$_{28}$Br$_{3.20}$C$_{l0}$O$_{1.60}$P$_{0.80}$Pd$_{0.80}$; FW. 659.63; crystal system Orthorhombic; space group Pbca; a 9.5537(3), b 16.1939(5), c 36.712(1) Å; V/ Å^3 5679.8(3); Z 10; D$_c$/g.cm^{-3} 1.928; μ/mm^{-1} 6.359; Tmax/Tmin; F(000) 3216; crystal size/mm; θ limit/° 1.11 to 26.99; index $-12 \le$ h ≤ 12, $-20 \le$ k ≤ 20, $-46 \le$ l ≤ 27; collected refl. 37129; independent refl. 6190; R$_{int}$ 0.0392; Obs. Refl. [I $> 2(\sigma)$I] 4979; data/restr./param. 6190/0/301; goodness of fit 1.126; R/wR (I $> 4(\sigma)$I) 0.0267, 0.0633; R/wR (all data) 0.0421, 0.0758; $\Delta\rho_{max}$, $\Delta\rho_{min}$./eÅ^{-3} 0.647, -1.071. CSD reference code: 614696.

   3: Emp. Formula C$_{36}$H$_{66}$Br$_2$P$_2$Pd; FW. 827.05; crystal system Triclinic; space group P–1; a 9.9411(3), b 10.1338(2), c 10.7239(3) Å, α 112.736(1), β 109.434(1), γ 92.169(1)°; V/Å^3 922.26(4); Z 1; D$_c$/g.cm^{-3} 1.489; μ/mm^{-1} 2.780; Tmax/Tmin; F(000) 428; crystal size/mm; θ limit/° 2.22 to 27.00; index $-12 \le$ h ≤ 12, $-12 \le$ k ≤ 12, $-13 \le$ l ≤ 13; collected refl. 27711; independent refl. 4029; R$_{int}$ 0.0261; Obs. Refl. [I $> 2(\sigma)$I] 3864; data/restr./param. 4029/7/187; goodness of fit 1.050; R/wR (I $> 4(\sigma)$I) 0.0156, 0.0405; R/wR (all data) 0.0165, 0.0410; $\Delta\rho_{max}$, $\Delta\rho_{min}$/eÅ^{-3} 0.400, -0.241. CSD reference code: 614695.

23. NMR data: Because the palladium species is reactive, only a general indication of chemical shifts can be reported. Typical ^{1}H spectra for HTrop and HTropBr$_3$ protons are in the same region as for aromatic protons. Most notably, HTropBr$_3$ has a singlet for 2 protons in the range 8.2–8.4 ppm.

24. Bruker, 1998, SADABS (Version 2004/1), Bruker AXS Inc., Madison, WI USA.

25. G.M. Sheldrick, 1997, SHELXS-97 and SHELXL-97. University of Gottingen, Germany.

26. H. Shimanouchi, Y. Sasada, Acta Cryst., B29 (1973) 81.
27. this work paragraph 3.1.
28. K. Kubo, E. Yamamoto, A. Mori, Acta Cryst., C57 (2001) 611.
29. K. Tanaka, R. Nagahiro, S. Ohba, M. Eishima, Tetrahedron Lett., 42 (2001) 925.
30. G. Steyl, A. Roodt, Acta Cryst., C59 (2003) o525.
31. N. Sanna, F. Ramondo, L. Bencivenni, J. Mol. Struct., 318 (1994) 217.
32. H. Rostkowska, L. Lapinski, M.J. Nowak, L. Adamowicz, Int. J. Quantum Chem., 90(3) (2002) 1163.
33. J.R. Chipperfield, S. Clark, J. Elliott, E. Sinn, Chem. Comm., (1998) 195.
34. J.M. Elliot, J.R. Chipperfield, S. Clark, S.J. Teat, E. Sinn, Inorg. Chem., 41 (2002) 293.
35. T. Tsuji, H. Sekiya, Y. Nishimura, A. Mori, H. Takeshita, N. Nishiyama, Acta Cryst., C47, 1991, 2428.
36. G.R. Desiraju, T. Steiner, "The Weak Hydrogen Bond in Structural Chemistry and Biology", IUCr, Oxford University Press, Oxford UK, 2001.
37. CSD Version 5.27, May Update 2006.
38. G. Steyl, Acta Cryst., E61 (2005) m1860.
39. G. Steyl, Acta Cryst., E62 (2006) m650.
40. G. Steyl, Acta Cryst., E62 (2006) m1324.
41. G. Steyl, Acta Cryst., E62 (2006) m974.
42. M.L. Clarke, A.G. Orpen, P.G. Pringle, E. Turley, Dalton Trans. (2003) 4393.
43. J. Tsuji, "Palladium Reagents and Catalysts: New Perspectives for the 21st Century", 2004, Wiley, Chichester, PO19 8SQ, UK.

CHAPTER 16

THEORETICAL AND EXPERIMENTAL MODELS OF MOLECULES ILLUSTRATED WITH QUANTUM-CHEMICAL CALCULATIONS OF ELECTRONIC STRUCTURE OF H_2CN_2 ISOMERS

J.F. OGILVIE AND FENG WANG

Abstract: To compare aspects of theoretical and experimental models of molecules, we employ the results of quantum-chemical calculations on diazomethane and six structural isomers with formula H_2CN_2; significant deficiencies of both models impede comparison between a calculated value of a property and a corresponding value deduced from experiment

INTRODUCTION

When, about year 1957, Coulson was reputed to have remarked to the effect that "the objective of quantum chemistry is to paint a picture, not to take a photograph", conditions regarding both theoretical methods and the practice of calculation were much more primitive than those at present. During the past half century, quantum-chemical computations – as an application of quantum-mechanical principles and procedures for the calculation of molecular electronic structure – have evolved from being a tedious manual task, on which only a few expert and mathematically minded chemists would embark, to become a routine computation for which several computer programs, some even free of cost, are available; operation of these programs requires little understanding of either details of the models or algorithms or even the nature of the calculations in relation to their prospectively fundamental quantum-mechanical underpinning. Not only do the innumerable publications accumulated in this field include many examples in which their authors have focused on artefacts of quantum-mechanical methods, such as wave functions in wave mechanics or their purported constituent orbital components, but also terms that originated in a questionable mathematical basis have pervaded general chemical usage in a

J.C.A. Boeyens and J.F. Ogilvie (eds.), Models, Mysteries and Magic of Molecules, pp. 341–364.
© 2008 *Springer.*

purely qualitative and loose manner devoid of mathematical significance, such as the description of a tetrahedrally 'ligated' carbon atom as being of type "sp^3". The practice of quantum chemistry is almost invariably built on a model of atomic orbitals and their combinations that serve as a basis of calculations of electronic structure with atomic nuclei at chosen positions, but commonly the latter relative positions are varied for the purpose of finding a minimum energy of the ensemble of electronic and nuclear particles. As we demonstrate here, this model proves practical for the purpose of estimating molecular properties that might be compared with deductions from experiment, even though such a comparison be imperfect, as we discuss here also. In the present work we hence undertake not merely to calculate some properties of selected molecular species for actual or prospective comparison with apparently related experimental data but also to appraise the scene of those calculations so as to illuminate both the underlying models and possible pitfalls in such comparisons.

For this purpose we have selected molecular species in a set specified according to a chemical formula H_2CN_2; this merely pentatomic molecular entity comprising atomic centres of three types is remarkable in implying a possible existence of multiple chemical compounds, some already characterized and others prospective, even if likely unstable. All species result from structural isomerism – a varied topology or order of putative connectivity of relative spatial locations of atomic centres. Among thirteen such plausible isomers, those with both a nearly collinear, arrangement of massive atoms and a dihydrogenic moiety are diazomethane H_2CNN, cyanamide H_2NCN, and isocyanamide or N-aminoisonitrile H_2NNC; with hydrogenic atomic centres farther separated, carbodiimide HNCNH and nitrilimine HCNNH are characterized to some extent, but further possibilities CN(H)NH, NC(H)NH, HCN(H)N and HCN(H)N, each bearing a hydrogenic atomic centre attached at the middle member of the skeleton, are unknown experimentally. Isomers with a cyclic skeleton include diazirine or 1,1-diazirine c-H_2CN_2 that is stable and well characterised, and 1,2-diazirine or isodiazirine c-HCN(H)N not yet identified from experiment, apart from c-C(NH)$_2$ and c-CNNH$_2$ that are unknown and likely to be even less stable under conventional conditions in a laboratory. The geometric structures of the seven isomers upon which we have undertaken calculations are depicted in Figure 16-1.

The diversity of these isomers and the significant variability of the chemical properties and reactivity of the corresponding chemical substances have attracted much attention from the point of view of calculations of molecular structure. Prompted by a qualitative consideration of the interconversion of isomers [1], Hart [2] embarked on a major comparison of geometric structures through innovative calculations on five acyclic isomers; to explore the bonding and reactivity of these isomers, he employed for this purpose a basis set of gaussian lobes, and calculated molecular orbitals in a self-consistent field that were transformed to canonical localised molecular orbitals. Apart from routine calculations on an individual or a few isomers, Moffat [3], Thomson and Glidewell [4], Guimon et al. [5], Boldyrev et al. [6], Kawauchi et al. [7] and Maier et al. [8] performed increasingly sophisticated calculations on multiple isomers.Although Hart [2], for lack of automatic optimization

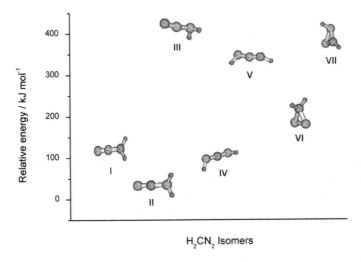

H_2CN_2 Isomers

Figure 16-1. Depiction of geometric structures of isomers of H_2CN_2 showing the relations between those structures and the relative energies according to data in Table 16-1; the molecules are diazomethane I, cyanamide II, isocyanamide III, carbodiimide IV, nitrilimine V, 1,1-diazirine VI and 1,2-diazirine VII

of geometric structural parameters, applied a manual method for this purpose, subsequent authors reported their estimates of optimized bond lengths and interbond angles of various isomers.

Several calculations of geometric structure of these isomers have thus been reported, but there has been little effort devoted to estimate the electric and magnetic properties and to their comparison. For instance, the vibrational polarizability describes the polarization of a molecule due to a displacement of the atomic centres from their relative equilibrium positions [9]; this quantity, which is represented with a tensor of second order having up to six independent components of which some might be zero by symmetry, governs the intensity in the vibrational Raman spectrum. As the value of this polarizability depends on the frequency of light that serves to excite the Raman scattering, a calculation for comparison with experiment might take this factor into consideration. Such vibrational polarizability causes a net contribution to the total molar polarization that is smaller than the electronic polarizability with all atomic nuclei in their equilibrium positions, but the ratio of magnitudes of these polarizabilities varies with the nature of the molecular structure. We have calculated these quantities that are difficult to measure, particularly with chemical compounds as inconvenient to handle as diazomethane and most isomers. Whereas the electric dipolar moment is commonly deduced from splitting of lines in pure rotational transitions with a Stark effect involving an externally applied electric field, estimates of the rotational g factor and the magnetizability result from application of a Zeeman effect through a magnetic field on similar transitions. Although only an anisotropic component of magnetizability is experimentally evaluated in this way, it is convenient to compute the isotropic magnetizability and even the diamagnetic and paramagnetic contributions thereto that arise through the use of *London atomic orbitals* in the calculation [10]. For polyatomic molecules, the rotational *g* factor is represented by a tensor of second rank with generally six independent components. These properties are difficult to measure through a direct experiment, but their calculation presents no particular difficulty when each property is related to some aspect of the distribution of atomic nuclei and their associated electronic density.

The progressive improvement of both computer hardware and software during these five decades during which calculation of these molecular properties has become feasible has enabled much improved estimates of not only structural parameters but also other properties that might be experimentally measured directly or deduced from spectral or other experiments. As an example of the present capability of calculations of molecular electronic structure, in the present work we have conducted calculations of various electric and magnetic properties of H_2CN_2 in seven structural isomers with a basis set at a uniformly high level, for comparison with existing experimental data or for prediction for subsequent experiments; because these electric and magnetic properties can depend sensitively on molecular geometries, we specify also those geometries, and related spectral data. We interpret all these results in the light of our present understanding of the nature of the models underlying both calculations and experimental measurements.

CALCULATIONS

For all calculations of molecular electronic structure we utilized software Dalton 2.0 [11] to implement numerical solution of Schrödinger's equation and to estimate molecular properties. The electronic energy and properties we calculated with wave functions according to a basis set, denoted *aug-cc-pVTZ*, devised by Dunning and coworkers [12], involving a self-consistent field of the type complete active space and multiple configurations. The number of electrons in active shells associated with functions for one electron in a selected set varied with the particular isomer; these numbers of electrons in active shells and numbers of active orbitals, respectively, for each isomer follow: diazomethane, 4, 6; cyanamide, 10, 10; isocyanamide, 4, 6; carbodiimide, 12, 10; nitrileimine, 8, 9; 1,1-diazirine, 6, 10; 1,2-diazirine, 10, 9. The same contractions, denoted (6s3p2d|4s3p2d) for H and (11s6p3d2f|5s4p3d2f) for C and N, were employed in all calculations, in total 224 primitive and 184 contracted gaussian basis functions.

Such a basis set combines well with coupled-cluster wave functions to tend to converge in a consistent and predictable manner towards limits of the basis set and the theory. Calculation of the rotational *g* tensor and magnetizability involved use of rotational London orbitals [10]. Optimization, first order in derivatives of energy with respect to internuclear distances, yielded all reported geometric structures of arrangements of atomic centres; we calculated all reported molecular properties at these optimized geometries.

RESULTS

In several tables we present the results from our calculations. One large table contains values of parameters describing electric and magnetic properties and moments of inertia and rotational parameters of optimized structures for which such comparison is convenient. Succeeding tables contain parameters for optimized geometric structures, their vibrational wavenumbers and intensities, with experimental data for comparison if available; because the most pertinent structural parameters vary according to a particular isomer, and because the symmetries of the isomers involve four distinct point groups – C_{2v}, C_2, C_s and C_1, each with its fundamental vibrational modes in separate classes, we present one such table for each individual structural isomer, except combining content for cyanamide and isocyanamide into one table.

In each case the results are applicable to a net electrically neutral molecular species of formula $^1H_2{}^{12}C^{14}N_2$ with unit spin multiplicity and for the electronic ground state. Because our calculation pertains formally to a single molecule with coordinate axes fixed in the molecular frame oriented according to the inertial axes, we express our results generally in terms of molecular quantities and with SI units but unified atomic mass unit; to facilitate comparison with results reported elsewhere, in a few cases we present additional data converted with appropriate factors.

In an order emphasizing the structural relations, Figure 16-1 depicts the optimized geometric structures of seven isomers with chemical formula H_2CN_2 – diazomethane I, cyanamide II, isocyanamide or N-aminoisonitrile III, carbodiimide IV, nitrilimine V, diazirine VI and 1,2-diazirine VII. With one column for each isomer in the same order, Table 16-1 presents the results of calculations of electric and magnetic and some spectral properties of isolated molecules in their optimized structures that are directly comparable in this way. The first row contains the total energy of the molecular system, including relativistic corrections of types mass velocity and Darwin; the second and third rows indicate the energy of each isomer relative to that of the most stable isomer, cyanamide, per molecule and per mole respectively, appropriately rounded.

Of rows in the next group for electric properties, the fourth shows the total electric dipolar moment. The next five rows present an analysis of net electronic populations associated with each atomic centre, according to an atomic polar tensor [13]; each value listed represents the net alteration of electronic population associated with a particular atomic centre through its participation as a constituent of the particular molecule in a specific isomeric form. To distinguish the two hydrogenic atomic centres if they lie in chemically inequivalent positions, H_a is nearer a carbon atom than H_b; likewise if the two nitrogens have inequivalent positions N_a is nearer a carbon atom than N_b. The next six rows present elements of a symmetric

Table 16-1. Energies, electric dipolar moments, net atomic populations, vibrational polarizabilities and mean vibrational molecular polarization, magnetizability and contributions thereto, isotropic g tensor and nuclear and electronic paramagnetic and diamagnetic contributions thereto, principal moments of inertia and rotational parameters calculated for $^1H_2{}^{12}C^{14}N_2$ in seven structural isomers

property	diazomethane	cyanamide	isocyanamide	carbodiimide	nitrilimine	1,1-diazirine	1,2-diazirine
total energy/hartree + 148	−0.1584322	−0.2010834	−0.0508985	−0.1741517	−0.0716112	−0.1382128	−0.0694224
relative energy/10⁻¹⁸ J	18.6	0	65.5	11.7	56.4	27.4	57.4
relative energy/kJ mol⁻¹, including residual energy	104	0	394	67	333	164	340
electric dipolar moment p/10⁻³⁰ C m	5.654	13.637	14.061	7.346	7.425	5.791	9.884
atomic population on C	−0.413	0.333	0.011	0.925	−0.600	0.129	0.261
atomic population on H_a	0.118	0.195	0.198	0.321	0.244	0.017	0.103
atomic population on H_b	0.118	0.195	0.198	0.321	0.192	0.017	0.125
atomic population on N_a	0.773	−0.408	−0.127	−0.784	0.641	−0.082	−0.255
atomic population on N_b	−0.596	−0.315	−0.280	−0.784	−0.576	−0.082	−0.234
vibrational polarizability α_{aa}/10⁻⁴¹C² m² J⁻¹	3.432	4.956	1.491	23.876	⋯	0.293	1.793
vibrational polarizability α_{ab}/10⁻⁴¹C² m² J⁻¹	0	0	0	0	⋯	0	−0.031
vibrational polarizability α_{ac}/10⁻⁴¹C² m² J⁻¹	0	3.480	−1.477	−1.755	⋯	0	−0.026
vibrational polarizability α_{bb}/10⁻⁴¹C² m² J⁻¹	0.145	0.316	6.242	2.958	⋯	0.958	1.559
vibrational polarizability α_{bc}/10⁻⁴¹C² m² J⁻¹	0	0	0	0	⋯	0	0.256
vibrational polarizability α_{cc}/10⁻⁴¹C² m² J⁻¹	21.035	4.060	2.505	0.204	⋯	0.094	0.511
mean vibrational molecular polarization/10⁻³⁰ m³	5.673	2.151	2.360	6.232	⋯	0.310	0.890
magnetizability, isotropic/10⁻²⁹ JT⁻²	−36.55	−45.91	−43.27	−46.82	−39.39	−31.77	−37.34
magnetizability, diamagnetic/10⁻²⁹ JT⁻²	−180.71	−188.42	−179.00	−183.71	−178.45	−140.13	−146.49
magnetizability, paramagnetic/10⁻²⁹ JT⁻²	144.16	142.51	135.73	136.89	139.06	108.36	109.15
isotropic g tensor	−0.3195	0.1561	0.1445	0.0484	−0.2184	−0.1821	−0.1115
nuclear contribution to g	0.7179	0.7031	0.6922	0.6933	0.6875	0.5932	0.5952
diamagnetic contribution to g	−0.0401	−0.0401	−0.0461	−0.0414	−0.0456	−0.0576	−0.0577
paramagnetic contribution to g	−0.9973	−0.5069	−0.5016	−0.6035	−0.8603	−0.7177	−0.6490
inertial parameter I_a/10⁻²⁰ u m²	1.807222	1.701204	1.759552	1.386995	1.421473	12.387741	12.202547
inertial parameter I_b/10⁻²⁰ u m²	45.216582	50.258732	46.189584	49.003341	45.955446	21.381566	23.954082
inertial parameter I_c/10⁻²⁰ u m²	47.023803	51.332726	47.110536	49.277031	46.217609	30.32326	34.43216
rotational parameter A/m⁻¹	932.7925	990.9233	958.0635	1215.4068	1185.9267	136.0832	138.1484
rotational parameter B/m⁻¹	37.2820	33.5417	36.9466	34.4010	36.6825	78.8419	70.3748
rotational parameter C/m⁻¹	35.8491	32.8399	35.7831	34.2099	36.4745	55.6931	48.959

tensor for vibrational electric dipolar polarizabilities [9] relative to principal inertial axes; for these calculations we assumed a static polarizability corresponding to zero frequency. The following row indicates the corresponding mean vibrational contribution to the total molecular electric dipolar polarization.

The next seven rows pertain to magnetic properties of each isomer in its optimized structure. The molecular magnetizability is the factor of proportionality that yields the magnetic dipolar moment induced in a freely rotating molecule subjected to an external magnetic field; this magnetizability is a tensorial quantity, of which the isotropic magnetizability is one third the trace of this tensor, calculated as a sum of diamagnetic and paramagnetic contributions [10] that are listed separately. For a free molecule there is likewise a tensor for the rotational g factor, which measures the extent to which a magnetic dipolar moment arises from the rotation of the molecule about its centre of mass; again obtained as one third the trace of the g tensor, the isotropic value listed is the sum of nuclear and electronic contributions, of which the latter have diamagnetic and paramagnetic components, all of which are listed separately. Of the last six rows, three contain the moments of inertia about the principal rotational axes, and another three rows present the corresponding rotational parameters.

Six further tables, one for each structural isomer of $^1H_2{}^{12}C^{14}N_2$ investigated except for cyanamide and isocyanamide combined into one table, present the calculated values of parameters pertaining to the optimized geometric structure of the atomic centres, as lengths of inferred chemical bonds or the smallest internuclear distances, and interbond angles, in a set sufficient to define unambiguously the geometric arrangement of atomic centres according to a specified point group. These tables include also the calculated wavenumbers of fundamental vibrational modes within specified symmetry classes, the corresponding calculated intensities, the calculated wavenumbers scaled by 0.95, and pertinent data from experiment for structure, wavenumber and intensity where available.

DISCUSSION

Relation Between Calculation and Experiment

Before discussing in detail the numerical results of our computational work, we describe the theoretical and computational context of the present calculations: apart from deficiencies of models employed in the analysis of experimental data, we must be aware of the limitations of both theoretical models and the computational aspects. Regarding theory, even a single helium atom is unpredictable [14] purely mathematically from an initial point of two electrons, two neutrons and two protons. Accepting a narrower point of view neglecting internal nuclear structure, we have applied for our purpose well established software, specifically Dalton in a recent release 2.0 [9], that implements numerical calculations to solve approximately Schrödinger's temporally independent equation, thus involving wave mechanics rather than quantum mechanics in other forms. The trial wave function for only the electrons is composed from basis functions in a chosen set, described as *aug-cc-pVTZ*, implying three gaussian functions to represent each 'valence' orbital, correlation-consistent and augmented with diffuse and polarization functions; although in principle a further set at the maximum theoretical level within Dalton 2.0, described as *aug-cc-pV6Z*, implying use of six gaussian functions analogously, is available for H, C and N atomic centres, the accessible computing resources precluded such use for all isomers: to maintain a common level of basis set for all isomers, we accepted the former set. In each basis function, the coefficient of a coordinate in an exponent is fixed, whereas the coefficient of each gaussian function in a fixed set is adjusted to yield a minimum energy of the selected molecular system. As those basis functions serve to mimic wave functions of an atom with one electron, further procedures serve to take into account, in a necessarily incomplete and approximate manner, repulsion and correlation between electrons. A further calculation of perturbational type has as its objective partially to take into account relativistic effects, of types mass velocity and Darwin, not encompassed directly in solution of Schrödinger's equation; for atoms H, C and N these relativistic corrections are small, and vary little between the various structural isomers. The wave function that results from this calculation is an artefact of this approach to our application of quantum mechanics, for which reason we refrain from discussing any aspect of this artefact, or of its even more artefactual constituent basis functions; a molecular distribution of electronic density, which is in principle an observable quantity, can clearly not be decomposed uniquely into exponential or gaussian functions in a finite sum. As the criterion for an optimal wave function is the minimum total energy, both with optimized coefficients of basis functions and with optimized geometry, the resulting electronic density is subject to error reflecting an incomplete basis set; additional error results from an incomplete account of electronic correlation. Various molecular properties are customarily related to the electronic density and to the arrangement of the atomic nuclei, but a particular property might be sensitive to that density in particular regions of the effective molecular volume, such as near specific nuclei; for this and other reasons, the quality of calculation of each such property varies according to its nature, and some values of properties would then inevitably be nearer experimental values of these properties than others.

Apart from use of experimental values of atomic – rather than nuclear – and electronic masses and of electric charges, the basis of this calculation has an empirical component. The calculation is certainly not made genuinely from first principles or *ab initio*, firstly because the composition of the basis set is predetermined, by those who have published this basis set [12] and by the authors of Dalton software [11] who have incorporated it, according to its success in reproducing experimentally observable quantities and other calculated properties. Secondly, the solution of Schrödinger's equation is based on a separation of electronic and nuclear motions, essentially with atomic nuclei fixed at relative positions, which is a further empirical imposition on the calculation; efforts elsewhere to avoid such an arbitrarily distinct treatment of subatomic particles, even on much simpler molecular systems, have proved only partially successful [15–17], at a greatly increased cost and complication of such calculation. With account taken concurrently of both nuclear and electronic motions, Schrödinger's equation has only recently been solved exactly for the hydrogen and other atoms [18]. A practical advantage of a separation of electronic and nuclear motions is that, on that basis, methods and algorithms have been well developed to estimate diverse molecular properties, even those such as the rotational g factor that partially transcend that approximation of separate treatment of electronic and nuclear motions [10]. According to that separate treatment, all molecules with the same formula and total charge represent local minima on a single hypersurface of potential energy, provided that all crossings of surfaces are avoided when a complete hamiltonian is applied. Although, to achieve a local minimum on an hypothetical hypersurface of potential energy in nine spatial dimensions applicable to the prospective motion of atomic nuclei, we undertook optimization of the relative coordinates of nuclear particles, with internuclear separations varied within an apparently small range from a specified initial conformation, there is within the progress of the calculation an implicit or explicit suppression of the seeking of a global minimum of energy; we become thereby able to distinguish these seven structural isomeric forms, among further possible, but thermodynamically less likely, isomers. As structural isomers of H_2CN_2 numbering six have been demonstrated to be sufficiently durable and stable to be characterized experimentally according to spectrometric measurements of various kinds, the separate treatment of electronic and nuclear motions that yields the corresponding classical structures in terms of geometric arrangements of atomic nuclei seems acceptable in the region of at least six of those minima on that putative hypersurface of potential energy; in regions not near those minima, such as those near the location of a *transition structure* that possesses one or more imaginary vibrational frequencies, such a separate treatment is questionable. Thirdly, a proper quantum-mechanical calculation is subject to requirements of indistinguishability of all identical particles, not just electrons; in the context of the present calculations there is neglect of permutation symmetry of the two atomic nuclei of both hydrogen, ^1H, and nitrogen, ^{14}N, that is contrary to that fundamental requirement of quantum mechanics.

As is typical of computer programmes for conventional calculations of molecular electronic structure and a resulting geometric conformation of relative nuclear positions and molecular properties, Dalton [11] provides no estimate of error or uncertainty in those internuclear separations or properties resulting from either numerical error or approximations in the method. Despite our use of the same basis set, *aug-cc-pVTZ*, for calculations on each structural isomer, there remains latitude in the conduct of the calculation according to the concept of a complete active space. That basis set is the largest for which our calculations, on the available hardware and within the limitations of Dalton, are practicable under the present conditions. Altering the number of electrons in that active space yields slight and apparently significant variations in internuclear distances and other descriptors of molecular properties of the types that we present in the several tables. Although we have accepted as the best wave function of each isomer the one among our extensive, but not exhaustive, tests that yielded the least total molecular energy, there is in general no monotonic trend of values of molecular properties, such as bond lengths or wavenumbers of characteristic vibrational modes, to converge to a limit well defined in the progress toward a converged total energy; the variation theorem is inapplicable to this process. Variation of the basis set would doubtless yield further slight variations of resultant molecular descriptors. As an exploratory test on only diazomethane, we made a calculation with the same basis set but with a further approximation of density functionals; the structural parameters and the properties, of those that the calculation permitted, varied slightly from those values obtained without this approximation. The total duration of a calculation with density functionals, but without all properties, was about two fifths that without this approximation.

Despite one's intent to compare calculated results with corresponding parameters deduced from experimental data, these quantities inherently defy direct comparison. In particular, the calculated structural parameters for an optimized structure pertain to a geometric arrangement of atomic centres in their relative equilibrium locations according to a local minimum on an hypothetical hypersurface of potential energy that is an artefact of a separate treatment of electronic and nuclear motions. Experimental data pertain either to these small molecules in particular quantum states, which have a completely indefinite geometric structure so effectively no extension in space or time and no correspondence to a classical structure [19], or to ensembles of molecules averaged over internal states occupied at a temperature of a particular experiment. For one to deduce accurately from experimental data, for instance from spectral data for transitions between discrete molecular states, the equivalent theoretical function of potential energy of a free molecule as a dependence on internuclear separations that is independent of nuclear masses, one must take into account the fact that electrons fail to follow perfectly the putative nuclear motions conventionally described as vibrational and rotational. To enable one to derive from spectral data a function purely for potential energy, an elimination of the effect of finite nuclear masses requires not only extensive spectral measurements on multiple isotopic variants but also application of corrections described as adiabatic and nonadiabatic, encompassed within adiabatic terms and rotational and vibrational g factors; this process, and in some cases also the effects of finite nuclear volumes, has been satisfactorily implemented for diatomic molecular species [20], but remains largely impracticable for polyatomic molecules; Dalton [11] nevertheless provides procedures for vibrational averaging based on an harmonic approximation. A comparison between geometric parameters obtained through a calculated molecular electronic structure and those from experiment must thus be necessarily rough. For other than an optimized structure, and particularly when the total electronic energy much exceeds that associated with the residual energy, known also as *zero-point energy*, within the most stable electronic state, the approximation of separate treatment of electronic and nuclear motions is subject to failure because multiple electronic states are likely to have comparable energies under those conditions, whether or not Dalton software indicates such a possibility.

For an optimized structure, the total electronic energy that Dalton yields, including effects of repulsion between atomic nuclei, is formally applicable to an isolated single molecule in its hypothetical equilibrium condition, which is neither a spectral nor a thermodynamic state; this energy is expressed most appropriately in SI units as joules, or aJ, per molecule. Even with an addition of residual energy, calculated from half the sum of vibrational wavenumbers of fundamental modes – harmonic or otherwise, such an energy still applies purely to an isolated molecule. Mere multiplication by Avogadro's constant to yield a nominal energy per mole is grossly misleading because a molar quantity implies a macroscopic sample in some state of aggregation. Under standard thermodynamic conditions, the state of aggregation varies with the structural isomer – diazomethane and diazirine are vapours at 300 K whereas cyanamide is an involatile solid – hence subject to particular conditions of at least pressure and temperature, and encompasses all energies of intermolecular interactions; such energies vary considerably with the intrinsic properties of individual molecules, such as the extent of polarity or molecular electric dipolar moment. Although these energies of intermolecular interactions under conditions applicable to standard thermodynamic states seem minuscule by comparison with total electronic energies, they become significant by comparison with small differences between energies of ideal molecular structural isomers, and likewise vary with the nature of each isomer. The variations in residual energy among these isomers are small but significant.

The electric dipolar moment derived from the Stark effect on molecular spectral transitions pertains formally to an expectation value of charge displacement over domains of internuclear separations and pertaining to particular quantum states, rather than representing that charge displacement for internuclear distances in a particular fixed set corresponding to an equilibrium conformation. Just as an atom within a molecule is poorly defined, the net atomic charge associated with a particular atomic centre is poorly defined both theoretically and experimentally: there are various possible methods both to calculate this quantity and to measure it, such as through diffraction by xrays and electrons, but the fundamental problem remains that in a molecule other than one with a single atomic nucleus there is no atom – only atomic nuclei and their associated electrons; any partition of total electronic charge is consequently arbitrary. Cioslowski's approach nevertheless yields estimates of redistribution of electronic charge on formation of a molecule, such as those listed in Table 16-1, that exhibit satisfactory properties [13]; although one might seek to apply such data to reinforce conventional notions of chemical binding and to indicate prospective modes of chemical reactivity, we refrain from such speculation. Reported in Table 16-1, these estimates described as atomic populations are first derivatives of the electric dipolar moment of a molecule with respect to cartesian coordinates [13], which differ significantly from any attempted integration of electronic charge in a chosen volume about one nucleus minus the nuclear charge or atomic number.

Like electric dipolar moment, the magnetic and other electric, properties of molecules deduced from spectral experiments pertain inevitably to expectation values over domains of internuclear separations; the calculated values, listed in Table 16-1, might however serve as approximate guides to what experiments might provide. The rotational parameters deduced from experiment are supposed to be inversely proportional to inertial parameters on the basis of a classical interpretation of a rigid assemblage of atomic nuclei and their directly associated electrons according to separate atomic masses. The distinction between calculated moments of inertia, and rotational parameters derived therefrom, and the corresponding experimental quantities, suffers from the same fundamental impediment as the parameters of geometric structures. In all cases only a rough comparison is formally practicable, but the trends of deviations between calculated and measured properties might provide guidance for the conduct and interpretation of future experiments.

Vibrational mode pose a particular dilemma for comparison of calculated and experimental quantities. The latter are derivable with great relative precision, from about one part in 10^3, in the worst cases of vibration-rotational bands with unresolved contours for a compound among the present seven observed as a gaseous sample, to about one part in 10^7, such as for band ν_7 of carbodiimide [21] in Table 16-4 for which a full rotational analysis proved practicable. Calculations of molecular electronic structure in Dalton and similar computational procedures yield wavenumbers according to a parabolic dependence of small displacements from the equilibrium conformation, which pertain thus to hypothetical harmonic vibrations. Any cross section of a postulated hypersurface of potential energy accurately deduced from experimental data according to a classical model lacks an exactly parabolic profile, even near a local minimum of energy of which our calculations confirm seven for these isomers of H_2CN_2; for this reason there is a systematic deviation between directly calculated vibrational wavenumbers and those measured for centres of vibration-rotational bands in infrared spectra or Raman scattering. To take into account this condition, an empirical method of adjusting the calculated wavenumbers involves scaling, by a value typically ≈ 0.95; such a value has formal justification for application to stretching modes involving hydrogen atomic centres [22], but for other modes is merely a convenient factor. To facilitate comparison between experimental and calculated wavenumbers, we apply this scaling factor in the tables for individual structural isomers. The variability of calculated wavenumbers associated with particular fundamental vibrational modes seems to be as much as ten per cent depending on the conditions of the calculation apart from the particular basis set, namely the number of electrons in the active space. Measurement of absolute intensities of vibration-rotational bands of polyatomic molecules is difficult, and the results are hence typically much less precise than for wavenumbers of band centres; for instance, about one part in ten has been achieved for diazomethane [23], whereas for some diatomic molecules [20] one part in 10^2 is typically achievable. Incompletely resolved separate lines assigned to individual vibration-rotational transitions and overlapping bands of separate vibrational modes complicate these measurements for polyatomic molecules. As presented in Tables 16-2–16-7, calculated intensities are obtained from dipolar gradients of atomic centres, also known as

Table 16-2. Calculated and experimental, structural and vibrational parameters of diazomethane (point group C_{2v}): calculated and measured lengths of bonds and interbond angles; calculated wavenumbers and intensities of fundamental modes, scaled wavenumbers, and experimental wavenumbers and intensities[a]

	calc.				exp.	
internuclear distance/10^{-10} m, C-H	1.064				1.075	
internuclear distance/10^{-10} m, C-N	1.304				1.300	
internuclear distance/10^{-10} m, N-N	1.113				1.139	
interbond angle/deg, H-C-H	126.2				126	
	ν/m^{-1}	I/10^{-21} m	ν/m^{-1}	ν/m^{-1}	I/10^{-21} m	
vibrational mode	calc.	calc.	scaled	exp.	exp.	
ν_1, a$_1$	334084	20.59	317380	307710	7.6	
ν_2, a$_1$	210522	897.74	199996	210157.7	284.5	
ν_3, a$_1$	153722	66.75	146036	141333	19.9	
ν_4, a$_1$	117275	1.80	111411	117700	5.1	
ν_5, b$_1$	57590	26.35	54711	56819		
ν_6, b$_1$	39225	208.38	37264	40887		
ν_7, b$_2$	347807	1.81	330417	318450		
ν_8, b$_2$	117355	6.55	111487	110900	0.2	
ν_9, b$_2$	41596	0.87	39516	41579	58.9	
residual energy/hc	709588		674109	672945		

[a] For references to sources of experimental data, see the text.

net atomic charges or the atomic polar tensor listed in Table 16-1; as these charges vary appreciably with the nature of the particular basis set and other aspects of the calculation, there is no expectation of great absolute accuracy of the resulting intensities, but relative values might serve for comparison with experimental data.

The extent of information from experiment about the quantitative aspects of molecular geometry and properties varies with the particular structural isomer; 1,2-diazirine is not yet positively detected from experiments, whereas much information about chemical and physical properties is known about diazomethane and 1,1-diazirine, with other isomers in intermediate conditions. Those two specified isomers are not the most stable, but the inversion motion of the amino moieties in both cyanamide and isocyanamide greatly impedes efforts to define structural parameters of these molecules from their spectra, apart from the more difficult conditions in working with these involatile substances. As little or no experimental thermochemical data are available for these isomers, we compare our order of relative stabilities with those of Hart [2], of Kawauchi et al. [7] and of Moffat [3]. As a concession to such comparison on a molar basis, we present in Table 16-1 our total energies including the scaled residual energy on a molar basis; Figure 16-1 depicts these energies for the seven selected structural isomers. All these authors agree that cyanamide is most stable, with carbodiimide next except for Hart [2]; other than our finding isocyanamide to be least stable, our order of stability of the

Table 16-3. Calculated and experimental, structural and vibrational parameters of cyanamide and isocyanamide (both point group C_s): calculated and measured lengths of bonds and interbond angles; calculated wavenumbers and intensities of fundamental modes, scaled wavenumbers, and experimental wavenumbers[a]

	cyanamide calc.	exp.	isocyanamide calc.
internuclear distance/10^{-10} m, N-H	1.016	1.001	0.998
internuclear distance/10^{-10} m, C-N	1.355	1.346	
internuclear distance/10^{-10} m, C-N	1.164	1.1645	1.162
internuclear distance/10^{-10} m, N1-N2	2.518	2.506	1.355
interbond angle/deg, H-N-H	109.5		109.7
interbond angle/deg, N-C-N or N-N-C	177.4		169.7

vibrational mode	ν/m^{-1} calc.	$I/10^{-21}$ m calc.	ν/m^{-1} scaled	ν/m^{-1} exp.	ν/m^{-1} calc.	$I/10^{-21}$ m calc.	ν/m^{-1} scaled
ν_1, a'	376526	74.73	357700	342000	372518	26.04	353892
ν_2, a'	242694	115.84	230559	227000	231498	98.50	219923
ν_3, a'	177238	70.90	168376	157500	180127	43.04	171121
ν_4, a'	111059	11.34	105506	105500	112634	89.15	107002
ν_5, a'	69328	359.94	65862	71386	98279	130.42	93365
ν_6, a'	31683	8.71	30099	38000	26130	2.29	24824
ν_7, a''	386150	107.09	366843	348000	381627	62.70	362548
ν_8, a''	128174	0.82	121765	72000	144691	11.01	137458
ν_9, a''	32353	0.50	30735		10769	4.82	10231
residual energy/hc	777602		738722		779136		740179

[a] For references to sources of experimental data, see the text.

Table 16-4. Calculated and experimental, structural and vibrational parameters of carbodiimide (point group C_2): calculated and measured lengths of bonds and interbond angles; calculated wavenumbers and intensities of fundamental modes, scaled wavenumbers, and experimental wavenumbers[a]

	calc.		exp.	
internuclear distance/10^{-10} m, N-H	0.997		1.0039	
internuclear distance/10^{-10} m, C-N	1.233		1.2247	
interbond angle/deg, H-N-C	116.7		119.1	
interbond angle/deg, N-C-N	171.5		171.6	
angle/deg, HN..NH	88.5		89.35	
	ν/m^{-1}	I/10^{-21} m	ν/m^{-1}	ν/m^{-1}
vibrational mode	calc.	calc.	scaled	exp.
ν_1, a	380037	65.95	361035	349800
ν_2, a	123820	0.10	117629	128500
ν_3, a	102125	23.93	97019	
ν_4, a	78250	106.49	74338	
ν_5, a	53179	0.04	50520	53700
ν_6, b	379817	241.55	360826	
ν_7, b	217176	1017.52	206317	210470.47
ν_8, b	97604	864.55	92724	89000
ν_9, b	52298	137.13	49683	53700
residual energy/hc	742153		705045	

[a] For references to sources of experimental data, see the text.

Table 16-5. Calculated and experimental, structural and vibrational parameters of nitrilimine (point group C_1): calculated and measured lengths of bonds and interbond angles; calculated wavenumbers and intensities of fundamental modes, scaled wavenumbers, and experimental wavenumbers[a]

	calc.			
internuclear distance/10^{-10} m, C-H	1.069			
internuclear distance/10^{-10} m, C-N	1.192			
internuclear distance/10^{-10} m, N-N	1.270			
internuclear distance/10^{-10} m, N-H	1.006			
interbond angle/deg, H-C-N	128.5			
interbond angle/deg, C-N-N	167.6			
interbond angle/deg, N-N-H	106.6			
	ν/m^{-1}	I/10^{-21} m	ν/m^{-1}	ν/m^{-1}
vibrational mode	calc.	calc.	scaled	exp.
ν_1, a	365122	21.77	346866	325010
ν_2, a	339346	17.12	322379	314250
ν_3, a	206186	574.09	195877	203270
ν_4, a	146618	162.63	139287	127810
ν_5, a	117182	18.27	111323	118750
ν_6, a	92333	55.35	87716	79210
ν_7, a	80640	0.89	76608	60650
ν_8, a	50203	28.52	47693	46140
ν_9, a	40669	18.01	38636	
residual energy/hc	719149		683192	

[a] For references to sources of experimental data, see the text.

Table 16-6. Calculated and experimental, structural and vibrational parameters of 1,1-diazirine (point group C_{2v}): calculated and measured lengths of bonds and interbond angles; calculated wavenumbers and intensities of fundamental modes, scaled wavenumbers, and experimental wavenumbers[a]

	calc.		exp.	
internuclear distance/10^{-10} m, C-H	1.069		1.0803	
internuclear distance/10^{-10} m, C-N	1.493		1.4813	
internuclear distance/10^{-10} m, N-N	1.181		1.228	
interbond angle/deg, H-C-H	120.2		120.5	
interbond angle/deg, N-C-N	46.6		48.98	
	ν/m^{-1}	I/10^{-21} m	ν/m^{-1}	ν/m^{-1}
vibrational mode	calc.	calc.	scaled	exp.
ν_1, a_1	328440	15.58	312018	302325
ν_2, a_1	197881	57.99	187987	162300
ν_3, a_1	159980	2.85	151981	145916.06
ν_4, a_1	101627	3.36	96546	99180
ν_5, a_2	102244	0	97132	96270
ν_6, b_1	104115	57.32	98909	96730
ν_7, b_1	83864	21.77	79671	80713.95
ν_8, b_2	340757	21.09	323719	313190
ν_9, a_2	119200	6.87	113240	112491.43
residual energy/hc	769054		730102	704558

[a] For references to sources of experimental data, see the text.

Table 16-7. Calculated structural and vibrational parameters of 1,2-diazirine (point group C_1): calculated lengths of bonds and interbond angles; calculated wavenumbers and intensities of fundamental modes, and scaled wavenumbers

	calc.		
internuclear distance/10^{-10} m, C-H$_a$	1.070		
internuclear distance/10^{-10} m, C-N$_a$	1.390		
internuclear distance/10^{-10} m, C-N$_b$	1.254		
internuclear distance/10^{-10} m, N$_a$-N$_b$	1.749		
internuclear distance/10^{-10} m, N-H$_a$	1.029		
interbond angle/deg, H-C-N	136.7		
interbond angle/deg, H-N-N	108.3		
interbond angle/deg, N-C-N	82.7		
	ν/m^{-1}	I/10^{-21} m	ν/m^{-1}
vibrational mode	calc.	calc.	scaled
ν_1, a	338430	2.40	321509
ν_2, a	331928	2.05	315332
ν_3, a	171791	3.84	163201
ν_4, a	137330	9.11	130464
ν_5, a	129595	70.05	123115
ν_6, a	111071	46.51	105517
ν_7, a	98527	68.03	93601
ν_8, a	81647	47.30	77565
ν_9, a	45464	7.51	43191
residual energy/hc	722891		686746

other isomers agrees with that of Kawauchi et al. [7] Our estimate of the energy of carbodiimide relative to cyanamide is larger than the value $15.1 \, \text{kJ mol}^{-1}$ [24] deduced from experiments in which both species appeared to coexist in thermodynamic equilibrium. Vincent and Dykstra [25] calculated the difference of energies of cyanamide and isocyanamide to be $220 \, \text{kJ mol}^{-1}$; that difference, smaller than ours, might reflect their less sophisticated level of calculation. Of five acyclic isomers, Hart [2] found isocyanamide second in stability; his order of the other four acyclic isomers is the same as ours.

As little or no experimental information is available for electric and magnetic properties other than electric dipolar moment, we make a general comparison here. With regard to electrical properties in Table 16-1, all structural isomers have electric dipolar moments in a moderate range, which facilitates detection of these species through pure rotational transitions in microwave spectra; for a species not yet so observed, 1,2-diazirine, the rotational parameters provided in this table enable rough prediction of frequencies and intensities of such transitions. The vibrational polarizability for the various structural isomers has a generally positive sign, but the magnitudes of components of this tensor vary considerably. For nitrilimine we omit from this table calculated values of vibrational polarizability, and the corresponding vibrational contribution to total molecular polarization, because these calculated values have uncharacteristically large magnitudes; for only this isomer the number of vibrational modes resulting from the calculation is eleven, including two spurious modes with small wavenumbers corresponding essentially to rotational motion; such effects, resulting from slight deficiencies of calculations of molecular electronic structure, are known to corrupt estimates of vibrational polarizability. Use of other active spaces for nitrilimine failed to eliminate these deficiencies.

For the rotational g tensor, the nuclear contribution to the isotropic value varies among isomers in a small range $0.68 - 0.72$ for acyclic isomers but is essentially the same, at 0.594, for the two cyclic isomers. In all cases both the diamagnetic and paramagnetic electronic contributions are negative; the magnitude of the diamagnetic component, in a small range $0.04 - 0.06$, is much smaller than the magnitude of the paramagnetic component, which varies in a range $0.5 - 1$ for these compounds at calculated equilibrium geometries. Depending on the relative magnitudes of electronic and nuclear contributions, the net effect, reflected in the isotropic g value, is therefore positive or negative. Wilson et al. [26] reported generally negative values of diagonal components of this tensor for many small molecular species, with differences between calculated and experimental values typically less than 3 per cent. As a result of their calculations for 61 compounds with a further density-functional approximation, Wilson et al. concluded [26] that calculations of such magnetic properties are generally reliable; that conclusion is expected to be applicable to our results of analogous calculations without that approximation, provided that these calculations involve a satisfactory basis set and sufficient account of electronic correlation. Like the isotropic rotational g factor, the isotropic magnetizability has diamagnetic contributions, in all cases negative and near $-1.8 \times 10^{-27} \, \text{J T}^{-2}$ for acyclic isomers or near $-1.4 \times 10^{-27}\,\mathrm{J\,T}^{-2}$ for cyclic isomers; the paramagnetic contributions are in all cases positive and near $1.4 \times 10^{-27}\,\mathrm{J\,T}^{-2}$ for acyclic isomers and $1.1 \times 10^{-27}\,\mathrm{J\,T}^{-2}$ for cyclic isomers. The net effect is consequently in a range/$10^{-28}\,\mathrm{J\,T}^{-2}$ from -3.1 to -4.6.

For other properties we discuss results separately for the various structural isomers of $^{1}\mathrm{H}_{2}{}^{12}\mathrm{C}^{14}\mathrm{N}_{2}$.

Diazomethane

For this species we constrained our calculation to retain a planar conformation belonging to point group C_{2v}, in accordance with experimental evidence; the calculations on that basis yielded no imaginary vibrational wavenumbers that would indicate a decreased symmetry. Although such a constraint of symmetry is superfluous in such a calculation, its presence greatly diminishes the duration of calculation and the extent of storage space for integrals. The calculated electric dipolar moment, listed in Table 16-1, is comparable with the experimental quantity, $5.0 \times 10^{-30}\,\mathrm{C\,m}$ [27], but no experimental values are available for comparison with our magnetic quantities. The rotational parameters that we calculated, according to Table 16-1, differ slightly from the corresponding experimental quantities/m^{-1} for the vibrational ground state, $A = 910.5603$, $B = 37.7108452$ and $C = 36.1757609$ [28]. In view of the qualifications, stated above, about a general comparison between calculated and experimental geometric parameters, the calculated lengths of chemical bonds are reasonably similar to experimental values, compared in Table 16-2. The scaled values of vibrational wavenumbers are generally nearer the experimental quantities [29–31] than the unscaled values. According to a single calculation for diazomethane using density functionals, those resulting vibrational wavenumbers would clearly benefit from a different scaling factor; there seems to be no decisive gain of accuracy of prediction from use of such density functionals, although that approach might decrease somewhat the total duration of calculation. The calculated intensities of vibrational transitions in fundamental modes of class a_1 have magnitudes in the same order as the experimental values [23], but for modes of class b_2 the disagreement is great; as all these experimental values were obtained from spectra at only moderate resolution, their reliability is questionable.

Cyanamide and Isocyanamide

Because their structures, differing only in the orientation of the cyano moiety with respect to the amino moiety, are similar, these two isomers that belong to the same point group, C_s, we conveniently consider together. According to our calculations, cyanamide and isocyanamide represent the most and least stable, respectively, of the seven selected isomers, and are the two most polar molecules, reflecting the natures of their constituent amino and cyano moieties. The atomic nuclei of both cyanamide and isocyanamide undergo a motion, inversion at the nitrogen atom of the amino moiety, of large amplitude, with which is associated a double minimum in the function for potential energy; for this reason the rotational parameters in the vibrational ground state comprise two sets, one for each state of the symmetry pair [21, 32]. Much more experimental information exists for cyanamide than for isocyanamide.

For cyanamide, the components of electric dipolar moment/10^{-30} C m parallel to inertial axes are $p_a = 14.14$ and $p_c = 3.03$ [33], slightly larger than our calculated values $p_a = 13.33$ and $p_c = 2.878$; the total moment from experiment is 14.46×10^{-30} C m, correspondingly larger than our calculated value in Table 16-3. We compare our calculated rotational parameters with the effective rotational parameters/m^{-1} for the vibrational ground state − $A = 1041.19325$, $B = 33.78923475$ and $C = 32.90918052$ [34], which are all a little larger than the calculated values. Our calculations indicate that both these isomers have a non-linear spine, consistent with experiment for both structural isomers; those calculated internuclear distances, in Table 16-3, agree satisfactorily with the values deduced from experiment for cyanamide.

For isocyanamide, spectral data of insufficient isotopic variants have been obtained to enable the derivation of experimental structural parameters; although application of the Stark effect to split spectral lines of a single isotopic species would suffice to yield values of the total electric dipolar moment and its two non-zero components, this experiment has not been reported. We compare our calculated rotational parameters with the effective rotational parameters/m^{-1} for the vibrational ground state $-A = 654.831$, $B = 34.16822$ and $C = 33.05467$ [35], from a mean of corresponding parameters for the two inversion states of least energy; the large difference between calculated and experimental values of A results from the large amplitude of wagging motion in these inversion states that is not taken into account in our calculation involving static relative nuclear positions. Our calculated values of other spectral and structural properties of isocyanamide are likely as near prospective experimental values as their counterparts for cyanamide.

Carbodiimide

According to data from microwave spectra of carbodiimide [24], the electric dipolar moment/10^{-30} C m is 6.34, somewhat smaller than the calculated value in Table 16-1; the rotational parameters/m^{-1} are $A = 1265.02346$, $B = 34.5803859$ and $C = 34.5775333$ [24], also larger than the calculated values. In Table 16-4, one highly accurate vibrational wavenumber is known, for mode ν_7 from spectra of gaseous samples [21], but other data emanate from spectra in solid phases [36]; the intensity that we calculated for the specified vibrational mode is large, consistent with the corresponding band being most readily observed in a spectrum of a gaseous sample, but two intensities, both small, are calculated for vibrational modes that seem to have been observed for solid samples, in which molecules are subject to significant intermolecular interactions.

Nitrilimine

As the only experimental observations of this isomer, apart from a possibly ambiguous signal in a mass spectrum [37], are vibrational absorption features attributed to its dispersion in solid argon [8], no measured properties correspond to any calculated result in Table 16-1 for this species, but Table 16-5 includes those vibrational data. The calculated angle between C—N and N—N bonds is approximately the same as for isocyanamide; the structure of nitrilimine differs from that of isocyanamide through transfer of one hydrogenic atomic centre from the terminal N to the terminal C. The angle between the C—H and N—H bonds is calculated to be 92°, similar to the angle between the two N—H bonds of carbodiimide.

1,1-Diazirine and 1,2-Diazirine

Of these two cyclic structural isomers, for diazirine the electric dipolar moment is known from the Stark effect in microwave spectra [38] to be 5.3×10^{-30} C m, somewhat smaller than the calculated value in Table 16-1. All calculated net atomic charges for diazirine have small magnitudes. The rotational parameters/m^{-1} for the vibrational ground state – $A = 136.59901$, $B = 78.9478949$ and $C = 55.7923163$ [39] – are comparable with the calculated values in Table 16-1. Some fundamental vibrational modes, as listed in Table 16-6, have wavenumbers well defined directly from experiment [40], whereas the wavenumber of ν_5 for the CH$_2$ twisting mode, being infrared inactive, is estimated through a force field fitted to many data for isotopic variants [40]. The wavenumber of ν_2 for the N=N stretching mode is overestimated in our calculation, consistent with a length of this bond smaller than from experiment. The scaled wavenumbers for the C—H stretching modes are still appreciably larger than the direct experimental values, but seem comparable with 'harmonic band centres' from the same fitted force field [39]; the scaled wavenumbers associated with the other six fundamental modes are comparable with the experimental data. As the energy of 1,2-diazirine is only slightly greater, according to Table 16-1, than that of nitrilimine, which has been formed by photolysis of a suitable precursor in solid argon [8], there is a prospect that 1,2-diazirine might likewise be prepared and stabilized in such an environment. We find, in Table 16-7, the length of a C—N bond at 1.75×10^{-10} m to be atypically large, in accordance with other large values in the first calculation [3] and subsequent work, but there is no indication through imaginary vibrational wavenumbers that the converged geometry pertains to a transition structure. There is a possibility of two geometrical isomers of this cyclic molecule, with both hydrogens on the same or different sides of the ring defined by the massive atoms; because the hydrogen attached to carbon is near the plane of that ring, the difference of energies is likely to be small.

CONCLUSION

By means of these pragmatic calculations on structural isomers of formula $^1H_2{}^{12}C^{14}N_2$, we have shown examples of the information of kinds that one might derive with contemporary software for quantum-chemical calculations of molecular electronic structure, namely some electric and magnetic properties, beyond the optimized geometric structures of equilibrium nuclear conformations and wavenumbers of vibrational modes. For this purpose we have applied a particular computer program, although one expects that our results would be closely reproduced with alternative computer programs and the same level of theory, providing that all users have conformed to appropriate conventions. To compare with experimental data is difficult because in all cases the quantities that one can calculate for a single sample species differ from what might be observed directly from experiments, and corrections of neither experimental data nor calculated results for equilibrium structures of polyatomic molecules are sufficiently developed to facilitate such comparison in a routine manner for such polyatomic molecules. In general, the experimental and theoretical models are intrinsically somewhat incompatible. Even for these small molecules, encompassing only five atomic nuclei of small atomic number and their 22 associated electrons, these calculations, and by implication analogous calculations on other systems involving more than two nuclei and two electrons, remain partially an art rather than an exact science, and a guide to prospective information that critical experiments might provide, rather than a standard to which experiments must aspire. This guidance is nevertheless helpful in the planning of experiments and in their interpretation; within the formal limitations of the circumstances of the generation of these results as we have discussed, their relative trends provide qualitative, or at best semi-quantitative, information about properties, structural or other, according to a conventional and classical notion of molecular structure.

These calculations have all been based on an orbital model, even though eventually each orbital became replaced with, and approximated by, three gaussian functions. Even for methods involving density functionals, an orbital basis is still essential because no alternative method of calculating required integrals has been devised. In principle, rather than using a function of form $e^{-\alpha R}$ like that for an s orbital or $e^{-\beta R^2}$ as a gaussian function, one might use functions of forms γ (R_0-R) or $\delta R(R_0-R)$ as products with Heaviside functions; although many of these functions, with varied values of parameters γ, δ and R_0 corresponding to coefficients α or β in the exponential functions, would undoubtedly be required to replace a particular exponential function, the tedious numerical integrations would be entirely or largely eliminated – indeed most might be replaced by simple algebraic formulae subject to highly efficient computation. The point is that the power of these calculations of molecular structure and properties, attested by the moderate agreement between calculated and experimental values in the preceding tables, reflects the enormous capacity of contemporary computer processors and memories rather than any particular resemblance between members of a basis set and solutions of Schrodinger's equation for an atom with one electron; such a solution defines an orbital. The latter solution, which may be made mathematically exact and algebraic, depends on a function for potential energy that comprises precisely only an electrostatic attraction between a single electron and an atomic nucleus, or even multiple atomic nuclei in a specified geometric arrangement. For even two electrons in the field of a single nucleus, a solution of such an exact and algebraic nature is impossible – the 'many-body problem'; the direct application of orbitals, derived for a system of one electron, to systems with multiple electrons is formally illogical, because one thereby takes into account all attractions to one or other nucleus but neglects the repulsion between each two electrons. In an actual calculation with standard software, such as Dalton, the latter deficiency is approximately remedied first by use of a self-consistent field and then by further account of electronic correlation according to various methods. Such remedies would be equally applicable in the use of these functions of forms γ $(R_0–R)$ and δR $(R_0–R)$ with their Heaviside factors. No particular effects of the results of the latter calculations as molecular properties would hence be attributable to orbitals because no orbitals would be involved in the calculation: there need be no attempt to mimic orbitals in the generation of the latter functions – one would simply proceed according to an entirely numerical protocol with the variation theorem until attaining the desired convergence and numerical accuracy. Although orbitals play an enormous – excessive – role in the teaching and practice of contemporary pure chemistry, they are, and remain, artefacts of a particular approach to quantum mechanics. Matrix mechanics has never been sufficiently developed to be competitive with wave mechanics for application to such electronic systems – perhaps we can blame the physicists of a century ago for their ignorance of linear algebra relative to their knowledge of differential equations, but an effort to rectify that imbalance in the development of matrix mechanics seems unforthcoming in the foreseeable future. There have even been efforts to rationalize purported direct experimental observations of orbitals [41], which appear futile because of their lack of physical existence [42]. Despite the demonstrated equivalence of matrix mechanics and wave mechanics in a context of pioneer quantum mechanics [43], Dirac proved Schrodinger's formulation to be grossly deficient with respect to Heisenberg's formulation at a more profound level of theory [44]. Quantum mechanics must be considered to be not a chemical, not even a physical, theory – where is the physics in considering momentum to be, or to be represented by, a matrix according to matrix mechanics, or to be or to be represented by $-i\hbar\ \partial/\partial q$ according to wave mechanics, or to be or to be represented by a difference of creation and destruction operators devised by Dirac? – but a collection of mathematical methods, more or less consistent and of varied degree of sophistication and complication, to calculate some property or quantity – most meaningfully an observable property – but only approximately except for the most prototypical and simple systems of only indirect chemical interest. Notwithstanding such reservations, to model the electronic density in molecules one can profitably apply orbitals, or their representatives, in calculations of a sort with conventional software that we here apply to these isomers of diazomethane.

Since Coulson's time, substantial progress has been made towards approaching a quantitative status of results of quantum-chemical calculations within the scope of a separate treatment of electronic and nuclear motions as a working model. As the present results indicate, much further work is required to attain that objective, such as improving and extending both basis sets and the software that incorporates these functions up to the still expanding limits of computer hardware, improving the algorithms for taking into account the purported vibrational motion – not merely in an harmonic approximation, increasing the range of molecular properties accessible within a particular calculation, and refining both experimental and theoretical models to improve their compatibility.

ACKNOWLEDGEMENTS

JFO thanks Dr. R. D. Harcourt, Dr. R. Kobayashi and Dr. D. J. D. Wilson for helpful advice and information, and Professors M. and B. P. Winnewisser for providing relevant publications; Swinburne University of Technology and its Faculty of Information and Communication Technologies through a Research and Development Grant supported his visit to Melbourne to conduct this project, and the Australian Partnership for Advanced Computing provided use of the National Supercomputing Facilities.

REFERENCES

1. J. F. Ogilvie, J. Mol. Struct. 3, 513–516 (1969).
2. B. T. Hart, Aust. J. Chem. 26, 461–476, 477–488 (1973).
3. J. B. Moffat, J. Mol. Struct. 52, 275–280 (1979).
4. C. Thomson and C. Glidewell, J. Comput. Chem. 4, 1–8 (1983).
5. C. Guimon, S. Khayar, F. Gracian, M. Begtrup and G. Pfister-Guillozo, Chem. Phys. 138, 157–171 (1989).
6. A. I. Boldyrev, P. von R. Schleyer, D. Higgins, C. Thomson and S. Kramarenko, J. Comput. Chem. 13, 1066–1078 (1992).
7. S. Kawauchi, A. Tachibana, M. Mori, Y. Shibusa and T. Yamabe, J. Mol. Struct. 310, 255–267 (1994).
8. G. Maier, J. Eckwert, A. Bothur, H. P. Reisenauer and C. Schmidt, Liebigs Ann. Chem.1041–1053 (1996).
9. D. Rinaldi, M. F. Ruiz-Lopez, M. T. C. Martins Costa and J. L. Rivail, Chem. Phys. Lett. 128, 177–181 (1986).
10. J. F. Ogilvie, J. Oddershede and S. P. A. Sauer, Adv. Chem. Phys. 111, 475–536 (2000).
11. Dalton, a computer program for molecular electronic structure, release 2.0 (2005); cf. http://www.kjemi.uio.no/software/dalton/dalton.html.
12. T. van Mourik and T. H. Dunning, Int. J. Quantum Chem. 76, 205 (2000) and references therein.
13. J. Cioslowski, J. Amer. Chem. Soc. 111, 8333–8336 (1989).
14. A. B. Pippard, Contemp. Phys. 29, 399–405 (1988).
15. I. L. Thomas, Phys. Rev. 185, 90–94 (1969) and related papers.
16. M. Cafiero and L. Adamowicz, Chem. Phys. Lett. 387, 136–141 (2004).
17. B. T. Sutcliffe and R. G. Woolley, Chem. Phys. Lett. 408, 445–447 (2005).
18. K. Sodeyama, K. Miyamoto and H. Nakai, Chem. Phys. Lett. 421, 72–76 (2006).
19. R. G. Woolley, Adv. Phys. 25, 27–52 (1976).

20. J. F. Ogilvie and J. Oddershede, Adv. Quantum Chem. 48, 253–318 (2005).
21. M. Birk and M. Winnewisser, Chem. Phys. Lett. 123, 386–389 (1986).
22. J. F. Ogilvie, Spectrochim Acta, 23A, 737–750 (1967).
23. M. Khlifi, P. Paillous, P. Bruston and F. Raulin, Icarus, 124, 318–328 (1996).
24. M. Birk, M. Winnewisser and E. A. Cohen, J. Mol. Spectrosc. 136, 402–445 (1989).
25. M. A. Vincent and C. E. Dykstra, J. Chem. Phys. 73, 3838–3842 (1980).
26. D. J. D. Wilson, C. E. Mohn and T. Helgaker, J. Chem. Theory Comput. 1, 877–888 (2005).
27. A. P. Cox and J. Sheridan, Nature, 181, 1000–1001 (1958).
28. E. Schafer and M. Winnewisser, J. Mol. Spectrosc. 97, 154–164 (1983).
29. C. B. Moore and G. C. Pimentel, J. Chem. Phys. 40, 329–341, 342–356 (1964).
30. J. Vogt, M. Winnewisser, K. Yamada and G. Winnewisser, Chem Phys 83, 309–318 (1984) and references therein.
31. L. Nemes, J. Vogt and M. Winnewisser, J. Mol. Struct. 218, 219–224 (1990).
32. E. Schafer, M. Winnewisser and J. J. Christiansen, Chem. Phys. Lett. 81, 380–386 (1981).
33. R. D. Brown, P. D. Godfrey and B. Kleibomer, J. Mol. Spectrosc. 114, 257–273 (1985).
34. R. D. Brown, P. D. Godfrey, M. Headgordon, K. H. Weidemann and B. Kleibomer, J. Mol. Spectrosc. 130, 213–220 (1988).
35. M. Winnewisser and J. Reinstaedtler, J. Mol. Spectrosc. 120, 28–48 (1986).
36. S. T. King and J. H. Strope, J. Chem. Phys. 54, 1289–1295 (1971).
37. N. Goldberg, A. Fiedler and H. Schwarz, Helv. Chim. Acta, 77, 2354–2362 (1994).
38. L. Pierce and V. Dobyns, J. Amer. Chem. Soc. 84, 2651–2652 (1962).
39. M. Bogey, M. Winnewisser and J. J. Christiansen, Can. J. Phys. 62, 1198–1216 (1984).
40. B. P. Winnewisser, A. Gambi and M. Winnewisser, J. Mol. Struct. 320, 107–123 (1994) and references therein.
41. W. H. E. Schwarz, Angew. Chem. Int. Ed. 45, 1508–1517 (2006).
42. J. F. Ogilvie, J. Chem. Ed. 67, 280–289 (1990) and references therein.
43. H. Primas, Chemistry, Quantum Mechanics and Reductionism, second edition, Berlin, Germany: Springer-Verlag (1983) and references therein.
44. P. A. M. Dirac, Nature, 203, 115–116 (1964).

CHAPTER 17

RECURRENT COMPLEX MODULES INSTEAD OF MOLECULES IN NON-MOLECULAR COMPOUNDS
Describing and Modelling Modular Inorganic Structures

GIOVANNI FERRARIS* AND MARCELLA CADONI

Abstract: To compensate for a lack of molecules, several inorganic crystal structures are describable in terms of complex modules that occur in various crystalline compounds. The structures that share common modules can be correlated via series, and their investigation belongs to a new branch of crystallography named modular crystallography. The crystal structures of members of a series so called polysomatic are typically based on at least two common modules; these modules are stacked differently in various members of the series. If the modules shared by the members differ to some extent, the merotype and plesiotype series are defined. To illustrate the basic concepts of modular crystallography, examples of series based on modules of the perovskite structure that occur in layered organic-inorganic hybrids and superconductors are presented. An unknown crystal structure can be modelled if relations with known members of a series are established. Examples of structure modelling for the silicates nafertisite, bornemanite, seidite-(Ce), kalifersite and tungusite are presented with the related series

INTRODUCTION

Crystalline molecular substances (e.g. sulfur) are rare among natural (minerals) and synthetic inorganic materials; most structures of these compounds are instead based on atoms in infinite arrays. In structures of this type, strong bonds tend to be ubiquitous and not confined to molecules. To overcome the lack of molecules to serve as *natural units* to describe a crystal structure, since the beginning of inorganic crystal chemistry, coordination polyhedra have been assumed as simple building units of non-molecular crystal structures (hereafter, inorganic crystal structures). These coordination polyhedra bear anions at their vertices, which are coordinated to a central cation. Aggregations of SiO_4 tetrahedra (hereafter Si-tetrahedra) to classify structurally the silicate minerals have long been used (see below), but a systematic

365

J.C.A. Boeyens and J.F. Ogilvie (eds.), Models, Mysteries and Magic of Molecules, pp. 365–390.

use of aggregations of coordination polyhedra as complex building units (modules) to describe inorganic structures is a recent practice. With its rapid expansion, a new branch of crystallography named *modular crystallography* has been initiated [1, 2].

The development of modular crystallography has been greatly stimulated by the following facts.

1. Polytypes are described as structures that differ mainly in the stacking of a particular layer along a direction [3]; for reasons of space, polytypy is not discussed in this chapter.

2. Following the first case treating the polysomatic series of biopyriboles [4], structures of several groups have been described as members of the same polysomatic series (see below).

3. The number of known complex modules that occur in distinct structures is increasing. Limited to oxygenated compounds, the following modules, in which the name applied to the module corresponds to that of the mineral species in which it occurs, have been reported [2]: bastnäsite, brucite, corundum, epidote, gehlinite, gibbsite, huanghoite-(Ce), lomonosovite, lorenzenite, mica, nacaphite, nasonite, nolanite, palmierite, perovskite, pyrochlore, pyroxene, rutile, seidozerite, schafarzikite, silinaite, spinel, synchisite, talc, topaz and zhonghuacerite-(Ce).

4. A modular interpretation of these structures proves useful according to several aspects, such as (i) to model unknown structures, because complex building modules can play a role like that of molecules; (ii) to establish correlations between properties and structure that might indicate paths to synthesize new materials; (iii) to interpret structural defects that lead to modifications of the chemical composition.

In the survey of modularity reported in Chapter 1 of [2], a general description of the modular interpretation of crystal structures has been extensively discussed. In this chapter we consider only planar modules, and the examples are devoted to show the polysomatic, merotype and plesiotype aspects (see below) of some series that are important either methodologically or for their technological and mineralogical relevance. The possibility of exploiting the modularity to model unknown crystal structures is illustrated with some examples.

POLYSOMATIC SERIES

The members of a polysomatic (signifying many bodies [4]) series are crystal structures based on the varied stacking of the same modules A, B, C, ... With reference to the case of two modules, A and B, a polysomatic series has a general formula $A_m B_n$, in which the running indices m and n represent the number of modules A and B contained in the asymmetric unit of a crystal structure. For a particular pair (m,n), distinct stacking sequences of A and B might occur, each corresponding to a separate structure. An unambiguous description of a member $A_m B_n$ must therefore show the complete sequence of modules: e.g. for $A_2 B_3$ the sequences $ABABB$ and $ABBAB$ represent distinct structures. All structures corresponding to a pair (m, n)

have the same chemical composition and correspond to polymorphs. For a given sequence, separate polytypes might occur if two adjacent modules can be stacked in distinct ways.

The chemical compositions of the members of a polysomatic series $A_m B_n$ are collinear with the composition of A and B. The periodicities (cell parameters) that are parallel to the building layers must be approximately the same (or multiples) in separate layers; otherwise the dimensional matching at the interface between two modules becomes impossible. Finally, the cell parameter corresponding to the stacking direction is a multiple of the sum of the thicknesses of the building modules A and B.

MEROTYPE SERIES

In a merotype ($\mu\acute{\epsilon}\rho o\varsigma =$ part) series [2, 5], the crystal structures of all members are based on one or more common modules that alternate with modules that, instead, are typical of each structure. Examples are given below.

PLESIOTYPE SERIES

In a plesiotype ($\pi\lambda\eta\sigma\acute{\iota}o\varsigma =$ near) series [2, 5], only the main features of the building modules and the principles of connection are preserved, but details of both the chemical composition and the configuration differ in the members of the series. Examples are given below, including mero-plesiotype series in which merotypy and plesiotypy occur at the same time.

THE ROOTS OF POLYSOMATISM

The term polysomatism has been introduced [4] to correlate the crystal structures of biopyriboles, i.e. micas (*bio*tite), *py*roxenes and amphi*boles*, all of which are based on modules of micas and pyroxenes (see below). The concept was latent in the literature as proved by the common description reported for the phyllosilicates (layer silicates). The description and classification of these silicates has long been based either on tetrahedral-octahedral (*TO*) or tetrahedral-octahedral-tetrahedral (*TOT*) layers (Figure 17-1). In these layers, T is a sheet of Si-tetrahedra in which each tetrahedron shares three oxygen atoms with three neighbouring tetrahedra; O is a sheet of octahedra that is delimited by close-packed oxygen atoms in two planes. Both sheets ideally display hexagonal symmetry that promotes polytypy in the crystal structures based on them. T and O sheets are clamped together via sharing of the non-bridged oxygen atoms of the T sheet. Stacking of *TO* sheets form the so-called 1:1 layer silicates; stacking of *TOT* sheets with cations (micas), mixed cations and water molecules (e.g., smectites) or a further octahedral sheet (chlorites) causes to form the so-called 2:1 layer silicates (Figure 17-1). According to the nomenclature of polysomatism, the group corresponds to a merotype series with the *TOT* layer as a fixed module and a variable module represented by the interlayer content.

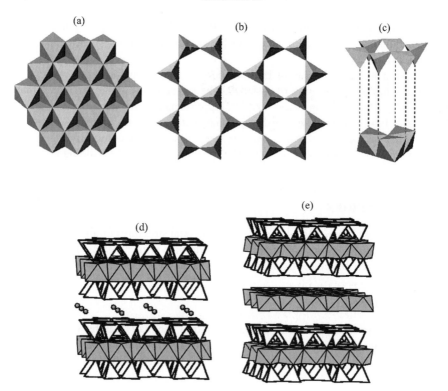

Figure 17-1. Octahedral O (a) and tetrahedral T (b) sheets and their connection (c) to form TOT layers as occur, e.g. in micas (d) and chlorites (e)

Biopyriboles (Figure 17-2) represent a first and now classic example of polyso-matic structures established by Thompson [4], who showed that the structures of micas, pyroxenes and amphiboles share, according to various ratios, the same modules of mica (M) and pyroxene (P) and that are members of a polysomatic series M_mP_p. The description of biopyriboles as members of a polysomatic series, a series belonging to the wider category of a homologous series [2], and the concomitant discovery of chain-width defects in pyroxenes, which were readily interpreted according to the proposed scheme of building series, proved definitively the predictive power of polysomatism in terms of structural characterization and modelling. The chain-width defects detected in pyroxenes were discovered to be slices of the freshly discovered multiple-chain-width biopyriboles jimthompsonite and chesterite [6], which were materializing theoretically possible members of the M_mP_p series. The modelling of carlosturanite [7] and of other modular struc-tures reviewed elsewhere [8] represents some early successes of polysomatism. These successful examples opened a prolific route as shown in comprehensive reviews [1, 2].

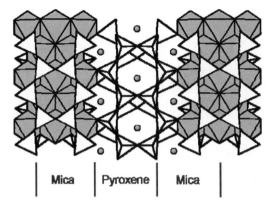

| Mica | Pyroxene | Mica | |

Figure 17-2. Projection along [100] of the crystal structure of an amphibole and its modular slicing as modules of mica and pyroxene

DESCRIBING MODULAR STRUCTURES

Several modular structures and related series are described in the section devoted to the modelling of crystal structures. To underline the utility of a modular description of the crystal structures (when practicable, of course), we first provide examples of compounds based on perovskite modules and important for materials science.

Modules of Perovskite

The perovskite-type structure offers to materials science some of the most versatile and important building modules. Many members of the perovskite family and their derivative structures play a fundamental role in several technical applications because they display diverse properties in such fields as electromagnetism, optics, catalysis, composite and porous materials.

Perovskite s.s. corresponds to the mineral $CaTiO_3$. Ideally, its crystal structure is cubic ($Pm\bar{3}m$, $a_c \sim 3.8$ Å) and is generally represented (Figure 17-3) by emphasizing the octahedral polyhedra surrounding the small cation Ti^{4+}. The octahedra share all their corners (oxygen atoms) to form a framework in which large dodecahedral (cubo-octahedral) sites have at their centre the large cation Ca^{2+}. In the ideal cubic structure of ABX_3 perovskites, the bond lengths of $A-X$ (dodecahedral site) and $B-X$ (octahedral site) are $a_c/(2)^{1/2}$ and $a_c/2$, respectively. The cations A and B might charges other than 2+ and 4+, respectively, in the archetype $CaTiO_3$: what must be kept constant and equal to three times the charge of X is the sum of the positive charges; for example, for $X = O$, $La^{3+}Cr^{3+}O_3$ and $Na^{1+}W^{5+}O_3$ are known. Apart from oxygen, X can be a halide.

The structure of perovskite type is the archetype of numerous synthetic and natural compounds with stoichiometry ABX_3 and their derivative structures. The latter are formally obtained from the archetype through mechanisms such as anion or

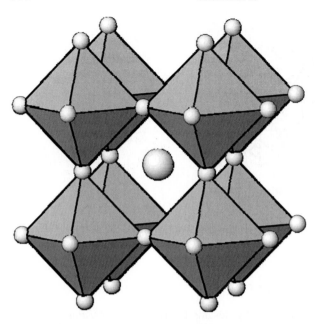

Figure 17-3. Structure of an ideal cubic perovskite represented as packing of corner-sharing octahedra. The large circle represents a cation in dodecahedral coordination at the centre of the unit cell

cation deficiency and splitting into independent subsets of crystallographic position sets that would be equivalent in the ideal cubic structure [9].

Modules of perovskite-type (abbreviated to perovskite) crystal structure alternating with other structural modules occur in several crystalline materials that are known as hybrid or intergrowth perovskites. Three-dimensional (3D – the octahedra share corners in three non-coplanar directions such that layers of thickness many octahedra are formed), two-dimensional (2D – the octahedra share corners in two directions such that layers of thickness one octahedron are formed), one-dimensional (1D – the sharing of octahedral corners develops along one direction only such that rows of octahedra result) and zero-dimensional (0D – only isolated octahedra occur) perovskite layers are known. In the 0D and 1D layers, the positions of the isolated octahedra (0D) and rows of octahedra (1D) are supposed to match those of the octahedral framework in the perovskite structure.

The 3D and 2D fragments can be cut according to various orientations from a perovskite structure (Figure 17-4): layers perpendicular to directions [001] (most common), [110] and [111] are known [10].

High-Temperature Superconductors

The crystal structures of several high-temperature superconductors (HTSC), particularly in the cuprate family, are describable as built of modules of perovskite together

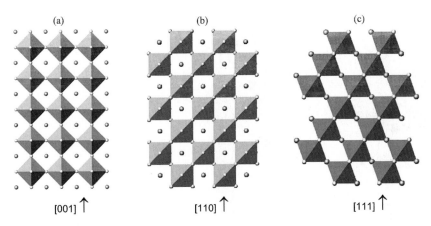

Figure 17-4. Projection of the perovskite structure to show the stacking of the octahedral sheets along directions [001] (a) [110] (b) and [111] (c); projections are in planes (100), (001) and ($1\bar{1}0$), respectively

with other modules that might range from structures like that of sodium chloride to sheets of either cations or anions only and even to organic molecules [9, 11–13].

The family of HTSC compounds with general formula $Tl_mBa_2Ca_{n-1}Cu_nO_{2n+m+2}$ ($m = 1, n = 1$ to 5; $m = 2, n = 1$ to 4), in which Tl, Ba and Ca are replaceable with other cations, offers typical examples of polysomatism based on four sheets (modules) conventionally indicated as $A = TlO$, $B = BaO$, $C = CuO_2$ and $D = Ca$. All these sheets have the same thickness, $\sim 3.1\,\text{Å}$. The CuO_2 sheet formally corresponds to a section of a perovskite cut through the octahedrally coordinated Cu atom; a sheet of this type and the pyramidal and octahedral modules of perovskite obtained on combining the sheets are shaded in Figure 17-5

Conventionally, the superconductors that we treat are labelled with four digits in a set that correspond, in order, to the stoichiometric coefficients of Tl, Ba, Cu and Ca. Figure 17-5 shows the known structures for $m = 1$. In these structures the Bravais lattice is primitive P, and the number of sheets that occur in one unit cell corresponds to the sum of four stoichiometric coefficients: 4, 6, 8, 10 and 12 for $n = 1, 2, 3, 5$ and 5, in order. The structures are tetragonal $P4/mmm$ and the value of cell parameter a, $\sim 3.85\,\text{Å}$, corresponds to twice the length of the Cu—O bond. From member n to $n+1$, cell parameter c increases by about $3.1\,\text{Å}$ that corresponds to the thickness of the added sheet. Explicitly, $c = 9.54$, 12.72, 15.92, 19.00 and $22.21\,\text{Å}$ for $n = 1$ [14], 2 [15], 3 [16], 4 [17] and 5 [18], respectively.

Randomly stacked defects of the building modules have been observed; these defects, which are common in a modular structure and alter the stoichiometry, are suggested [19] to be a possible source of anomalies noted in the superconducting behaviour of single crystals belonging to compounds of the series here described.

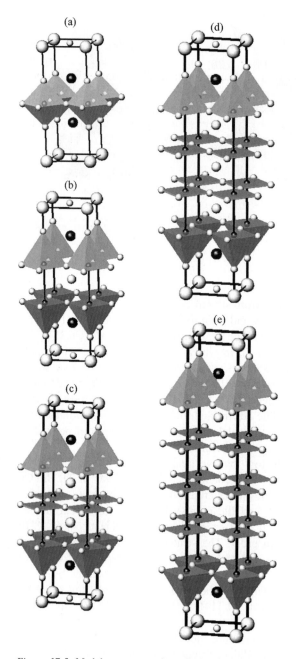

Figure 17-5. Modular representation of crystal structures of superconductors with general formula $TlBa_2Ca_{n-1}Cu_nO_{2n+3}$ for $n = 1$ (a) 2 (b) 3 (c) 4 (d) and 5 (e). Large and small black spheres represent Ba and Cu atoms, respectively; large, intermediate and small blank spheres represent Tl, Ca and O atoms, respectively

Organic-Inorganic Hybrid Perovskites

The crystal structures of many organic-inorganic layered hybrids consist of perovskite modules, which act as complex anion, and intercalated layers of organic molecules, which act as cation [20]. The combination of the properties of these organic and inorganic parts allows engineering materials that show useful magnetic, electrical and optical characteristics. An investigation of these materials in series might evince correlations between structure and properties and guide the synthesis of tunable functional materials.

The tailoring of layered organic-inorganic perovskite materials is obtainable on playing on various parameters of the basic ABX_3 perovskite structure: (a) the nature of A (dodecahedral) and B (octahedral) cations; (b) the nature of the X anion (either oxygen or halides); (c) the orientation and thickness of the perovskite layer. For a given orientation and composition of the perovskite layer, series of organic-inorganic layered perovskites can be built in at least three ways: (i) keeping fixed the thickness of the perovskite layer and altering the organic interlayer; (ii) keeping fixed the organic interlayer and altering the thickness of the perovskite layer; (iii) varying the thickness of both layers in the case that the organic layer can be incremented by polymerization. In case (i), for each chemical composition, thickness and orientation of the perovskite module a merotype series might be obtained on inserting various organic modules. Cases (ii) and (iii) represent polysomatic series.

The polysomatic series of general formula $(RNH_3)_2A_{n-1}B_nX_{3n+1}$ provides an attractive example of hybrid compounds based on (001) perovskite layers [20] (Figure 17-6). The running index n indicates the number of octahedral sheets that form the perovskite module. For $n=1$, the thickness of the perovskite layer is only one octahedron, and the A cation, which would occupy the dodecahedral cage in the corresponding 3D perovskite structure, is obviously absent; $(RNH_3)_2$ represents the interlayer organic part. For $R = C_4H_9$ (butyl), $A = (CH_3NH_3)^+$, $B = Sn^{2+}$ and $X = I^-$ five compounds with $1 \leq n \leq 5$ are known, and the general formula given above becomes $(C_4H_9NH_3)_2(CH_3NH_3)_{n-1}Sn_nI_{3n+1}$. This polysomatic series [10] illustrates effectively the engineering of materials that can be realized as a function of n. The compound for $n = 1$ is a semiconductor with a fairly large band gap; on increasing n, the resistivity of the material decreases, with a transition to metallic behaviour for $n \geq 3$; the material for $n = \infty$, $(CH_3NH_3)SnI$, is a tridimensional perovskite with the properties of a p-type metal having a small density of carriers.

$A'_2A_nB_nX_{3n+2}$ is the general formula for hybrid compounds in a family based on perovskite layers (110) (Figure 17-7). For example, for the polysomatic series $[NH_2C(I) = NH_2]_2(CH_3NH_3)_nSn_nI_{3n+2}$, methylammonium, (CH_3NH_3), and iodoformamidinium, $[NH_2C(I) = NH_2]$, occupy the dodecahedral site and the interlayer, respectively [21]. (111) layers of the perovskite structure occur in compounds of general formula $A'_2A_{n-1}B_nX_{3n+3}$ [10] (Figure 17-8).

Because the layers have varied boundaries, the stoichiometry of the perovskite module varies in the cuts $A_{n-1}B_nX_{3n+1}$, $A_nB_nX_{3n+2}$, and $A_{n-1}B_nX_{3n+3}$ for (001), (110) and (111) layers, respectively. Except the (110) cut, the dodecahedral cation

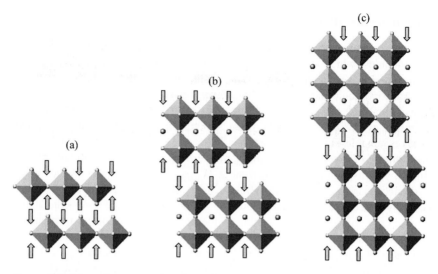

Figure 17-6. Schematic representation of crystal structures of organic-inorganic layered materials based on 1- (a), 2- (b) and 3-octahedra (c) thick (001) layers of perovskite alternating with an organic cation represented by arrows. See Figure 17-4(a)

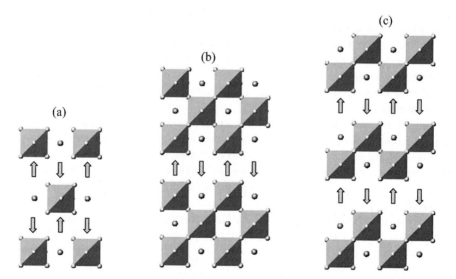

Figure 17-7. Schematic representation of crystal structures of organic-inorganic layered materials based on 1- (a) 2- (b) and 3-octahedra (c) thick (110) layers of perovskite alternating with an organic cation represented by arrows. See Figure 17-4 (b)

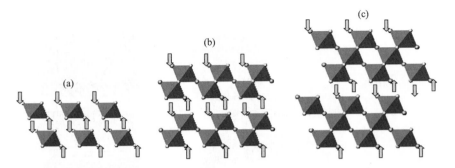

Figure 17-8. Schematic representation of crystal structures of organic-inorganic layered materials based on 1- (a) 2- (b) and 3-octahedra (c) thick (111) layers of perovskite alternating with an organic cation represented by arrows. See Figure 17-4 (c)

A is absent in the layer compounds of thickness one octahedron. This layer consists of isolated (0D) and corner-sharing octahedra (1D) in the (111) and (100) cuts of perovskite, respectively.

MODELLING MODULAR STRUCTURES

In the following sections we provide examples of modelling unknown inorganic modular structures and discuss the related series. Further examples are available elsewhere [2]; unfortunately, often in the literature the inputs from concepts of modular crystallography are not explicitly quoted. The numbers of modular structures are increasing rapidly, also because layered materials are increasingly a matter of intensive research [22]. As stated above, as the description of these structures is typically not given in terms of modularity, they can escape a search on keywords.

Most examples of modelled structures given in the following sections concern microporous mineral phases [23], a category of minerals that are generally poorly crystallized and are, consequently, unsuitable for a determination of crystal structure via diffractometric single-crystal methods [24].

MODELLING NAFERTISITE

A model of the crystal structure of nafertisite (*nfr*)(Figure 17-9) $\{(Na, K, \Box)_4$ $(Fe^{2+}, Fe^{3+}, \Box)_{10}[Ti_2O_3Si_{12}O_{34}](O, OH)_6$; $A2/m$, $a = 5.353$, $b = 16.176$, $c = 21.95$ Å, $\beta = 94.6°\}$ was obtained [25] by comparison with the crystal structures of bafertisite [26] [27] [28] (*bft*) (Figure 17-9) $\{Ba_2(Fe, Mn)_4[Ti_2O_4Si_4O_{14}](O, OH)_2$; $P2_1/m$, $a = 5.36$, $b = 6.80$, $c = 10.98$ Å, $\beta = 94°\}$ and astrophyllite [29, 30] (*ast*) (Figure 17-9) $\{(K, Na)_3(Fe, Mn)_7[Ti_2O_3Si_8O_{24}](O, OH)_4$; $P\bar{1}$, $a = 5.36$, $b = 11.63$, $c = 11.76$ Å, $\alpha = 112.1$, $\beta = 103.1$, $\gamma = 94.6°\}$.

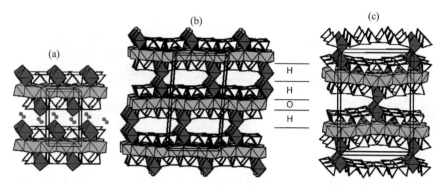

Figure 17-9. Perspective view of crystal structures of bafertisite (a) astrophyllite (b) and nafertisite (c); *H* and *O* sheets are shown

The raw structure model of nafertisite was obtained on applying the principles of the polysomatic series and modifying the structures of bafertisite and astrophyllite after noting the following.

1. Bafertisite, astrophyllite and nafertisite have a common value $a \sim 5.4\,\text{Å}$, which matches the value of a for micas.

2. $(b_{ast} - b_{bft}) \sim 1/2(b_{nfr} - b_{bft}) \sim 4.7\,\text{Å}$ corresponds to $b/2$ in mica.

3. The structures of bafertisite and astrophyllite are based on *HOH* (*H* for hetero, see below) layers that are comparable with the *TOT* layer of the phyllosilicates from which they differ mainly because of a periodic insertion of Ti-octahedra in the *T* sheet.

4. Ignoring details, an astrophyllite layer differs from a bafertisite layer only in having an additional mica-like module *M* sandwiched between two (010) modules that are built from Ti- and Si-polyhedra (bafertisite-like module *B*).

5. The difference in composition between bafertisite and astrophyllite and between astrophyllite and nafertisite is roughly the same, and corresponds to $(A, \square)(Y, \square)_3[Si_4O_{10}](OH, O)_2$. This difference is comparable to the composition of micas, in which *A* and *Y* represent interlayer alkaline or alkaline-earth cations and octahedral cations of the *O* sheet (see above), respectively.

6. $(d_{002})_{nfr} = 10.94\,\text{Å}$ matches the thickness of one structural layer in both bafertisite and astrophyllite.

The structural model of nafertisite was obtained on doubling the mica-like module *M* in the *HOH* layer of astrophyllite.

The Polysomatic Series of the Heterophyllosilicates

The successful modelling of the crystal structure of nafertisite, according to the principles of modular crystallography, indicated [25, 31, 32] to correlate via a polysomatic series the group of titanium silicates of which the structures are based

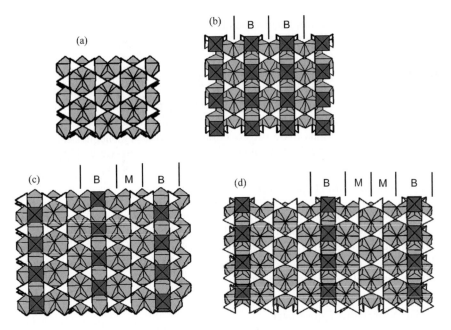

Figure 17-10. Orthogonal projection of a phyllosilicate *TOT* layer (a) and of *HOH* layers typical of bafertisite (b) astrophyllite (c) and nafertisite (d); *M* and *B* modules are shown

on *TOT*-like layers and to introduce the term *heterophyllosilicate*. In the members of the heterophyllosilicate polysomatic series, a row of Ti polyhedra (or replacing cations, e.g. Nb; hereafter, Ti only is used for short) periodically substitutes a row of disilicate tetrahedra within the *T* tetrahedral sheet of a mica *TOT* layer; the octahedral *O* sheet is instead maintained (Figure 17-10). *HOH* layers are thus obtained in which *H* stands for *hetero* to indicate the presence of rows of 5- or 6-coordinated Ti in the modified *T* sheet. The slice of mica-derived structure, which contains Ti polyhedra and has composition $B = A_2 Y_4 [Ti_2(O)_4 Si_4 O_{14}](O, OH)_2$, is conventionally called bafertisite-type *B* module, as mentioned above. The unmodified part of the *TOT* layer comprises modules $M = AY_3 [Si_4 O_{10}](O, OH)_2$, each being one silicate chain wide. The resulting polysomatic series is formally represented with the formula $B_m M_n$.

The $B_m M_n$ heterophyllosilicate polysomatic series has a chemical composition $A_{2+n} Y_{4+3n} [Ti_2(O')_{2+p} Si_{4+4n} O_{14+10n}](O'')_{2+2n}$. In this formula, atoms belonging, even in part, to the *H* sheet are shown in brackets. *A* and *Y* represent large (alkali) interlayer cations and octahedral cations of the *O* sheet, respectively; O' (bonded to Ti) and O'' (belonging to the octahedral sheet only) can be oxygen, OH, F or H_2O; the $14 + 10n$ oxygen atoms are bonded to Si. The value of p (0, 1, 2) depends on the configuration around Ti, which can be either five- or six-coordinate.

Except the case $n = 0$, which corresponds to micas, HOH layers of the following three types are known [24, 32] (Figure 17-10).

$(HOH)_B$ Bafertisite-Type Layer

Bafertisite represents the member B_1M_0 of the B_mM_n polysomatic series. The corresponding $(HOH)_B$ layer is the most versatile among the known HOH layers, being able to sandwich a various more or less complex interlayer contents; an updated list of more than 30 minerals containing this layer is given elsewhere [33].

$(HOH)_A$ Astrophyllite-Type Layer

Relative to the $(HOH)_B$ layer, in a $(HOH)_A$ astrophyllite-type layer an M module is present between two B modules. The titanosilicates based on a $(HOH)_A$ layer represent the members B_1M_1 of the heterophyllosilicate B_mM_n polysomatic series and differ from each other only by polytypy (i.e. stacking of the layers) and the chemical nature of the A and Y cations [34, 35].

$(HOH)_N$ Nafertisite-Type Layer

Relative to the $(HOH)_B$ layer, in a $(HOH)_N$ nafertisite-type layer two M modules are present between two B modules. Nafertisite and caryochroite [36] are the only two known titanosilicates based on the $(HOH)_N$ layer and that correspond to the members B_1M_2 of the heterophyllosilicate B_mM_n polysomatic series; they differ from each other mainly by the chemical composition of the O sheet and the water content in the interlayer.

The Bafertisite Mero-Plesiotype Series

$(HOH)_B$-bearing compounds are known in two groups: the götzenite group [37], in which Ca is at the centre of the hetero-polyhedra and the HOH layers are strongly interlinked, and the complex bafertisite series [32] in which the $(HOH)_B$ layer alternates with diverse interlayer contents. The bafertisite series is represented by the general formula $A_2\{Y_4[Ti_2(O')_{2+p}Si_4O_{14}](O'')_2\}W$, in which: $[Ti_2(O')_{2+p}Si_4O_{14}]^{r-}$ and $\{Y_4[Ti_2(O')_{2+p}Si_4O_{14}](O'')_2\}^{q-}$ are two complex anions that correspond to the H sheet and the HOH layer, respectively. Besides the already defined A, Y, O', O'' and p symbols, W represents further interlayer content.

In terms of merotypy and plesiotypy, the bafertisite-type structures belong to a mero-plesiotype series [38]. This series is merotype because the HOH module is constantly present in the crystal structure of all members, whereas a second module, namely the interlayer content, is peculiar to each member. The series has also a plesiotype character because the (i) chemical nature and coordination number of the Ti and Y cations and (ii) the linkage between the H and O sheets are variable. The two H sheets sandwiching an O sheet either face each other via the same type of

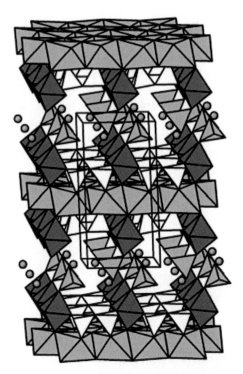

Figure 17-11. Perspective view along [100] of the crystal structure of vuonnemite

polyhedra (i.e. the heteropolyhedra face each other) or show a relative shift (i.e. a heteropolyhedron faces a tetrahedron) [39, 40], thus realizing two topologies. The *HOH* layer of two kinds occurring in vuonnemite, $(HOH)_V$ (Figure 17-11), and in bafertisite, $(HOH)_B$ (Figure 17-9), are considered [41] typical examples of the two topologies. Besides, as mentioned above, the coordination number of the Ti cations can be either five (square pyramid) or six (octahedron).

Perspectives of Technological Applications

Because of the structural parallelism described above, in principle the *HOH* layers of the heterophyllosilicates could serve like the *TOT* layers of the phyllosilicates to synthesize mesoporous pillared materials and, in general, layered materials, a perspective that has been recently investigated but not yet realized [33, 42].

MODELLING BORNEMANITE

Electron diffraction data obtained along two orientations enabled the X-ray powder diffraction (XRPD) pattern of bornemanite to be indexed and the following crystal data to be obtained: space group $I11b$; lattice parameters $a = 5.498$,

$b = 7.120$, $c = 47.95\,\text{Å}$, $\gamma = 88.4°$ [43]. Ideally the formula of bornemanite is $BaNa_3\{(Na, Ti)_4[(Ti, Nb)_2O_2Si_4O_{14}](F, OH)_2\}PO_4$ [44].

The values of the cell parameters and the presence of the complex anion $[(Ti,Nb)_2O_2Si_4O_{14}]$ in the chemical formula support a strong analogy between bornemanite and members of the bafertisite series. A comparison of the cell parameters shows that $c/2$ $(23.97\,\text{Å})$ of bornemanite corresponds to the sum in thickness of one lomonosovite-like module $(14.5\,\text{Å})$ [45] and one seidozerite-like module $(8.9\,\text{Å})$ [46] (Figure 17-12). Disregarding isomorphic substitutions (such as Ba for Na and Nb for Ti), half the sum of the crystal-chemical formulae of lomonosovite and seidozerite, $[Na_8\{(Na_2Ti_2)[Ti_2O_2Si_4O_{14}](O, F)_2\}(PO_4)_2 + Na_2\{(Na, Mn, Ti)_4[(Na, Ti, Zr)_2O_2Si_4O_{14}]F_2\}]/2$, corresponds well to the simplified crystal-chemical formula of bornemanite given above. On the basis of these

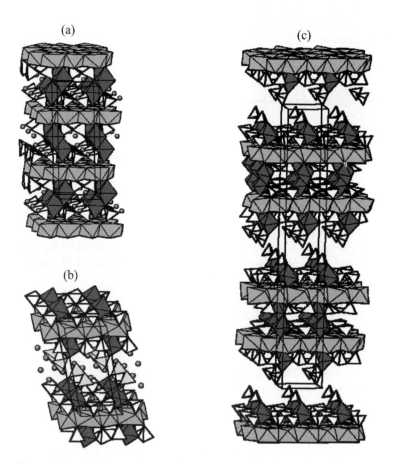

Figure 17-12. Perspective view along [100] of crystal structures of seidozerite (a) lomonosovite (b) and bornemanite (c); in bornemanite the interlayer cations are not shown

indications, a structure model for bornemanite has been built based on alternating seidozerite-like and lomonosovite-like modules (Figure 17-12). In practice, the structure of bornemanite is describable as a [001] stack of bafertisite-type heterophyllosilicate layers in which the lomonosovite and seidozerite interlayer contents alternate.

Modelling Kalifersite

The chemical composition and crystal data of kalifersite [47] $\{(K, Na)_5(Fe^{3+})_7[Si_{20} O_{50}](OH)_6.12H_2O; P\bar{1}, a = 14.86, b = 20.54, c = 5.29\,\text{Å}, \alpha = 95.6, \beta = 92.3, \gamma = 94.4°\}$ are comparable to those of the following microporous clay minerals: sepiolite [48] $\{Mg_8[Si_{12}O_{30}](OH)_4.12H_2O; Pncn, a = 13.40, b = 26.80, c = 5.28\,\text{Å}\}$; palygorskite [49, 50] $\{Mg_5[Si_8O_{20}](OH)_2.8H_2O$; two polytypes are known: $C2/m, a = 13.337, b = 17.879, c = 5.264\,\text{Å}, \beta = 105.27°; Pbmn, a = 12.672, b = 17.875, c = 5.236\,\text{Å}\}$. The structures of sepiolite and palygorskite (Figure 17-13) are based on a framework of chessboard-connected TOT ribbons that correspond to slabs of the TOT layer of the phyllosilicates (Figure 17-1). These ribbons develop along [001] and delimit [001] channels. Along the [010] direction, the $(TOT)_S$ ribbon of sepiolite is one chain wider than that, $(TOT)_P$, of palygorskite. This feature requires for sepiolite a value of b about $9\,\text{Å}$ longer than that of palygorskite, i.e. about $4.5\,\text{Å}$ per added T chain.

A model of the crystal structure of kalifersite has been obtained [47] (Figure 17-13) after realizing the following modular relationships with sepiolite and palygorskite.

(i) Kalifersite, sepiolite and palygorskite have similar values of their parameters a and c; the parameter c is along the fibrous direction of these silicates and its value corresponds to the periodicity of a pyroxene chain.

(ii) The b value of kalifersite is intermediate between that of palygorskite and sepiolite.

(iii) The $\{[Si_{20}O_{50}](OH)_6\}^{26-}$ silicate anion of kalifersite corresponds to the sum of those of sepiolite, $\{[Si_{12}O_{30}](OH)_4^{16-}$, and palygorskite, $\{[Si_8O_{20}](OH)_2\}^{10-}$.

A satisfactory structure model for kalifersite was conclusively based on a 1:1 chessboard arrangement of $(TOT)_P$ and $(TOT)_S$ [001] ribbons.

The Polysomatic Series of Palysepioles and Related Structures

Palygorskite (P), and sepiolite (S) are the end members of a modular series named *palysepiole (paly*gorskite + *sepiol*ite) polysomatic series P_pS_s [47]; kalifersite is the P_1S_1 member. Falcondoite [51] and loughlinite [52] differ from sepiolite only in the composition of the O sheet in the TOT ribbons; the same situation holds for yofortierite [53] and tuperssuatsiaite [54] in comparison to palygorskite.

Raite [55]$\{Na_3Mn_3Ti_{0.25}[Si_8O_{20}](OH)_2.10H_2O; C2/m, a = 15.1, b = 17.6, c = 5.290\,\text{Å}, \beta = 100.5°\}$ has a crystal structure that consists of a palygorskite-like framework, but the channel content differs substantially from that of palygorskite

(a)

(b)

(c)

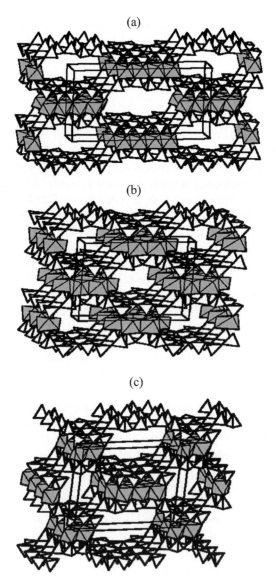

Figure 17-13. Perspective view along [100] of crystal structures of sepiolite (a) palygorskite (b) and kalifersite (c). The content of the channels is not shown

in the O sheet not being interrupted. Thus, raite and palygorskite share only the *TOT* building ribbon and are in a merotype relationship. In the structure of intersilite [56] {(Na,K)Mn(Ti,Nb)Na$_5$(O,OH)(OH)$_2$[Si$_{10}$O$_{23}$(O,OH)$_2$].4H$_2$O; $I2/m$, $a = 13.033$, $b = 18.717$, $c = 12.264$ Å, $\beta = 99.62°$}, sepiolite-like ribbons partially overlap along [010] because of tetrahedral inversions within the same ribbon.

Because the sepiolite framework is substantially modified, intersilite is in a plesiotype relationship with the palysepiole polysomatic series.

A Three-Module Series

A description of the palysepiole polysomatic series and related structures offers the opportunity to mention the main members of the lintisite series of which the crystal structures are based on three building modules. One of these modules is a cut of silinaite [57] (NaLiSi$_2$O$_5$.2H$_2$O; $C2/c$, $a = 14.383$, $b = 8.334$, $c = 5.061$ Å, $\beta = 96.6°$), a crystal structure that shows a chess-board arrangement of channels and ribbons comparable with that of palysepioles. Modules of silinaite (S) have been used [58] as one of three modules needed to describe the modularity in the lintisite group (Figure 17-14). The other two modules are slabs (L) of the structures of lorenzenite [59]

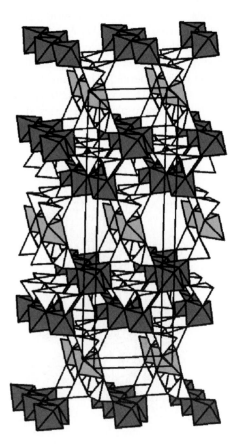

Figure 17-14. Perspective view along [001] of the crystal structure of lintisite. The content of the channels is not shown

{Na$_4$Ti$_4$(Si$_2$O$_6$)$_2$O$_6$; *Pnca*, $a = 14.487$, $b = 8.713$, c $= 4\,5.233$ Å} and of the hypothetic zeolite-like phase (Na,K)Si$_3$AlO$_8$.2H$_2$O (*Z*) [60]. Besides the three structures mentioned above, the members *LS* lintisite [61] [Na$_3$LiTi$_2$(Si$_2$O$_6$)$_2$O$_2$.2H$_2$O; *C2/c*, $a = 28.583$, $b = 8.600$, $c = 5.219$ Å, $\beta = 96.6°$] and *LZ* vinogradovite [62] {Na$_4$LiTi$_4$(Si$_2$O$_6$)$_2$[(Si, Al)$_4$O$_{10}$]O$_4$.(H$_2$O, Na, K)$_3$; *C2/c*, $a = 24.50$, $b = 8.662$, $c = 5.211$ Å, $\beta = 100.15°$} have been recognized [58] to be components of a family based on the modules *L*, *S* and *Z*.

MODELLING SEIDITE-(CE) – THE RHODESITE MERO-PLESIOTYPE SERIES

The first input to modelling the microporous crystal structure of seidite-(Ce) [63] (Figure 17-15) {Na$_4$(Ce, Sr)$_2${Ti(OH)$_2$(Si$_8$O$_{18}$)}(O, OH, F)$_4$.5H$_2$O; *C2/c*, $a = 24.61$, $b = 7.23$, $c = 14.53$ Å, $\beta = 94.6°$} came on comparing its chemical composition and crystal data with those of miserite [64] [KCa$_5$(Si$_2$O$_7$)(Si$_6$O$_{15}$)(OH)F; $P\bar{1}$, $a = 10.100$, $b = 16.014$, $c = 7.377$ Å, Å, $\alpha = 96.41$, $\beta = 111.15$, $\gamma = 76.57°$] and noting that the cell parameters of miserite (*m*) and seidite-(Ce) (*s*) are related as follows: $a_s \cong 2a_m$, $c_s \cong b_m$, $b_s \cong c_m$. Seidite-(Ce) was discovered to share its silicate module with rhodesite [65] (Figure 17-15) {K$_2$Ca$_4$[Si$_8$O$_{18}$(OH)]$_2$.12H$_2$O; *Pmam*, $a = 23.416$, $b = 6.555$, $c = 7.050$ Å} and related structures (see below) with which it forms a mero-plesiotype series. The

(a)   (b)

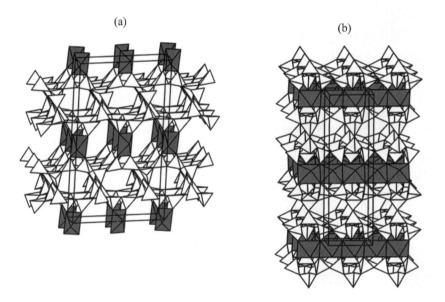

Figure 17-15. Perspective view along [010] of crystal structures of seidite-(Ce) (a) and rhodesite (b). The content of the channels is not shown

following description of this series clarifies also the modular concepts that inspired the modelling of the structure of seidite-(Ce).

The (100) layer comprising eight-membered channels delimited by corner-sharing Si-tetrahedra that occurs in seidite-(Ce) is present also in the following structures, besides rhodesite: macdonaldite [66] $\{Na_4(Ce,Sr)_2\{Ti(OH)_2(Si_8O_{18})\}(O,OH,F)_4$. $5H_2O$; $Cmcm$, $a = 14.081$, $b = 13.109$, $c = 23.560\,\text{Å}\}$; delhayelite [67] $\{K_7Na_3Ca_5[Si_7AlO_{19}]_2 F_4Cl_2$; $Pmmn$, $a = 24.86$, $b = 7.07$, $c = 6.53\,\text{Å}\}$; hydrodelhayelite [68] $\{K_2Ca_4[Si_7AlO_{17} (OH)_2]_2.6H_2O$; $Pnm2_1$, $a = 6.648$, $b = 23.846$, $c = 7.073\,\text{Å}\}$; monteregianite-(Y) [69] $\{K_2Na_4Y_2[Si_8O_{19}]_2.10H_2O$; $P2_1/n$, $a = 9.512$, $b = 23.956$, $c = 9.617\,\text{Å}$, $\beta = 93.85°\}$; AV-9 [70] $\{K_2Na_4Eu_2[Si_8O_{19}]_2.10H_2O$; $C2/m$; $a = 23.973$, $b = 14.040$, $c = 6.567\,\text{Å}$, $\beta = 90.35°\}$. In all these compounds, the silicate double layer alternates with a sheet of cations that show mainly an octahedral coordination. The various *octahedral* sheets are shown in Figure 17-16. In seidite-(Ce) the octahedral sheet consists of isolated octahedra (Figure 17-15).

In conclusion, seidite-(Ce) and the mentioned rhodesite-like compounds form a mero-plesiotype series. Nearly the same (plesiotype aspect) double silicate layer occurs in all members and alternates with a variable octahedral module (merotype aspect).

MODELLING TUNGUSITE – THE REYERITE MEROTYPE SERIES

The crystal structure of tungusite[71] $\{[Ca_{14}(OH)_8](Si_8O_{20})(Si_8O_{20})_2[(Fe^{2+})_9 (OH)_{14})]$; $P\overline{1}$, $a = 9.714$, $b = 9.721$, $c = 22.09\,\text{Å}$, $\alpha = 90.13$, $\beta = 98.3$, $\gamma = 120.0°\}$ was modelled by comparison with the crystal structures (Figure 17-17) of reyerite [72] $[Ca_{14}(Na, K)_2Si_{22}Al_2O_{58}(OH)_8.6H_2O$; $P\overline{3}$, $a = 9.765$, $c = 19.067\,\text{Å}]$ and gyrolite [73] $[Ca_{16}NaSi_{23}AlO_{60}(OH)_8.14H_2O$; $P\overline{1}$, $a = b = 9.74$, $c = 22.40\,\text{Å}$, $\alpha = 95.7$, $\beta = 91.5$, $\gamma = 120.0°]$. In particular, the following aspects were considered.

1. The difference $(Ca,Na)O_2.8H_2O$ between the chemical compositions of reyerite and gyrolite is accounted for by the presence in gyrolite of a partially filled (001) octahedral X sheet that is sandwiched between two centrosymmetrically related (001) S_1OS_2 layers [72, 73]. The O sheet consists of edge-sharing Ca-octahedra; the S_1 and S_2 sheets are built as six-membered rings of tetrahedra pointing upwards and downwards in the ratios 1:1 and 3:1 for S_1 and S_2, respectively.

2. The ideal composition of tungusite differs from that of gyrolite by the presence of six divalent cations, which complete the (001) X sheet of which the dimensions $(a \sim b \sim 9.72\,\text{Å}$, $\gamma = 120°$, thickness about 2.8 Å) correspond to those of a 3 x 3 trioctahedral sheet.

Fedorite [73] $[K_2(Ca_5Na_2)Si_{16}O_{38}(OH,F)_2.H_2O$; $P\overline{1}$, $a = b = 9.67$, $c = 12.67\,\text{Å}$, $\alpha = 102.2$, $\beta = 71.2$, $\gamma = 120.0°]$ is also based on O and S_2 modules (Figure 17-17). Reyerite, gyrolite and tungusite contain the same module OS_2 that, instead, stands alone in the crystal structure of fedorite. These minerals, together with other

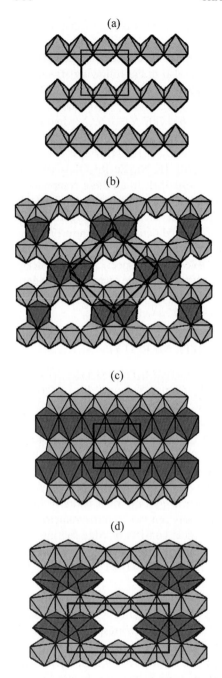

Figure 17-16. Orthogonal projection on (001) of the "octahedral" sheets of rhodesite
(a) monteregianite-Y (b) delhayelaite (c) and AV-9 (d)

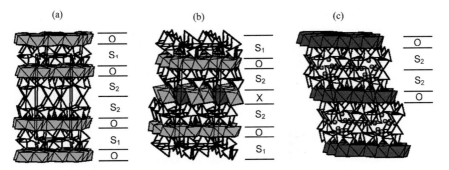

Figure 17-17. Perspective view of crystal structures of reyerite (a) gyrolite (and tungusite) (b) and fedorite (c)

phases [2], are describable as members of a merotype series of which the fixed module OS_2 might either stand alone or be intercalated by one or more modules. The case of only one intercalated module is represented by the presence of S_1 in reyerite; the case of two intercalated modules (S_1 and X) is represented by gyrolite and tungusite. The members of the reyerite merotype series might show a complex polytypy [73, 74].

CONCLUSIONS

Between molecular and non-molecular crystal structures (practically the totality of inorganic crystalline materials), there is a strong configurational asymmetry consisting in the lack of molecules as natural building units in structures of the second type. To overcome this asymmetry, which is a drawback both to describe and, even more, to model inorganic structures, the branch of modular crystallography has evolved recently [2] on systematically considering, among other features, the recurrence of some complex modules in various structures. A complex module is based on several traditional coordination polyhedra that, since the beginning of inorganic structural science, have served to describe crystal structures. The structures based, either completely or partially, on the same modules can be correlated to form series in which features such as collinear variation of chemical composition and cell parameters are observed. These features are exploitable in processes such as modelling unknown structures that show the same characteristics and synthesizing new members of a series in search of enhanced useful properties (materials engineering).

The utility of considering, when applicable, crystal structures as based on complex modules that occur in varied structures, has been demonstrated on describing inorganic materials in several series. In particular, the method has been applied to modelling unknown modular crystal structures. The method is useful when a direct solution of the crystal structure is impracticable, e.g. because suitable single crystals are not available and because information from the powder diffraction pattern is poor relative to the complexity of the crystal structure. The low crystallinity of the modelled structures added further problems to the use of the XRPD patterns even to refine moderately the structural model. The only possible refinement has been based on the distance least-square (DLS) program [75] that allows one to optimize distances and angles of a structural model.

The validation of the modelled structures has been achieved on comparing calculated and observed XRPD patterns, besides matching the calculated and experimental chemical compositions and crystal data, including the space group. To initiate a search for known crystal structures that show structural features comparable to those of the unknown structure, the latter data are already indispensable.

ACKNOWLEDGMENTS

Research financially supported by MIUR (Rome, Italy; PRIN project "Minerals to materials: crystal chemistry, microstructures, modularity, modulations"). We are grateful to the Editors of this book for the critical reading of the stext.

REFERENCES

1. Merlino S (1997) Modular aspects of minerals, EMU Notes in Mineralogy, vol. 1. Eötvös University Press, Budapest Hungary.
2. Ferraris G, Makovicky E, Merlino S (2004) Crystallography of Modular Materials. IUCr Monographs in Crystallography, Oxford University Press, Oxford UK.
3. Guinier A, Bokij GB, Boll-Dornberger K, Cowley JM, Ďurovič S, Jagodzinski M, Krishna P, DeWolff PM, Zvyagin BB, Cox DE, Goodman P, Hahn Th, Kuchitsu K, Abrahams SC (1984) Acta Crystallogr A40:399.
4. Thompson, JB (1978) Am Mineral 63:239.
5. Makovicky E (1997) EMU Notes on Mineralogy 1:315.
6. Veblen DR, Buseck PR (1979) Am Mineral 64:687.
7. Mellini M, Ferraris G, Compagnoni R (1985) Am Mineral 70:773.
8. Veblen DR (1991) Am Mineral 76:801.
9. Mitchell RH (2002) Perovskites: Modern and ancient. Almaz Press, Thunder Bay Canada.
10. Mitzi DB (2001) J Chem Soc Dalton Trans p.1.
11. Raveau B, Michel C, Hervieu M, Groult D (1991) Crystal chemistry of high TC superconducting copper oxides. Springer-Verlag, Berlin Germany.
12. Shekhtman VS (1993) The real structure of high-T_c superconductors. Springer-Verlag, Berlin Germany.
13. Vainshtein BK, Fridkin VM, Indenbom VL (1994) Structure of Superconductors in: Structure of Crystals (2nd edition). Springer-Verlag, Berlin Germany.
14. Matheis DP, Snyder RL (1990) Powder Diffraction 5:8.
15. Kolesnikov NN, Korotkov VE, Kulakov MP, Lagvenov GA, Molchanov VN, Muradyan LA, Simonov VI, Tamazan RA, Shibaeva RP, Shchegolev IF (1989) Physica C 162:1663.
16. Morosin B, Gingley DS, Schirber JE, Venturini EL (1988) Physica C 156:587.
17. Ogborne DM, Weller MT (1994) Physica C 230:153.
18. Weller MT, Pack MJ, Knee CS, Ogborne DM, Gormezano A (1997) Physica C 282:849.
19. Leonyuk L, Babonas G-J, Maltsev V, Vetkin A, Rybakov V, Reza A (1999) J Crystal Growth 198–199:619.
20. Mitzi DB, Field CA, Harrison WTA, Guloy AM (1994) J Chem Soc, Dalton Trans p.1.
21. Mitzi DB, Wang S, Field CA, Chess CA, Guloy AM (1995) Science 267:1473.

22. Auerbach SM, Carraio KA, Dutta PK (2004) Handbook of Layered Materials. Marcel Dekker, New York USA.
23. Ferraris G., Merlino S (2005) Micro and mesoporous mineral phases, Reviews of Mineralogy and Geochemistry, vol 57. Mineralogical Society of America, Washington DC USA.
24. Ferraris G, Gula A (2005) Rev Mineral Geochem 57:69.
25. Ferraris G, Ivaldi G, Khomyakov AP, Soboleva SV, Belluso E, Pavese A (1996) Eur J Mineral 8:241.
26. Guan YaS, Simonov VI, Belov NV (1963) Dok Akad Nauk SSSR 149:1416.
27. Pen ZZ, Shen TC (1963) Scientia Sinica 12:278.
28. Rastsvetaeva RK, Tamazyan RA, Sokolova EV, Belakovskii DI (1991) Sov Phys Crystallogr 36:186.
29. Shi N, Ma Z, Li G, Yamnova NA, Pushcharovsky DYu (1998) Acta Crystallogr B54:109.
30. Woodrow PJ (1967) Acta Crystallogr 22:673.
31. Ferraris G, Khomyakov AP, Belluso E, Soboleva SV (1997) Proc 30th Inter Geol Congress 16:17.
32. Ferraris G (1997) EMU Notes in Mineralogy 1:275.
33. Ferraris G, Bloise A, Cadoni M. (2007) Micropor Mesopor Mater, doi:10.1016/j.micromeso.2007.02.036.
34. Piilonen PC, Lalonde AE, McDonald AM, Gault RA, Larsen AO (2003) Can Mineral 41:1.
35. Piilonen PC, McDonald AM, Lalonde AE (2003) Can Mineral 41:27.
36. Kartashov PM, Ferraris G, Soboleva SV, Chukanov NV (2006) Can Mineral 44:1331.
37. Christiansen CC, Rønsbo JG (2000) N Jb Mineral Mh p. 496.
38. Ferraris G, Ivaldi G, Pushcharovsky DYu, Zubkova N, Pekov IV (2001) Can Mineral 39:1307.
39. Sokolova E, Hawthorne FC (2004) Can Mineral 42:797.
40. Christiansen CC, Makovicky E, Johnsen ON (1999) N Jb Mineral Abh 175:153.
41. Nèmeth P, Ferraris G, Radnóczi G, Ageeva OA (2005) Can Mineral 45:973.
42. Ferraris G (2006) Solid State Phenomena 111:47.
43. Ferraris G, Belluso E, Gula A, Soboleva SV, Ageeva OA, Borutskii BE (2001) Can Mineral 39:1667.
44. Men'shikov YuP, Bussen IV, Goiko EA, Zabavnikova NI, Mer'kov AN, Khomyakov AP (1975) Zapiski Vserossiyskogo Mineralogicheskogo Obschestva 104:322.
45. Belov NV, Gavrilova GS, Solov'eva LP, Khalilov AD (1978) Sov Phys Dokl 22:422.
46. Simonov VI, Belov NV (1960) Sov Phys Crystallogr 4:146.
47. Ferraris G, Khomyakov AP, Belluso E, Soboleva SV (1998) Eur J Mineral 10:865.
48. Brauner K, Preisinger A (1956) Tsch Mineral Petrogr Mitt 6:120.
49. Artioli G, Galli E (1994) Mater Science Forum 166–169:647.
50. Chiari G, Giustetto R, Ricchiardi G (2003) Eur J Mineral 15:21.
51. Springer G (1976) Can Mineral 14:407.
52. Fahey JJ, Ross M, Axelrod JM (1960) Am Mineral 45:270.
53. Perrault G, Harvey Y, Pertsowsky R (1975) Can Mineral 13:68.
54. Càmara F, Garvie LAJ, Devouard B, Groy TL, Buseck PR (2002) Am Mineral 87:1458.
55. Pushcharovsky DYu, Pekov IV, Pluth J, Smith J, Ferraris G, Vinogradova SA, Arakcheeva AV, Soboleva SV, Semenov EI (1999) Crystallogr Rep 44:565.
56. Yamnova NA, Egorov-Tismenko YuK, Khomyakov AP (1996) Crystallogr Rep 41:239.
57. Grice JD (1991) Can Mineral 29:363.
58. Merlino S, Pasero M (1997) EMU Notes in Mineralogy 1:297.
59. Sundberg MR, Lehtinen M, Kivekäs R (1987) Am Mineral 72:173.
60. Smith JV (1977) Am Mineral 62:703.
61. Merlino S, Pasero M, Khomyakov AP (1990) Z Kristallogr 193:137.
62. Rastsvetaeva RK, Simonov VI, Belov NV (1968) Sov Phys Dok 12:1990.
63. Ferraris G, Belluso E, Gula A, Khomyakov AP, Soboleva SV (2003) Can Mineral 41:1183.
64. Scott JD (1976) Can Mineral 14:515.
65. Hesse KF, Liebau F, Merlino S (1992) Z Kristallogr 199:25.
66. Cannillo E, Rossi G, Ungaretti L, Carobbi SG (1968) Atti Accad Naz Lincei, Classe Sci Fisi 45:399.
67. Cannillo E, Rossi G, Ungaretti L (1970) Rend Soc Ital Mineral Petrol 26:63.

68. Ragimov KG, Chiragov MI, Mamedov KS, Dorfman MD (1980) Dokl Akad Nauk Azerbaid SSR 36:49.
69. Ghose S, Sen Gupta PK, Campana CF (1987) Am Mineral 72:365.
70. Ananias D, Ferreira A, Rocha J, Ferreira P, Rainho JP, Morais C, Carlos LD (2001) Am Chem Soc 123:5735.
71. Ferraris G, Pavese A, Soboleva SV (1995) Mineral Mag 59:535.
72. Merlino S (1988) Mineral Mag 52:247.
73. Merlino S (1988) Mineral Mag 52:377.
74. Zvyagin BB (1997) EMU Notes in Mineralogy 1:345.
75. Baerlocher Ch, Hepp A, Meier WM (1978) Manual DLS-76. A Program for the simulation of crystal structures by geometric refinement. ETH, Zurich Switzerland.

CHAPTER 18

MODELS FOR ISOMERIC BISPIDINE
COMPLEXES – ACCURATE PREDICTION VERSUS
THOROUGH UNDERSTANDING

PETER COMBA AND MARION KERSCHER

Abstract: Bispidine transition-metal complexes possess several unique properties. We suggest these
to be derived primarily from the rigid ligand backbone, which enforces specific geometries
on the metal center and also produces a large elasticity in the coordination geometries. The
unexpected trend of complex stabilities along the series of the first transition-metal row,
isomerism of various types and the reactivities of the iron systems towards oxygen activation are reviewed on the basis of the structures enforced by the bispidine ligands; various
computational approaches are discussed, which might assist a thorough understanding of
the various molecular properties

Key words: molecular mechanics, DFT, ligand-field theory, oxygen activation, Jahn-Teller distortion,
distortional isomer

INTRODUCTION

Bispidine-derived ligands (bispidine = 3,7-diazabicyclo[3.3.1]nonane; the ligands
discussed here are shown in Chart 1) are rigid and, with respect to the two tertiary
amine donors, highly preorganized. Additional donors at C2 and C4, and at N3 and
N7, (see Chart 1 for nomenclature) lead to tetra-, penta- and hexadentate as well
as to dinucleating ligands with the rigid diazabicyclic bispidine and various less
preorganized and more flexible pendant donor groups. Several experimentally determined structures of metal-free ligands are given in Figure 18-1. Rearrangement of
the pendant donor groups is generally a low-energy process; L^{10} in Figure 18-1 is an
example that supports this assumption and is the only highly preorganized multiden-
tate bispidine-type ligand [1]. The synthesis and properties of many bispidine-type
ligands [2] and their transition-metal complexes have been described [3]. Applications of bispidine complexes range from metal ion selectivity, [4, 5] nuclear
medicinal chemistry [6] and molecular magnetism, [7] to mechanistic biomimetic

391

J.C.A. Boeyens and J.F. Ogilvie (eds.), Models, Mysteries and Magic of Molecules, pp. 391–409.
© 2008 *Springer*.

Chart 1.

L^n

$[M(L^1)X_2]^{n+}$

subsituents at 1,3,5,7,9 omitted

ligand	R	R′	R″
L^1	⌇CH$_3$	⌇CH$_3$	pyridin-2-yl
L^2	⌇CH$_3$	⌇CH$_3$	6-methylpyridin-2-yl
L^3	⌇CH$_3$	⌇CH$_3$ / pyridin-2-ylmethyl	1-methylimidazol-2-yl
L^4	⌇CH$_3$	⌇CH$_3$	pyridin-2-yl
L^5	pyridin-2-ylmethyl	pyridin-2-ylmethyl	pyridin-2-yl
L^6	pyridin-2-ylmethyl	2-(pyridin-2-yl)ethyl	pyridin-2-yl
L^7	⌇CH$_3$	6-methylpyridin-2-ylmethyl	quinolin-2-yl
L^8	⌇CH$_3$	2-(pyridin-2-yl)ethyl	quinolin-2-yl
L^9	⌇CH$_3$		pyridin-2-yl
L^{10}	⌇CH$_3$	2-hydroxybenzyl	pyridin-2-yl
L^{11}	⌇CH$_3$	⌇CH$_3$	2-methoxyphenyl

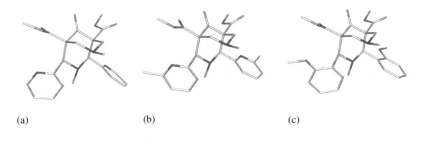

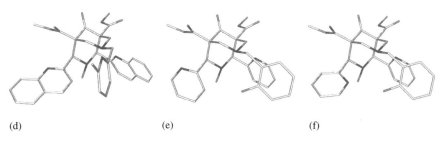

(a)   (b)   (c)

(d)   (e)   (f)

Figure 18-1. Plots of the X-ray structures of bispidone ligands (a) L^1, (b) L^2, (c) L^{11}, (d) L^8, and (e), (f) L^{10}

chemistry, [8–11] bleaching [12, 13] and oxidation catalysis [14–20]. All these applications are based on the specific structural properties of bispidine transition-metal complexes, enforced by the rigid bispidine backbone, the shape of the potential-energy surface of the complexes, and the resulting electronic properties of the complexes [3].

The main features of bispidine complexes are (i) the coordination geometries that are derived from *cis*-octahedral with two sterically and electronically distinct sites for the binding of substrates, which in pentadentate ligand systems generate isomeric complexes with disparate properties (Figure 18-2), [3, 21] and (ii) nearly flat potential-energy surfaces of the complexes with various shallow minima; i.e. whereas the ligands are rigid, the coordination geometries are elastic and typically there are various nearly degenerate isomers with strikingly disparate properties [21]. Three unique examples of isomerism observed in bispidine complexes are given in Figure 18-3.

An interesting example of a molecular property derived from the structure of the complexes and therefore enforced by the bispidine ligands is that of the complex stabilities reported in Table 18-1. [5] The copper(II) stability constant of the tetradentate ligand L^1 is about 10^7 times that of the methylated tetradentate ligand L^2. This effect has been explained with a change of structure (coligand *trans* to N3 (L^1) vs. *trans* to N7 (L^2) and a concomitant change of the Jahn-Teller axis with the result of a partial quenching of the elongation due to the tight five-membered

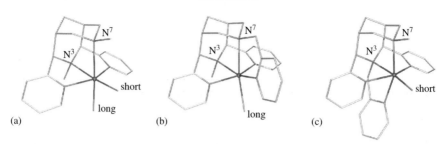

Figure 18-2. Plots of the structures of bispidine metal complexes with the tetradentate ligand L^1 (a) with two sterically and electronically different sites for substrate coordination, and with the isomeric pentadentate ligands L^4 (b) and L^5 (c), which have one of those sites blocked by a pendant pyridine

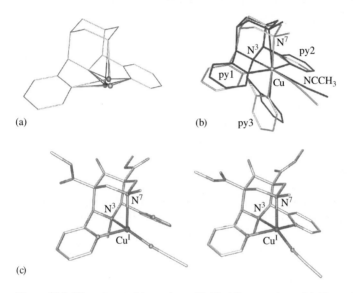

Figure 18-3. Three types of isomerism with bispidine complexes (a) The variation of the metal ion site in the bispidine cavity, [25] (b) Jahn-Teller isomerism, [22] and (c) distortional isomerism of the copper(I) complexes [39]

chelate rings, which involve the pendant pyridine groups) and is paralleled by a difference in the reduction potential about 300 mV (this correlation is due to similar stability constants of the copper(I) complexes) [5]. With an additional pyridine donor the pentadentate ligand L^4 leads, as expected, to a copper(II) complex stability, which is about 100 times that of L^1. However, the isomeric pentadentate ligand L^5 has a less stable copper(II) complex by about 0.001, and is even less stable than the complex with the tetradentate ligand L^1. The unexpected changes in copper(II) stability along the series of ligands have been interpreted with a change of the direction of the Jahn-Teller elongation (see Figure 18-3b) [22–24]. The interesting observation, therefore, is that the complex stabilities with cobalt(II), nickel(II) and

Table 18-1. Complex stabilities ($logK_{ML}$) of metal ion bispidine complexes in aqueous solution or acetonitrile, $\mu = 0.1\,M$ (KCl) [5]

$logK_{ML}$		L^1	L^2	L^4	L^5	L^6
Cu(I)	a)	5.61±0.32	5.48±0.32	5.69±0.22	6.29±0.43	4.97±0.52
	b)	9.5	8.9	9.6	8.8	7.9
Cu(II)		16.56±0.05	9.60±0.07	18.31±0.12	15.66±0.03	16.28±0.10
	c)	12.5	7.1	15.3	14.5	16.2
Co(II)		5.46±0.05	–	6.23±0.05	13.69±0.05	7.30±0.06
Ni(II)		–	7.50±0.09	6.10±0.08	9.54±0.06	5.02±0.07
Zn(II)		11.37±0.01	–	8.28±0.05	13.57±0.04	9.18±0.05
Li(I)	d)	2.7±0.09	3.9±0.10	–	3.65±0.09	3.70±0.08

a) by NMR titration, in MeCN.

b) calculated in H_2O $\left(\ln \frac{K_{Cu^I}}{K_{Cu^{II}}} = \frac{E^\circ \cdot n \cdot F}{RT} \right)$.

c) calculated in CH_3CN, see b).

d) by UV-V is titration.

zinc(II) also fail to follow the naively expected trends (see Table 18-1). A possible interpretation is that similar structural distortions are observed for all metal ions, i.e. the "Jahn-Teller effects" mentioned above are not primarily electronic effects but are enforced by the ligand; this interpretation is supported experimentally by structural data of bispidine-zinc(II) complexes (d^{10}) [5].

The aim of the present communication is to discuss possible reasons for these and other distortions and to describe models with which one can predict the corresponding structures. The correlation of the ensuing structures with molecular properties is also discussed, and answers are sought to the question whether an accurate structural prediction might help to understand specific distortions. In addition, some fascinating properties and applications of bispidine complexes are presented and models are discussed, which assist their thorough understanding.

STRUCTURAL ANALYSES

The rigidity of the bispidine ligands and the elasticity of the coordination spheres of these complexes are demonstrated in Figure 18-4 [25]. Whereas the ligand backbone, apart from small changes of the torsional angle about the $C2,C4$—$C_{pyridine}$ bond, has a constant shape (Figure 18-4a), the metal ions adopt various positions in the bispidine cavity. From experimental structures it appears that there are positions of two types, one with M-N7>M-N3 and the other with M-N7<M-N3; see Figure 18-4b,c (the third structure (black metal center, Cu^{II}) also has M-N7>M-N3 and is due to an additional electronic perturbation, i.e. a pseudo-Jahn-Teller distortion).

A molecular-mechanics scan of the potential-energy surface of $[M(L^1)]^{n+}$ indicates that the two distortional isomers (M-N7<>M-N3) are nearly degenerate for cobalt(II) (see Figure 18-5); this effect emerges also from experimental structural studies [25]. The lower-energy, nearly flat curve in Figure 18-5 indicates that the

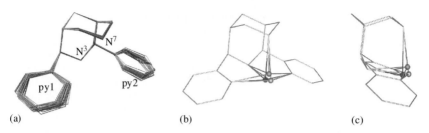

Figure 18-4. (a) Overlay plot of the ligand portion of 40 X-ray structures of complexes with ligand L¹;
(b), (c) average structure of (a) with three metal centers (Cu^{II}: dark, Cu^{I}: middle, Mn^{II}: light) included
in their crystallographically determined sites [25]

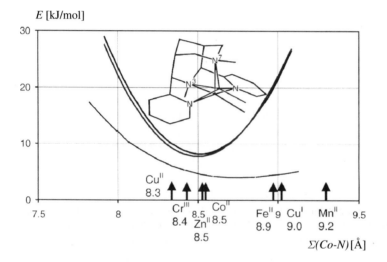

Figure 18-5. Cavity shape and size (hole size) curves of L¹ (lowest energy curve, no metal-donor
interaction terms included), and of the two isomers of [Co(L¹)]²⁺ with the cobalt-bispidine interaction
terms included; the arrows are averaged sums of observed M—L bond distances [25]

ligand is unselective with respect to the metal ion size. Except for small metal ions,
there is little strain induced in the ligand when a metal ion becomes coordinated,
and for the larger metal ions (M-N $\geq$ 2.05 Å) the steric energy is nearly independent
of the metal ion.

Similar effects are observed for penta- and hexadentate bispidine ligands, and
the shape of the corresponding potential-energy surfaces is essentially identical, i.e.
flat and with a steep rise of energy for small metal ions [4]. A thorough molecular
mechanics analysis has been made for complexes of ligand L⁶; the experimental
and computed structural data together with an analysis of the cavity shape (hole
size calculations, Figure 18-6) have been used to interpret the complex stabilities
(see also Table 18-1 above) [4]. The interesting observation is that the stability

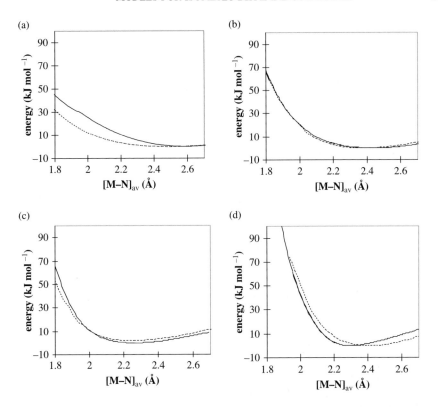

Figure 18-6. Cavity shape and size (hole size) curves of bispidine ligands (molecular mechanics, no metal center dependent energy terms included; broken lines are with sum constraints, solid lines are with an approach with individual, asymmetric variations of all six M—N bonds) of (a) L^1, (b) L^4, (c) L^5 and (d) L^6

of the cobalt(II) complex is larger than that of nickel(II), in contrast to expectations derived from the Irving-Williams series behavior [26, 27]. Similar trends are observed with the tetra- and pentadentate bispidine ligands [5]. Whereas molecular mechanics accurately reproduces the structures and allows one to understand qualitatively the stability constants, a thorough interpretation must include also electronic effects; this has not yet been achieved.

Of particular interest are the structural effects (pseudo-Jahn-Teller distortions) of the copper(II) complexes. For tetragonal copper(II) complexes the theoretical basis might be visualized with a Mexican-hat potential (see Figure 18-7) [28, 29]. The stabilization (E_{JT}, Figure 18-7a) due to an elongation might occur along one of three molecular axes, which produces three isomeric structures of disparate stability (Figure 18-7b). The first examples of such 'Jahn-Teller isomers' were recently isolated and structurally characterized [22–24]. Some available experimental structures are presented in Figure 18-8. All these structures are accurately

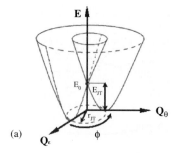

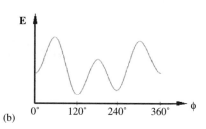

Figure 18-7. (a) The Mexican hat potential energy surface for copper(II) in an octahedral ligand field, and (b) a cross section through the warped rim (dotted line in (a)) of the Mexican hat potential [28]

reproduced or predicted with molecular mechanics [30] (the model used (LFT-MM) involves a parameterization scheme that, in addition to the conventional bonded and non-bonded interactions, includes a ligand-field-based term) [31, 32].

The structure of the copper(II) complex of L^7 is also accurately predicted (Figure 18-8g). This structure has long bonds to q1 (2.311 Å) and q2 (2.269 Å) and short bonds to N3 (1.953 Å), N7 (2.095 Å) and py (1.927 Å); viz. the elongation is along q_1-Cu-q_2 and there is a ligand missing *trans* to N7, in the plane of the strongly bound ligands, perpendicular to the Jahn-Teller axis [23]. Although this unique structure is accurately predicted, the computational result (molecular mechanics) does not help to understand the fundamental reason behind it. A possible qualitative interpretation is that the bispidine ligand enforces a geometry on the complex, which leads to an electronic configuration on copper(II) that destabilizes the bonding on the N7-Cu-X_A axis (stabilization of the corresponding d orbital; X_A is the coligand *trans* to N7). The six-membered chelate ring involving N7 and py for some reason compresses the Cu–N7 bond, and the bond *trans* to N7 is therefore weakened ("trans-influence"; holohedrized symmetry [28]).

This interpretation is supported, qualitatively again, by the experimental structure of the similar cobalt(III) complex of L^9 (Figure 18-9), which provides a rare example of a pentacoordinate, square-pyramidal high-spin cobalt(III) complex [33]. Our preliminary interpretation of this structure is that effects similar to those described above for the copper(II) complexes produce a low-energy d_{z^2} orbital in the hexacoordinate precursor of the cobalt(III) complex shown in Figure 18-9; this structure presumably has a low-spin electronic ground state and, due to the small ligand field and the low-lying d_{z^2} orbital, has a small energy barrier to a high-spin excited state. The latter is Jahn-Teller active and, due to the ligand-enforced small Co-N7 distance, dissociates the ligand *trans* to N7 for reasons similar to those given above for the copper(II) complex (holohedrized symmetry). All these qualitative interpretations obviously require a thorough study of the electronic structures. Preliminary DFT and LF-DFT calculations (ligand-field analysis by DFT calculations, see Chapter by M. Atanasov, P. Comba, C. Daul, and F. Neese) support this idea [34].

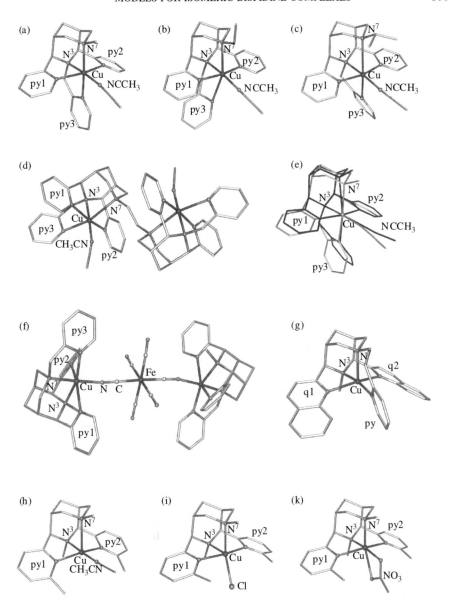

Figure 18-8. Structural plots of bispidine-copper(II) complexes with various directions of the Jahn-Teller elongation (given in brackets; hydrogen atoms and substituents at C1, C5 and C9 omitted) (a)–(d) $[Cu(L^5)(NCCH_3)]^{2+}$ and derivatives (a) (py1-Cu-py2), (b) (N7-Cu-py3), (c) (N7-Cu-py3), (d) (N7-Cu-py3), (e) overlay plot of the chromophores of (a) and (d), (f) $\{Fe(CN)_6[Cu(L^4)]_2\}^+$ (py1-Cu-py2), (g) $[Cu(L^7)]^{2+}$ (q1-Cu-q2), (h) $[Cu(L^2)(NCCH_3)]^{2+}$ (Cu-N7), (i) $[Cu(L^2)(Cl)]^+$ (Cu-N3) and (k) $[Cu(L^2)(NO_3)]^+$ (py1-Cu-py2)

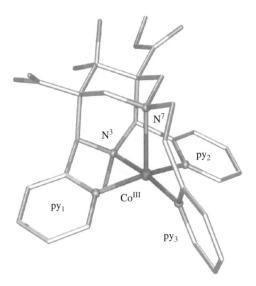

Figure 18-9. Plot of the X-ray structure of $[Co(L^9)]^{3+}$ (one of the two ester groups of the ligand is hydrolyzed)

An important feature in the quantum-mechanical analysis [34] is a misdirection of the M—N7 bond. This effect is due to the tight five-membered chelate rings involving N3, the pyridine (or quinoline) groups and the metal ion, and is also due to the constant and small N3···N7 distance (~2.9 Å) of the rigid bispidine backbone, [3, 25] i.e. there is decreased overlap between the N7 lone pair and the metal d_{z^2} orbital. There is hence a ligand-induced elongation along M—N7 and therefore a high degree of complementarity of these ligands for copper(II); this effect provides a reason for the large copper(II) stability constants (in particular with L^1, see Table 18-1 above). Why then are there isomeric structures with short Cu—N7 bonds and an elongation along the pyridine groups (see Figure 18-8 above, structures a, f, g, k)? An interesting observation is that in those structures the angular geometry about C2 and C4 is unstrained in both isomers (nearly ideal tetrahedral angles). The conversion from one isomer to the other is best described by a tilt about the C2···C4 axis, followed by a rearrangement of the torsional angles about C2, C4—$C_{pyridine}$ (see Figure 18-10). The two isomers are thus expected to have similar energies.

DYNAMICS

The copper(I) complexes of the tetradentate ligands L^1 and L^2 occur in three structural forms; two have been characterized by single-crystal X-ray diffraction and are distortional isomers (bond-stretch isomers [35–37], see Figure 18-11) [3, 38, 39].

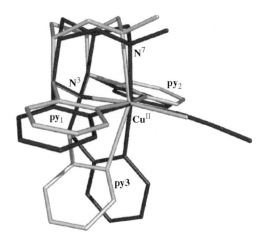

Figure 18-10. Overlay plot of the molecular cations $[Cu(L^5)(NCCH_3)]^{2+}$ and $[Cu(L^5)(Cl)]^+$ (Jahn-Teller elongation along py1-Cu-py2 and N7-Cu-py3, respectively)

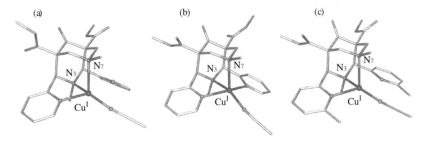

Figure 18-11. Plots of the complex cations of various isomers of bispidine-copper(I) complexes, determined by X-ray crystallography (hydrogen atoms omitted (a) $[Cu(L^1)(NCCH_3)]^+$, (b) $[Cu(L^1)(NCCH_3)]^+$ and (c) $[Cu(L^2)(NCCH_3)]^+$[39]

With L^1 both isomers have been crystallized; the tetracoordinate compound is yellow, the pentacoordinate is red (Figure 18-11a,b). The fact that both are observed indicates that they have similar energies, and in solution fluxionality is expected and observed through ^{1}H-NMR spectra. An interesting structural detail is that the rotation of the pendant pyridine rings differs in the two reported structures of the tetracoordinate forms with L^1 and L^2. The structure of the complex with L^2 (Figure 18-11c) seems to represent an intermediate between the tetra- and pentacoordinate forms, isolated and characterized for the L^1-based complex (Figure 18-11a,b) with the pyridine group still partially bonded.

From the ^{1}H-NMR spectra it follows that the solution behavior of $[Cu(L^1)(NCCH_3)]^+$ differs from that of $[Cu(L^2)(NCCH_3)]^+$ (see Figure 18-12) [39]. With L^1 the spectra indicate a symmetrical five-coordinate structure up to ambient temperature. The line broadening at higher temperatures

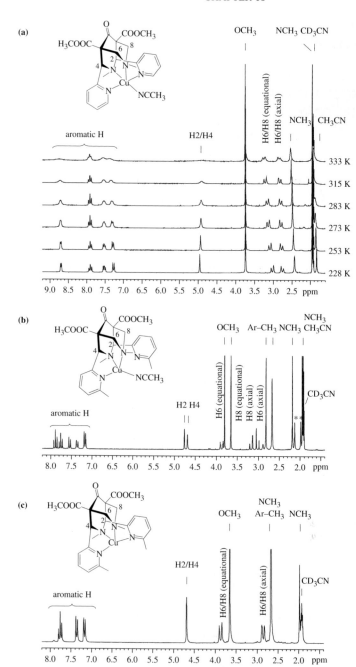

Figure 18-12. ^{1}H-NMR spectra (200 MHz, CD$_3$CN) of the isomeric copper(I) complexes of the tetraden-tate bispidine ligands L^1 and L^2 (a) temperature dependent spectra of [Cu(L^1)(NCCH$_3$)]$^+$, (b) ambient temperature spectrum of the unsymmetrical four-coordinate form of [Cu(L^2)(NCCH$_3$)]$^+$ and (c) ambient temperature spectrum of [Cu(L^2)]$^+$ [39]

indicates that the two enantiomeric four-coordinate forms are in a rapid equilibrium (Figure 18-11a). From the spectrum at ambient temperature of the L^2–based complex (Figure 18-12b) it emerges that the four-coordinate complex is more stable with the methylated ligand. Figure 18-12c shows a second form of the L^2-derived complex, which is assigned to a highly symmetrical four-coordinate structure [39].

The interconversion between the various structural forms involves bond making and bond breaking which molecular mechanics is unable to model, but the entire behaviour of both complexes is satisfactorily simulated with DFT calculations [39]. Both the computed structures and energetics agree with the solid-state structural and the solution ^{1}H-NMR experimental data (see Table 18-2). For L^1 the five-coordinate complex is slightly more stable and for L^2 the four-coordinate structure is preferred. Also, for the L^2-based system a four-coordinate structure with both pyridine groups coordinated to copper(I) (see the structure proposed on the basis of the NMR spectra, Figure 18-12c) is optimized and is only about 5 kJ/mol less stable than the unsymmetrical four-coordinate species with an acetonitrile donor coordinated to the copper(I) center. For L^1 the corresponding structure is approximately 15 kJ/mol less stable.

Table 18-2. Selected calculated and experimental geometric parameters for the $[Cu(L^n)(CH_3CN)]^+$ complexes, n = 1, 2. Experimental values are given in italics, * indicates that the structure is a transition state, ** is the 4-coordinate complex without $NCCH_3$ [39]

Ligand	L^1				L^2			
Parameters	5-coord	*5-coord*	4-coord	*4-coord*	5-coord*	4-coord	*4-coord*	4-coord**
Cu-N3	2.418	*2.292*	2.228	*2.203*	2.166	2.223	*2.203*	2.217
Cu-N7	2.288	*2.188*	2.234	*2.160*	2.203	2.224	*2.184*	2.200
Cu-Npy1	2.249	*2.169*	2.225	*2.066*	2.632	2.186	*2.096*	2.019
Cu-Npy2	2.249	*2.247*	2.745	*3.118*	2.632	3.188	*2.897*	2.019
Cu-NCCH$_3$	1.993	*1.936*	1.935	*1.873*	1.918	1.920	*1.900*	–
angles(°)								
N7-Cu-N3	82.46	*83.15*	84.58	*83.77*	86.56	84.91	*83.63*	86.59
N3-Cu-Nac	160.71	*154.95*	146.85	*134.16*	147.05	139.64	*150.09*	–
torsions angles (°)								
N3-C-C-Npy1	37.26	*38.39*	39.69	*41.53*	46.58	41.18	*35.69*	37.29
N3-C-C-Npy2	–37.26	*–39.94*	–44.54	*–61.15*	46.58	–55.60	*–45.44*	37.29
Relative energies (kJ/mol)								
Solvated by MeCN	0.00	–	1.98	–	3.08	0.00	–	5.05

REACTIVITY

There is a rich bispidine-iron chemistry that involves applications of the iron(II) complexes as bleaching catalysts, [12, 13] of L^n / Fe^{II} / H_2O_2 systems ($n=1,4,5$) in non-heme iron oxidation catalysis [14, 16–18, 20, 40] and as a mechanistic model for the Fenton reaction [15]. Based on much published data on iron systems with other ligands [41–47], metastable iron(III) and iron(IV) intermediates with bispidine ligands have been characterized spectroscopically (UV-vis-NIR, EPR, Raman, Mössbauer); the interconversion between the various forms has been studied with time-dependent spectroscopy (stopped-flow experiments), and catalytic organic transformations (primarily the oxidation of cyclooctene to the corresponding epoxide and 1,2-diol products) have been investigated by product analyses and extensive ^{18}O-labelling experiments [14, 15, 20]. All putative oxidants, which have been found in systems with various other ligands, have been identified in the iron bispidine chemistry [14, 15]. An interesting feature is that the two pentadentate ligands have disparate reactivities. Whereas the L^5-based complex has stable iron(III) and iron(IV) intermediates, the L^4-based system is catalytically much more active; however, mechanistically and in terms of product distributions the two systems are similar [14, 20]. The catalyst with the tetradentate bispidine ligand L^1 shows a separate behavior. No high-valent iron species were detected spectroscopically (except in reactions with alkyl peroxides [40]), and the different product distribution indicates another catalytic mechanism [48]. An exciting result with the pentadentate ligand L^5-based iron complex is that the reaction with H_2O_2 in aqueous solution directly and without an iron(III) intermediate produces the iron(IV) oxo complex, which is stable in aqueous solution [15]. This result is relevant to the still disputed mechanism for the Fenton reaction. The resulting mechanistic scenario involving the iron-based chemistry without addition of a substrate is presented in Chart 2.

With cyclooctene as the substrate, the pentadentate ligand-based systems yield epoxide and diol products in ratios that depend strongly on the reaction conditions.

Chart 2

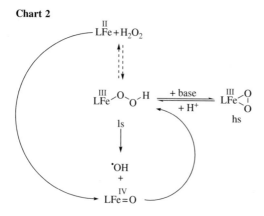

The more active L^4-based catalyst that is among the most active iron-based systems is unique for several reasons: e.g. dependence of the product distribution on the atmosphere (ambient or argon), formation of diol products with pentadentate ligand systems, formation of *cis* and *trans* diol products. The novel mechanism shown in Chart 3 was proposed on the basis of the experimental data (except for the direct reactions from iron(II) to iron(IV) and the reaction from the oxo-iron(IV) complex to the epoxide product; these are small modifications reflecting the theoretical studies, see below) [20]. The L^1-based catalyst belongs to the most active iron-based catalyst systems and has unique properties, i.e. both oxygen atoms in the diol product, which is selectively *cis*-configurated, arise directly from H_2O_2. The mechanistic implications of this result appear in Chart 4 [48, 49].

Computational methods were used to support the various mechanisms derived from experiment, and to obtain detailed information about the series of intermediates and transition states; these studies are primarily based on DFT calculations. All possible pathways and all possible spin states of the intermediates and transition states have been considered [16–18, 48, 49]. As an example, Figure 18-13 shows the major results related to the oxidation of cyclooctene by $[(L^4)$ Fe=O$]^{2+}$ (see also Chart 3) [17]. In summary, the computational methods have been used to study the structures and electronic properties of the iron(III) intermediates, [14] and of

Chart 3

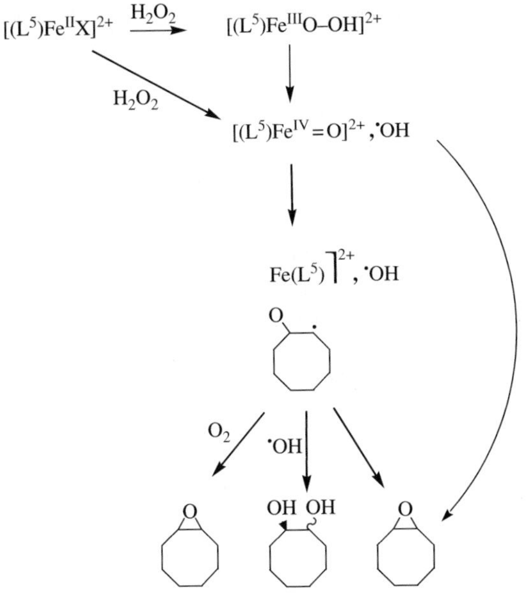

$[(L^5)Fe^{II}X]^{2+} \xrightarrow{H_2O_2} [(L^5)Fe^{III}O-OH]^{2+}$

H_2O_2

$[(L^5)Fe^{IV}=O]^{2+}$, $^{\bullet}OH$

$Fe(L^5)\rceil^{2+}$, $^{\bullet}OH$

O_2 / $^{\bullet}OH$

OH OH

Chart 4

$(L)–Fe^{II}—X \xrightarrow{+H_2O_2} (L)–Fe^{II} \underset{^{\bullet\bullet\bullet}OH}{\cdots O} \longrightarrow (L)–Fe^{IV}=O$

OH

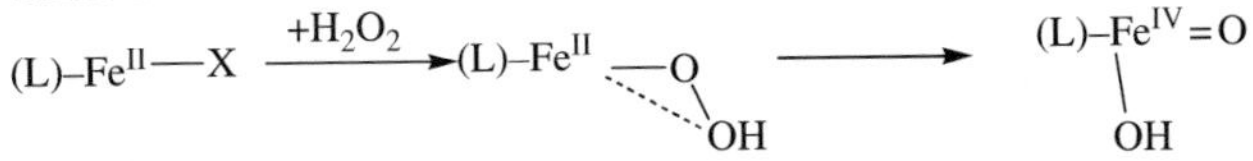

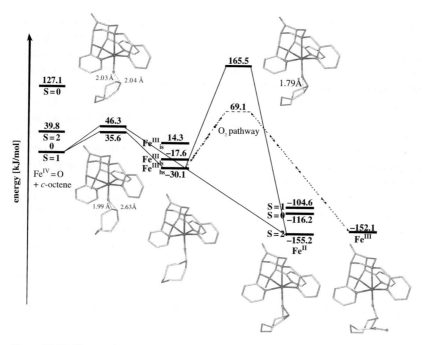

Figure 18-13. The reaction profile (DFT) of the oxygenation of cyclooctene by $[(L^4) Fe=O]^{2+}$ (see Chart 2) [17]

the ferryl products [18]. The structure of the high-spin peroxo iron(III) complex of the pentadentate ligands L^4 and L^5 was found to be seven-coordinate on the basis of the experimentally observed and the computed Raman transitions [14]. From the thorough analysis of the ferryl complexes of L^1, L^4 and L^5 it emerges that the bispidine-ligand-enforced structure (see section on the structural aspects above), the *trans* influence of the Fe=O bond and Jahn-Teller effects in the high-spin iron(IV) electronic configuration are in a subtle equilibrium and are three major effects that determine the structure and electronic properties of these highly reactive oxidants [18]. The study of the oxidation reaction with H_2O_2 of the bispidine-iron(II) complexes with the pentadentate ligands L^4 and L^5 revealed that the question whether there is a direct conversion to the ferryl complexes (O—O heterolysis vs. homolysis) depends strongly on the spin state. [16] In a similar study with the tetradentate bispidine L^1 a dihydroxo iron(IV) complex was found as an initial product (see also Chart 4), [49] which was subsequently established to be relevant to the oxidation of alkenes [48].

CONCLUSION

We have shown that bispidine coordination chemistry is exceptionally rich with many unique properties of the complexes, and with possible applications in various areas. We attribute the specific and unique properties of the bispidine transition-metal complexes to the rigid diazaadamantane-derived bispidine backbone and the asymmetry with respect to the tertiary amines N3 and N7, due to the pendant donors at C2 and C4, producing *cis*-octahedral coordination geometries with two sterically and electronically distinct sites for substrate coordination. Important features are the rigid and highly preorganized bispidine backbone (Figure 18-4) and the elasticity of the coordination geometry (Figures 18-4 and 18-6). The basic conclusion of the observations and interpretations discussed here is that the rigid bispidine ligands enforce specific geometries on the transition-metal sites, which therefore adopt specific electronic configurations, and these configurations might lead to additional structural perturbations. The resulting geometric and electronic structures are the basis for the observed unique thermodynamic properties, reactivities and spectra of the bispidine complexes. A combination of steric and electronic effects and the difficulty to separate them makes some molecular properties difficult to understand thoroughly.

The discussion of the data reviewed here indicates that some observations are best interpreted, at least qualitatively, on the basis of the experimental structural data. The modeling of structural features is generally accurate with approaches based on molecular mechanics. This success includes systems in which electronic effects obviously contribute to the specific distortions as these are generally included in the parameterization of the force fields, [50] especially in electronically doped molecular-mechanics approaches (see section on the modeling with LFT-MM). A thorough interpretation of observed structures, accurately simulated with force-field methods, might need, however, a quantum-mechanical approach to achieve a fundamental understanding of the specific distortions (see Figures 18-8 and 18-9). Molecular properties might be interpreted on the basis of simple correlations with structural data and, if these are not available from experimental data, molecular mechanics might yield accurate structural models. For a thorough interpretation of reaction mechanisms, in which bond making and bond breaking are important, quantum-mechanical approaches are required (see sections on dynamics and reactivity).

ACKNOWLEDGEMENT

Generous financial support by the Deutsche Forschungsgemeinschaft (DFG) of the work reviewed here is gratefully acknowledged.

REFERENCES

1. Comba, P.; Tarnai, M.; Wadepohl, H., *work in progress.*
2. Mannich, C.; Mohs, P., *Chem. Ber.* **1930,** B63, 608.
3. Comba, P.; Kerscher, M.; Schiek, W., *Prog. Inorg. Chem.,* **2007,** 55, 613.
4. Bleiholder, C.; Börzel, H.; Comba, P.; Ferrari, R.; Heydt, A.; Kerscher, M.; Kuwata, S.; Laurenczy, G.; Lawrance, G. A.; Lienke, A.; Martin, B.; Merz, M.; Nuber, B.; Pritzkow, H., *Inorg. Chem.* **2005,** 44, (22), 8145.

5. Born, K.; Comba, P.; Ferrari, R.; Kuwata, S.; Lawrance, G. A.; Wadepohl, H., *Inorg. Chem.*, **2007**, 46, 458.

6. Comba, P.; Kerscher, M.; Pietzsch, H.-J.; Spies, H.; Stephan, H.; Juran, S. *Radioaktive Metallkomplexe von Bispidinderivaten und deren Verwendung für die nuklearmedizinische Diagnostik und Therapie sowie Verfahren zur Herstellung radioaktiver Metallkomplexe von Bispidinderivaten.* Deutsches Patent 2006.

7. Atanasov, M.; Busche, C.; Comba, P.; Rajaraman, G., *in preparation.*

8. Börzel, H.; Comba, P.; Katsichtis, C.; Kiefer, W.; Lienke, A.; Nagel, V.; Pritzkow, H., *Chem. Eur. J.* **1999**, 5, 1716.

9. Börzel, H.; Comba, P.; Pritzkow, H., *J. Chem. Soc., Chem. Commun.* **2001**, 97.

10. Börzel, H.; Comba, P.; Hagen, K. S.; Kerscher, M.; Pritzkow, H.; Schatz, M.; Schindler, S.; Walter, O., *Inorg. Chem.* **2002**, 41, 5440.

11. Born, K.; Comba, P.; Daubinet, A.; Fuchs, A.; Wadepohl, H., *J. Biol. Inorg. Chem.,* **2007**, 12, 36.

12. Börzel, H.; Comba, P.; Hage, R.; Kerscher, M.; Lienke, A.; Merz, M., Patent WO 0248301 A1 (2002-06-20), Patent US 2002/014900 (2002-10-17) *Ligand and complex for catalytically bleaching a substrate.*

13. Comba, P.; Koek, J. H.; Lienke, A.; Merz, M.; Tsymbal, L., Patent WO 03/104234 A1 (2003-12-18). *Ligand and complex for catalytically bleaching a substrate.*

14. Bukowski, M. R.; Comba, P.; Limberg, C.; Merz, M.; Que Jr., L.; Wistuba, T., *Angew. Chem. Int. Ed.,* **2004**, 43, 1283.

15. Bautz, J.; Bukowski, M.; Kerscher, M.; Stubna, A.; Comba, P.; Lienke, A.; Münck, E.; Que Jr, L., *Angew. Chem. Int. Ed.,* **2006**, 45, 5681.

16. Anastasi, A. E.; Lienke, A.; Comba, P.; Rohwer, H.; McGrady, J. E., *Eur. J. Inorg. Chem.,* **2007**, 65.

17. Comba, P.; Rajaraman, G., *Inorg. Chem.,* submitted.

18. Anastasi, A.; Comba, P.; McGrady, J.; Lienke, A.; Rohwer, H. *Inorg. Chem.,* **2007**, 46, 6420.

19. Comba, P.; Kuwata, S.; Tarnai, M.; Wadepohl, H., *J. Chem. Soc., Chem. Commun.* **2006**, 2074.

20. Bukowski, M. R.; Comba, P.; Lienke, A.; Limberg, C.; Lopez de Laorden, C.; Mas-Balleste, R.; Merz, M.; Que Jr., L., *Angew. Chem. Int. Ed.,* **2006**, 45, 3446.

21. Comba, P.; Schiek, W., *Coord. Chem. Rev.* **2003**, 238–239, 21.

22. Comba, P.; Hauser, A.; Kerscher, M.; Pritzkow, H., *Angew. Chem. Int. Ed.* **2003**, 42, (37), 4536.

23. Comba, P.; Lopez de Laorden, C.; Pritzkow, H., *Helv. Chim. Acta* **2005**, 88, 647.

24. Comba, P.; Martin, B.; Prikhod'ko, A.; Pritzkow, H.; Rohwer, H., *Comptes Rendus Chimie* **2005**, 6, 1506.

25. Comba, P.; Kerscher, M.; Merz, M.; Müller, V.; Pritzkow, H.; Remenyi, R.; Schiek, W.; Xiong, Y., *Chem. Eur. J.* **2002**, 8, 5750.

26. Irving, H.; Williams, R. J. P., *Nature* **1948**, 162, 746.

27. Irving, H.; Williams, R. J. P., *J. Chem. Soc.* **1953**, 3192.

28. Figgis, B. N.; Hitchman, M. A., *Ligand Field Theory and its Applications.* Wiley-VCH: Weinheim, New York, 2000; p. 354.

29. Bersuker, I. B., *Chem. Rev.* **2001**, 101, 1067.

30. Benz, A.; Comba, P.; Deeth, R. J.; Kerscher, M.; Wadepohl, H., *work in progress.*

31. Deeth, R. J.; Hearnshaw, L. J. A., *J. Chem. Soc., Dalton Trans.* **2005**, 3638.

32. Deeth, R. J.; Hearnshaw, L. J. A., *J. Chem. Soc., Dalton Trans.* **2006**, 1092.

33. Comba, P.; Lawrance, G. A.; Wadepohl, H.; Wunderlich, S., *work in progress.*

34. Atanasov, M.; Comba, P., *work in progress.*

35. Stohrer, W.-D.; Hoffmann, R., *J. Am. Chem. Soc.* **1972**, 74, 779.

36. Parkin, G., *Chem. Rev.* **1993**, 93, 887.

37. Rohmer, M. M.; Bénard, M., *Chem. Soc. Rev.* **2001**, 30, 340.

38. Börzel, H.; Comba, P.; Hagen, K. S.; Katsichtis, C.; Pritzkow, H., *Chem. Eur. J.* **2000**, 6, 914.

39. Born, K.; Comba, P.; Kerscher, M.; Rohwer, H., *work in progress.*

40. Bautz, J.; Comba, P.; Que Jr., L., *Inorg. Chem.* **2006**, 45, 7077.

41. Meunier, B., *"Metal-Oxo and Metal-Peroxo Species in Catalytic Oxidation".* Springer: Berlin, 2000; Vol. 97.

42. Meunier, B., *Biomimetic Oxidations Catalyzed by Transition Metal Complexes*. Imperial College Press: 2000.
43. Que Jr., L.; Ho, R. Y. N., *Chem. Rev.* **1996,** 96, 2607.
44. Solomon, E. I.; Brunold, T. C.; Davis, M. I.; Kensley, J. N.; Lee, S.-K.; Lehnert, N.; Neese, F.; Skulan, A. J.; Yang, Y.-S.; Zhou, J., *Chem. Rev.* **2000,** 100, 235.
45. Costas, M.; Mehn, M. P.; Jensen, M. P.; Que Jr., L., *Chem. Rev.* **2004,** 104, 939.
46. Collins, T. J., *Acc. Chem. Res.* **2002,** 35, 782.
47. Chen, K.; Costas, M.; Que Jr., L., *J. Chem. Soc., Dalton Trans.* **2002,** 672.
48. Bautz, J.; Comba, P.; Lopez de Laorden, C.; Mentzel, M.; Rajaraman, G., *Angew. Chem.*, in press.
49. Comba, P.; Rajaraman, G.; Rohwer, H., *Inorg.Chem. 2007, 46, 3826.*
50. Boeyens, J. C. A.; Comba, P., *Coord. Chem. Rev.* **2001,** 212, 3.

CHAPTER 19

THE LIGAND-FIELD PARADIGM

Insight into Electronic Properties of Transition-metal Complexes Based on Calculations of Electronic Structure

MIHAIL ATANASOV, PETER COMBA, CLAUDE A. DAUL
AND FRANK NEESE

Abstract: An overview and a critical comparison of contemporary models to describe and to predict electronic multiplet structures and the spectroscopic behavior of transition-metal complexes with open d-shells is given in relation to experimental data including d-d absorption and ESR spectra. A ligand-field density-functional theory (LFDFT) predicts these properties with a success similar to *ab initio* approaches, such as the spectroscopy oriented configuration-interaction method, and better than time-dependent density-functional theory applied to open shell systems. Using well characterized systems, from classical coordination compounds [FeO_4^{2-}, CrX_6^{3-} (X = F,Cl), CoL_6^z(z = −3, L = CN^-; z = 2 and 3, L = H_2O)] to Fe^{IV} macrocyclic compounds with biochemical and catalytic activity, it is shown that LFDFT is able also to characterize larger systems and subtle effects such as those from surrounding influences and the second coordination sphere

INTRODUCTION

An intrinsic feature of transition-metal ions is the localized character of their 3d electrons; this property is preserved to a great extent in their complexes, but modified by covalency. One can hence formulate a metal-ligand interaction in these compounds as being mainly ionic and interpret a metal-ligand bond as a donor-acceptor bond. In simple terms, a ligand donates electrons into the empty valence shell of the transition-metal ion – the partly filled 3d and the empty 4s and 4p shells, which leads to metal-centred antibonding and ligand-centred bonding molecular orbitals (MO) (except for the case of metal-ligand π-backbonding where the metal-based t_{2g}-orbitals in an octahedral complex are bonding with the ligands), but these interactions are supposed to be weak enough to be treated with perturbation theory. Electronic transitions associated with absorption and emission spectra in the visible

411

J.C.A. Boeyens and J.F. Ogilvie (eds.), Models, Mysteries and Magic of Molecules, pp. 411–445.
© 2008 *Springer*.

region are located within the many-electron states that originate from a well defined d^n-configuration of the transition-metal ion. Ligand-to-metal (LMCT) as well as metal-to-ligand (MLCT) charge transfer transitions are not comprised in this manifold and need a different treatment. All these features, which are mostly consistent with the interpretation of experiment, define what we call a *Werner-type* complex.

An approach broadly used to interpret the properties of these complexes has been ligand-field theory (LFT), i.e. various models beginning with classical crystal-field theory (CFT) and extending to the angular-overlap-model parameterization of the ligand field [1, 2]. Various extensions of the latter model, such as the *cellular-ligand-field* (CLF) formulation [1] include effects of low symmetry, such as s-d mixing, orbital-phase coupling and misdirected ligand fields due to bent bonds; these concepts have found applications to various transition-metal complexes and their structural and electronic properties [1, 2].

With the development of current theories of electronic structure, such as multi-reference ab initio approaches and in particular density-functional theory (DFT), the appealing ligand-field approach became displaced by these approaches theoretically well justified but chemically less transparent and less readily analyzed and interpreted (but see Ref. [3][a] as an exception). In contrast, the parametric structure of ligand-field theory and the need of experimental data to allow adjustment of these parameters makes this model a tool for interpretation rather than for predictions of electronic properties of transition-metal complexes. A proposed DFT-supported ligand-field theory (LFDFT) enables one to base the determination of LF parameters solely on DFT calculations from first principle [3–6]. This approach is equally suitable to predict electronic transitions [4] and, with appropriate account of spin-orbit coupling [5], one is able to calculate with satisfactory accuracy g- and A-tensor parameters [6].

In this review we focus on applications of the LFDFT method to complexes of 3d metals and compare these with both spectral data from well documented sources [1, 2] and other theoretical methods, such as the spectroscopically oriented CI (SORCI) method [7] and time-dependent density-functional theory (TDDFT) [8–17]. In particular we intend to emphasize applications to systems with atypical electronic properties of interest for bio-inorganic chemistry and homogeneous catalysis. These systems include macrocyclic amines of Fe in various oxidation states, which are of interest for enzymatic and catalytic reactivity both for redox and bond-breaking reactions.

Under the second topic of '*Ligand-field Theory and its Extensions*' we describe the basic concepts behind the various versions of LFT – the angular-overlap model (AOM) and its extensions. In the section named '*The physical background: conditions for the applicability of the ligand-field approach*' we sketch briefly the theoretical foundation and limits of applicability of the effective-hamiltonian approach with special attention to electronic multiplets. In the 'theory' section, we describe various approaches in current calculations of electronic structure, such as LFDFT, SORCI and TDDFT, with the various 'applications' detailed in the following section, before an outlook for further developments.

LIGAND-FIELD THEORY AND ITS EXTENSIONS

LFT originated as a purely electrostatic model – crystal-field theory (CFT) [18], in which d-electronic multiplets of transition metals are perturbed by ligands as point charges or point dipoles. The CF operator (Equation 1) acts within the space of Slater determinants (SD) composed of purely d-spin-orbitals in which two-electron energies are taken into account with the Coulomb operator and one-electron energies with a crystal-field potential (v_{CF}), the first and second terms in Equation 1, respectively.

$$(1) \qquad H_{CF} = \sum_{i<j} 1/r_{ij} + v_{CF}$$

Explicit use of these operators, involving bond lengths and ligand charges (dipoles) (for v_{LF}) and a basis set for d or f orbitals was made, but the approach was unable to account for the spectrochemical series and the nephelauxetic effect. In two seminal papers [19, 20] Van Vleck showed that metal-ligand covalency yields leading contributions to the multiplet splittings in a transition-metal complex and governs their optical and magnetic properties; he proposed LFT as isomorphous to CFT in which the operators of the LF hamiltonian,

$$(2) \qquad H_{LF} = \sum_{i<j} G(i,j) + v_{LF}$$

are taken as effective, and ligand-field splittings are described in terms of parameters adjusted from experiment. In Equation 2, $G(i,j)$ is an effective, or screened, Coulomb operator, and v_{LF} is an effective ligand-field operator, represented within a basis of d or f orbitals by 5×5 or 7×7 LF matrices, respectively.

From CFT to LFT, one proceeds from free-ion two-electron energies, expressed in terms of Racah parameters B_0 and C_0, to screened parameters $B < B_0$ and $C < C_0$, which take into account the nephelauxetic effect.

In contrast to the more general LFT approach that defines global (i.e. symmetry-adapted) parameters, the angular-overlap model (AOM) [21–24] uses local contributions to the LF potential that arise from interactions of the metal d orbitals with valence orbitals of the individual ligating atoms (ligators). For each properly aligned metal-ligand pair, possessing linear local symmetry, the AOM introduces two d-orbital energy parameters – e_σ for a σ ($dz^2 - p_z$) and e_π for a π ($d_{zx} - p_x$, $d_{yz} - p_y$) bond. Contributions from all ligands in the coordination sphere of the metal are added to yield a 5×5, generally off-diagonal (in the case of low-symmetry), ligand-field matrix V_{ij}, which thus represents an additive potential:

$$(3) \qquad V_{ij} = \sum_{l=1}^{N_L} \sum_{\lambda=\sigma,\pi} F_{\lambda i}(\theta_l, \varphi_l, \psi_l) F_{\lambda j}(\theta_l, \varphi_l, \psi_l) e_{\lambda l}$$

$F_{\lambda i}$ and $F_{\lambda j}$ are described with eulerian angles θ_l, φ_l and ψ_l, accounting for the geometry (position) of each ligand l. A generalization of the AOM by Gerloch et al - the cellular-ligand-field model [25–27] – was based on an assumption that metal-ligand covalent bonding is due mainly to metal 4s, 4p and ligand p and s functions. Thus, according to the CLF one presupposes that 3d orbitals (electrons) are spectators (probes) of the electron density due to the remaining electrons in the bonding and non-bonding orbitals (lone pairs, χ). By spherically averaging the potential V due to this density – $<V>$, one arrives at mean 'd' orbitals and mean 'd'-orbital energies. In a second step, the deviation of V from spherical symmetry, $V' = V - <V>$, serves to define $e_{\lambda l}$ as

$$(4) \qquad e_{\lambda l} \approx\; <d_\lambda|H|d_\lambda> + \sum_\lambda <d_\lambda|H'|\chi_\lambda><\chi_\lambda|H'|d_\lambda> /(\varepsilon_d - \varepsilon_\chi)$$

in which $H = V + T$ and $H' = V' + T$, with T the kinetic energy operator; ε_d and ε_χ are the energies of orbitals d and χ, respectively. The first term (Equation 4), called the static term, is formally analogous to the electrostatic matrix element in CFT. The second term is called the dynamic contribution to $e_{\lambda l}$. Arguments have shown that the dynamic term is larger than the static one, especially when $\lambda = \pi$ relative to $\lambda = \sigma$ [26]. Extensive reviews on the application of CLF theory as a rationale for AOM have been published [1, 27], but calculations following the recipes of the CLF are still lacking. DFT calculations show that metal d orbitals in Werner-type complexes are strongly involved in bonding, in most cases considerably more strongly than the metal 4s and 4p orbitals [28]. Thus, whereas interpretations based on CLF must be corrected because of these new circumstances, the basic message of Equation 4 is still vital, as has been shown by further developments [3].

The Physical Background: Conditions for the Applicability of the Ligand-Field Approach

We focus here on systems comprising atoms of transition metals and ligands, which can be bridging or terminal. In calculations of electronic structure, one directly recognizes antibonding molecular orbitals as being dominated by metal d or f functions, which are partly filled, and bonding orbitals dominated by ligand AO, which are fully occupied. Following Löwdin [29] we write the Schrödinger equation $H\psi = E\psi$ in a discrete representation based on the use of a complete orthonormal set $\Phi = \{\Phi_k\}$, and introduce the hamiltonian matrix $\mathbf{H} = \{H_{kl}\}$ and the column vector $\mathbf{c} = \{c_k\}$ using the relations:

$$(5) \qquad H_{kl} = \langle \Phi_k | H | \Phi_l \rangle, c_k = \langle \Phi_k | \Psi \rangle,$$

$$(6) \qquad \Psi = \sum_k c_k \Phi_k$$

$$(7) \qquad \mathbf{H}\mathbf{c} = E\mathbf{I}\mathbf{c}$$

$\mathbf{I}$ - the identity matrix.

We subdivide the system into two parts, one built from metal nd orbitals (d) and another composed of valence metal $(n+1)$s and $(n+1)$p and ligand functions (v). Then the eigenvalue problem (Equation 7) can be represented in the form of the pseudo-eigenvalue Equation 8, completely restricted to the d-subspace, with the explicit form of $\mathbf{H_{dd'}}'$ given by Equation 9. $\mathbf{H_{dd'}}$ is a $N_d \times N_d$ matrix. Thus, for a d^2 system, for instance, $N_d = 45$. $\mathbf{H_{dv}}$ ($\mathbf{H_{vd}}$) are rectangular $N_d \times N_v (N_v \times N_d)$ matrices, where N_v is, in principle, infinite. No approximation is inherent in Equations 8–10. Solution of the secular determinant equation (Equation 10) yields N_d eigenvalues of Equation 7 with eigenvectors expressed as (finite) combinations of the sub-basis $\Phi = \{\Phi_k\}$. The representation given by Equations 8–10

(8) $\mathbf{H'_{dd'}c_d} = E\mathbf{I_d c_d}$

(9) $\mathbf{H'_{dd'}} = \mathbf{H_{dd'}} + \mathbf{H_{dv}}(E\mathbf{I_v} - \mathbf{H_{vv'}})^{-1}\mathbf{H_{vd}}$

(10) $|\mathbf{H'_{dd'}} - E\mathbf{I_{dd'}}| = 0$

provides the physical background of ligand-field theory. The matrix $\mathbf{H_{dd'}}$ represents the purely electrostatic effect on the metal d orbitals by the surrounding ligand nuclei and the valence-electron distribution excluding the d electrons, subject to the conventional description in terms of a *crystal-field theory* applied to TM impurities in crystals. As orbitals of subsystems d and v are orthogonal to each other, the $\mathbf{H_{dd'}}$ matrix also incorporates important exchange (Pauli) repulsion terms, which have been shown to be proportional to squares of the corresponding overlap integrals, allowing formulation of the ligand field as a pseudopotential [30, 31]. The second term in Equation 9 is energy dependent. For diagonal $\mathbf{H_{dd'}}'$, perturbation expansions (presupposing that $|H_{vv}-H_{dd}| >> |H_{vd}|$) allow one to identify E with the corresponding diagonal element of $\mathbf{H_{dd'}}$. The second term in Equation 9 reflects metal-ligand covalency (charge transfer) and is the subject of parameterization according to the angular overlap model [2]. Earlier analysis based on Equation 9 (Equation 4) was purely theoretical, attempting to place a correct context and limits of applicability of the ligand-field approach within the main body of quantum chemistry [25]. In the section titled *The LFDFT – an attempt to revive LFT*, we describe a practical scheme to deduce the matrix $\mathbf{H_{dd'}}'$ from DFT and to apply it directly to the calculation of d^n-electronic multiplets.

The effective hamiltonian representation of equations [8–10] is readily extensible to systems with more than one TM. The $\mathbf{H'_{dd'}}$ matrix for such cases contains terms that account for d-electron delocalization (via the second term in Equation 9) from one metal to another – indirectly via the bridging ligands or directly via the corresponding off-diagonal terms of $\mathbf{H_{dd'}}$ [33]. This interaction produces magnetic-exchange coupling, which is reviewed elsewhere [34, 35].

Equation 9 is applicable both within a sub-space of one-electron states (orbitals) and to the many-electron states resulting from the redistribution of all d electrons within the active d-orbital subspace. Their treatment requires both two-electron repulsion and one-electron integrals. The general scheme described below allows one to deduce also these integrals from DFT.

Theory

We present three new developments that we compare critically against experimental data – LFDFT [3, 4, 36, 40], SORCI [7] and TDDFT for spin-open-shell transition-metal complexes [14].

The LFDFT – an attempt to revive LFT

According to the third topic of this article, '*The physical background: conditions for the applicability of the ligand-field approach*' one can in principle reduce the complicated many-electron problem of a TM complex to the subspace of configurations built from electron replacements within MO dominated by TM d functions without approximation. As follows from Equation 9, the general matrix element for such a hamiltonian differs, however, for each electronic state, being dependent on energy (E). If d-d electron states are well separated from charge-transfer states one can substitute for a given d-electronic state (i) the variable energy E in Equation 9 by its diagonal element (H_{didi}) and use perturbation theory to various orders. Being parametric, LFT implies a separate parameter set for each state within the LF domain, but it is an intrinsic feature of LFT to use the *same set of parameters* – B, C and the 5×5 LF matrix – for each state belonging to the manifold of a given d^n configuration. Accordingly, averaging of one and two-electron matrix elements must be conducted beforehand to make this condition possible. A second feature of LFT is that, different from SCF theories – HF or Kohn-Sham (KS) DFT, electron repulsion between d electrons is treated explicitly, implying the need to define and to calculate average d orbitals in a common set in terms of which matrix elements of both $G(i,j)$ and v_{LF} (Equation 2) become evaluated. There occurs in LFT no such evaluation, but parameters are introduced that must be fitted from experiment. In principle, having common orbitals, matrix elements of $G(i,j)$ and v_{LF} can be explicitly calculated and used in a further step for configuration interaction within the subspace of SD belonging to the d^n configuration (complete active space, CAS). One can then obtain expectation values of all operators, including vibronic coupling, but this result has not yet been achieved. Instead, we proposed a general and user-oriented recipe [2, 3] based on DFT, comprising the following steps. Rather than presenting a general theory (see [3, 4] for details), we take as an example tetrahedral FeO_4^{2-} with Fe^{VI} in a d^2 configuration. The geometry adopted in such a calculation is crucial. As the method is based on DFT, it is reasonable to take a DFT-optimized geometry, but GGA functionals that are more suitable for energy calculations yield TM—L bonds that are too long compared with experiment. We thus recommend use of the simple LDA-VWN functional when optimizing the geometries of such complexes, or one can simply adopt the experimental geometry. In a second step, one identifies from a preliminary DFT calculation the MO dominated by d orbitals that we call here LF orbitals (LFO). This step can be immediately done, and, if 'd MO' are recognizable as being prevailed by d functions, one can proceed. 'd orbitals' must be well separated from ligand orbitals for LFT to work, implying that, beginning with an ionic d^n configuration, the intermixing of CT states is weak. A spin-restricted (with same orbitals for different spins) SCF-DFT calculation of the average d^n configuration (AOC) is made, on specifying an equal occupation $n/5$ on each LFO. The KS orbitals using this AOC are best suited for a treatment in which interelectronic repulsion is – as in LFT – approximated by screened atomic-like Racah parameters B and C for spherical symmetry; we thus assign any energetic effect due to deviation of the LF orbitals from this symmetry to contributions to the one-electron matrix $\mathbf{v_{LF}} = \{h_{\mu\nu}\}$ (μ, $\nu = 1$ to 5). From these LFO, the energies of all 45 SD are calculated on maintaining the electron density frozen (without SCF, spin-unrestricted). This recipe is also consistent with the prerequisites of the LF approach; here orbital relaxation is taken into account only at the level of averaging the electron density to provide proper LF orbitals, whereas all SD energies for subsequent LF treatment are calculated without SCF iterations. In this respect the proposed procedure differs from a MCSCF or CASSCF Ansatz. The energies of all 45 SD for the bare FeO_4^{2-} are listed in Table 19-1, with expressions for the energy of every SD in terms of parameters B, C and $10 Dq$. We thus obtain a system of 45 linear equations with three unknown parameters. An arbitrary shift of all DFT energies compared to the LF ones for each SD energy must be considered because of varied gauge origins. This shift is effected on introducing an additional regression coefficient to yield a system of 45 over-determined linear equations (Equation 11), in which $\mathbf{X}$ stores LF parameters, $\mathbf{E}$ the energies of the SD, and $\mathbf{A}$ the coefficients of the linear relation between $\mathbf{E}$ and $\mathbf{X}$. This system is solved in a least-square sense (Equation 12) to obtain B, C and $10 Dq$. For the bare FeO_4^{2-} anion we obtain $B = 285$, $C = 1533$ and $10 Dq = 11369$ cm^{-1}. A comparison

(11) $$\mathbf{E} = \mathbf{AX}$$

(12) $$\mathbf{X} = (\mathbf{A^T A})^{-1} \mathbf{A^T E}$$

between the SD energies calculated using these values of B, C and $10 Dq$ with DFT data shows a remarkable consistency between the DFT formalism and the LFT parameterization (Figure 19-1). This comparison also shows that $10 Dq$ obtained in this way is near (within an error less than 2%) the KS orbital energy difference $\varepsilon^{KS}(t_2) - \varepsilon^{KS}(e)$.

When all LF parameters are evaluated, they can be introduced into a favoured LF program [37–39] to yield all multiplet energies and expectation values of all operators for comparison with experiment. In the case of d^2 FeO_4^{2-} with tetrahedral symmetry, energy matrices can be written explicitly (Table 19-2); the role of single excitation (for the $^1T_2$ and $^3T_1$ terms) and double excitations (for $^1A_1$, ^1E, $^1T_2$ and $^3T_1$) is important – we return to this point when we look at applications later in the article. According to this procedure, both dynamical correlation (via the DFT

Table 19-1. DFT values for energies/eV of all 45 Slater determinants (SD) due to the d^2 configuration of Fe in tetrahedral FeO_4^{2-} and their ligand-field expressions in terms of interelectronic repulsion B and C and the cubic ligand-field splitting parameter $10\,Dq$

Energy	SD	expression	Energy	SD	expression	Energy	SD	expression
−1.20944	$z2^+\ z2^-$	$12B+3C$	−0.55996	$z2^-\ xy^+$	$4B+C+10Dq$	1.56350	$yz^+\ yz^-$	$12B+3C+20Dq$
−2.25626	$z2^+\ x2\text{-}y2^+$	0	−0.83723	$z2^-\ xy^-$	$10Dq$	0.66546	$yz^+\ xz^+$	$3B+20Dq$
−2.02183	$z2^+\ x2\text{-}y2^-$	$4B+C$	−1.21542	$x2\text{-}y2^+\ x2\text{-}y2^-$	$12B+3C$	0.92390	$yz^+\ xz^-$	$6B+C+20Dq$
−0.52452	$z2^+\ yz^+$	$9B+10Dq$	−0.73462	$x2\text{-}y2^+\ yz^+$	$3B+10Dq$	0.66546	$yz^+\ xy^+$	$3B+20Dq$
−0.35645	$z2^+\ yz^-$	$10B+C+10Dq$	−0.49523	$x2\text{-}y2^+\ yz^-$	$6B+C+10Dq$	0.92390	$yz^+\ xy^-$	$6B+C+20Dq$
−0.52452	$z2^+\ xz^+$	$9B+10Dq$	−0.73462	$x2\text{-}y2^+\ xz^+$	$3B+10Dq$	0.92390	$yz^-\ xz^+$	$6B+C+20Dq$
−0.35645	$z2^+\ xz^-$	$10B+C+10Dq$	−0.49523	$x2\text{-}y2^+\ xz^-$	$6B+C+10Dq$	0.66546	$yz^-\ xz^-$	$3B+20Dq$
−0.83723	$z2^+\ xy^+$	$10Dq$	−0.42621	$x2\text{-}y2^+\ xy^+$	$12B+10Dq$	0.92390	$yz^-\ xy^+$	$6B+C+20Dq$
−0.55996	$z2^+\ xy^-$	$4B+C+10Dq$	−0.29169	$x2\text{-}y2^+\ xy^-$	$12B+C+10Dq$	0.66546	$yz^-\ xy^-$	$3B+20Dq$
−2.02183	$z2^-\ x2\text{-}y2^+$	$4B+C$	−0.49523	$x2\text{-}y2^-\ yz^+$	$6B+C+10Dq$	1.56350	$xz^+\ xz^-$	$12B+3C+20Dq$
−2.25626	$z2^-\ x2\text{-}y2^-$	0	−0.73462	$x2\text{-}y2^-\ yz^-$	$3B+10Dq$	0.66546	$xz^+\ xy^+$	$3B+20Dq$
−0.35645	$z2^-\ yz^+$	$10B+C+10Dq$	−0.49524	$x2\text{-}y2^-\ xz^+$	$6B+C+10Dq$	0.92390	$xz^+\ xy^-$	$6B+C+20Dq$
−0.52452	$z2^-\ yz^-$	$9B+10Dq$	−0.73463	$x2\text{-}y2^-\ xz^-$	$3B+10Dq$	0.92390	$xz^-\ xy^+$	$6B+C+20Dq$
−0.35645	$z2^-\ xz^+$	$10B+C+10Dq$	−0.29169	$x2\text{-}y2^-\ xy^+$	$12B+C+10Dq$	0.66546	$xz^-\ xy^-$	$3B+20Dq$
−0.52452	$z2^-\ xz^-$	$9B+10Dq$	−0.42622	$x2\text{-}y2^-\ xy^-$	$12B+10Dq$	1.56350	$xy^+\ xy^-$	$12B+3C+20Dq$

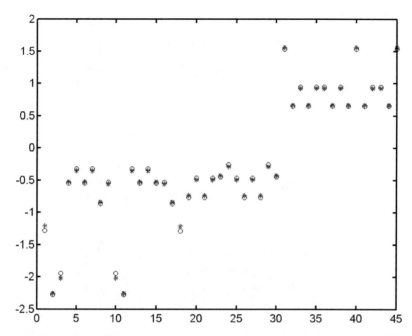

Figure 19-1. Calculated (DFT-LDA, stars) and reproduced (using best fit $B = 285\,cm^{-1}$; $C = 1533\,cm^{-1}$ and $10\ Dq = 11369\,cm^{-1}$, circles) energies of all 45 Slater determinants of bare tetrahedral FeO_4^{2-} [R(Fe—O) = 1.675 Å]

Table 19-2. Matrices for the energies of multiplets of a tetrahedral d^2 complex[a]

d^2

$^1A_1(^1G,^1S)$

$$\begin{array}{cc} t_2^2 & e^2 \\ \begin{bmatrix} 10B+5C+20Dq & \sqrt{6}(2B+C) \\ \sqrt{6}(2B+C) & 8B+4C \end{bmatrix} \end{array}$$

$^1E\ (^1D,^1G)$

$$\begin{array}{cc} t_2^2 & e^2 \\ \begin{bmatrix} B+2C+20Dq & -2\sqrt{3}B \\ -2\sqrt{3}B & 2C \end{bmatrix} \end{array}$$

$^1T_2(^1D,^1G)$

$$\begin{array}{cc} t_2^2 & t_2e \\ \begin{bmatrix} B+2C+20Dq & 2\sqrt{3}B \\ 2\sqrt{3}B & 2C+10Dq \end{bmatrix} \end{array}$$

$^3T_1\ (^3F,^3P)$

$$\begin{array}{cc} t_2^2 & t_2e \\ \begin{bmatrix} -5B+20Dq & 6B \\ 6B & 4B+10Dq \end{bmatrix} \end{array}$$

$t^2e\ ^1T_1(^1G)\ 4B+2C+10Dq$
$t^2e\ ^3T_2(^3F)\ -8B+10Dq$
$e^2\ \ ^3A_2(^3F)\ -8B$

[a] Multiplets of transition-metal ions in crystals, S.Sugano,Y.Tanabe, H.Kamimura, Academic Press, NY USA, 1970; J.S.Griffith, The Theory of Transition-Metal Ions, Cambridge, University Press, 1971.

functional) and non-dynamical correlation (via CI) are taken into account. Results for FeO_4^{2-} are listed in Table 19-3.

We consider the general case of a d^n complex without symmetry. For such a complex the LF matrix is off-diagonal with 15 independent matrix elements and the values of B and C. These quantities are evaluated as follows. The energies of SD from the DFT calculation, $E(\Phi_I)$ $(I = 1, \ldots, N_{SD})$, are given by

$$(13) \quad E(\Phi_I^{KS}) = (\Phi_I^{KS}|H|\Phi_I^{KS}) = a_{I1}\varepsilon_1^{LFO} + a_{I2}\varepsilon_2^{LFO} + a_{I3}\varepsilon_3^{LFO} + a_{I4}\varepsilon_4^{LFO}$$
$$+ a_{I5}\varepsilon_5^{LFO} + a_{I6}B + a_{I7}C + a_{I8}$$

The energies of LFO, denoted by ε_i^{LFO} $(i = 1, \ldots 5)$, B, C, and the parameter a_{I8} absorbing the arbitrariness of the gauge origin are obtained from a least-square fit (Equations 11, 12) as already described. To obtain the 5×5 matrix $\mathbf{v}_{LF}$ we use the fact that ε_i^{LFO} are near the corresponding KS-orbital energies ε_i^{KS} $(i = 1, \ldots 5)$. We then use the corresponding KS eigenvectors to reconstruct matrix $\mathbf{v}_{LF}$ from ε_i^{LFO} $(i = 1, \ldots, 5)$. From the components of the eigenvector matrix $\mathbf{V}$ (Equation 14) built from such columns one takes only the components corresponding to the d functions. We denote the square matrix composed of these column vectors by $\mathbf{U}$. As $\mathbf{U}$ is neither normalized nor orthogonal, we introduce an overlap matrix $\mathbf{S}$ defined in Equation 15 and use Löwdin's symmetric orthogonalization scheme to

Table 19-3. Comparison between theoretical and experimental wavenumbers/cm^{-1} of d-d transitions of FeO_4^{2-}[a]

Transition	LFDFT		SORCI[d]		TDDFT		exp.
	bare[b]	solv.[c]	bare	solv.	LDA-VWN	SAOP	
$^3A_2 \rightarrow {}^1E$	5304	5227	7678	7822	4904	9484	6215[e]
$^3A_2 \rightarrow {}^1A_1$	9666	9473	11548	11512	8888	13041	9118[e]
$^3A_2 \rightarrow {}^3T_2$	11369	11365	11505	11368	12412	14872	12940[e]
$^3A_2 \rightarrow {}^3T_1$	14468	14145	13918	13782	13493	15735	17700[e]
MUSE	1566	1618	2278	2363	1569	2772	–
MUSE($^3A_2 \rightarrow {}^3\Gamma$)	2402	2565	2608	2745	2368	1948	–
MUSE($^3A_2 \rightarrow {}^1\Gamma$)	730	672	1946	1981	770	3596	–

[a] R(Fe—O) = 1.675 Å from LDA (COSMO) geometry optimization.
[b] B = 285; C = 1533, 10Dq = 11369 cm^{-1}.
[c] B = 252; C = 1622, 10Dq = 11365 cm^{-1}.
[d] CAS(6,5) reference; a fit of B,C and 10Dq to this values yields B = 204(202), C = 2703(2731), 10Dq = 11505(11368) cm^{-1} without(with) solvent.
[e] values taken from Brunold TC, Hauser A, Guedel HU (1994) J.Lumin. 59:321.

obtain an equivalent set of orthogonal eigenvectors (**C**, Equation 16). We identify these vectors with eigenfunctions of the effective LF

$$
\begin{array}{ccccc}
\varepsilon_1{}^{KS} & \varepsilon_2{}^{KS} & \varepsilon_3{}^{KS} & \varepsilon_4{}^{KS} & \varepsilon_5{}^{KS} \\
\cdots & \cdots & \cdots & \cdots & \cdots
\end{array}
$$

$$
(14) \qquad V = \begin{array}{c} d_{xy} \\ d_{yz} \\ d_{z^2} \\ d_{xz} \\ d_{x2-y2} \end{array}
\begin{bmatrix}
U_{11} & U_{12} & U_{13} & U_{14} & U_{15} \\
U_{21} & U_{22} & U_{23} & U_{24} & U_{25} \\
U_{31} & U_{32} & U_{33} & U_{34} & U_{35} \\
U_{41} & U_{42} & U_{43} & U_{44} & U_{45} \\
U_{51} & U_{52} & U_{53} & U_{54} & U_{55}
\end{bmatrix}
$$

$$
\begin{array}{ccccc}
\cdots & \cdots & \cdots & \cdots & \cdots
\end{array}
$$

$$
(15) \qquad \mathbf{S = U^T . U}
$$

$$
(16) \qquad \mathbf{C = US^{-1/2}}
$$

hamiltonian $\mathbf{v_{LF}}$ that we seek as

$$
(17) \qquad \varphi_i = \sum_{\mu=1}^{5} c_{\mu i} d_\mu \ (i = 1\text{--}5)
$$

and the corresponding eigenvalues $(\varphi_i|V_{LF}|\varphi_i)$ with ε_i^{LFO} $(i = 1, \ldots 5)$:

$$
(18) \qquad \varepsilon_i^{LFO} = (\varphi_i|v_{LF}|\varphi_i)
$$

The 5×5 matrix $\mathbf{V_{LF}}$ is then given by

$$
(19) \qquad \mathbf{V_{LF} = C.E.C^T} = \{v_{\mu\nu}\} = \Big\{\sum_i c_{\mu i}\varepsilon_i^{LFO} c_{\nu i}\Big\}
$$

E is the diagonal matrix built from ε_i^{LFO}; $\mathbf{E} = \mathrm{diag}\{\varepsilon_i^{KS}\}$. The matrix V_{LF} is obtained in a general form with no assumption (such as in CFT or AOM); it is particularly suitable in cases in which application of CFT or AOM is hindered because of the high level of parameterization.

Within the same formalism we have shown how to derive spin-orbit coupling energies [5], g-tensors, hyperfine-coupling tensors [6] and zero-field splittings [40].

The LFDFT formalism has been implemented with the aid of the Amsterdam density-functional (ADF) package [41] that allows one to define precisely orbital occupation and thus to calculate all energies needed. Diverse exchange-correlation functionals have been used and tested. Numerical details are given elsewhere [3–6] (see also 'Applications').

The SORCI method

The SORCI method has been developed in Refs. [7, 42] and a detailed description together with an evaluation of its performance for the prediction of d-d multiplets in transition metal hexaquo complexes has been provided in ref [3][a]. Briefly, SORCI is a simplified multireference *ab initio* CI variant which combines the concepts of variation/perturbation theory, difference dedicated CI [43, 44], individual selection [45] and natural orbitals in an attempt to provide a balanced and unbiased description of a number of low-lying excited states of interest. Here we provide additional evidence that this is indeed the case by computing the multiplet energies of a number of anionic transition metal complexes. The results for cationic, octahedral hexaquo complexes have been quite encouraging with average errors in the transition energies of $2000 \, cm^{-1}$ and less [3].[a]

All computations in this section have been performed using the ORCA program package [46], a DFT BP86 functional and a basis set of triple zeta quality (TZV(P)). In all MRCI computations the T_{sel}, T_{pre} and T_{nat} tresholds were set to 10^{-6}, 10^{-4} and 10^{-5} Eh, respectively which are the default values of the method. A CAS(n,5) has been chosen in all cases, unless otherwise specified. This choice is motivated, of course, by ligand field theory and provides a suitable model space over which the many-particle wavefunctions are expanded.

TDDFT for open-shell systems

TDDFT [47], [8] has had great success in calculation of excitation energies in closed-shell systems in many organic and inorganic molecules [10, 11, 48, 54] and even in molecules containing transition metals [54–56]. Except for systems with long-range charge-transfer excitations [57], it has been efficient and satisfactorily accurate. Its extension to open-shell systems developed [58–60] to include spin [14] and space [17] open shells in transition-metal complexes. The theorem of Runge and Gross [47] provided a theoretical basis of TDDFT on considering the time evolution of electron density $\rho(\mathbf{r}_1,t)$ under the influence of an external potential $\nu_{ext}(\mathbf{r}_1,t)$. Given a wave function $\Psi(\mathbf{x}_1,\mathbf{x}_2, \ldots ,\mathbf{x}_N,t)$, $\mathbf{x}_i=[\mathbf{r}_i,\mathbf{s}_i]$ at time $t = 0$, at all subsequent times $t > 0$ the density $\rho(\mathbf{r}_1,t)$ determines the potential uniquely up to an additive function that depends purely on time. In turn $\nu_{ext}(\mathbf{r}_1,t)$ uniquely determines $\Psi(\mathbf{x}_1,\mathbf{x}_2, \ldots ,\mathbf{x}_N,t)$, which can therefore be considered to be a functional of the time dependent density $\Psi([\rho],t)$. Hence a unique mapping is established

$$(20) \qquad \rho(\mathbf{r}_1, t) \Leftrightarrow \nu_{ext}(\mathbf{r}_1, t) \Leftrightarrow \Psi(\mathbf{x}_1, \mathbf{x}_2, \ldots , \mathbf{x}_N, t)$$

Most implementations of TDDFT to an open-shell system use an spin-unrestricted approach, because orbital-energy differences concerned with partially occupied orbitals are generally too small in a spin-restricted approach, and the orbital-energy difference in DFT is the leading term in the electron-excitation energy.

The master equations of the TDDFT method for open-shells follow (see [13] for a derivation):

(21) $\Omega F_I = \omega_I^2 F_I$

(22) $\Omega_{ia\sigma, jb\tau} = \delta_{\sigma\tau}\delta_{ij}\delta_{ab}(\varepsilon_{a\sigma} - \varepsilon_{i\sigma})^2 + 2\sqrt{\varepsilon_{a\sigma} - \varepsilon_{i\sigma}}K_{ia\sigma, jb\tau}\sqrt{\varepsilon_{b\tau} - \varepsilon_{j\tau}}$

(23) $K_{ia\sigma, jb\tau} = K_{ai\sigma, bj\tau}^{Coul} + K_{ai\sigma, bj\tau}^{XC}$

(24) $K_{ai\sigma, bj\tau}^{Coul} = \int d\mathbf{r} \int d\mathbf{r}' \psi_{a\sigma}^*(\mathbf{r})\psi_{i\sigma}(\mathbf{r})\frac{1}{|\mathbf{r}-\mathbf{r}'|}\psi_{b\tau}^*(\mathbf{r}')\psi_{j\tau}(\mathbf{r}')$

(25) $K_{ai\sigma, bj\tau}^{XC} = \int d\mathbf{r} \int d\mathbf{r}' \psi_{a\sigma}^*(\mathbf{r})\psi_{i\sigma}(\mathbf{r})f_{XC}^{\sigma\tau}(\mathbf{r}, \mathbf{r}', \omega)\delta(\mathbf{r} - \mathbf{r}')\psi_{b\tau}^*(\mathbf{r}')\psi_{j\tau}(\mathbf{r}')$

with $f_{XC}^{\sigma\tau}(\mathbf{r}, \mathbf{r}', \omega)$ denoting the exchange-correlation kernel :

(26) $f_{XC}^{\sigma\tau}(\mathbf{r}, \mathbf{r}', \omega) = \frac{\delta v_{XC}^\sigma(\mathbf{r}, \omega)}{\delta \rho_\tau(\mathbf{r}', \omega)} = \frac{\delta^2 E_{XC}}{\delta \rho_\sigma(\mathbf{r}, \omega)\delta \rho_\tau(\mathbf{r}', \omega)}$

Within the adiabatic approximation $f_{XC}^{\sigma\tau}(\mathbf{r}, \mathbf{r}', \omega)$ is substituted by the frequency-independent derivative $f_{XC}^{\sigma\tau}(\mathbf{r}) = \frac{\delta v_{XC}^\sigma(\mathbf{r})}{\delta \rho_\tau(\mathbf{r}')}$; in Equations 22–26, a,b and i,j are indices for virtual and occupied orbitals, respectively, σ, τ denote the spin, and ε and ψ are KS orbital energies and MO. Spin-unrestricted TDDFT has been implemented in the response module of the ADF program package [41, 61], consistent with earlier implementations for closed systems. The Davidson algorithm [62] serves to calculate the lowest excitation energies as the lowest eigenvalues ω^2 of Equation 21. ADF makes full use of symmetry, decreasing significantly the computational effort and assisting an interpretation of the results. In the calculation of excitation energies, the adiabatic LDA (ALDA) approach has been used with the LDA-VWN functional and, alternatively, of the *statistical average of orbital potentials* (SAOP) model potential. In calculating energies of spin-flip transitions we applied the Tamm-Dancoff approximation [52]. Note that the equations given above are only valid in the case that no nonlocal potentials enter into the DFT equations. In the presence of, e.g. Hartree-Fock exchange, the equations to be solved are somewhat more complex [11][b] but can be approached along the same lines.

Applications

Classical coordination compounds

Electronic transitions in FeO_4^{2-}   Tetrahedral FeO_4^{2-} is characterized by a e^2 ground-state configuration producing $^3A_2$ (the ground state) and ^1E and $^1A_1$(spin-flip) excited states. These states interact with states of the same symmetry stemming from the t_2^2 doubly excited configuration. The singly excited $e^1 t_2^1$ configuration gives rise to $^3T_2$ and $^3T_1$ states, and the latter can interact with one more $^3T_1$ state from the doubly excited t_2^2 configuration. The electronic spectrum (Table 19-3) displays two sharp lines at 6215 and 9118 cm^{-1} corresponding to the ^1E and $^1A_1$ transitions from the ground state and two broad bands centered at 12940 and 17700 cm^{-1} due to spin-allowed transitions to the $^3T_2$ and $^3T_1$ excited states. All these features are well reproduced by LFDFT for both bare and charge-compensated species, and the same is valid for the SORCI method; mean unsigned errors (MUSE given separately for the spin-allowed, for spin-forbidden and for both in Table 19-3) are comparable for the two methods, being slightly in favor of the LFDFT result. There is a bias for the LFDFT and SORCI methods, correspondingly to underestimate (overestimate) energies of spin-forbidden transitions. This effect is easily understood in terms of LFT showing (cf. Table 19-2) a stronger dependence of these transitions on B and particularly on C (see Table 19-2); the latter becomes calculated smaller in DFT but greater in SORCI relative to the value obtained using a direct fit of LFT expressions (Table 19-2) to the spectrum ($B = 375$, $C = 1388$ cm^{-1}; compare with Table 19-3). In the case of SORCI the main reason for the overestimation must be blamed on basis set deficiencies. Since the low-spin states have, in general, larger contributions from dynamic electron correlation it would be necessary to use much more extensively polarized basis sets in order to obtain more accurate results than the ones presented here. Nevertheless, the agreement one obtains with experiment, even with these very small basis sets used here and that can easily be used in studies on larger molecules, might be considered as remarkable and is associated with the focus of the method on energy differences rather than total energies. For the sake of comparison in Table 19-3, we included TDDFT values for the four transition wavenumbers using two functionals. The comparison with experiment is worse than for LFDFT and SORCI, but still acceptable, in particular when using the simple LDA-VWN functional, although large deviations for the spin-forbidden transitions are encountered with the SAOP functional.

Transition-metal cyanide complexes LFDFT (LDA-VWN) calculations of electronic transitions of red $Fe(CN)_6^{3-}$ and yellow $Co(CN)_6^{3-}$ have been published [4]. Here we reproduce the results for $Co(CN)_6^{3-}$ and compare these from SORCI and TDDFT in Table 19-4. From the t_{2g}^6 closed-shell $^1A_{1g}$ ground state, the spin-allowed $t_{2g}^6 \rightarrow t_{2g}^5 e_g^1$ transitions to $^1T_{1g}$ and $^1T_{2g}$ are observed at 31950 and 38600 cm^{-1}, respectively. SORCI impressively reproduces the wavenumbers of these transitions, but LFDFT less satisfactorily, the values of 10 Dq and the transition wavenumber to $^3T_{1g}$ being exaggerated by about 4000–5000 cm^{-1}. The wavenumber of the latter transition is less by 2254 cm^{-1} in SORCI than from experiment, but there is a large discrepancy with experiment in the TDDFT calculation; deviations are extreme – as large as 10097 cm^{-1} for the $^1A_{1g} \rightarrow {}^3T_{2g}$ transition – when using the SAOP functional.

Table 19-4. Comparison between theoretical and experimental wavenumbers/cm^{-1} of d-d transitions of $Co(CN)_6^{3-a}$

Transition	LFDFT[b]	SORCI[c]	TDDFT		exp.
			LDA-VWN	SAOP	
$^1A_{1g} \rightarrow {}^1T_{1g}$	35107	31014	34792	39317	31950[d]
$^1A_{1g} \rightarrow {}^1T_{2g}$	40900	38874	37333	41680	38600[d]
$^1A_{1g} \rightarrow {}^3T_{1g}$	29845	23716	31542	36067	25970[e]
$^1A_{1g} \rightarrow {}^3T_{2g}$	32721	28191	31615	36123	–
MUSE	3110	1155	3227	5224	
MUSE($^1\Gamma \rightarrow {}^1\Gamma$)	2728	605	2054	6848	
MUSE($^1\Gamma \rightarrow {}^3\Gamma$)	3875	2254	5572	10097	

[a] $R(Co-C) = 1.899$ Å $R(C-N) = 1.172$ Å in all calculations.
[b] $B = 387$; $C = 2573$, $10\ Dq = 37180\ cm^{-1}$.
[c] CAS(6,5) reference.
[d] values adopted from Alexander J J, Gray H B (1968) J.Am. Chem. Soc. 90: 4260 as averages over band energy maxima measured in $K_3Co(CN)_6$ (32100, 38500 cm^{-1}) and n-$Bu_4NCo(CN)_6$ (31800, 38700).
[e] value reported from Miskowski, VM; Gray HB, Wilson RB, Solomon EI, Inorg.Chem. (1979), 18, 1410.

Electronic transitions in CrF_6^{3-} and $CrCl_6^{3-}$ Optical spectra of CrF_6^{3-} and $CrCl_6^{3-}$ anions with a $^4A_{2g}(t_{2g}^3)$ ground-state configuration are characterized by broad bands due to spin-allowed transitions to the $^4T_{2g}$ and $^4T_{1g}$ $(t_{2g}^2e_g^1)$ and the $^4T_{1g}(t_{2g}e_g^2)$ excited states, and sharp lines due to spin-forbidden (spin-flip) transitions – ^2E_g, $^2T_{1g}$ and $^2T_{2g}$ belonging to the ground-state configuration. Experimental transition wavenumbers as deduced from band maxima and wavenumbers of sharp lines for CrF_6^{3-} and $CrCl_6^{3-}$ are listed in Tables 19-5 and 19-6, respectively, with computed values. Focusing on the LFDFT data, we notice a large increase of the wavenumber of $^4A_{2g} \rightarrow {}^4T_{2g}$ from a bare to a charge-compensated CrF_6^{3-} species, thus improving the agreement with experiment. This effect is less pronounced for the less ionic $CrCl_6^{3-}$ complex. Not only $10\ Dq$ but also the values of B and to a less extent C are affected by the solvent. As follows from MUSE values, both LFDFT and SORCI give similar results, but deviations of spin-forbidden transitions deviate in opposite directions relative to experiment, being smaller and larger for the two methods respectively. The same comments concerning the basis set as in the case of FeO_4^{2-} apply here. For the TDDFT results we state a total failure to describe the spectrum of CrF_6^{3-} and, to a lesser extent, for $CrCl_6^{3-}$ - wavenumbers of spin-allowed transitions are calculated here much greater than experimental values; TDDFT predicts only one of the two $^4A_{2g} \rightarrow {}^4T_{1g}$ transitions at roughly half the transition wavenumber between the $^4T_{1g}(a)$ and $^4T_{1g}(b)$ states. The underlying reason is that these two transitions correspond to nominal $t_{2g}^3 \rightarrow t_{2g}^2e_g^1$ and $t_{2g}^3 \rightarrow t_{2g}^1e_g^2$ excitations, and strongly admix via Coulomb-repulsion terms. The second transition corresponds to a double excitation from the ground state and is ignored with TDDFT.

Table 19-5. Comparison between theoretical and experimental wavenumbers/cm^{-1} of d-d transitions of CrF$_6^{3-}$ [a]

Transition	LFDFT[a]		SORCI[d]	TDDFT[e]	exp.[f]
	bare[b]	solvated[c]			
$^4A_{2g}(t_{2g}^3)$	0	0	0	0	0
$^4T_{2g}(t_{2g}^2e_g^1)$	13569	15480	15629	25550	15200
$^4T_{1g}(t_{2g}^2e_g^1)$	19443	21949	22338	27953	21800
$^4T_{1g}(t_{2g}^1e_g^2)$	30339	34346	35105	–	35000
$^2E_g(t_{2g}^3)$	12497	13002	18254	34720	16300
$^2T_{1g}(t_{2g}^3)$	13044	13601	18988	34792	16300
$^2T_{2g}(t_{2g}^3)$	18628	19568	25392	42656	23000
MUSE(total)	3347	1752	1351	14614	–
MUSE($^4A_{2g} \rightarrow {}^4\Gamma_g$)	2883	361	357	8252	–
MUSE($^4A_{2g} \rightarrow {}^2\Gamma_g$)	3810	3143	2345	18856	–

[a] R(Cr—F) = 1.957 Å from LDA (COSMO) geometry optimization.
[b] B = 605; C = 2694, 10 Dq = 13569 cm^{-1}.
[c] B = 657, C = 2731, 10 Dq = 15480 cm^{-1}.
[d] CAS(6,5) reference, COSMO solvent model.
[e] SAOP functional, all-electron calculation, COSMO; spin-flip transitions have been calculated within the Tamm-Dancoff approximation.
[f] taken from Allen GC; El-Sharkawy AM, Warren KD (1971) Inorg.Chem. 10:2538.

Table 19-6. Comparison between theoretical and experimental d-d transition wavenumbers/cm^{-1} of CrCl$_6^{3-}$ [a]

Transition	LFDFT[a]		SORCI[d]	TDDFT[e] (SAOP)	exp.[f]
	bare[b]	solvated[c]			
$^4A_{2g}(t_{2g}^3)$	0	0	0	0	0
$^4T_{2g}(t_{2g}^2e_g^1)$	13023	13599	14054	21154	12800
$^4T_{1g}(t_{2g}^2e_g^1)$	18018	18480	19366	22517	18200
$^4T_{1g}(t_{2g}^1e_g^2)$	28445	29412	31834	–	~ 29000
$^2E_g(t_{2g}^3)$	10887	10790	17224	12420	14430
$^2T_{1g}(t_{2g}^3)$	11308	11182	17671	12460	15010
$^2T_{2g}(t_{2g}^3)$	16402	16345	23000	19316	20600
MUSE(total)	320	497	2185	3703	–
MUSE($^4A_{2g} \rightarrow {}^4\Gamma_g$)	3814	3908	1751	6336	–
MUSE($^4A_{2g} \rightarrow {}^2\Gamma_g$)	2067	2202	2618	1948	–

[a] R(Cr—Cl) = 2.335 Å – from a LDA(COSMO) geometry optimization.
[b] B = 493; C = 2406, 10 Dq = 13022 cm^{-1}.
[c] B = 473, C = 2412, 10 Dq = 13600 cm^{-1}.
[d] CAS(6,5) reference, COSMO solvent model.
[e] SAOP functional, all-electron calculation, COSMO; spin-flip transitions have been calculated within the Tamm-Dancoff approximation.
[f] taken from Schwartz RW (1976) Inorg.Chem. 15:2817.

Table 19-7. Comparison between wavenumbers/cm^{-1} of d-d transitions of low-spin Co(H$_2$O)$_6^{3+}$ calculated with various approaches (bare cation, and explicit solvent, LFDFT, see Figure) and with experiment

Electronic state[a]	Co(H$_2$O)$_6^{3+}$			[Co(H$_2$O)$_6^{3+}$][H$_2$O]$_{12}$	exp.[f]
	LFDFT[b]	SORCI[c]	TDDFT[d]	LFDFT[e]	
^{1}A$_{1g}$	0	0	0	0	–
^{1}T$_{1g}$	15370	15670	16528	15669	16600
^{1}T$_{2g}$	24537	23600	22017	24726	24900
^{3}T$_{1g}$	9212	5257	9156	10271	–
^{3}T$_{2g}$	13782	10779	9380	14798	–
^{5}T$_{2g}$	9660	–	–	12649	–
MUSE	796	1115	1478	552	

[a] symmetry notations are given in O$_h$.
[b] this work, R(Co—O) = 1.873 Å from LDA geometry optimization; PW91 functional was taken for the calculations of the energy multiplets; 10 Dq = 16994, B = 775, C = 2781 cm^{-1}.
[c] Neese F, Petrenko T, Ganyushin D, Olbrich G, Coord.Chem.Rev. 251 (2007) 288–327, CAS(6,5) reference space, addopted treshold values for the natural orbital populations (T$_{nat}$), multireference perturbation theory (T$_{sel}$) and reference space (T$_{pre}$) selections are 10^{-5}, 10^{-6} and 10^{-4}, respectively.
[d] SAOP model functional used; extended model cluster with geometry speciefied in Figure 19-2.
[e] R(Co—O) = 1.896 Å; 10Dq = 17102, B = 750, C = 2415 cm^{-1}, this work.
[f] taken from C.K. Jørgensen, Absorption Spectra and Chemical Bonding, Pergamon Press, Oxford UK,1962.

Electronic transitions in Hexa-Aqua Complexes of Co(III) and Co(II) The blue complex Co(H$_2$O)$_6^{3+}$ has a low-spin ^{1}A$_{1g}$(t$_{2g}^6$) ground state. From the many possible transitions only the spin-allowed ones to ^{1}T$_{1g}$ and ^{1}T$_{2g}$ (t$_{2g}^5$e$_g^1$) are observed in the spectrum (Table 19-7, Figure 19-2). The wavenumbers of two transitions are well

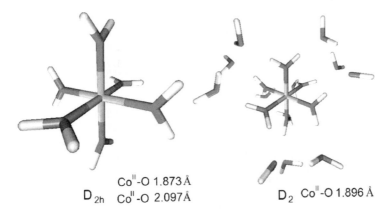

Co^{III}-O 1.873Å
D$_{2h}$ Co^{II}-O 2.097Å

D$_2$ Co^{III}-O 1.896 Å

Figure 19-2. Co(H$_2$O)$_6^{z+}$ (z = 3, 2) and {[Co(H$_2$O)$_6$]12H$_2$O}$^{3+}$ model clusters for LFDFT calculation of energies of d-d transitions; Co—O bond lengths are given (from LDA-VWN) (see color plate section)

reproduced with all three methods compared here. As expected for a closed-shell system, the TDDFT method works satisfactorily in this case. We studied also with LFDFT the effect of solvent water, which we took into account using COSMO (not shown in Table 19-7) and an explicit solvent adopting an extended-cluster model with 12 additional H_2O molecules completing a second coordination sphere (Figure 19-2). Except the spin-forbidden transition (not observed) and in contrast to CrF_6^{3-}, no essential solvent effect was established.

The pink high-spin complex $Co(H_2O)_6^{2+}$ possesses a $^4T_{1g}(a)$ ground state. The spectrum shows three bands at 8100, 16000 and, split, 19400–21550 cm^{-1} corresponding to $^4T_{2g}$, $^4A_{2g}$ and $^4T_{1g}(b)$ excited states; a further sharp line at 11300 cm^{-1} has been assigned to $^4T_{1g}(a) \rightarrow {}^2E_g$. Neglecting the possible trigonal splitting (vibronic activity) of the $^4T_{1g}$ ground state (but notice a splitting by 2000 cm^{-1} of the third spin-allowed band), we calculated the wavenumbers of electronic transitions using LFDFT and compare them with the SORCI result in Table 19-8. Because of the both space- and spin-open-shell character of the $^4T_{1g}$ ground state of $Co(H_2O)_6^{2+}$, a TDDFT calculation was impracticable. A serious attempt was made to extend the TDDFT method to such cases [17]. Both the LFDFT and SORCI results reproduce the experimental spin-allowed transitions (Table 19-8), but, if the assignment of the only observed spin-forbidden transition is correct, its wavenumber is reproduced less satisfactorily by theory – as expected, it is calculated too large (small) by the SORCI (LFDFT) methods (Table 19-8).

Table 19-8. Comparison between wavenumbers/cm^{-1} of d-d transitions of high-spin $Co(H_2O)_6^{2+}$ calculated with various approaches and with experiment

Electronic state[a]	LFDFT[b]	SORCI[c]	exp.[d]
$^4T_{1g}$	0	0	–
$^4T_{2g}$	7755	6630	8100
$^4A_{2g}$	16617	14313	16000
$^4T_{1g}$	19548	19970	19400
			21550
^2E_g	6224	13130	11300
MUSE	1546	1389	–
MUSE($^4T_{1g} \rightarrow {}^4\Gamma_g$)	370	1242	–
MUSE($^4T_{1g} \rightarrow {}^2E_g$)	5076	1830	

[a] symmetry notations are given in O_h.

[b] this work, R(Co—O) = 2.097 Å from LDA geometry optimization; PW91 functional was taken for the calculations of the energy multiplets; 10 Dq = 8862, B = 860, C = 3013 cm^{-1}.

[c] F. Neese, T. Petrenko, D. Ganyushin, G. Olbrich, Coord. Chem. Rev. 251(2007) 288–327, CAS(6,5) reference space, addopted treshold values for the natural orbital populations (T_{nat}), multireference perturbation theory (T_{sel}) and reference space (T_{pre}) selections are 10^{-5}. 10^{-6} and 10^{-4}, respectively.

[d] taken from C.K. Jørgensen, Absorption Spectra and Chemical Bonding, Pergamon Press, Oxford UK, 1962.

Special cases

In focusing on the d-localized states and presenting the LF potential as an additive quantity, classical LFT is approximate and thus must be regarded as a model rather than as a genuine theory. Cases were thereby encountered that were not reproduced (explained). These cases are essentially s-d mixing, bent (misdirected) metal-ligand bonds and phase-coupled ligators. For cases such as these, extensions of the conventional theory have been made [2]. Even in situations in which LF fails, it was used diagnostically, thus leading one to create essential chemical concepts. As LFDTF is free from the assumption of the additivity of the LF potential and provides the LF matrix *as it is*, it is readily extensible to include not only d but other metal-centered orbitals. We can thus apply the method to discuss once more, at a higher level, all these concepts.

s-d Mixing in square planar complexes   Cu(II) forms tetragonal amine CuN_4XY complexes with coplanar CuN_4 moieties and distant X, Y ligands consistent with its Jahn-Teller unstable ground state. As follows from AOM energy expressions (Equation 27), Cu(II) possesses a $^2B_{1g}$ ground state, assuming D_{4h} symmetry,

$$(27.1) \quad e(b_{1g}) = 3e_\sigma(N)$$

$$(27.2) \quad e(a_{1g}) = e_\sigma(N) + e_\sigma(X) + e_\sigma(Y)$$

$$(27.3) \quad e(b_{2g}) = 4e_\pi(N)$$

$$(27.4) \quad e(e_g) = 2e_\pi(N) + e_\pi(X) + e_\pi(Y)$$

with $^2A_{1g}$, $^2B_{2g}$ and ^2E_g excited states arising from electronic transitions from the $a_{1g}(dz^2)$, $b_{2g}(d_{xy})$ and $e_g(dxz,yz)$ to the $b_{1g}(dx^2-y^2)$ orbital, respectively. For the compound $Na_4[Cu(NH_3)_4][Cu(S_2O_3)_2]$, which represents the closest approach to a CuN_4 square planar chromophore [63], we neglect contributions from X and Y in Equation 27; as N possesses no orbital of π-symmetry in the plane, one obtains for the wavenumbers of the $^2B_{1g} \rightarrow {}^2A_{1g}$ and $^2B_{1g} \rightarrow {}^2E_g$, $^2B_{2g}$ transitions $2e_\sigma(N)$ and $3e_\sigma(N)$. We compare these with the values reported experimentally, i.e. 18400 and 19200 cm^{-1}, respectively [63]. According to the AOM, the ratio of these transitions should be 2/3, which should further increase if weak coordination to X and Y is taken into account. It follows that the energy of d_{z^2} is calculated anomalously small by the AOM. In D_{4h} symmetry the d_{z^2} and 4s orbitals of Cu have the same symmetry and can thus mix. To account for this mixing, Smith introduced one additional negative energy term ($-4e_{sd}$) in Equation 27.2 [64, 65] with the following expression as given by second-order perturbation theory:

$$(28) \quad e_{sd} = \frac{\langle 3d|h|4s\rangle^2}{E_{4s} - E_{3d}}$$

Accordingly, the spectrum was fitted, yielding values $e_\sigma(N) = 6400$ and $e_{sd} = 1400\,cm^{-1}$ for the two parameters. Since then, using this approach one is able to rationalize the spectra of many square-planar Cu(II) complexes [1]. Hyperfine-coupling tensors due to the 3/2 nuclear spin of Cu subsequently supported this interpretation; ESR spectra of tetragonally compressed octahedral Cu(II) complexes unambigously manifested direct spin density at the nuclei of Cu due to d_{z2}–4s mixing via the Fermi contact contributions to the A-tensor [66]. Computed energies of the d-d transition wavenumbers of $Cu(NH_3)_4^{2+}$ with various methods are compared with experiment in Table 19-9. Both LFDFT and the SORCI method, as well as the direct ΔSCF calculation, are mutually consistent; they compare satisfactorily with experiment. Theoretical values are systematically larger than experimental ones; probably the effect of a weak axial coordination in the solid cannot be excluded. As seen from Equations 27, a small axial perturbation can affect significantly the d-d transitions – the wavenumbers of both $d_{z2} \to d_{x2-y2}$ and $d_{xz,yz} \to d_{x2-y2}$ transitions are expected to decrease. In support of this, wavenumbers of band maxima in the spectrum of the aqueous solution are red-shifted relative to the solid. One can readily *translate* the d-d transition wavenumbers into $e_\sigma(N)$, $e_\pi(N)$ and e_{sd} parameter values, included in Table 19-9. Parameter $e_\pi(N)$ has a non-zero value; its origin has been discussed previously [65]. d-d transitions calculated using TDDFT show

Table 19-9. Theoretical and experimental (solid and aqueous solution) wavenumbers/cm^{-1} of d-d transitions in $Cu(NH_3)_4^{2+}$

Electron transition[a]	LFDFT	DFT-ΔSCF	SORCI		TDDFT			exp.	
			CAS (17,10)[e]	CAS (9,5)	LDA VWN	SAOP	solid[c]	aqueous[d] solution	
$^2B_1 \to {}^2B_2$	17477	16370	18469	17340	34341	39494	–	14033	
$^2B_1 \to {}^2A_1$	20727	19431	20566	20082	32744	37768	18400	16533	
$^2B_1 \to {}^2E$	19840	19009	21130	20324	34220	39744	19200	17501	
MUSE[b]	1484	611	2048	1403	14682	19956	–	–	
$e_\sigma(N)$	7401	7216	7930	7769	11366	13331	6400	–	
$e_\pi(N)$	1182	1320	1330	1492	–60	125	1400	–	
e_{sd}	1481	1250	1176	1136	2503	2776	–	–	

[a] symmetry notations in the C_{4v} point group with a hole on dx^2-y^2, dz^2, dxz,yz and dxy leading to ground $^2B_1$ and the excited $^2A_1$, ^2E and $^2B_2$ states, respectively; R(Cu— N) = 2.049 Å from LDA-VWN geometry optimization.
[b] energies corresponding to band maxima in the solid state spectrum is taken as reference.
[c] reported in $Na_4[Cu(NH_3)_4][Cu(S_2O_3)_2].H_2O$ which represents the closest approach to a genuine CuN_4 square planar chromophore, see Hathaway BJ, Stephens FS, J Chem Soc (A) 884(1970).
[d] from Neese F, Magn.Res.in Chem.2004, 42, S187–S198 and cited references; other data not listed in the table are: $^2B_1 \to {}^2E$ ligand-to-metal charge transfer transition at 44196 cm^{-1}, $g\bot = 2.047$. $g|| = 2.241$; $A||^{Cu}$(MHz) (–) 586, $A\bot^{Cu}$(MHz) –(68) $A||^N$(MHz) (+)39.1 $A\bot^N$(MHz) (+) 31.7.
[e] in addition to the 9 – 5d and the 4 (e+b$_1$+a$_1$) orbitals the 4s on the metal has been included in the active space; $T_{Nat} = 10^{-5}$; $T_{Sel} = 10^{-5}$.

a strong deviation from experiment with both the LDA-VWN functional and the SAOP model potential; d-d transition wavenumbers exceed experimental values by as much as $15000\,\mathrm{cm}^{-1}$ and $20000\,\mathrm{cm}^{-1}$, respectively. This result is common for TDDFT on application to complexes of Cu(II) [67].

Bent bonds (misdirected valency) A common practice in AOM considerations, mainly for the sake of simplicity, is to assume that metal(M)–ligand(L) bonds are classified of σ- and π-types in a linear ($C_{\infty v}$) M–L molecule (L=F or Cl, for example); this pseudo-symmetry is preserved in a ML_n complex. One thus postulates that the LF matrix is diagonal in the local frame with two parameters e_σ and e_π characterizing each M–L bond— (δ-bonding is ignored, Equation 29). This situation can alter greatly for composite ligands such as macrocyclic amines, in which carbon atoms attached to each N ligator can affect its lone pairs, to prevent them from being optimally aligned for σ bonding with the metal. In the simplest case of a local pseudo-symmetry C_s, (Figure 19-3, Equation 29),

$$V(C_{\infty v}) = \begin{array}{c} \begin{array}{ccc} d_{z2} & d_{xz} & d_{yz} \end{array} \\ \begin{bmatrix} e_\sigma & 0 & 0 \\ 0 & e_\pi & 0 \\ 0 & 0 & e_\pi \end{bmatrix} \end{array}$$

$$\text{(29)} \qquad \Rightarrow V(C_s) = \begin{array}{c} \begin{array}{ccc} d_{z2} & d_{xz} & d_{yz} \end{array} \\ \begin{bmatrix} e_\sigma & e_{\sigma\pi x} & 0 \\ e_{\sigma\pi x} & e_{\pi x} & 0 \\ 0 & 0 & e_{\pi y} \end{bmatrix} \end{array}$$

$$\Rightarrow V(C_1) = \begin{array}{c} \begin{array}{ccc} d_{z2} & d_{xz} & d_{yz} \end{array} \\ \begin{bmatrix} e_\sigma & e_{\sigma\pi x} & e_{\sigma\pi y} \\ e_{\sigma\pi x} & e_{\pi x} & e_{\pi x\pi y} \\ e_{\sigma\pi y} & e_{\pi x\pi y} & e_{\pi y} \end{bmatrix} \end{array}$$

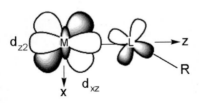

Figure 19-3. Model for rigid bent bonding; the ligand atom forms strong bonds to R but weaker ones to a transition metal. The figure illustrates how bent σ and π bonds between the metal (M) and donor atoms (L) require a local ligand-field matrix element connecting the $d_\sigma(dz^2)$ and the $d\pi(d_{xz})$ orbitals. The bent geometry implies a positive (negative) $e_{\sigma\pi}$ when the σ-lone pair points towards the negative(positive) quadrant of the xz plane

this condition affects the nominal values of e_σ and e_π and creates an off-diagonal matrix element $e_{\sigma\pi}$ that accounts for the $\sigma(dz^2)-\pi(d_{xz})$ mixing. Liehr [68] discovered this phenomenon and Gerloch et al. [69–71] introduced it into the LFT formalism and applied this extended model to rationalize spectra and bonding in many TM complexes [72]. A problem of applying broadly this concept was the numerous model parameters thereby introduced; thus the general case of a M—L bond with no symmetry yields six independent parameters $(V(C_{\infty v}) \rightarrow V(C_s) \rightarrow V(C_1))$, Equation 29). This concept is worthy of reconsideration using LFDFT, being free from any assumption in constructing the LF matrix. We take again as example $Cu(NH_3)_4{}^{2+}$ and let the trigonal axis of each NH_3 ligand deviate from the Cu—N bond direction. We introduce the angle (α) for the tilting of each of the four NH_3 ligands in such a way that the C_{4v} symmetry is conserved (Figure 19-4). In doing so, we eliminate mixing between $\sigma(b_{1g}, a_{1g})$ and $\pi(e_g)$ orbitals (the $e_{\sigma\pi}$ term in $V(C_s)$, Equation 29). This allows us to analyze the effect of the bent bonds on only the diagonal terms. The dependence of the wavenumbers of d-d transitions is depicted in Figure 19-5. A parameterization using Equation 27, but setting zero values for energies due to X and Y, allows one to quantify the effect of bending as reflected by the dependence of e_σ, e_π and e_{sd} on the angle α (Figure 19-6). The wavenumbers of all transitions decrease with increasing α (Figure 19-5), which is mainly due to a reduced M—L σ-overlap.

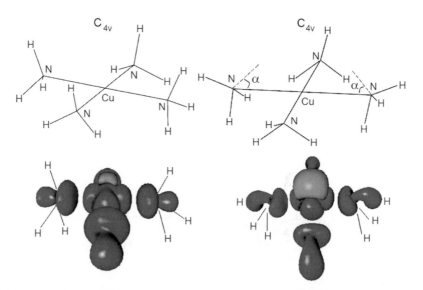

Figure 19-4. Model for $Cu(NH_3)_4{}^{2+}$ taken to illustrate the effect of bent bonding; no bending (left); an extreme case of bent bonding ($\alpha = 45°$) (right). Note that (see caption to Figure 19-3) the $\sigma - \pi$ mixing caused by each pair of trans NH_3 ligands cancel; there is no $\sigma(b_1, a_1) - \pi(e)$ mixing in C_{4v} symmetry (see color plate section)

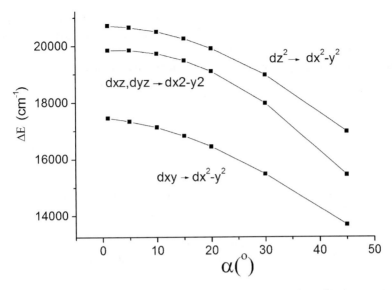

Figure 19-5. The dependence of the energies of d-d transitions in $Cu(NH_3)_4^{2+}$ on the geometric parameter α quantifying the concerted bending of all four NH_3 ligands with conservation of the C_{4v} symmetry

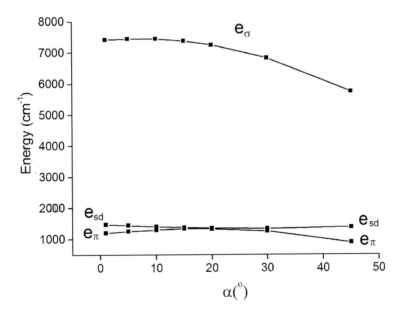

Figure 19-6. Dependence of the AOM parameters e_σ, e_π and e_{sd} on the geometric parameter α quantifying the concerted bending of all four NH_3 ligands with conservation of the C_{4v} symmetry

For angles $\alpha < 20°$, the e_π parameter increases, whereas e_σ and e_{sd} decrease with increasing α (Figure 19-6). To diminish the number of parameters, approximations were used – Equation 30, e^o_σ- the σ-antibonding energy for ideally aligned orbitals – expressing e_σ, e_π and $e_{\sigma\pi}$ in terms of the angle α estimated from structural data and assuming a given hybridization on the donor

$$\begin{aligned} e_\sigma &= \cos^2 \alpha . e^o_\sigma \\ e_\pi &= \sin^2 \alpha . e^o_\sigma \\ e_{\sigma\pi} &= \sin \alpha \cos \alpha . e^o_\sigma \end{aligned} \tag{30}$$

atom (rigid model - Figure 19-3) [73]. A closer look at Figure 19-6 and the numerical values of e_σ, e_π (Table 19-9) shows that their dependence on α is less steep than expected based on Equations 30; apparently a re-hybridization on the donor atoms occurs on bending to maximize metal-ligand overlap.

Formation of bent bonds is also reflected in diagrams of orbital contours using the spin density as a probe (Figure 19-4); spin density accumulates on H – one of each NH_3 that becomes more coplanar with the MN_4 plane. The transfer of spin density from the metal to atoms from the second coordination sphere, induced by bent bonding, might have important consequences for the reactivity in macrocyclic compounds.

Orbital phase coupling (Orgel effect) in complexes with π-conjugate bis-bidentate ligands Frontier orbitals in chelate ligands are part of an extended network. In such cases perturbation from bonding and antibonding delocalized ligand π-MO, rather than separate and independent ligand functions, governs the respective LF potential and LF splittings. Orgel predicted [74] this effect, which Ceulemans

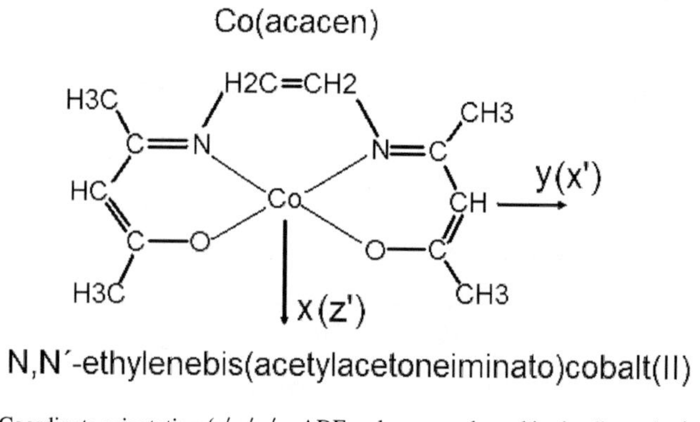
Co(acacen)
H2C=CH2
H3C
CH3
C=N
N=C
HC
Co
CH
y(x')
C—O
O—C
H3C
CH3
x(z')
N,N′-ethylenebis(acetylacetoneiminato)cobalt(II)

Figure 19-7. Coordinate orientation (x′,y′,z′ – ADF and x,y,z – adopted in the discussion) and schematic geometry of the Coacacen complex; the geometry taken for the calculation is that given by X-ray crystallographic data, reported by F.Cariati, F.Morazzoni, C.Busetto, G.Del Piero, A.Zazzetta, J.Chem.Soc. Dalton Trans. (1976) 342

et al [75–77] later incorporated in AOM. We consider here as an example the planar complex Coacacen, acacen = N,N'-ethylenebis (acetylacetoneiminato), with a d^7 configuration on Co(II) (Figure 19-7). For this complex, and assuming a square-planar geometry, d_{xz} and d_{yz} are expected not to split, being both affected (destabilized - L→M π-donation or stabilized - M→L π-back donation) due to π-overlap with the ligand to the same extent ($2e_\pi$). Inspection of the electronic structure of the free ligand shows that its HOMO π orbital is in phase, whereas its LUMO π^* is out of phase. According to the coordinate orientation of Figure 19-7 and $C_{2v}(y)$ symmetry, the $b_2(\pi)$ ligand orbital causes a destabilization of d_{yz} (π-donation), whereas the $a_2(\pi^*)$ ligand orbital causes a stabilization of the d_{xz} orbital (π-back donation). The parameters $e_{\pi s}$ and $e_{\pi s}'$ quantify the effect (Figure 19-8). A DFT calculation lends full support to this splitting pattern. In Figure 19-9 we present the energies of the five d-orbitals, as calculated using LFDFT, and their contour plots as well as AOM expressions. Based solely on the AOM, as many as six parameters are required to compute the orbital splitting, but there are only four splitting energies: the parameters $e_{\pi s}$ and $e_{\pi s}'$ cannot be determined independently, and the s-d mixing accounted for by the e_{sd} energy can not be neglected. The large degree of parameterization prevents use of the model for predictive purposes, but it assists an interpretation of the results. The parameters from a full LFDFT calculation (Table 19-10) have been used to calculate all multiplets, the lowest ones being listed in Table 19-11, in satisfactory agreement with experimental data. Using a

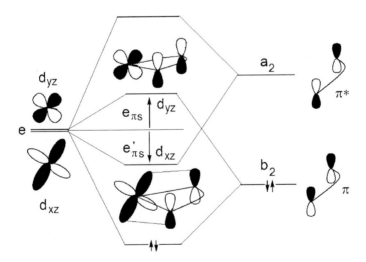

Figure 19-8. Orbital-energy diagram showing the interaction between the out-of-plane d_π-orbitals $d_{xz,yz}$ and the HOMO and LUMO of a five-membered double-bonded bidentate ligand. Symmetry labels a_2 and b_2 refer to the $C_{2v}(y)$ point group, i.e.

	$C_2(y)$	$\sigma_v(xy)$	$\sigma_v(yz)$
$a_2(d_{xz})$	+1	−1	−1
$b_2(d_{yz})$	−1	−1	+1

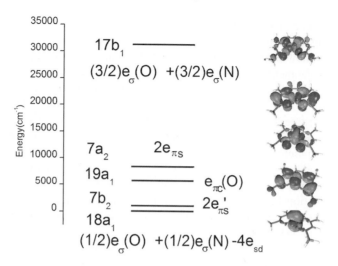

Figure 19-9. MO energy diagram for the d-based orbitals of Coacacen as determined by LFDFT (see color plate section)

Table 19-10. Non-empirically determined parameters used in the calculation of the ESR parameters and multiplet structure of Co(acacen)

B	$512\pm53\,\mathrm{cm}^{-1}$
C	$3118\pm225\,\mathrm{cm}^{-1}$
$(x'y'\|h_{lf}\|x'y')$	$-1071\pm407\,\mathrm{cm}^{-1}$
$(y'z'\|h_{lf}\|y'z')$	$6308\pm407\,\mathrm{cm}^{-1}$
$(z2'\|h_{lf}\|z2')$	$5052\pm407\,\mathrm{cm}^{-1}$
$(x'2-y'2\|h_{lf}\|x'2-y'2)$	$3731\pm407\,\mathrm{cm}^{-1}$
$(z2'\|h_{lf}\|x'2-y'2)$	$2771\pm407\,\mathrm{cm}^{-1}$
$(x'z'\|h_{lf}\|x'z')$	$-24003\pm407\,\mathrm{cm}^{-1}$
ζ	$460\,\mathrm{cm}^{-1}$
k	0.77

formalism elaborated elsewhere [6], we calculated also the *g*-tensor values, compared with those from a ZORA calculation and experimental ones in Table 19-12. A large in-plane anisotropy with $g_{xx} > g_{yy}$ was measured and reproduced in both independent calculations. This anisotropy stems from the large splitting between the d_{yz} and d_{xz} orbitals because of the Orgel effect. An alternative explanation of this anisotropy was based on the mechanism of bent bonding [69], which the present results fail to support.

Table 19-11. Mutiplet splitting energies determined by the LFDFT method (GGA functional, frozen core approximation)

	LFDFT	exp.
$^2A_2$	0.0	–
$^2A_1$	4665	–
$^2B_2$	7036	4000
$^2A_1$	10885	8000
$^4B_2$	13021	–
$^4A_1$	12835	–.
$^4B_2$	14694	–

Table 19-12. g-tensor values of [Co(acacen)] determined by LFDFT and spin-orbit restricted ZORA calculation and compared with experiment

	ZORA		LFDFT-GGA		Exp.
	LDA	GGA	A	B	
g_{xx}	2.85	2.76	3.21	2.80	2.92/3.26
g_{yy}	1.89	1.93	1.87	1.94	1.90 ± 0.03
g_{zz}	1.91	1.92	1.87	2.11	2.00 ± 0.02
g_{iso}	2.22	2.20	2.28	2.32	

A: two states model 97% $| dyz^1 dxy^2,{}^2A_2 > +3%|dz^{2\ 1}dxy^2,{}^2A_1 >$.
B: full calculation, $g_{iso} = (g_{xx}+g_{yy}+g_{zz})/3$.
Exp.values: range of values, because of strong dependence on the host lattice.

Complexes of Biological and Catalytic Interest:Fe(IV) oxo-complexes

High-valent iron-oxo intermediates are commonly invoked in catalytic cycles of mononuclear iron enzymes that activate O_2 to effect metabolically important oxidative transformations. Catalytic pathways of many mononuclear non-heme iron enzymes are proposed to involve high-valent iron-oxo intermediates as the active oxidizing species. Two isomeric pentadentate bispidine Fe(II) complexes (bispidine = 3,7 – diazabicyclo 1,3,3,nonane) in the presence of H_2O_2 are catalytically active for the epoxidation and 1,2-dihydroxylation of cyclooctene [78, 79]. Spectral and mechanistic studies indicate that in all these cases a Fe(IV) = O intermediate is responsible for the catalytic process [80]. As theoretical methods based on A Fe(IV) = O oxo complex [Fe(IV)O(TMC)(NCCH₃)]²⁺ (complex I, Figure 19-10) has been isolated and characterized by X-ray crystallography, Mössbauer measurements [81] and MCD spectra [82]. The X-ray structure shows that Fe is hexacoordinate forming a strong Fe=O bond and weaker bonds to four equatorial amine N-donors and to the terminal N from CH_3CN (Figure 19-10). Spectral data have been interpreted in terms of a ground state with $S = 1$ and a large zero-field splitting $D = 29$ (3) cm^{-1} has been reported [81]. As theoretical methods based on

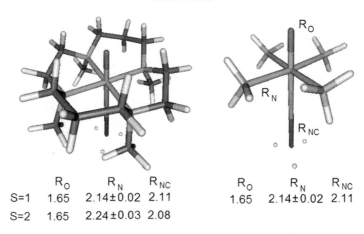

	R_O	R_N	R_{NC}		R_O	R_N	R_{NC}
S=1	1.65	2.14±0.02	2.11		1.65	2.14±0.02	2.11
S=2	1.65	2.24±0.03	2.08				

Figure 19-10. Fe(IV) = O complexes studied in this work: [FeO(TCM)(CH$_3$CN)]$^{2+}$ (complex I, left) - the geometry of the S = 1 species has been adopted from a BP86 geometry optimization, Ref. [82]; the geometry of the S = 2 species is given for comparison, note that R$_O$ and R$_{NC}$ alter little, but bond lengths to equatorial ligands R$_N$ strongly increase on going from S = 1 to S = 2; [FeO(NH$_3$)$_4$(CH$_3$CN)]$^{2+}$ (complex II, right) – the geometry of the first coordination sphere was taken from complex I (S = 1) but N have been saturated by H – atoms (see color plate section)

DFT and ab initio are used in all mechanistic investigations aiming to understand reaction mechanisms, it is tempting to use this complex to test these approaches. To this end, we undertook a LFDFT calculation of the d-d spectral transitions; the results are compared with those from SORCI and TDDFT in Table 19-13. Using LFDFT and a DFT geometry-optimized structure (LDA of VWN plus gradient corrections, Becke88 for exchange, and Perdew 86 for correlation) we reproduced the observed d-d transitions. A MO-energy diagram for the MO dominated by Fe(IV) 3d functions appears in Figure 19-11.

The d_{z2} and $d_{xz,yz}$ orbitals are strongly destabilized by σ and π Fe=O antibonding, leading to a $b_2^2 e^2$ ground-state configuration and two unpaired electrons on the π-antibonding d_{xz} and d_{yz} orbitals. The d_{x2-y2} orbital forms weaker sigma bonds with the equatorial amines; it is calculated to lie about 22000 cm^{-1} above the lowest $b_2(d_{xy})$ orbital. Thus the large stabilization of the $^5A_1$ high-spin state by exchange forces ($\sim$22000 cm^{-1}) is nearly completely canceled by the d_{x2-y2}-d_{xy} orbital-energy difference when promoting one electron from d_{xy} to the empty d_{x2-y2} orbital. As a result, the two states $^3A_2$ and $^5A_1$ are found to have similar energy. With the parameter set provided by LFDFT and using the S = 1 optimized geometry, $^3A_2$ is calculated to be about 1000 cm^{-1} below $^5A_1$. Already a small downward energy shift of d_{x2-y2}, due to elongation of the equatorial Fe—N bonds, can yield a change to a $^5A_1$ ground state (compare with Figure 19-10, S = 2 geometry) [83]. As expected, simple ΔSCF calculations as well as TDDFT tolerate the $^3A_2$ ground state, whereas SORCI, which was found generally to exaggerate the values of B and C, produces the opposite effect – a $^5A_1$ ground term is calculated with

Table 19-13. Theoretical and experimental term energies/cm^{-1} (D_{4h} symmetry notation) of the $S = 1$ form of trans-Fe(O)(NCCH$_3$)TMC^{2+} (I) and LFDFT values for the Fe(O)(NCCH$_3$)(NH$_3$)$_4^{2+}$ (II) model complex

Electr. state	LFDFT^g		SORCI	ΔSCF	TDDFT (SAOP)	Assignment	exp(polariz.) (I)
	(I)	(II)	(I)	(I)	(I)		
$^3A_2$	0	1264	3226	0	0	xy^{2}xz^1 yz^{1a}	–
$^5A_1$	911	0	0	6488	8146	xy → x^2–y^2	–
$^1B_2$	10036	10233	–		6170	spin-flip^b	–
$^5B_1$	10629	11645	–		13759	xy → z^2	–
$^1B_1$	10979	11968	–		6314	spin-flip^b	–
^3E	11968	12323	–	10486	9472	xy → xz, yz	~10500(xy)
^3E	15680	15441	–	12668	13726	xz, yz → z^2, x^2–y$^{2\ d}$	13000(xy)
$^1A_1$	16924	17241	10484		11589	spin-flip^c	–
$^3A_1$	19347	17575	–	12570	18372	xy → x^2–y^2	17000(z)
^5E	19359	17837	–		–	xy(xz, yz) → z^2, x^2–y$^{2\ e}$	–
^1E	20360	20284	9436		13070	xy → xz, yz, x^2–y$^{2\ d}$	–
$^3B_2$	22012	21157	–		f	xy→x^2–y^2	–
^3E	22184	21884	–	17848	18739	xz, yz → z^2, x^2–y$^{2\ d}$	–
$^3B_1$	22599	23380	–		19662	xy → x^2–y^2	–
$^3A_1$	23799	23897	–		–	xy → x^2–y^2	–
$^3A_2$	26525	25955	–	23948	–	xy → z^2, x^2–y$^{2\ d}$	25000(z)

^{a} ground state configuration.
^{b} spin-change within the ground state configuration.
^{c} spin-change within the ground state configuration with significant contributions from excited statecon-figurations.
^{d} – substantial multiconfigurational character.
^{e} - two-electron transition.
^{f} not found in the energy range below the ligand-to-metal - charge transfer excitations.
^{g} calculated using the reported $S = 1$ DFT geometry optimized structure (A. Decker, J.-U. Rohde; L. Que, Jr., E.I. Solomon, J. Am. Chem. Soc. 2004, 126, 5378, see also Supporting Information) showing negligible deviation from a D_{4h} symmetry; (I) diagonal ligand field (v) matrix elements $<z^2|v|z^2>$, $<x^2 - y^2|v|x^2 - y^2>$, $<xz|v|xz> = <yz|v|yz>$ and $<xy|v|xy>$, and B and C values /cm^{-1} are 31678, 21960, 13230, 0, 786, 3485 (I) and 33237, 20005, 14732, 0, 1071, 3250 (II).

this method. With a BP8-optimized structure, the equatorial ligand field must be somewhat underestimated as the length of the Fe=O bond is predicted accurately, whereas the Fe–N(eq) bonds are calculated to be longer than from experiment, thus underestimating the σ-interaction and the energy of the σ-antibonding $b_1(dx^2–y^2)$ orbital. Equatorial Fe–N(eq) bonds appear shorter with a LDA geometry, similar to the X-ray structure; thus $b_1(dx^2–y^2)$ is calculated higher in energy, and, as the splitting of the $b_2(dxy)$ and $b_1(dx^2–y^2)$ MO determines the position of the $^5A_1$ state relative to $^3A_2$, the SORCI calculations give a more realistic result with such a geometry: $^5A_1–^3A_2 = -564$ (LDA) and -968 (X-ray) cm^{-1}, respectively [84]. The order relative to experiment remains reversed. As expected, the B3LYP value, 3065 cm^{-1} (not listed in Table 19-13), for the splitting between $^5A_1$ and $^3A_2$, using

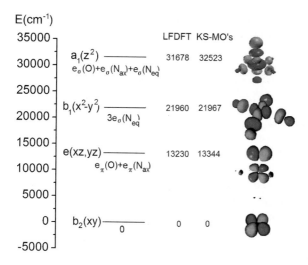

$E(cm^{-1})$

35000 —

30000 — $a_1(z^2)$ ——————   31678   32523
            $e_\sigma(O)+e_\sigma(N_{ax})+e_\sigma(N_{eq})$

25000 —

20000 — $b_1(x^2-y^2)$ ——————   21960   21967
            $3e_\sigma(N_{eq})$

15000 —

10000 — $e(xz,yz)$——————   13230   13344
            $e_\pi(O)+e_\pi(N_{ax})$

5000 —

0 — $b_2(xy)$——————   0   0
            0

-5000 —

LFDFT KS-MO's

Figure 19-11. MO energy diagram for complex I (Figure 19-10) calculated with LFDFT; the energies of LFDFT orbitals are compared with those resulting from the average-of-configuration KS calculation. Orbital contours (pertaining to values of the density of 0.05 a.u.) are plotted, but atoms are omitted for clarity (see color plate section)

a BP86 or B3-LYP optimized structure, is smaller than the BP86 one, $4516\,cm^{-1}$, after geometry relaxation, to be compared also with the value without relaxation – $6488\,cm^{-1}$ – in Table 19-13.

Comparing the results of Table 19-13, we state that both LFDFT and TDDFT and the simple ΔSCF calculation yield results of satisfactory quality relative to experiment. Initial SORCI calculations have been attempted but based on the results of Ref. [83] on a small model as well as our preliminary results the entire system indicate that the size of basis set, what would be needed in order to obtain the quintet-triplet ordering correctly, is too extensive for the calculation to be feasible with the present implementation of the SORCI method. Thus algorithmic improvements are necessary in order to routinely tackle molecules of the size of $[Fe(IV)O(TMC)(NCCH_3)]^{2+}$.

In Table 19-13 we list the calculations for the cluster, in which, without altering the geometry (complex I, BP86), we saturated each amine N donor with protons. Such a cluster has been used in model calculations previously [83]. $^5A_1$ is stabilized here by $1264\,cm^{-1}$ with respect to $^3A_2$, as expected for NH_3 being a weaker donor than amine N (but a SORCI calculation yields a much larger stabilization, $5348\,cm^{-1}$). One can conclude that for a realistic model one should take the actual geometry of complex I into account.

To test the LFDFT results, we have also calculated the zero-field splitting (ZFS) D. Using standard ligand-field arguments the following equation has been derived [83, 84]

$$(31) \quad D(^3A_2) \cong \varsigma_o^2 \left[\frac{1}{3} \frac{\alpha_{xy}^2 \alpha_{x2-y2}^2}{\Delta[^5A_1(b_2 \to b_1)]} + \frac{\alpha_{xz,yz}^4}{\Delta[^1A_1(e \to e)]} \right.$$
$$\left. + \frac{1}{4} \left(\frac{\alpha_{xz,yz}^2 \alpha_{xy}^2}{\Delta[^3E(b_2 \to e)]} + \frac{\alpha_{xz,yz}^2 \alpha_{x2-y2}^2}{\Delta[^3E(e \to b_1)]} + \frac{3\alpha_{xz,yz}^2 \alpha_{z2}^2}{\Delta[^3E(e \to a_1)]} \right) \right]$$

in which ς_o is the $Fe^{4+}(3d)$ spin-orbit coupling parameter (ZORA value, 629 cm^{-1}), α_i represents the fractional iron character in the Fe-3d based MO i and Δ are transition wavenumbers to the indicated excited states. The excited states that enter Equation 31 are defined with

$$|^3A_2\rangle = |b_2^+ b_2^- e_x^+ e_y^+|$$
$$|^3E(b_2 \to e)\rangle = |b_2^+ e_x^+ e_x^- e_y^+|; |b_2^+ e_x^+ e_y^+ e_y^-|$$
$$(32) \quad |^3E(e \to b_1)\rangle = |b_2^+ b_2^- e_x^+ b_1^+|; |b_2^+ b_2^- e_y^+ b_1^+|$$
$$|^3E(e \to a_1)\rangle = |b_2^+ b_2^- e_x^+ a_1^+|; |b_2^+ b_2^- e_y^+ a_1^+|$$
$$|^5A_1(b_2 \to b_1)\rangle = |b_2^+ e_x^+ e_y^+ b_1^+|$$
$$|^1A_1(e \to e)\rangle = \frac{1}{\sqrt{2}}(|b_2^+ b_2^- e_x^+ e_x^-| + |b_2^+ b_2^- e_y^+ e_y^-|)$$

Furthermore, from our LFDFT analysis we estimate

$$(33) \quad \alpha_{xy}^2 = 0.955; \alpha_{xz,yz}^2 = 0.588; \alpha_{x2-y2}^2 = 0.611; \alpha_{z2}^2 = 0.678$$

and for the average spin-orbit parameter in the complex,

$$(34) \quad \zeta = \frac{1}{5}(\alpha_{xy}^2 + 2\alpha_{xz,yz}^2 + \alpha_{x2-y2}^2 + \alpha_{z2}^2)\varsigma_o = 0.684\varsigma_o = 430 \text{ cm}^{-1}$$

With these data a LFDFT full-CI calculation yields $D(^3A_2) = 26$ cm^{-1} in agreement with the experimental value (29 cm^{-1} [81]). As follows from Equation 31–33, the largest contribution to D (80%) results from the mixing of the $^3A_2$ ground term with the nearest $^5A_1$ state.

CONCLUSIONS AND OUTLOOK

We have compared three approaches – LFDFT, SORCI ab initio and purely DFT methods (mostly TDDFT and ΔSCF) – for their ability to reproduce multiplet structures of ten transition-metal complexes with open d-shells and well documented optical d-d spectra. We found that in most cases the former two theories work comparably well, whereas TDDFT for open-shells was deficient, yielding wavenumbers of electronic transitions that considerably exceed experimental ones. Although both LFDFT and SORCI yield energies of spin-allowed transitions in agreement with experiment, there is an essential bias concerning spin-forbidden transitions – LFDFT underestimates but SORCI overestimates of their energies – at least as long as one does not use extensive basis sets. Seeking an explanation, we state that DFT exaggerates metal-ligand covalency and thus yields smaller B and C values, whereas SORCI in combination with small basis sets turns the effect in an opposite direction. In the case of close electronic states of different spin multiplicity, the LFDFT and SORCI methods thus tend to tolerate low- and high-spin states respectively.

The applicability of LFDFT, like LFT itself, is rooted in an effective hamiltonian theory that states that, in principle, it is possible to define precisely a hamiltonian for a sub-system such as the levels of a transition metal in a transition-metal complex or a solid. This condition is possible, because in Werner-type complexes the metal-ligand bond is mostly ionic and as such allows one to take a spectroscopically justified preponderant electronic d^n or f^n configuration as well defined; ligand-to-metal and metal-to-ligand charge-transfer states are well separated from excitations within this configuration.

We placed most emphasis on LFDFT, which is a non-empirical approach to a theory (LFT) regarded as borne out by experiment. It is thus a simple model rather than a theory, in which atomic-like quantities, such as parameters of inter-electronic repulsion – there are two such, B and C, for nd-transition metals, or three $-(F_2, F_4, F_6)$ for rare earths and actinides – and the spin-orbit coupling energy are treated in spherical symmetry as atomic-like quantities implying an averaging over the d^n or f^n configuration. All chemical aspects are then considered on a one-electron basis and are hidden in the matrix elements of the 5×5 (3d) or 7×7 (4f or 5f) ligand-field potential. These matrix elements and their orbital-energy splittings after application of the many-electron treatment as described under '*The LFDFT – an attempt to revive LFT*' (i.e. when taking the ε_i^{LFO} in Equation 19) are near those that result from just substituting the Kohn-Sham energies ε_i^{KS} in Equation 19. A simple average-of-configuration DFT calculation thus gives a clue about the structure of the ligand field and the energy splittings of the d- or f-orbitals. Keeping in mind that LFDFT values of B and C are underestimated with current DFT exchange-correlation potentials and that they vary less strongly from one complex to another being affected by nephelauxetic reduction in a well known way [85], one can take more realistic values for these parameters, either calculating them explicitly (using explicitly LF orbitals) or adopting them from those of the free ion after a proper nephelauxetic reduction. A similar approach to LFT has been proposed [86, 87]. No such restrictions apply to SORCI which is a general purpose method and performs as well on small molecules or conjugated pi-systems as on d-d multiplet energies. Its performance for charge-transfer transitions in transitions metal complexes has, however, not been assessed.

The approach to the parameters of the ligand field is not restricted to the DFT method. The procedure in '*The LFDFT – an attempt to revive LFT*' can instead be combined with any method for electronic structure, supposing that it is able to yield (after some averaging) reasonable LF orbitals and effective LF matrix elements. Thus DFT failed badly for rare earths [32, 88, 89]. For that purpose we applied a spectroscopically adjusted extended Hückel model [90]. Alternatively, one can proceed beyond LFT and take average natural atomic orbitals resulting from an average multi-configurational procedure, such as SORCI, and calculate all matrix elements of both one- and two-electron operators.

The LFDFT procedure for single nuclear complexes is readily extensible to polynuclear magnetic clusters, as has been already shown in calculations of isotropic exchange-coupling energies in exchange-bridged dimers [34, 35]. The approach has been applied to calculate and to predict anisotropic exchange-coupling parameters in cyanide-bridged transition-metal complexes [91]. This approach has great predictive power in molecular magnetism.

REFERENCES

1. A.J. Bridgeman and M. Gerloch, Progr.Inorg.Chem. 45(1996) 179–281.
2. T. Schönherr, M. Atanasov and H.Adamsky, In: A.B.P. Lever (ed) Comprehensive Coordination Chemistry II, From Biology to Nanotechnology, Fundamentals, Vol. 1, Section 2.36, Elsevier, Amsterdam Netherlands, 2003, p. 443–455.
3. (a) F. Neese, T. Petrenko, D. Ganyushin and G. Olbrich, Coord.Chem.Rev. 251(2007) 288–327.
   (b) M. Atanasov, C.A. Daul and C. Rauzy, Chem.Phys.Lett. 367(2003) 737–746.
4. M. Atanasov, C. Daul and C. Rauzy, Struct. and Bonding, 106(2004) 97–125.
5. M. Atanasov, C. Rauzy, P. Bättig and C. Daul, Int. J. Quantum Chem. 102(2005) 119–131.
6. C.Daul, C.Rauzy, M.Zbiri, P.Baettig, R.Bruyndonckx, E.J.Baerends and M.Atanasov, Chem.Phys.Lett. 399(2004), 433–439.
7. F. Neese, J. Chem. Phys. 119(2003), 9428–9443.
8. E.K.U. Gross and W. Kohn, Adv. Quantum Chem. 21(1990) 255.
9. E.K.U. Gross, J.F. Dobson and M. Petersilka, In: R. F. Nalewajski (ed) Density Functional Theory, Springer Series: Topics in Current Chemistry, Springer, Berlin Germany, 1996.
10. M.E. Casida, In: D.P. Chong (ed) Recent advances in density functional methods, Vol.1, World Scientific, Singapore, 1995, p. 155.
11. (a) R. Bauernschmitt and R. Ahlrichs, Chem. Phys. Lett. 256(1996) 454.
    (b) R. Bauernschmitt and R. Ahlrichs, J. Chem. Phys. 104(1996) 9047–9052.
12. S.J.A. Van Gisbergen, J.G. Snijders and E.J. Baerends, J. Chem. Phys. 103(1995) 9347.
13. A. Rosa, G. Ricciardi, O. Gritsenko and E.J. Baerends, Struct.Bond. 112(2004) 49–116.
14. F. Wang and T. Ziegler, Mol. Phys. 102(2004) 2585.
15. A. Dreuw, M. Head-Gordon, Chem. Rev. 105(2005) 4009–4037.
16. F. Wang, T. Ziegler, E. van Lenthe, S.J.A. Van Gisbergen and E.J. Baerends, J. Chem. Phys. 122(2005) 204103.
17. M. Seth and T. Ziegler, J. Chem. Phys. 123(2005) 144105.
18. H. Bethe, Ann. d. Physik, 3(1929) 165.
19. J.H. Van Vleck, J. Chem. Phys. 3(1935) 803–806.
20. J.H. Van Vleck, J. Chem. Phys. 3(1935) 807–813.
21. C.K. Jørgensen, R. Pappalardo and H.-H. Schmidtke, J. Chem. Phys. 39(1963) 1422.
22. H.-H. Schmidtke and Z. Naturforsch. 19a(1964) 1502–1510.
23. C.E. Schäffer and C.K. Jørgensen, Mol. Phys. 9(1965) 401–412.

24. C.E. Schäffer, Struct. Bond. 5(1968) 68–95.
25. M. Gerloch, J.H. Harding and R.G. Woolley, Struct. Bond. 46(1981) 1–46.
26. R.G. Woolley, Mol. Phys. 42(1981) 703–720.
27. M. Gerloch and R.G. Woolley, Progr. Inorg. Chem. 31(1984) 371–446.
28. M. Atanasov, C.A. Daul and E. Penka Fowe, Monatshefte für Chemie, 136(2005) 925–963.
29. P.-O. Löwdin, In: C.H. Wilcox (ed) Perturbation Theory and its Applications in Quantum Mechanics, Wiley, New York USA, 1966, p. 255–294.
30. C.J. Ballhausen and J.P. Dahl, Theor. Chim. Acta, 34(1974) 169.
31. C.J. Ballhausen, Molecular Electronic Structures of Transition Metal Complexes, McGraw-Hill, New York USA, 1979, pp. 53–54.
32. M. Atanasov, C. Daul, H.U. Güdel, T.A. Wesolowski and M. Zbiri, Inorg. Chem. 44(2005) 2954–2963.
33. M. Atanasov and H.-H. Schmidtke, Chem. Phys. 124(1988) 205.
34. M. Atanasov and C.A. Daul, Chem. Phys. Lett. 379(2003) 209.
35. M. Atanasov and C.A. Daul, Chem. Phys. Lett. 381(2003) 584.
36. M. Atanasov and C.A. Daul, Chimia, 59(2005) 504–510.
37. H. Adamsky, T. Schönherr and M. Atanasov, In: A.B.P. Lever (ed) Comprehensive Coordination Chemistry II, From Biology to Nanotechnology, Vol. 2, Elsevier, Amsterdam Netherlands, 2003, p. 661–664; http://www.aomx.de
38. A.J. Bridgeman, In: A.B.P. Lever (ed) Comprehensive Coordination Chemistry II, From Biology to Nanotechnology, Fundamentals, Vol. 2, Elsevier, Amsterdam Netherlands, 2003, p. 669–672; A.R. Dale, M.J. Duer, N.D. Fenton, M. Gerloch, M. Jones and R.F. McMeeking, CAMMAG5, University of Cambridge, 2001, available by contacting Dr. A.J. Bridgeman, University of Hull, UK, E-mail: a.j.bridgeman@hull.ac.uk.
39. J. Bendix, In: A.B.P. Lever (ed) Comprehensive Coordination Chemistry II, From Biology to Nanotechnology, Vol. 2, Elsevier, Amsterdam Netherlands, 2003, p. 673–676; Ligfield.
40. M. Atanasov and C.A. Daul, C.R. Chimie 8(2005) 1421–1433.
41. A. Berces et al, ADF2004.01; SCM, Theoretical Chemistry, Vrije Universiteit, Amsterdam, The Netherlands, 2004. Available from: http://www.scm.com/
42. For a clear account of the method including $[Cu(NH_3)_4]^{2+}$ as a completely worked out example see: F.Neese, Magn.Res.Chem. 42(2004) S187–S198.
43. J. Miralles, J.P. Daudey and R. Caballol, Chem. Phys. Lett. 198(1992), 555.
44. J. Miralles, O. Castell, R. Caballol and J.P. Malrieu, Chem. Phys. 172(1993), 33.
45. (a) B. Huron, J.P. Malrieu and P. Rancurel, J. Chem. Phys. 58(1973) 5745. (b) R.J. Buenker and S.D. Peyerimhoff, Theoret. Chim. Acta, 35(1974) 33. (c) M. Hanrath, B. Engels, Chem. Phys. 225(1997) 197.
46. F. Neese, ORCA, an ab-initio, density functional and semiempirical program package, Max-Planck Institute for Bioinorganic Chemistry, Mülheim an der Ruhr, Germany, 2005.
47. E. Runge and E.K.U. Gross, Phys. Rev. Lett. 52(1984) 997.
48. C. Jamorski, M.E. Casida and D.R. Salahub, J. Chem. Phys. 104(1996) 5134.
49. R.E. Stratmann, G.E. Scuseria and M.J. Frisch, J. Chem. Phys. 109(1998) 8218.
50. D.J. Tozer and N.C. Handy, J. Chem. Phys. 109(1998) 10180.
51. S. Hirata and M. Head-Gordon, Chem. Phys. Lett. 302(1999) 375.
52. S. Hirata and M. Head-Gordon, Chem. Phys. Lett. 314(1999) 291.
53. S.J.A. Van Gisbergen, J.G. Snijders and E.J. Baerends, Comput. Phys. Commun. 118(1999), 119.
54. S.J.A. Van Gisbergen, J.A. Groeneveld, A. Rosa, J.G. Snijders and E.J. Baerends, J. Phys. Chem. A 103(1999) 6835.
55. A. Rosa, E.J. Baerends, S.J.A. Van Gisbergen, E. Van Lenthe, J.A. Groeneveld and J.G. Snijders, J. Am. Chem. Soc. 121(1999) 10356.
56. A. Rosa, G. Ricciardi, E.J. Baerends and S.J.A. Van Gisbergen, J. Phys. Chem. A, 105(2001) 3311.
57. A. Dreuw, L.J. Weisman and M. Head-Gordon, J. Chem. Phys. 119(2003) 2943.
58. E. Broclawik and T. Borowski, Chem. Phys. Lett. 339(2001) 433.

59. B. Dai, K. Deng, J. Yang and Q. Zhu, J. Chem. Phys. 118(2003) 9608.

60. V.N. Nemykin and P. Basu, Inorg. Chem. 42(2003) 4046.

61. G.te Velde, F.M. Bickelhaupt, E.J. Baerends, C. Fonseca Guerra, S.J.A. Van Gisbergen, J.G. Snijders and T. Ziegler, J. Comput. Chem. 22(2001) 931–967; http://www.scm.com/

62. E.R. Davidson, J. Comput. Phys. 17(1975) 87.

63. B.J. Hathaway and F. Stephens, J. Chem. Soc. (A) 1970, 884–888.

64. D.W. Smith, Inorg. Chim. Acta, 22(1977) 107.

65. D.W. Smith, Struct. Bond.(Berl) 35(1978) 87–118.

66. M.A. Hitchman, R.G. McDonald and D. Reinen, Inorg. Chem. 25(1986) 519–522.

67. F. Neese, J. Biol. Inorg. Chem. 11(2006) 702–711.

68. A.D. Liehr, J. Phys. Chem. 68(1964) 665–772.

69. R.J. Deeth, M.J. Duer and M. Gerloch, Inorg. Chem. 26(1987) 2573–2578.

70. R.J. Deeth, M.J. Duer and M. Gerloch, Inorg. Chem. 26(1987) 2578–2582.

71. R.J. Deeth and M. Gerloch, Inorg. Chem. 26(1987) 2582–2585.

72. M.J. Duer, N.D. Fenton and M. Gerloch, Int. Rev. Phys. Chem. 9(1990) 227–280.

73. D. Reinen, M. Atanasov and S.-L. Lee, Coord. Chem. Rev. 175(1998) 91–158.

74. L.E. Orgel, J. Chem. Soc. 1961, 3683.

75. A. Ceulemans, M. Dendooven and L.G. Vanquickenborne, Inorg. Chem. 24(1985) 1153.

76. A. Ceulemans, M. Dendooven and L.G. Vanquickenborne, Inorg. Chem. 24(1985) 1159.

77. A. Ceulemans, R. Debuyst, F. Dejehet, G.S.D. King, M. Vanhecke and L.G. Vanquickenborne, J. Phys. Chem. 94(1990) 105–113.

78. M.R. Bukowski, P. Comba, C. Limberg, M. Merz, L. Que, Jr, T. Wistuba, Angew. Chem. Int. Ed. 43(2004) 1283–1287.

79. M.R. Bukowski, P. Comba, A. Lienke, C. Limberg, C. Lopez de Laorden, R. Mas-Balleste, M. Merz, L. Que, Jr, Angew. Chem. Int. Ed. 118(2006) 3524.

80. P. Comba and G. Rajaraman, submitted for publication.

81. J.-U. Rohde, J.-H. In: M.H. Lim, W.W. Brennessel, M.R. Bukowski, A. Stubna, E.Münck, W. Nam and L. Que, Jr., Science 299(2003) 1037–1039.

82. A. Decker, J.-U. Rohde, L. Que, Jr. and E.I. Solomon, J. Am. Chem. Soc. 126(2004) 5378–5379.

83. F. Neese, J. Inorg. Biochem. 100(2006) 716–726.

84. J.C. Schöneboom, F. Neese and W. Thiel, J. Am. Chem. Soc. 127(2005) 5840–5853.

85. C.K. Jørgensen, Struct. Bond. 1(1966) 3–31.

86. C. Anthon, J. Bendix and C.E. Schäffer, Inorg. Chem. 42(2003) 4088.

87. C. Anthon, J. Bendix and C.E. Schäffer, Inorg. Chem. 43(2004) 7882.

88. A. Borel, L. Helm and C. Daul, Chem. Phys. Lett. 383(2004) 584.

89. L. Petit, A. Borel, C. Daul, P. Maldivi and C. Adamo, Inorg. Chem. 45(2006) 7382–7388.

90. M. Atanasov, C. Daul and H.U. Güdel, In: J. Leszczynski (ed) Computational Chemistry: Reviews of Current Trends, Vol.9, World Scientific, New Jersey USA, 2005, p. 153–194.

91. M. Atanasov, P. Comba and C.A. Daul, J. Phys. Chem. A, 2006, 110(2006) 13332–13340.

CHAPTER 20

THE HOLISTIC MOLECULE

JAN C.A. BOEYENS

Abstract: The quantum and classical views of a molecule are worlds apart, and only the latter has gained popularity in the chemical community. The classical model considers a molecule as a set of atoms connected through electron-pair bonds, structured according to simple valence rules. What is widely considered to be a quantum-mechanical molecular model is defined by electron-pair bonds directed by hybrid orbitals. By showing that orbital hybridization amounts to a simple rotation of coordinate axes and that a linear combination of hybrid orbitals violates the exclusion principle, all LCAO models are shown to reduce to the classical. The familiar classical concepts that feature in theories of chemical bonding have no operational meaning in quantum theory. Grafting these variables on a quantum-mechanical description of a molecule amounts to an abstraction that irreversibly breaks the holistic symmetry and conceals the molecule's quantum properties. A molecule is defined quantum-mechanically with a molecular wave function, a quantum potential and quantum torque. The observables associated with these functions are electron density (the unit operator), total energy, electronegativity and orbital angular momentum. No other derived concepts are needed to characterize a molecule or rationalize its chemical and physical properties. The universally valid holistic nature of quantum theory has the important implication that non-local interactions, mediated by the quantum potential, pervade the entire fabric of a molecule. Once this property is recognized, the mysterious attributes of molecules disappear. In principle, quantum potential plays a significant role elucidating chemical reactions, intramolecular rearrangements, crystal growth, polymorphism, polytypism, quasi-crystallinity, enzyme catalysis, allosteric effects, protein folding and the magic of DNA

INTRODUCTION

When Avogadro first mooted the notion of *molecule* there was no evidence for molecular structure or directed chemical bonding. Molecules, not atoms, were identified as the particles that feature in the various gas laws, without reference to internal structure, and assumed to be spherical.

447

J.C.A. Boeyens and J.F. Ogilvie (eds.), Models, Mysteries and Magic of Molecules, pp. 447–475.
© 2008 *Springer*.

When he suggested that the liaison between atoms that constitute a molecule could be formulated in terms of directed bonds, van't Hoff encountered serious opposition. It is all the more remarkable how his ideas later became so firmly embedded in chemical thinking that even the mighty quantum theory has so far failed to have a visible impact on this classical model. In fact, van't Hoff's model, although largely incompatible with quantum theory, managed to entrench itself in a guise that resembles that theory.

Quantum-mechanical analysis of a molecular system calls for the solution of a wave equation

$$(1)\qquad H\psi = E\psi$$

in which the molecular Hamiltonian operator, H, is a function of all electron and nuclear masses and coordinates only, as in an ideal gas. In this instance the Hamiltonian operator and its eigenfunctions are spherically symmetrical. Solutions that reflect a structured arrangement of the nuclei, as required by the van't Hoff model, can occur only under boundary conditions of broken symmetry, induced by an external field, such as a crystal field. The effect of an external field is alignment of eigenvectors in the field direction to reveal the symmetry that remains hidden in the field-free situation. This effect, commonly referred to as symmetry breaking, enables the observation of a three-dimensional molecular structure of the van't Hoff type. It is important to note that the eigenvalues of the Hamiltonian operator consist of the three quantities, *energy, orbital angular momentum* and the *components of angular momentum* in a three-dimensional coordinate system. Only the latter of these is affected by the external field. The components of angular momentum therefore represent the only quantum-mechanically defined molecular vectorial quantities fixed by a three-dimensional nuclear framework of a molecule. In particular, an observed three-dimensional structure is not necessarily a minimum-energy arrangement, but arises from the alignment of vectors to minimize resultant orbital angular momentum. Complete quenching of orbital angular momentum requires a symmetrical arrangement of atoms. When this cannot be achieved the result is a chiral, optically active molecule.

There is no quantum-mechanical evidence for spatially directed bonds between the atoms in a molecule. Directed valency is an assumption, made in analogy with the classical definition of molecular frameworks, stabilized by rigid links between atoms. Attempts to rationalize the occurrence of these presumed *covalent bonds* resulted in the notion of *orbital hybridization*, probably the single most misleading concept of theoretical chemistry. As chemistry is traditionally introduced at the elementary level by medium of *atomic orbitals*, chemists are conditioned to equate molecular shape with orbital hybridization, and reluctant to consider alternative models. Here is another attempt to reconsider the issue in balanced perspective.

Chemical readers with uncertain demand of mathematics, may decide at this point to skip the analysis of history, physics and mathematics of hybridization, and proceed directly to the summary that restates the mathematical results in words.

HYBRIDIZATION

The universally accepted argument that *explains* the structure of methane in terms of the well known scheme of sp^3 orbital hybridization derives from several statements and postulates originally formulated by Linus Pauling [1]. Some of these statements are repeated *verbatim* below.

1. There are three p orbitals (with $l = 1$) in each shell (beginning with the L shell) corresponding to the values -1, 0 and $+1$ for the magnetic quantum number m_l. [Henceforth represented by m.]
2. The Pauli exclusion principle can be expressed in the following way: *there cannot exist an atom in such a quantum state that two electrons within it have the same set of quantum numbers.*
3. There is one s orbital in each electron shell, with a given value of the total quantum number n; three p orbitals, corresponding to $m = -1$, 0, and $+1$, in each shell beginning with the L shell;
4. In other words, for the formation of an electron-pair bond two electrons with opposed spins and a stable orbital of each of the two bonded atoms are needed.
5. The carbon atom, nitrogen atom, and other first row atoms are limited to four covalent bonds using the four orbitals of the L shell.
6. ... *the three p orbitals are directed along the three Cartesian axes and will tend to form bonds in these directions.* [my italics.]
7. From the foregoing discussion it may be inferred that the quadrivalent carbon atom would form three bonds at right angles to one another and a fourth weaker bond (using the s orbital) in some arbitrary direction. This is, of course, not so; and, instead, it is found *on quantum-mechanical study of the problem* [my italics] that *the four bonds of carbon are equivalent and directed toward the corners of a regular tetrahedron,* as had been inferred from the facts of organic chemistry.
8. There are four orbitals in the valence shell of the carbon atom. We have described these as the $2s$ and the three $2p$ orbitals, ... These, are however, not the orbitals used directly in bond formation by the atom. (They are especially suited to the description of the free carbon atom; if quantum theory had been developed by the chemist rather than the spectroscopist it is probable that the tetrahedral orbitals described below would play the fundamental role in the theory, in place of the s and p orbitals.)
9. We assume that the radial parts of the wave functions ψ_s and ψ_{p_x}, ψ_{p_y}, ψ_{p_z}, are so closely similar that their differences can be neglected. The angular parts are

$$s = 1$$
$$p_x = \sqrt{3}\sin\theta\cos\phi$$
$$p_y = \sqrt{3}\sin\theta\sin\phi$$
$$p_z = \sqrt{3}\cos\theta$$

θ and ϕ being the angles used in spherical polar coordinates.

10. An equivalent set of tetrahedral bond orbitals, differing from these in orientation only, is

$$t_{111} = \frac{1}{2}(s + p_x + p_y + p_z)$$

$$t_{1\bar{1}\bar{1}} = \frac{1}{2}(s + p_x - p_y - p_z)$$

$$t_{\bar{1}1\bar{1}} = \frac{1}{2}(s - p_x + p_y - p_z)$$

$$t_{\bar{1}\bar{1}1} = \frac{1}{2}(s - p_x - p_y + p_z)$$

In sofar as these statements refer to single particles, most of them make quantum-mechanical sense. However, confusion sets in as soon as simultaneous reference is made to the behaviour of more than one electron. Statements 6–10 are typical of this kind. It is intended to assess the validity of these statements that together make the case for sp^3 hybridization, but first a brief summary of the degenerate set of wave functions that feature in the analysis.

The Wave Function $\psi(2, 1, m)$

The simplest form of sp^3 hybridization as generally postulated occurs for the carbon atom with the assumed $2s^2 2p^2$ electronic configuration. This notation derives from the solution of Schrödinger's equation for the hydrogen electron in the central field of a nuclear proton. The set of one-electron wave functions that occurs as the solution of the differential equation is characterized by the quantum numbers $n = 1, 2, \ldots, l = 0, 1, \ldots, (n-1)$, $m = -l, \ldots, +l$. Each wave function, $\psi(n, l, m)$, generally referred to as an *orbital*, describes the electron in one of two stationary states characterized by a further quantum number $m_s = \pm\frac{1}{2}$. The orbital is an eigenfunction of the motion with definite eigenvalues of total energy, total orbital angular momentum and orbital angular momentum in projection on an arbitrary direction, fixed in space, specified by integral values of the quantum numbers n, l and m. In hydrogen the electron may occur in any eigenstate, defined by a specific set of allowed quantum numbers n, l, m and m_s. It is mathematically convenient to formulate the central-field wave equation in spherical polar coordinates, as defined with reference to Figure 20-1 by

$$z = r\cos\theta \qquad \rho = r\sin\theta$$
(2)
$$x = r\sin\theta\cos\phi \qquad y = r\sin\theta\sin\phi$$

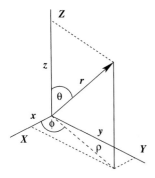

Figure 20-1. Transformation from Cartesian to spherical polar coordinates

The allowed normalized eigenfunctions and eigenvalues are given by:

(3) $$\psi_m(\phi) = e^{im\phi} f(\theta, r) \qquad L_z = m\hbar$$

(4) $$\psi_{lm}(\theta, \phi) = P_l^m(\cos\theta)e^{im\phi} \qquad L^2 = l(l+1)\hbar^2$$

(5) $$\psi_{nlm}(r, \theta, \phi) = R_{nl}(r)P_l^m(\cos\theta)e^{im\phi} \qquad E = -\frac{\mu e^4}{2\hbar^2(4\pi\epsilon_0)^2}\frac{1}{n^2}$$

For $n = 2$ the angular momentum functions are

(6) $$l = 0: \qquad Y_{0,0}(\psi_{2s}) = \frac{1}{\sqrt{4\pi}}$$

(7) $$l = 1: \qquad Y_{1,0}(\psi_0) = \sqrt{\frac{3}{4\pi}}\cos\theta = \sqrt{\frac{3}{4\pi}}\cdot\frac{z}{r}$$

(8) $$Y_{1,\pm1}(\psi_{\pm1}) = \mp\sqrt{\frac{3}{8\pi}}\sin\theta\, e^{\pm i\phi} = \mp\sqrt{\frac{3}{8\pi}}\cdot\frac{x\pm iy}{r}$$

Angular momentum

The three components of classical angular momentum are

(9) $$L_x = yp_z - zp_y$$

(10) $$L_y = zp_x - xp_z$$

(11) $$L_z = xp_y - yp_x$$

The product of any two components gives an expression that can be rearranged to show that, for instance

$$L_xL_y = (yp_z - zp_y)(zp_x - xp_z)$$
$$= (zp_x - xp_z)(yp_z - zp_y)$$
$$= L_yL_x$$

The two components are said to *commute*. Quantum-mechanically the components of linear momentum are represented by differential operators, such as $p \to -i\hbar\partial/\partial x$, to give operators

(12) $L_x = \frac{\hbar}{i}\left(y\frac{\partial}{\partial z} - z\frac{\partial}{\partial y}\right)$

(13) $L_y = \frac{\hbar}{i}\left(z\frac{\partial}{\partial x} - x\frac{\partial}{\partial z}\right)$

(14) $L_z = \frac{\hbar}{i}\left(x\frac{\partial}{\partial y} - y\frac{\partial}{\partial x}\right)$

In this case a *commutator* such as (L_x, L_y) is non-zero, in which

$$(L_x, L_y) = L_x L_y - L_y L_x$$

$$= -\hbar^2\left[\left(y\frac{\partial}{\partial z} - z\frac{\partial}{\partial y}\right)\left(z\frac{\partial}{\partial x} - x\frac{\partial}{\partial z}\right) - \left(z\frac{\partial}{\partial x} - x\frac{\partial}{\partial z}\right)\left(y\frac{\partial}{\partial z} - z\frac{\partial}{\partial y}\right)\right]$$

Because x, y and z directions are mutually orthogonal all mixed derivatives such as $\partial x/\partial y$ are equal to zero, and hence

(15) $(L_x, L_y) = -\hbar^2\left[y\frac{\partial}{\partial x} - x\frac{\partial}{\partial y}\right] = i\hbar L_z$

By cyclic permutation, $(L_y, L_z) = i\hbar L_x$, $(L_z, L_x) = i\hbar L_y$. The three components of angular momentum therefore do not commute.

The absolute value of the angular momentum (the total angular momentum) $|L|$ is defined by the relation

(16) $L^2 = L_x^2 + L_y^2 + L_z^2$

The commutation relations of L^2 with the components can be calculated directly, e.g.

$$(L^2, L_z) = (L^2 L_z - L_z L^2) = (L_x^2 + L_y^2)L_z - L_z(L_x^2 + L_y^2)$$

$$= i\hbar(L_x L_y + L_y L_x - L_y L_x - L_x L_y) = 0$$

The same calculation, repeated for (L^2, L_x); (L^2, L_y), shows that L^2 commutes with all components L_x, L_y and L_z individually and it is therefore possible to measure simultaneously L^2 and any single component of L. However, as these components do not commute amongst themselves, not more than one of these can be specified independently at a time. If L_z is considered known, L_x and L_y remain undefined within the limitation

(17) $L_x^2 + L_y^2 = L^2 - m^2\hbar^2$

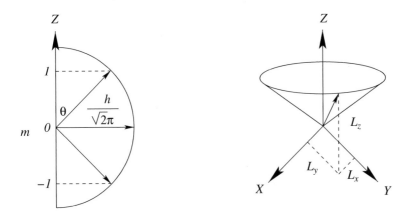

Figure 20-2. The orbital angular momentum vectors for various values of m. The vector L precesses about the Z-axis, so that L_z stays constant, but L_x and L_y are variable

The vector model The total angular momentum can be represented by a vector of length $L/\hbar = \sqrt{l(l+1)}$ with components m, up to $\pm l$, in the z-direction. The possible projections of $L = \sqrt{2}\hbar$ for $l = 1$, are shown in Figure 20-2. As L_x and L_y are undefined the angular momentum vector should be considered as randomly distributed over all possible allowed values of the azimuthal angle, consistent with the known projected value on the Z-axis, covering a cone with a vector angle given by $\cos\theta = m/\sqrt{l(l+1)}$. Although the component of angular momentum is quantized in only one direction, the direction of this chosen axis has no physical significance.

Rotation of axes

To prove the previous statement it is necessary to show that the same function is obtained by working in a coordinate system in which the axes have been rotated by an arbitrary amount [2]. As an illustration, consider the effect on the vector $r = L/\hbar$, for $l = 1$, of the rotation through an arbitrary angle β about the y-axis. The transformed coordinates are

$$z' = r\cos(\theta + \beta)$$
(18)
$$= z\cos\beta - x\sin\beta$$
$$x' = r\sin(\theta + \beta)$$
(19)
$$= x\cos\beta + z\sin\beta$$
(20) $\quad y' = y$
(21) $\quad r' = r$

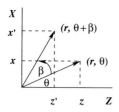

Figure 20-3. Rotation of a vector through an arbitrary angle β

From (7) and (8) the transformed wave functions for arbitrary β and for $\beta = 90°$ follow as:

$$(22) \qquad \psi_1' = \sqrt{\frac{3}{8\pi}} \frac{(x \cos\beta - z \sin\beta + iy)}{r} \rightarrow \sqrt{\frac{3}{8\pi}} \left(\frac{-z + iy}{r} \right)$$

$$(23) \qquad \psi_0' = \sqrt{\frac{3}{4\pi}} \frac{z \cos\beta - x \sin\beta}{r} \rightarrow \sqrt{\frac{3}{4\pi}} \left(\frac{-x}{r} \right)$$

$$(24) \qquad \psi_{-1}' = \sqrt{\frac{3}{8\pi}} \frac{x \cos\beta - z \sin\beta - iy}{r} \rightarrow \sqrt{\frac{3}{8\pi}} \left(\frac{-z - iy}{r} \right)$$

The coordinates expressed in terms of the eigenfunctions are

$$(25) \qquad x = \frac{r}{2\sqrt{3/8\pi}} (\psi_1 + \psi_{-1})$$

$$(26) \qquad iy = \frac{r}{2\sqrt{3/8\pi}} (\psi_1 - \psi_{-1})$$

$$(27) \qquad z = \frac{r}{\sqrt{3/4\pi}} \psi_0$$

Substituting back into (22)–(24) shows that the transformed wave functions are linear combinations of the eigenfunctions of L_z:

$$(28) \qquad \psi_1' = -\frac{1}{2}\psi_1(\cos\beta - 1) - \frac{1}{\sqrt{2}}\psi_0 \sin\beta - \frac{1}{2}\psi_{-1}(\cos\beta + 1)$$

$$(29) \qquad \psi_0' = \frac{1}{\sqrt{2}}\psi_1 \sin\beta + \psi_0 \cos\beta + \frac{1}{\sqrt{2}}\psi_{-1} \sin\beta$$

$$(30) \qquad \psi_{-1}' = -\frac{1}{2}\psi_1(\cos\beta + 1) + \frac{1}{\sqrt{2}}\psi_0 \sin\beta - \frac{1}{2}\psi_{-1}(\cos\beta - 1)$$

These transformations show that a particle with eigenfunction ψ_0, i.e. $L_z = 0$ in the old system may have eigenvalues $+1$, 0, or -1 after transformation.

In summary, when an atom has a definite value of L_z, it has indefinite values of L_x and L_y, but it has the latent ability to develop a definite, but completely unpredictable, value of either L_x or L_y, provided, for example, that it interacts with a suitably oriented Stern-Gerlach apparatus. In such a process, it would, of course, develop an indefinite value for L_z. As a generalization of this result, as equations (28–30) show, on arbitrary rotation[1], any given spherical harmonic $Y_l^m(\theta, \phi)$, becomes a linear combination of spherical harmonics with the same l, and coefficients that depend on the angle and direction of rotation.

The degenerate state

Theories of chemical bonding based on the properties of degenerate states with fixed l assume independent behaviour of the electrons in these states. In particular, for three electrons in the three-fold degenerate p-state with $l = 1$, they are assumed to have distinct values of m, without mutual interference. To make this distinction it is necessary to identify some preferred direction in which the components of angular momentum are quantized. By convention this direction is labeled as Cartesian Z. If the electrons share the degenerate p-state with parallel spins, they must share the same direction of quantization. This being the case, only one of the electrons can have the quantum number $m = 0$, characteristic of the real function (7).

The eigenfunctions for the other two electrons constitute a two-dimensional complex pair[2] (8). Any linear combination of these functions results in an equivalent set of three new orthogonal eigenfunctions of orbital angular momentum quantized in a new direction that depends on the transformation coefficient matrix, once again consisting of one real function and a complex pair. Polar plots of the probability distributions $|Y_1^m|^2$ are commonly shown as in Figure 20-4, which is a gross exaggeration of the density distribution.

The functions Y_l^m, in spherical polar coordinates, appear as solutions to Laplace's equation of degree l:

$$V_l = r^l Y_l^m(\theta, \varphi), \quad Y_l^m = P_l^m(\cos\theta)e^{im\varphi}$$

---

[1] Rotation about the Y-axis is readily generalized to arbitrary rotations and larger values of l [3].
[2] A solution of Laplace's equation

(31) $\nabla^2 V_l = 0,$

separated in Cartesian coordinates has the form

(32) $V_l = (ax + by + cz)^l$

(33)   with $a^2 + b^2 + c^2 = 0$

Because of (33) the three numbers a, b and c cannot be simultaneously real.

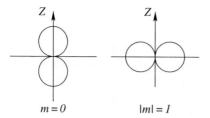

$m=0$ $\qquad$ $|m|=1$

Figure 20-4. Textbook examples of the degenerate p-functions shown as polar plots. These plots are to be imagined as having rotational symmetry about the Z-axis. For $|m| = 1$ the resulting distribution resembles a torus

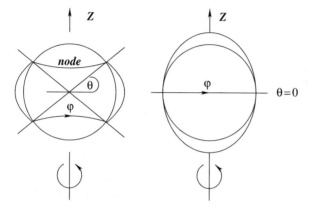

Figure 20-5. Spherical surface harmonics for $l = 1$, $m = \pm 1$ (left) and $m = 0$

For $r = 1$, $V_l = Y_l^m$ so that Y_l^m is the value of the solid harmonic in the surface of the unit sphere at points defined by the coordinates θ and φ, and hence are known as *surface harmonics* of degree l. The associated Legendre polynomials $P_l^m(\cos\theta)$ have $l - m$ roots. Each of them defines a nodal cone that intersects a constant sphere in a circle. These nodes, as shown in Figure 20-5, are in the surface of the sphere and not at $r = 0$ as assumed in the definition of atomic orbitals. Surface harmonics are obviously undefined for $r = 0$. The linear combinations $\psi_1 \pm \psi_{-1}$ define one real and one imaginary function directed along the X and Y Cartesian axes respectively, but these functions (denoted p_x and ip_y) are no longer eigenfunctions of L_z, but of L_x or L_y instead.

The question whether two real functions such as p_x and p_z can be defined simultaneously may appear not to be answered explicitly by the fact that L_x and L_z do not commute. The answer however, comes from the complementarity of angular momentum and rotation angle, of the same kind as the complementarity between linear momentum and position[3]. An electron can have definite angular momentum [2]

---

[3] Direct comparison [3] shows that, for given l, corresponding matrix components of position and momentum operators with respect to Y_l^m have the same dependence on m.

only when its wave function has the appropriate dependence on angle, i.e. $Y_l^m(\theta, \phi)$, analogous to the fact that it can have a definite momentum only when its wave function has an appropriate dependence on position, i.e. $\exp(ipx/\hbar)$. The uncertain rotation angle for electrons with $m = \pm 1$ therefore disallows rotation in a fixed plane perpendicular to either X or Y. The uncertainty nowhere implies that the angular dependence of electron density on m is any different from the m-dependence of angular momentum. It is inferred that the vectors p_x, p_y, p_z, like L_x, L_y and L_z do not commute and cannot be simultaneous solutions to (4).

Fact or Fiction?

The reason that all the arguments of the previous section are consistently ignored by chemists lies with the number of undergraduate textbooks that repeat the Pauling model as scientific fact. After fifty years, a tradition, established over generations of chemists, is no longer subject to scrutiny. It may therefore help to re-examine some of the original assumptions.

1. The "orbital" concept is not precisely defined and according to Coulson [4] is a relic from the Bohr-Rutherford planetary model of the atom:

The wave function may be said to describe the motion of the electron. In a rather loose way we could speak of it as describing an orbit, though of course we have now abandoned the hopeless attempt to follow the electron in its path; in fact we believe that even the idea of a path has no meaning. But on account of its relationship to the distribution of the electron we call it an *atomic orbital* a.o. For example, there are three *p*-type a.o.'s in which the boundary surface consists of two regions together resembling a "dumb-bell". There is a very marked directional character in these orbitals, which we exhibit by means of a suffix p_x, p_y, p_z.

Chemistry teachers are encouraged to accept this vague picture at face value and to believe that p_x, p_y, p_z occurs as a three-fold degenerate solution of Schrödinger's equation for the H electron. It does not.

3. Reading this statement together with 9 may create the impression of a one-to-one relationship between the so-called p_x, p_y and p_z orbitals and the three allowed values of m. Although Pauling [1] never makes this identification, authors of numerous textbooks fall into the trap. One author [5], after correctly defining the angular wave functions for $m = 0$ and $m = \pm 1$ as $(3/4\pi)^{1/2} \cos\theta$ and $(3/8\pi)^{1/2} \sin\theta \exp(\pm i\phi)$ respectively, proceeds to state without further proof:

There are actually three degenerate *p* wave functions, distinguished by their *m* values. They all have the same orbital shape but their symmetry axes are mutually at right angles to each other – hence their designation p_x, p_y and p_z according to whether their lobes are symmetrical about the *x*, *y* or *z* reference axis. There are only three linearly independent *p*-type orbitals – any similar orbital, pointing in a different direction, may be regarded as the superposition of certain amounts of p_x, p_y and p_z. This is done vectorially so that we may write

$$p = ap_x + bp_y + cp_z$$

where *a*, *b*, and *c* are the direction cosines of the angles between p_x, p_y and p_z respectively and the new orbital *p*.

As shown already, the linear combinations that define p_x and p_y are simple rotations of the coordinate axes, as, for instance, defined by (25). It follows that the functions p_x and p_y are both characterized by the quantum number $m = 0$ that defines zero component of angular momentum, now directed along the Cartesian X and Y axes respectively.

4. In conjunction with 7 this statement infers that p_x, p_y and p_z act simultaneously to form covalent bonds. These "orbitals" can never occur together. As soon as p_z is rotated into p_x or p_y the remaining two transform into an orthogonal complex pair. (Compare 22–24).

5–7. These statements are all based on the same false premises that the pair of real functions, p_x and p_y, are equivalent to the complex pair $\exp(\pm i\phi)$.

8. The statement in parentheses reflects a complete misreading of the hydrogen model. In a central (Coulomb) field the only angle-dependent solutions to the wave equation are the spherical harmonics. The only way in which to generate "tetrahedral orbitals" is by introducing boundary conditions that define an external field with tetrahedral symmetry. Such a procedure was outlined by Jahn and Teller [6] but has never been fully developed. It is not a derivation of molecular geometry, but requires prior knowledge of the molecular symmetry of interest.

10. As the labels imply, the t's provide the geometric definition of the four half body-diagonals of a cube, which define a tetrahedron. When the definitions of 9 are compared with Equation (2) the symbols p_x and p_y are seen to be the identification of Cartesian axes in polar coordinates. The sp^3 linear combination is thereby identified as a geometrical, rather than a quantum-mechanical, definition of the four assumed tetrahedral bonds of methane in polar coordinates.

Some of the statements are vague and misleading. Even apart from the fact that the three real p functions cannot be defined in a single coordinate system for the purpose of linear combination, a more serious objection is that each of them has the same magnetic quantum number $m = 0$. In addition they also have $n = 2$ and $l = 1$. The assumption therefore violates the Pauli exclusion principle, emphasized in statement 2. However, the sp^3 carbon has three electrons with $n = 2$, $l = 1$ and $m = 0$, and there are only two possible spin values, $m_s = \pm\frac{1}{2}$. The exclusion principle is widely recognized to be as ruthless as the second law of thermodynamics. The idea of sp^3 hybridization is therefore as ludicrous as perpetual motion.

Tetrahedral carbon

Orbital hybridization, like the Bohr model of the hydrogen atom in its ground state, is an effort to dress up a defective classical model by the assumption ad hoc of quantum features. The effort fails in both cases because the quantum-mechanics of angular momentum is applied incorrectly. The Bohr model assumes a unit of quantized angular momentum for the electron which is presumed to orbit the nucleus in a classical sense. Quantum-mechanically however, it has no orbital angular momentum. The hybridization model, in turn, spurns the commutation rules of quantized angular momenta and assumes the superposition of classical cartesian components in a quantum-like sense. This assumption violates the strictest of quantum rules and the end result is purely classical as demonstrated for methane.

Each of the three functions p_x, p_y, p_z is a linear combination of the eigenfunctions of L_z. Linear combinations of the three functions themselves must therefore also be linear combinations of these same eigenfunctions and solutions of (4). Howsoever physically inappropriate, it is still of interest to examine the mathematics of these linear combinations.

The functions p_x and p_y are defined as linear combinations of the ψ_1 and ψ_{-1} of (8), i.e.

$$(34) \qquad p_x \propto \psi_1 + \psi_{-1} = -\sqrt{\frac{3}{8\pi}} \sin \theta (e^{i\phi} + e^{-i\phi}) = -\sqrt{\frac{3}{2\pi}} \sin \theta \cos \phi$$

$$(35) \qquad p_y \propto \psi_1 - \psi_{-1} = \sqrt{\frac{3}{8\pi}} \sin \theta (e^{i\phi} - e^{-i\phi}) = i\sqrt{\frac{3}{2\pi}} \sin \theta \sin \phi$$

and

$$(36) \qquad p_z \propto \psi_0 = \sqrt{\frac{3}{4\pi}} \cos \theta = \sqrt{2}\sqrt{\frac{3}{2\pi}} \cos \theta$$

The widely accepted forms of these functions therefore correspond to:

$$(37) \qquad p_x = \sin \theta \cos \phi = -\sqrt{\frac{2\pi}{3}} (\psi_1 + \psi_{-1})$$

$$(38) \qquad p_y = \sin \theta \sin \phi = -i\sqrt{\frac{2\pi}{3}} (\psi_1 - \psi_{-1})$$

$$(39) \qquad p_z = \cos \theta = \frac{1}{\sqrt{2}} \sqrt{\frac{2\pi}{3}} \psi_0$$

These equations allow calculation of the linear combination

$$(40) \qquad p^3 = p_x + p_y + p_z = \sin \theta (\cos \phi + \sin \phi) + \cos \theta$$

$$(41) \qquad = -\sqrt{\frac{2\pi}{3}} \left[(1+i)\psi_1 + (1-i)\psi_{-1} - \frac{1}{\sqrt{2}} \psi_0 \right]$$

The vector p^3 defines a new quantization direction (40) of angular momentum as a linear combination (41) of the eigenfunctions ψ_0 and $\psi_{\pm 1}$, which is equivalent to a rotation of the axes (28–30). The s contribution has no effect on the direction.

To interpret this result it is important to note that by (26) the y-coordinate is complex and hence the angle of rotation, β of Equations (28–30), refers to a rotation with respect to the complex XY plane. When the coefficients of the equivalent linear combinations (29) and (41) are compared, it is seen that only ψ_0 has a real coefficient, which specifies the rotation angle of the Z-axis with respect to the complex plane that contains Y; $\cos\beta = 1/\sqrt{2}$, $\beta = 45°$. The absolute direction of p^3 (40) depends on the value of ϕ, as fixed by the direction of the Y-axis, which remains undefined in a complex plane. The linear combination therefore defines a vector that rotates at an azimuthal angle of $45°$ with respect to old Z, as Y rotates in the XY-plane. It may be fixed at any value of ϕ to define a new Z-direction, in terms of which, the degenerate set of p^3 functions satisfies (7) and (8). Nothing is gained. The vector p^3 simply specifies a new arbitrary direction of the polar axis. The original Z-direction is no longer defined.

Given this situation, it is almost trite to point out that the four t vectors of sp^3 belong to four equivalent, but different, linear combinations, each of them a member of a different set of three degenerate functions, and they never occur together as simultaneous solutions of (4) in a single coordinate system. A linear combination of these t functions, once more, is a solution of (4) and defines a quantization direction along the original Z-axis.

New solutions can be generated indefinitely by new linear combinations, amounting to an endless number of axial rotations without any progress beyond the definition and rotation of the three-fold degenerate set, defined by (7) and (8). In the solid angle of 4π the polar axis (Z) can be directed in infinitely many directions, each representing a linear combination of ψ_1, ψ_0 and ψ_{-1} of (7) and (8). In a chemical context the bottom line is that each linear combination (e.g. t_{111}) defines a specific orientation of the polar axis[4]. With this choice of axes there is no second polar axis and no other linear combination (e.g. $t_{1\bar{1}\bar{1}}$) that solves (4) in the same coordinate system. The tetrahedral carbon atom remains quantum-mechanically undefined.

The prescription [7], which defines the set of four tetrahedral orbitals, consists in writing the product states $sp_x p_y p_z$ as a Slater determinant:

$$
\begin{vmatrix}
s(1) & p_x(1) & p_y(1) & p_z(1) \\
s(2) & p_x(2) & p_y(2) & p_z(2) \\
s(3) & p_x(3) & p_y(3) & p_z(3) \\
s(4) & p_x(4) & p_y(4) & p_z(4)
\end{vmatrix}
$$

The purpose of this formulation is to obtain an anti-symmetric wave function for the four carbon valence electrons. The determinant, and hence the total wave function, changes sign with the interchange of any two electrons, as required. However, each of the individual product states, such as $s(1)p_x(2)p_y(3)p_z(4)$, is symmetrical and hence forbidden; i.e. nonexistent. The prescription would be valid in a classical environment, albeit with obscure meaning.

---

[4]There is nothing 'tetrahedral' about this direction, which is only defined with respect to an arbitrary Z.

The thermodynamic problem

The carbon atom is assumed here to be structured according to the one-electron Schrödinger solution for hydrogen. The L-shell with its four electrons therefore consists of the eigenstates defined by $Y_m^l = Y_0^0, Y_0^1, Y_1^1, Y_{-1}^1$, i.e. $2s_0$, $2p_0$, $2p_1$, and $2p_{-1}$. The p-functions occur as a degenerate set, with components of angular momentum, $L_z = m\hbar$. Consequently, excitation of a carbon atom from the ground state to the first excited state

$$2s^2 2p^2 \rightarrow 2s^1 2p^3$$

requires not only energy, but also angular momentum. The transition could be effected by absorption of a photon with an angular momentum component $\hbar$ [8] and the selection rule $\Delta l = \pm 1$, but not by thermal excitation. In hybridization theory this conservation of angular momentum is ignored. Despite the large difference in energy of the s and p-states, linear combinations of the four eigenfunctions are nevertheless assumed to produce alternative solutions to the wave equation, which define the allowed geometries of carbon compounds. This act of faith is not supported by the laws of physics nor the mathematical model. As long as a central-field potential applies, ANY linear combination of the relevant eigenfunctions redefines the SAME eigenfunctions as before, but with rotated coordinate axes. The only way to obtain a wave function for tetrahedral carbon is by formulating the differential equation with the appropriate potential and boundary conditions that define a tetrahedral ligand field[5]. To achieve this, the molecular geometry still has to be assumed, rather than predicted, and the electron density would not necessarily appear as four Lewis pairs.

Summary

The Schrödinger wave equation that describes the motion of an electron in an isolated hydrogen atom is a second-order linear differential equation that may be solved after specification of suitable boundary conditions, based on physical considerations. The solution to the equation, known as a *wave function* provides an exhaustive description of the dynamic variables associated with electronic motion in the central Coulomb field of the proton.

Solution of the equation is achieved by a standard mathematical procedure, known as *separation of variables*, such that the total wave function is obtained as the product of three sets of *eigenfunctions* with associated *eigenvalues* of energy

---

[5]The linear combination of atomic orbitals (LCAO) used as an interpolation method to derive the symmetry properties of energy bands in crystals [9] has as coefficients the values of plane waves, $\exp(ik.R)$, at the positions R of atoms in the unit cell. Rather than a LCAO the derived wave function therefore is a linear combination of Bloch sums with the full space-group symmetry of the crystal. Atomic orbitals, by contrast, are modes in a spherically symmetrical central Coulomb field.

and orbital angular momentum[6], quantified by three *quantum numbers*. A further variable, called *spin*, remains hidden, unless special care is exercised to ensure the correct transformation properties of the system [10]. It is quantified by two further quantum numbers.

The terminology and symbolism used to specify the various quantum numbers are not too informative. The numbers are known as the *principal (n), azimuthal (l), magnetic* (m_l), *spin (s)* and *magnetic spin* (m_s), quantum numbers. The first three are integers, such that, for one set of eigenfunctions, n is a positive number, l is always less than n and m_l has a total of $(2l+1)$ allowed values, clustered about zero. For $n = 2$ and $l = 1$ it follows that m_l has the three possible values $+1$, 0 and -1. The quantum numbers s and m_s have half-integer values. All electrons have $s = \frac{1}{2}$ and $m_s = \pm\frac{1}{2}$.

The total-energy eigenvalues of the electron are inversely proportional to the square of n. Each value of n is said to specify an *energy level*, and the energy of the electron remains the same while it stays at the same energy level, irrespective of the values of the other quantum numbers. *Sub-levels* with the same energy are known as *degenerate states* of the electron.

The allowed values of the various quantum numbers determine that each energy level has a fixed number of sub-levels. At the first level ($n = 1$) there is only one allowed state, known as an s state; in this case $1s$, having both l and m_l equal to zero. For $n = 2$ there are two possibilities, $l = 0$ or $l = 1$. In the first case the condition $l = m_l = 0$ defines the $2s$ state. The sub-level with $l = 1$ has $(2l+1) = 3$ possible states, known as p-states and is *three-fold* degenerate with $m_l = 1$, 0 or -1.

The quantum number l specifies the angular momentum of the electron in units of $\hbar$ (h bar), known as *Planck's constant*. In the presence of an applied magnetic field the component of angular momentum in the direction of the field is quantified by m_l as $L_z = m_l \hbar$. The subscript z refers to the convention of defining a right-handed set of Cartesian laboratory axes such that Z coincides with the direction of the magnetic field.

It is worth noting that the square of total angular momentum of the electron is a scalar quantity whereas L_z is a vector. For $l = 0$, in any s-state, including the $1s$ ground state of hydrogen, the electron has zero angular momentum. It is a curious fact that an electron in the p-state with $m_l = 0$ has non-zero angular momentum, $L = \sqrt{l(l+1)} \cdot \hbar = \sqrt{2}\hbar$, but zero component in the direction of an applied field (compare Figure 20-2). The only possible explanation [11] of this phenomenon is that the total angular momentum does not arise from rotation about an axis, but from another kind of rotation in a *spherical mode*.

An important property of quantum-mechanical angular momentum is the prediction and demonstration that the Cartesian components thereof do not commute[7]. This property is responsible for the fact that two components, such as L_z and L_x

---

[6]When the term *angular momentum* is used in the following without qualification, it always refers to orbital angular momentum.

[7]For non-commuting variables, a and b, the product $ab \neq ba$.

cannot be measured simultaneously. In order to measure L_x it is necessary to apply a magnetic field in the X-direction, which destroys all previous knowledge of L_z. It is the same property, exhibited by visible light, which can be polarized in only one direction at a time.

The square of an electronic eigenfunction has the same physical meaning as the intensity of a light beam and is interpreted as a measure of *charge density*. The angular distribution of the density depends on the spherical harmonic eigenfunctions of the angular momentum. A polar plot of such a function (Figure 20-4) for $m_l = 0$ is assumed to show extension of electron density in the direction of Z, whereas the functions $m_l = \pm 1$ describe identical[8] density distributions. In reality these charge distributions respectively resemble oblate and prolate ellipsoids centred in the XY-plane, and are shown schematically in Figure 20-5. There is no reason that the magnetic field should not line up with the laboratory X or Y axes. This choice has no effect on the charge distribution, except that the direction of elongation would be labeled X (or Y) and the ellipsoid defined in the YZ (or ZX) plane. It has become general practice in chemistry to refer to the presumed polar density distribution for $m_l = 0$, shown in Figure 20-4, as a p_z-orbital. Should the coordinate axes be relabeled, this entity should also be relabeled as either p_x or p_y. It is physically meaningless to specify two of these orbitals at the same time.

The power of quantum mechanics is revealed by experimental confirmation of the predicted spectroscopic properties of atomic hydrogen. The reasonable expectation of successfully extending the method to many-electron atoms and molecules has been thwarted by mathematical complexity. It has never been possible to solve the wave equation for the motion of more than one particle. The most complex chemical system that has been solved (numerically) is for the single electron in the field of two protons, clamped in place, to define the molecular ion H_2^+. In order to apply the methods of quantum mechanics to any atom or molecule, apart from H and H_2^+, it is necessary to apply approximation methods or introduce additional assumptions based on chemical intuition.

The first, reasonable assumption to be made, was that the energy and angular momentum eigenfunctions obtained in the hydrogen problem, could serve as a guideline to find approximate solutions for more complex atoms. As for hydrogen, it may therefore be assumed that, the lowest energy state would occur when all electrons of the atom are concentrated in the lowest $1s$ state. This assumption could soon be demonstrated by spectroscopic analysis to be unwarranted. A better assumption that predicts the distribution of electrons among energy levels of the hydrogen type, was formulated on the hand of what became known as Pauli's *exclusion principle*.

The way in which the exclusion principle determines the order of hydrogen-like energy-level occupation in many-electron atoms is by dictating a unique set of quantum numbers, n, l, m_l and m_s, for each electron in the atom. Application of

---

[8] The two eigenfunctions only differ in the sense of the rotation that defines the direction of the angular momentum vector.

this rule shows that the sub-levels with $l = 0, 1, 2$ can accommodate no more than 2, 6, 10 electrons respectively. In particular, no more than two electrons with $m_s = \pm\frac{1}{2}$, can share the same value of m_l. The idea of having three electrons of a carbon atom in so-called p_x, p_y and p_z orbitals, each with $m_l = 0$, is therefore forbidden by the exclusion principle.

Despite the fact that the physics of atomic structure therefore militates against a $p_x p_y p_z$ set of electrons on the same atom, it may be (and often is) argued that, since each of the three eigenfunctions, separately, solves the atomic wave equation, a linear combination of the three must likewise be a solution of the same equation[9]. Formation of such a linear combination is a purely mathematical procedure without any reference to electrons. It simply is a manipulation of three one-electron eigenfunctions and it is of interest to examine what physical meaning attached to the operation.

As each of the three functions p_x, p_y and p_z is a linear combination of the three eigenfunctions with $m_l = 1, 0, -1$, their linear combination is just another linear combination of these same eigenfunctions, only with different coefficients. Representing the eigenfunctions by the symbol F, the final linear combination of p_x, p_y, p_z, therefore has the algebraic structure

(42) $$LC = aF_1 + bF_0 + cF_{-1}$$

Analysis of this type of expression shows that LC is just another one-electron function with $m_l = 0$ or ± 1, directed at a special angle, that depends on the coefficients a, b and c. An infinite number of such linear combinations is possible, each defining another one-electron eigenfunction directed at one of an infinite number of angles, measured with respect to the original laboratory coordinate system. The important conclusion is that each linear combination corresponds to a new choice of axes. Selection of the polar axis along any Z always leaves p_x and p_y undefined as separate entities. In particular, there is no hope ever to simulate the tetrahedral structure of methane in terms of a linear combination of carbon electron eigenfunctions. That requires four linear combinations, each with a different polar axis, which is physically impossible.

CHEMICAL COHESION

The fact that molecular dissociation energies can be modelled in terms of interacting point charges does not imply that intramolecular interactions obey the laws of classical electrostatics. Assuming an equilibrium structure for H_2^+ as shown, the electrostatic interaction energy follows as

(43) $$\frac{3e^2}{4\pi\epsilon_0 r} = \frac{4.2 \times 10^3}{r} \quad \text{kJ mol}^{-1}$$

---

[9]This is a mathematical property of any linear second-order differential equation.

Equating this quantity with the measured dissociation energy 269 kJ mol^{-1}, predicts an interatomic distance 16 Å, an order of magnitude too large. On closer approach between the nuclei, the calculated energy increases without bound. The classical model therefore fails disastrously.

By comparison, numerical solution of the one-electron Schrödinger equation for clamped nuclei, predicts the correct dissociation energy and interatomic distance for H_2^+. These calculations show that the electrostatic interaction which stabilizes the molecule arises from quantum-mechanical charge distributions. Classical models, such as the Lewis or van't Hoff models, can therefore be rejected at the outset. Semi-classical models such as LCAO, are of the same kind, in view of the demonstrated classical nature of hybridization.

If H_2^+ is to serve as a model for quantum-mechanical simulation of chemical cohesion in general, certain aspects need more careful consideration. In order to find the minimum-energy interatomic distance it is necessary to repeat the calculation at a series of preselected values of r. The resulting binding-energy curve has a minimum at the observed interatomic distance. This is also the best result that can be anticipated for more complex molecules, i.e. a minimum-energy separation between all atoms, considered pairwise, without distinction between first and second nearest neighbours. For example, the ethane molecule, C_2H_6 will be modelled in terms of 1 C—C, 12 C—H and 120 H—H interactions. The final result is not a three-dimensional structure, but a *radial distribution* of nuclear density. This conclusion is consistent with the fact that minimization of energy, which is a scalar quantity, cannot generate conformational information. It is standard practice to call on chemical intuition at this point, in the form of an assumed classical, van't Hoff type, structure. This assumption corrupts the results of the calculation by forcing preconceived notions into the model. From there on the calculation has no theoretical value, apart from confirming the assumed classical structure.

The useful information generated by the H_2^+ calculation is that optimal nearest-neighbour separation relates to minimization of the energy of interaction between an electron and two nuclei. The interaction between the nuclei is said to be *mediated* by the *exchange* of an electron[10]. This electron exchange is a non-classical process, and the resulting structure has nothing in common with the assumed linear arrangement. It is important to realize that the idea of clamped nuclei is equivalent to the assumption of a classical linear molecule. Quantum-mechanically however, nuclear positions are defined as a probability density and at the optimal internuclear distance the molecular structure approximates a spherical shell of positive density within a negative charge cloud, shown in

---

[10]By analogy, electromagnetic interaction is said to be mediated by the exchange of a virtual photon.

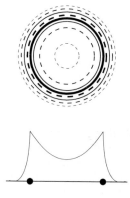

Figure 20-6. Spherical contours of charge densities

Figure 20-6 as a contour map of positive (solid lines) and negative densities. The two-dimensional diametrical section resembles the electron density for clamped nuclei.

For this $1s$ electron angular momentum plays no part [11] in the reconstruction. As a matter of fact, all homonuclear diatomic molecules probably have such spherical structures, assuming angular momentum vectors to cancel.

Non-local Interaction

If the charge distribution of Figure 20-6 appears bizarre, it is probably because protons are intuitively, but erroneously, considered to be less quantum mechanical than electrons. However, the interaction between electrons and nuclei in atoms, is the same as in molecules. The extranuclear electrons in atoms are prevented from penetrating into the nucleus under the influence of Coulombic attraction, by a repulsive force due to the quantum potential[12], [12],

$$(44) \qquad F_q = \frac{\partial V_q}{\partial r} = -\frac{\partial V}{\partial r}$$

Such a force balance is also responsible for establishing stationary states for molecules.

---

[11] Separation of the variables to enable solution of the electronic wave equation of H_2^+ requires clamping of the nuclei and hence imposing cylindrical symmetry on the system. The calculated angular momentum eigenfunctions are artefacts of this approximation and do not reflect the full symmetry of the quantum-mechanical molecule.

[12] The wave equation for continuous classical systems is non-linear. It differs from the Schrödinger equation in the appearance of the quantum-potential term, which linearizes the equation. All of the non-classical features of quantum-mechanics, such as linear superpositions, therefore arise from the quantum potential.

The quantum potential of a many-particle molecule is defined in terms of the radial wave function, R, as

$$(45) \qquad V_q = \sum_{j=1}^{n} -\frac{\hbar^2}{2R}\left(\frac{\nabla_j^2 R}{m_j}\right)$$

and is seen to depend on the quantum state of the entire system. Whereas R occurs in both numerator and denominator the quantum potential does not necessarily decay with distance. A *non-local connection* is said to exist between all particles in the system. If the system is disturbed in a localized region of three-dimensional space, the configuration space wave as a whole will respond and consequently all the particles making up the system are affected instantaneously.

The quantum potential connects all particles summed over in (45). In principle, the sum combines all particles in the universe into a single whole, and the corresponding wave function is said to be *holistic*. The holistic connection is broken whenever the wave function may be factorized as a product state. If the factorizable wave function is written in polar form,

$$(46) \qquad \psi(x_j, x_k) = Re^{iS/\hbar} = \psi_A(x_j)\psi_B(x_k)$$

the phase and amplitude functions are given by

$$(47) \qquad S(x_j, x_k) = S_A(x_j) + S_B(x_k)$$

$$(48) \qquad R(x_j, x_k) = R_A(x_j)R_B(x_k)$$

The quantum potentials are

$$(49) \qquad V_q = V_q^A(x_j) + V_q^B(x_k)$$

$$(50) \qquad V_q^A = -\frac{\hbar^2}{2R_A}\sum_j \frac{\nabla_j^2 R_A}{2m_j}$$

$$(51) \qquad V_q^B = -\frac{\hbar^2}{2R_B}\sum_k \frac{\nabla_k^2 R_B}{2m_k}$$

If these summands overlap the wave function is nonfactorizable and said to be *entangled*. It means that the motion of the two sectors, j and k, are correlated and this happens because they are physically connected by some classical interaction.

The individual sectors of a factorizable product-state system are *partially holistic* and appropriate to describe molecules. Each sector of this type has a characteristic quantum potential that keeps it together by balancing the classical potential. A molecule may therefore be viewed as a chemical entity made up of particles that move under the influence of a common quantum potential.

MOLECULAR CONFORMATION

Quantum-mechanical energy relationships that drive chemical interactions must inevitably lead to an unstructured probability distribution of matter. Neither can the total orbital angular momentum, which is observed as a squared quantity, dictate the three-dimensional conformation of a molecule. The components of angular momentum, however, are quantized in the direction of a polar molecular axis. As molecular fragments join up during chemical reaction, these components add vectorially, and any arrangement that minimizes the resultant angular momentum vector is favoured. The argument that this interaction is responsible for the symmetry breaking that generates molecular structure is supported by the demonstration that it correctly predicts the occurrence of geometrical and optical isomerism. Isomers of these types are stabilized by torsional barriers to rotation and molecular chirality respectively.

Torsional Rigidity

In the absence of a magnetic field, the polar axis for angular momentum on a free atom may be chosen in any direction without loss of generality. Interaction with another atom by electron exchange immediately changes this situation and the Z-direction for the diatomic system is established along the interatomic axis. In the case of a carbon atom interacting with a hydrogen atom, the electron density on carbon is polarized as shown in Figure 20-7 (a).

When two carbon atoms with residual angular momentum interact, coupling between the associated magnetic moments leads to an alignment which defines the polar axis in two possible ways, as shown in Figure 20-7 (b), both of them equally effective to quench the orbital angular momentum. If the two hydrogen atoms of a CH_2 molecule are assumed equivalent, the most likely distribution of hydrogen atoms is in the XY-plane, coincident with the concentration of the electron density, $(C2s^2 2p_1^1 2p_{-1}^1)$, that quenches the angular momentum, as shown in Figure 20-7(c,d).

When two CH_2 molecules combine, it seems reasonable to assume the superposition of two identical charge distributions of this kind, with a common Z-axis. An alternative is to assume that the two carbon atoms line up with respective electronic configurations of $C(p_z p_{+1})$ and $C(p_z p_{-1})$ as in 7(b)E. The azimuthal distribution of the angular momentum vectors of such a pair is in opposite sense as shown in Figure 20-7(e). The actual distribution of electron density in the ethylene molecule is conjectural. The obvious guess that anti-parallel $2p_z$ solutions quench the angular momentum, is favoured, but not confirmed, by chemical intuition.

Ethylene exhibits a non-steric barrier to rotation about the molecular axis. The conventional explanation ascribes this effect to the lateral π overlap of parallel p_z orbitals with spin pairing. The triple bond in acetylene is conventionally defined in terms of one sp σ-bond and two π-bonds at right angles to each other. Triple bonds should therefore be torsionally even more rigid than double bonds. The curious fact is that in dimetal systems, with the same type of triple bond as in acetylene, there are no geometrical isomers due to torsional rigidity [13].

The torsional barrier in ethylene is accounted for more logically by the energy required to twist the angular momentum vectors, which are anti-parallel in the planar arrangement, out of this alignment. In the case of acetylene, Figure 20-7(f), two CH molecules line up with the molecular axis along Z and electron pairs with respectice quantum numbers $m_l = \pm 1$ on the two carbon atoms according to 7(b)A. No torsional barrier is predicted.

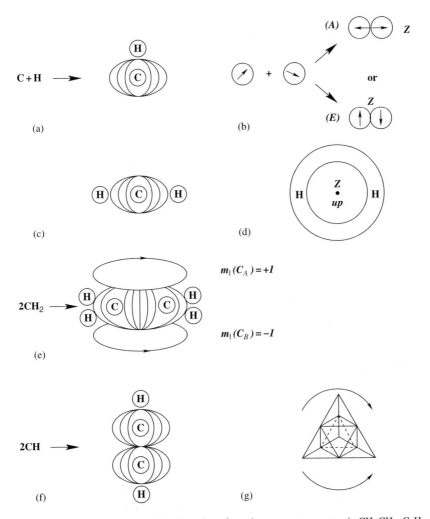

Figure 20-7. Diagrams to show the orientation of angular momentum vectors in CH, CH$_2$, C$_2$H$_4$, C$_2$H$_2$ and CH$_4$ molecules

Stereoisomerism

Optical activity is one of the best known, but least understood, phenomena of organic chemistry. Coupling between a polarized photon field and a molecule requires that the molecule has a magnetic moment, and hence residual angular momentum. The recognition that normal optical activity is always associated with chiral molecules, indicates that chirality is also characterized by non-zero angular momentum. Achiral molecules become optically active in an applied magnetic field, which confirms that an induced magnetic moment is responsible for this so-called Faraday effect. The demonstration that dissymmetric tetrahedral chiral molecules have non-zero angular momentum follows from an analysis of the structure of methane [11], shown in Figure 20-7(g).

The structure is obtained by assuming the four hydrogen atoms to be equivalent. If Z is defined by any H—C direction, the angular momentum is quenched only when two charges are seen to rotate in opposite sense on parallel planes, perpendicular to Z, $(m_l = \pm 1)$. By this argument four sets of parallel planes are finally required to surround the central carbon atom in a symmetrical fashion, because any of the four choices of Z must produce the same charge distribution. The eight planes intersect at the midpoints of the six lines that connect pairs of hydrogen atoms, to define a regular tetrahedron. When successive hydrogen atoms are replaced by different atoms, such as F, Cl or Br the octahedron becomes distorted, but the angular momentum remains quenched until all ligands are different and produce a chiral arrangement. Optical activity is fully accounted for [14].

CHEMICAL REACTION

To understand why and how chemical reactions happen it is necessary to consider also intermolecular interactions. It is only in the hypothetical case of an *ideal* gas that intermolecular interactions are totally absent. In all other systems they represent an important factor that affects molecular conformation, reactivity and stability. Whenever molecules co-exist in equilibrium it means that intermolecular forces are not sufficient to pull the molecules apart or together into larger aggregates. Equilibrium implies a balance of thermodynamic factors, and when these factors change, intermolecular interactions may overcome the integrity of a partially holistic molecule, and lead on to chemical reaction. Onset of the reaction is said to be controlled by an *activation energy barrier*. This barrier must clearly be closely allied to the quantum potential of the molecule.

Activation Energy

The link between activation energy and quantum potential is revealed by an atom under pressure. Uniform compression of an atom is simulated numerically by studying the energy eigenvalues as a function of the radial boundary condition

$$\lim_{r \to r_0} \psi(r) = 0$$

for $r_0 < \infty$. As r_0 decreases (interpreted as increasing pressure) all energy eigenvalues move to higher levels until the ground-state level coincides with the ionization level. At this point an electron becomes decoupled from the nucleus and finds itself in a hollow sphere. It is the point at which the classical potential has been overcome by the quantum potential. The electron has scaled the activation barrier and is said to be in the *valence state*.

Each atom has a characteristic ionization radius [15] and a characteristic value of valence-state quantum potential, identified with electronegativity [11]. It is only the artificial compression barrier that keeps the activated electron confined. In the real world an activated electron is free to interact with its environment and initiate chemical reaction. The activation is rarely caused by uniform compression and, more typically, is due to thermal, collisional, or catalytic activation.

The situation with molecules is no different. When a molecule enters the valence state it interacts with its environment. It is no longer the former partially holistic unit, and its quantum potential is redefined in the changing context. The molecule either breaks down into smaller wholes or it combines with other species to establish a larger whole. The two processes often occur concurrently to produce several reaction products.

In less violent encounters, such as crystallization, intermolecular interaction is just sufficient to establish a more extended whole, without completely destroying the partial molecular wholes. In a case such as this, interactions separate into strong intramolecular and weaker intermolecular interactions.

Intramolecular Rearrangement

The Beckmann rearrangement of ketoximes into acid amides, represented by the scheme:

consists of the trans interchange of a hydroxyl group with an aryl or alkyl group, also with retention of chirality [16]. No conceivable mechanism, based on the conventional view of covalent bonding, can possibly account for these observations, without breaking or making bonds. The accepted mechanism [17] simply states that

The migrating group does not become free but always remains connected in some way to the substrate.

To understand this, and other intramolecular rearrangements, it is necessary to give up the naïve notion of Lewis-type electron-pair bonds. The alternative is to view all interaction within a partially holistic molecule as mediated by its quantum potential and quantum torque. Both of these quantities are specified by the total molecular wave function. The necessary theory for this approach has not been worked out, and probably will not be, for as long as chemists cling to their outdated classical models.

There is no immediate need to abandon the conventional use of structural formulae as graphical models of molecular shape, provided it carries no connotation of electron pairs or hybrid orbitals. The extensive use of LCAO methods to simulate electron densities in molecules may be even harder to give up, but is also more misleading than the naïve model of localized Lewis pairs.

CONDENSED MATTER

Molecular quantum potential and non-local interaction depend on molecular size and the nature of intramolecular cohesion. Macromolecular assemblies such as polymers, biopolymers, liquids, glasses, crystals and quasicrystals are different forms of condensed matter with characteristic quantum potentials. The one property they have in common is non-local long-range interaction, albeit of different intensity. Without enquiring into the mechanism of their formation, various forms of condensed matter are considered to have well-defined electronic potential energies that depend on the nuclear framework. A regular array of nuclei in a structure such as diamond maximizes cohesive interaction between nuclei and electrons, precisely balanced by the quantum potential, almost as in an atom.

Band theory provides a picture of electron distribution in crystalline solids. The theory is based on nearly-free-electron models, which distinguish between conductors, insulators and semi-conductors. These models have much in common with the description of electrons confined in compressed atoms. The distinction between different types of condensed matter could, in principle, therefore also be related to quantum potential. This conjecture has never been followed up by theoretical analysis, and further discussion, which follows, is purely speculative.

Macromolecules

The mysterious behaviour of bio-macromolecules is one of the outstanding problems of molecular biology. The folding of proteins and the replication of DNA transcend all classical mechanisms. At this stage, non-local interaction within such holistic molecules appears as the only reasonable explanation of these phenomena. It is important to note that, whereas proteins are made up of many partially holistic amino-acid units, DNA consists of essentially two complementary strands. Non-local interaction in DNA is therefore seen as more prominent, than for proteins. Non-local effects in proteins are sufficient to ensure concerted response to the polarity and pH of suspension media, and hence to direct tertiary folding. The induced fit of substrates to catalytic enzymes could be promoted in the same way. Future analysis of enzyme catalysis, allosteric effects and protein folding should therefore be, more ambitiously, based on an understanding of molecular shape as a quantum potential response. The function of DNA depends even more critically on non-local effects.

Crystals

The properties of a crystal depend on its history. Factors such as chemical composition and connectivity, thermodynamic conditions and crystallization medium, all contribute to the eventual morphology and the likelihood of polymorphism, polytypism, twinning and other secondary interactions to occur. Different modifications of the crystalline material will be characterized by different values of quantum potential and relate to each other in the same way as isomeric molecules.

The process of crystal nucleation and growth is equivalent to a phase transition. The initial phase might be a gas, liquid, solution or solid (e.g. glass or another crystal) and the final phase need not be a crystal as traditionally defined. It could be a liquid crystal, a quasi-crystal, a polytype or some other defect solid. The phase transition proceeds *via* a *critical state*, which is intermediate between the two phases of the transition and holds the key to the understanding of crystal growth.

Seen as the spontaneous appearance of subcritical embryos, nucleation seems to defy the laws of thermodynamics which require an inverse relationship between solubility and cluster size. For the same reason, crystal growth in the absence of dislocations appears equally mysterious. Both of these processes are made intelligible by viewing the critical state as a state of hidden symmetry [11]. It is implied that supercritical clusters with long-range order exist in mother liquors at concentrations below critical. The mysterious agent that induces long-range order in fluids, solutions, plasmas, glasses and other defect solids, is most likely the quantum potential. Despite the absence of formal translational symmetry in these non-crystalline states, long-range order does occur as a result of non-local interaction, carried by the radial wave function as defined before in paragraph 3.1.

Polytypism

The same arguments may be used to simplify the logic behind the formation of crystals with complicated translational symmetry patterns, such as *polytypes*. Successive unit cells in a given crystallographic direction are found displaced from the sites predicted by normal symmetry translation. In the simple example of Figure 20-8 all of the orthorhombic unit cells (abc) are shifted by $b/3$ between adjacent layers. The primitive cell of the 3O polytype has $a' = 3a$, as shown in bold outline. The standard explanation to account for the long-range order in polytypes ascribes their formation to the operation of screw dislocations. This explanation is not without problems. If a given crystal develops around a single dislocation, it is hard to understand how cocrystallization of the large number of possible SiC polytypes are, for instance, prevented and why all embryos in a given batch should contain identical dislocations. The alternative of long-range forces is therefore not precluded and the only sensible long-range force probably arises from the quantum potential. Moreover, if the quantum potential is sensitive to the exact thermodynamic conditions of the critical state, a single polytype could well be expected to crystallize during any specific experiment.

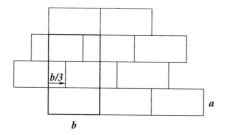

Figure 20-8. Unit cell of a 3O polytype based on regular displacement of the elementary orthorhombic cell of dimensions a × b × c

Quasicrystals

The long-range order found in quasicrystals is even more mysterious as it occurs independent of translational symmetry. A one-dimensional Fibonacci chain [18] illustrates non-periodic long-range order of the type observed in quasicrystals. The chain contains two types of interval, long (*L*), and short (*S*):

$$\boxed{\quad L \quad | \quad S \quad | \quad L \quad | \quad L \quad | \quad S \quad} \text{ etc}$$

It starts with *S* and *L* and grows by adding the previous (shorter) chain to the current one to produce the next:

$$S \quad L \quad LS \quad LSL \quad LSLLS \quad LSLLSLSL \quad LSLLSLSLLSLLS$$

It is the same algorithm that generates the numerical Fibonacci series

$$0,1,1,2,3,5,8,13,21 \ldots$$

The ratio of adjacent terms *i.e.* $1/1, 2/1, /3/2, \ldots$ converges to the irrational golden ratio, $\Phi = (1 + \sqrt{5})/2 = 1.618034\ldots$, which features prominently in the geometrical description of icosahedral quasicrystals.

When defined as a continued fraction

$$x = 1 + \cfrac{1}{1 + \cfrac{1}{1 + \cfrac{1}{1 + \cfrac{1}{1 + \ldots}}}} = 1 + \frac{1}{x}, \quad \text{i.e. } x = \Phi$$

the golden ratio exhibits the important property of self-similarity, best known from fractal structures, and also responsible for the long-range order in dodecahedral quasicrystals. This observation leads to the interesting conclusion that in the case of quasicrystals the quantum potential reveals the attribute of self-similarity in common with topology of space-time. Unpublished work of the author provides evidence that all long-range order and non-local interaction have their origin in this commonality.

REFERENCES

1. L. Pauling, *The Nature of the Chemical Bond*, 3rd ed., 1960, Cornell Univ. Press, Ithaca.

2. D. Bohm, *Quantum Theory*, 1951, Prentice-Hall, Englewood Cliffs, NJ.

3. H.A. Kramers, *Die Grundlagen der Quantentheorie*, 1938, Akademische Verlagsgesellschaft, Leipzig.

4. C.A. Coulson, *Valence*, 2nd. ed., 1961, Oxford University Press, London.

5. C.W.M. Cumper, *Wave Mechanics for Chemists*, 1966, Heineman Educational Books, London.

6. H.A. Jahn and E. Teller, *Stability of Polyatomic Molecules in Degenerate Electronic States, Proc. Roy. Soc. (Lond.)* Series A, 161 (1937) 220–235.

7. G.K. Vermalapalli, *Physical Chemistry*, 1993, Prentice-Hall, Englewood Cliffs.

8. J.C.A. Boeyens, *The Theories of Chemistry*, 2003, Elsevier, Amsterdam.

9. J.C. Slater and G.F. Koster, *Simplified LCAO Method for the Periodic Potential Problem, Phys. Rev.*, 1954 (94) 1498–1524.

10. J.C.A. Boeyens, *Understanding electron spin, J. Chem. Ed.*, 1995 (72) 412–415.

11. J.C.A. Boeyens, *New Theories for Chemistry*, 2005, Elsevier, Amsterdam.

12. P.R. Holland, *The Quantum Theory of Motion*, 1993, Cambridge University Press, Cambridge.

13. F.A. Cotton and R.A. Walton, *Multiple Bonds between Metal Atoms*, 1982, Wiley, New York.

14. J.C.A. Boeyens, *Quantum theory of molecular conformation, C.R. Chimie*, 2005 (8) 1527–1534.

15. J.C.A. Boeyens, *Ionization radii of compressed atoms, J. Chem. Soc. Faraday Trans.*, 1994 (90) 3377–3381.

16. J. Kenyon and D.P. Young, *Retention of asymmetry during the Curtius and the Beckmann change, J. Chem. Soc.*, 1941, 263–267.

17. J. March, *Advanced Organic Chemistry. Reactions, Mechanisms, and Structure*, 4th ed., 1992, Wiley, New York.

18. C. Janot, *Quasicrystals*, 1992, Clarendon Press, Oxford.

INDEX

INDABA
Berg-en-Dal, Kruger National Park
South Africa
20 - 25 August 2006

Figure 1-1. Mars: the epitome of the Indaba five theme – *mystery, models, magic and molecules*

Figure 1-2. Mars Oasis, Antarctica; a terrestrial niche for the survival of endolithic cyanobacteria in a *"limits of life"* situation; a translucent Beacon sandstone outcrop containing endolithic microbial communities on top of a dolerite sill. Dr David Wynn-Williams, Head of Antarctic Astrobiology at the British Antarctic Survey, Cambridge, is prospecting for endoliths. Reproduced from the book Astrobiology: The Quest for the conditions of Life, Eds: Gerda Horneck, Christa Baumstark-Khan, 2003, Springer. With kind permission of Springer Science and Business Media

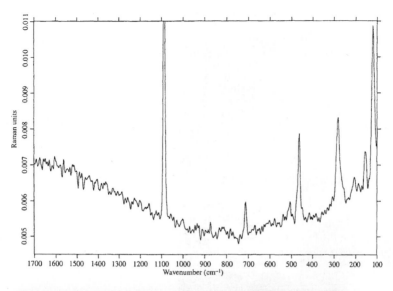

Figure 1-5. Chasmolith in feldspar, Lake Hoare, Antarctica: with the Raman spectrum of calcite, quartz and feldspar from the cyanobacterial colonisation zone

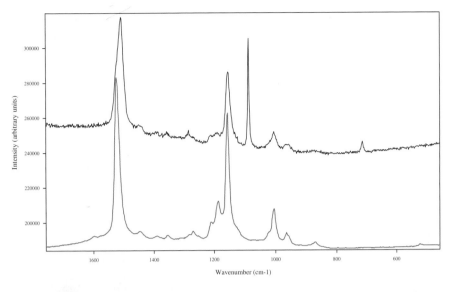

Figure 1-12. Novel endolith in a vacuole in a volcanic basalt lava matrix, Svalbard, Spitsbergen, Norwegian Arctic, showing two different carotenoids in admixture. Accessory photopigments are also identified in this system (not shown here)

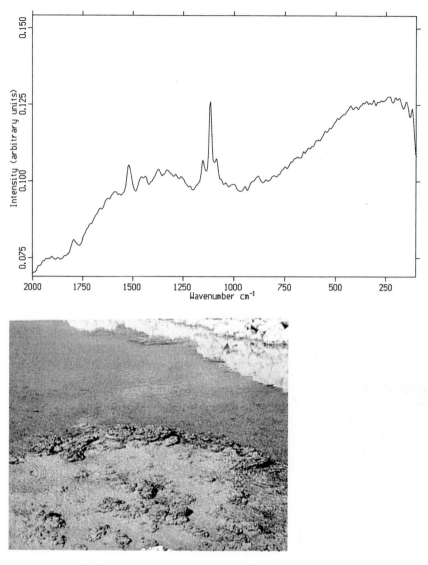

Figure 1-13. Salda Golu Lake, Turkey: hydromagnesite colonised stromatolites in a saltern. The FT-Raman spectrum of the cyanobacterial colonised zone in a stromatolite is shown

Figure 1-14. Juventus Chasmae, potential stromatolite region, on Mars. Reproduced from the Journal of the Geological Society, 156, Michael J. Russell et al., Search for signs of ancient life on Mars: expectations from hydromagnesite microbialites, Salda Lake, Turkey, 1999, with permission from the Geological Society

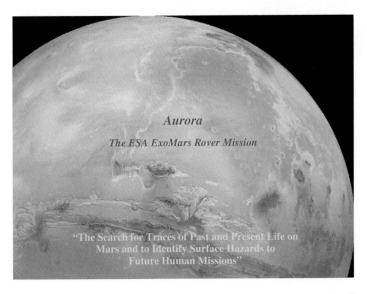

Figure 1-20. AURORA: the European Space Agency/NASA programme, incuding the ESA ExoMars mission

Figure 1-22. Symbolic depiction of the Mars mission phases, 2007–2033, involving a revisitation of manned lunar exploration before embarking upon the manned missions to Mars

Figure 2-1. The application of electroluminescence in lighting the display in this watch. An electric signal, caused by pressing the "light" button on the watch, excites the molecules in the display medium that cause luminescence upon their relaxation

Figure 2-2. A non-linear optical material, ammonium dihydrogen phosphate, displaying second-harmonic generation, the frequency doubling of light (infrared to blue). The origin of this physical phenomenon is entirely dependent on ionic displacement or molecular charge-transfer

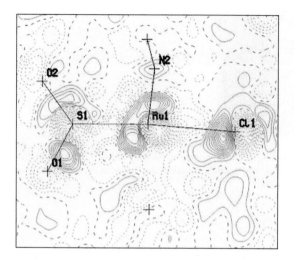

Figure 2-5. A "photo-difference" map showing the ground state (already modelled) depicted by the black lines together with the electron density associated with the light-induced $[Ru(SO_2)(NH_2)_4Cl]Cl$ complex[10, 11]. In this example, SO_2 is the photo-active ligand, undergoing a η^1-SO_2 (end-on) to η^2-SO_2 (side-bound) photoisomerisation. The sulfur atom and one oxygen of the η^2-SO_2 bound ligand are evident in this Fourier-difference map as the green feature and more diffuse green area, respectively, on the left of the figure (the other oxygen is not visible here as it lies out of the plane shown)

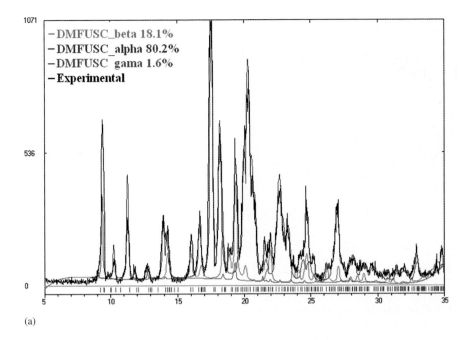

(a)

Figure 3-9. (a) Powder XRD of crystalline dimethyl fuschone **6** shows a mixture α (80%), β (18%) and γ forms (2%) at 30°C

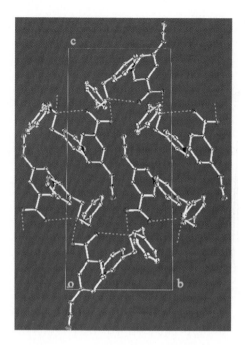

Figure 5-21. (a) Crystal structure of form II. The two azido groups are faced at the inversion center

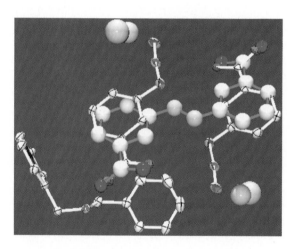

Figure 5-21. (b) molecular structure after the irradiation

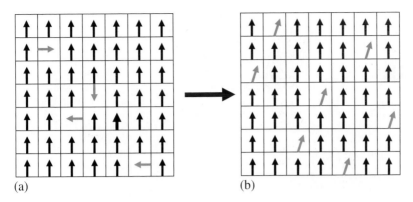

Figure 5-22. Excited molecules produced in the crystalline lattice. (a) at the early stage, (b) at the equilibrium state

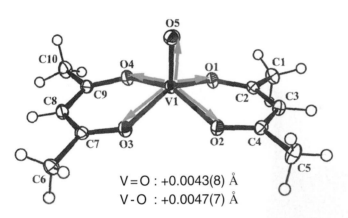

V=O : +0.0043(8) Å
V-O : +0.0047(7) Å

Figure 5-27. Elongation of V=O and V—O bonds at the excited state

Figure 8-1. (a) The natural blue coloration of the shell of lobster *Homarus gammarus* (b) the colour of cooked lobster. From Dr P Zagalsky with permission

Figure 8-3. (a) ASX in hexane at a dilution similar to that in lobster crustacyanin (left) (b) beta-crustacyanin (right). From Dr P Zagalsky with permission

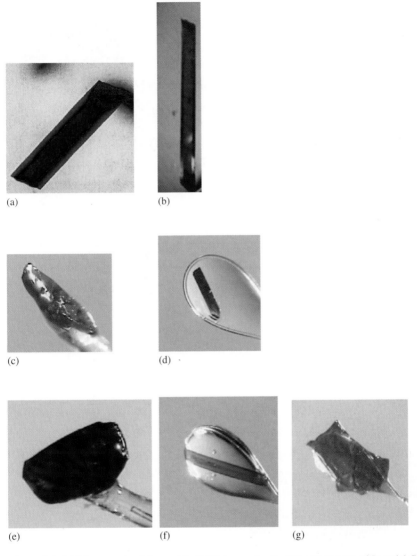

Figure 8-4. (a) Blue crystals of the protein β-CR (grown by Prof Naomi Chayen of Imperial College London; figure reproduced from Chayen (1998) with permission of the author and of IUCr Journals) and for comparison those of several unbound carotenoids (Bartalucci et al 2007) (b) chloroform solvate of ASX (c) pyridine solvate of ASX (d) unsolvated form of ASX (e) canthaxathin (f) zeaxanthin (g) β,β-carotene

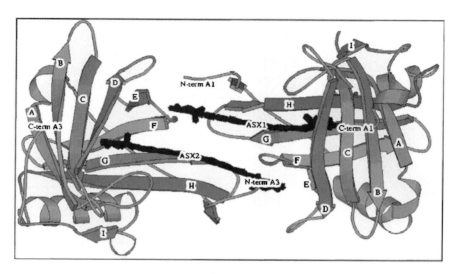

Figure 8-5. The structure of the β-CR at 3.2Å elucidated using X-ray crystallography. From Cianci et al 2002, with the permission of PNAS

Figure 8-6. A Velella velella (approx 1cm across) whose velallacyanin confers the distinctive blue colour on this sea creature (Zagalsky, 1985)

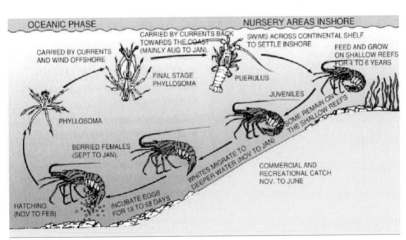

Figure 8-7. The Australian Western Rock lobster has a red phase (top) and a white phase (middle); reprinted from Comparative Biochemistry and Physiology, Part B 141 authors Wade et al (2005) entitled "Esterified astaxanthin levels in lobster epithelia correlate with shell colour intensity: Potential role in crustacean shell colour formation" pp. 307–313 Copyright (2005), with permission from Elsevier and the authors. The bottom schematic shows the life cycle; Figure from Wade 2005 with the permission of Dr. N. Wade and originally from the Western Australian Fisheries website http://www.fish.wa.gov.au/

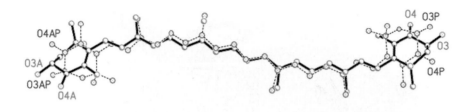

Figure 8-9. Conformation of unbound ASX-Cl (red) best overlay against protein bound ASX (blue); top is the view perpendicular to the plane of the polyene chain and bottom is the view edge on to the polyene chain. From Bartalucci et al (2007) with permission of IUCr Journals

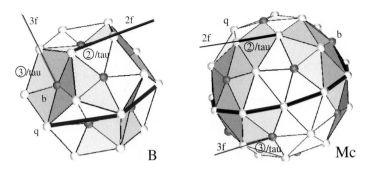

Figure 12-3. (*left*) A Bergman (B) and (*right*) a Mackay (Mc) polyhedron as an atomic decoration of the icosahedral tiling $\mathcal{T}^{*(2F)}$ in the model $\mathcal{M}(\mathcal{T}^{*(2F)})$ (see Ref.[3]). The standard length along the twofold directions is ②, $\tau^{-1}② = 2.96$ Å, and along the threefold direction is ③, $\tau^{-1}③ = 2.57$ Å, $\tau = (\sqrt{5}+1)/2$. By convention, twofold directions are presented in blue, threefold in yellow. A cut of B with a fivefold plane from Figure 12-18 is a pentagon, of edge length ② = 4.77 Å and a cut of Mc with the same plane is a decagon, of edge length $\tau^{-1}② = 2.96$ Å. Both polygons are marked with dark blue thick lines

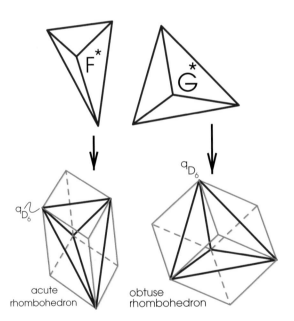

Figure 12-5. The tiling $\mathcal{T}^{(P)}$ can be derived from the F-phase tiling $\mathcal{T}^{*(2F)}$ upon replacing the two tetrahedra F* and G* by the acute and obtuse rhombohedra [6] respectively. By convention, fivefold directions are in red, twofold in blue

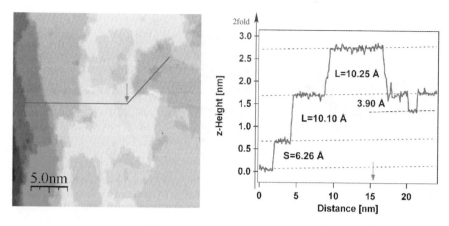

Figure 12-15. (*left*) $25 \times 25 \, nm^2$ STM image of a twofold terrace-stepped surface of i-AlPdMn. (*right*) Height profile measurement along the solid red line (N. B. the red line in this figure is NOT related to a fivefold direction) in the (*left*) image with the corresponding step heights. A single pit of 3.9 Å is detected

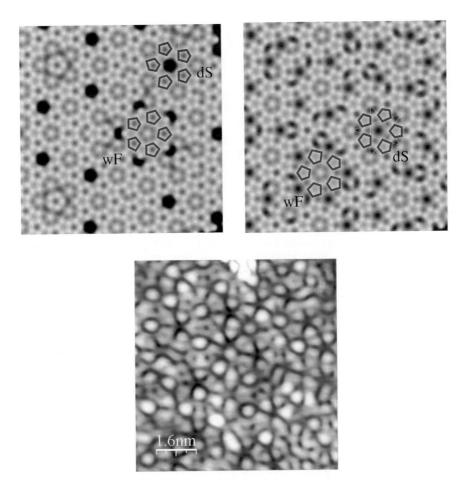

Figure 12-17. STM simulations of a fivefold surface on candidates for the bulk-terminating layers (*top-left*) on a (b,q,(a),q,b)-layer at $z = 47.96$ Å in the patch of $\mathcal{M}(\mathcal{T}^{*(2F)})$ (see Ref.[3]), (*top-right*) on a (q,b,b,q)-layer at $z = 50.00$ Å in the patch of $\mathcal{M}(\mathcal{T}^{*(2F)})$ (see Ref.[3]). Both images are 80×80 Å²; the vertical range is 1.5 Å. (*bottom*) An STM image of the real fivefold clean surface of i-AlPdMn, 80×80 Å²; the vertical range is 1.5 Å, $I = 100$pA, $U = 300$meV. The candidates for the observed fivefold symmetric local configurations, "white flower" (wF) and the "dark star" (dS) are marked on simulations of layers of both kinds. By convention, twofold directions are marked in blue. The wF and the dS are not marked on the STM image in this Figure, (*bottom*); these are marked in Figure 12-18 (*right*). The same local configurations were observed on the STM images of the fivefold surface of i-AlCuFe; see Figure 4(a) in Ref. [19]

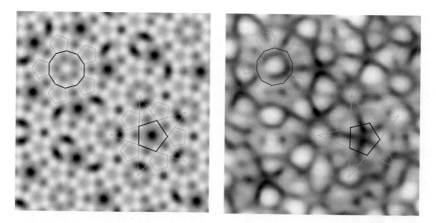

Figure 12-18. (*left*) An STM simulation of the fivefold termination $50 \times 50\,\text{Å}^2$ large, taken from Figure 12-17 (*top-right*); (*right*) An STM image of the real surface $50 \times 50\,\text{Å}^2$ large, taken from Figure 12-17 (*bottom*). By convention, twofold directions are in blue. The intersection of a Bergman polyhedron by the fivefold surface (*right*) and by the termination (*left*) are framed with a pentagon of edge length ② $= 4.77\,\text{Å}$, marked in dark blue. The intersection of a Mackay polyhedron is a decagon of edge length τ^{-1}② $= 2.96\,\text{Å}$ ($\tau = (\sqrt{5}+1)/2$), marked in dark blue. Compare the shape and the scale of the intersection to Figure 12-3

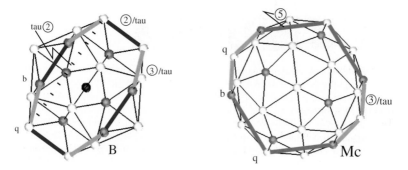

Figure 12-19. Bergman (B) and Mackay (Mc) polyhedra. On each an equator, orthogonal to a twofold direction, is marked. τ^{-1}② $= 2.96\,\text{Å}$, ② $= 4.80\,\text{Å}$, τ② $= 7.76\,\text{Å}$, τ^{-1}③ $= 2.57\,\text{Å}$, ⑤ $= 4.56\,\text{Å}$. By convention, if marked in color, fivefold directions are in red, twofold in blue, threefold in yellow

Figure 12-20. (*top-left*) Experimental STM image of the twofold surface of i-AlPdMn, $80 \times 80\,\text{Å}^2$ large, with 2.5 Å vertical range (the corrugation on the surface has a range 2.5 Å) and $I = 100$ pA and $U = 300$ meV. (*top-right*) STM simulation with gaussian convolution filter of the "sphere model", $80 \times 80\,\text{Å}^2$ large on a twofold termination in the model $\mathcal{M}(\mathcal{T}^{*(2F)})$ at $x = 49.3818\,\text{Å}$ (see Ref.[3]) with vertical range 2.5 Å. (*bottom-left*) The "sphere model" STM simulation of the terminating layer at $x = 49.3818\,\text{Å}$ (see Ref.[3]) $50 \times 50\,\text{Å}^2$. On it we mark in colour a Bergman (B) and a Mackay (Mc) cluster, cut by the termination on the equator (compare the shape and the scale to Figure 12-19). By convention, if marked in color, fivefold directions are in red, twofold in blue, threefold in yellow. (*bottom right*) STM simulation with gaussian convolution filter of the "sphere model" on a twofold termination in the model $\mathcal{M}(\mathcal{T}^{*(2F)})$ at $x = 49.3818\,\text{Å}$, $50 \times 50\,\text{Å}^2$ large with vertical range 1.5 Å

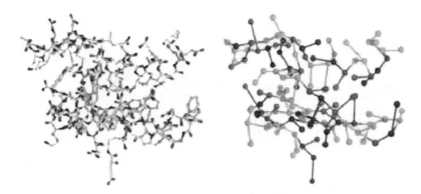

Figure 14-1. Atomic structure and reduced representation in terms of a graph of peaks of a protein chain (PDB file: 1B69). The peaks are colour-coded per amino-acid type. Visualizations of the atomic and reduced representations were obtained on combining the Swiss PDB Viewer (http://www.expasy.org/spdbv) and the ray-tracing POV-Ray (http://www.povray.org) programs

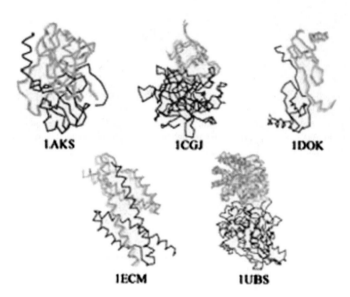

1AKS   **1CGJ**   **1DOK**

1ECM   **1UBS**

Figure 14-5. Cα trace superimpositions of the native solutions (blue and yellow) and the best candidates obtained with the GA in its 3000 / 200 / 60.0 / 0.1 configuration (green), for five complexes that composed our parameterization set – *1aks, 1cgj, 1dok, 1ecm,* and *1ubs*. Vizualizations were obtained on combining the Swiss PDB Viewer (http://www.expasy.org/spdbv) and the ray-tracing POV-Ray (http://www.povray.org) programs

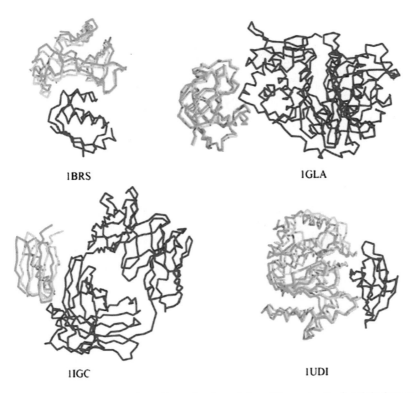

Figure 14-6. Cα trace superimpositions of native solutions (blue and yellow) and the best candidates obtained with the GA in its 3000 / 200 / 60.0 / 0.1 configuration (green), for four complexes that composed our validation set – *1brs (A and D chains), 1udi, 1igc, and 1gla.* Vizualizations were obtained on combining the Swiss PDB Viewer (http://www.expasy.org/spdbv) and the ray-tracing POV-Ray (http://www.povray.org) programs

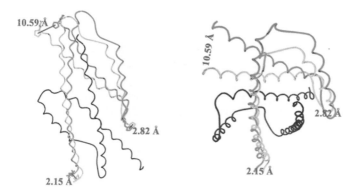

Figure 14-7. Ribbon representation (front and top views) of one proposed structure (*1ecm*) for which the interface with the partner (blue) is satisfactorily recognized by our GA (green), but which shows increasing divergence in the higher parts when compared with the crystalline structure (yellow). Distance calculations were performed on three amino acids, i.e. Asn5, His67, and His95, which present *rmsd* values 2.15, 2.82, and 10.59 Å, respectively. Vizualizations were obtained on combining the Swiss PDB Viewer (http://www.expasy.org/spdbv) and the ray-tracing POV-Ray (http://www.povray.org) programs

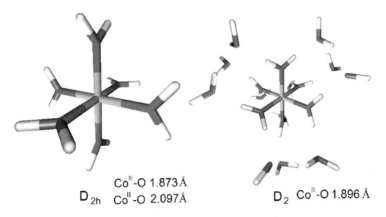

$$D_{2h} \quad \begin{array}{l} \text{Co}^{\text{III}}\text{-O } 1.873\,\text{Å} \\ \text{Co}^{\text{II}}\text{-O } 2.097\,\text{Å} \end{array} \qquad D_2 \quad \text{Co}^{\text{III}}\text{-O } 1.896\,\text{Å}$$

Figure 19-2. Co(H$_2$O)$_6^{z+}$ (z = 3, 2) and {[Co(H$_2$O)$_6$]12H$_2$O}$^{3+}$ model clusters for LFDFT calculation of energies of d-d transitions; Co—O bond lengths are given (from LDA-VWN)

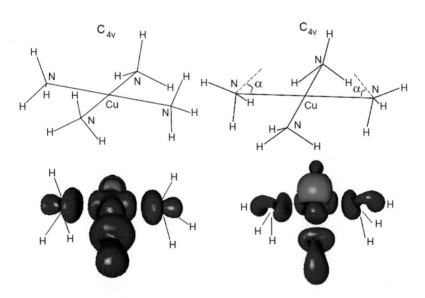

Figure 19-4. Model for Cu(NH$_3$)$_4^{2+}$ taken to illustrate the effect of bent bonding; no bending (left); an extreme case of bent bonding ($\alpha = 45°$) (right). Note that (see caption to Figure 19-3) the $\sigma - \pi$ mixing caused by each pair of trans NH$_3$ ligands cancel; there is no $\sigma(b_1, a_1) - \pi(e)$ mixing in C$_{4v}$ symmetry

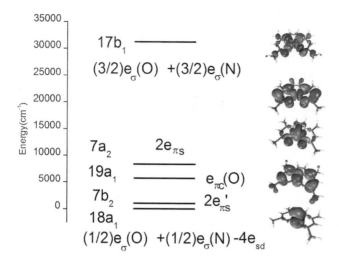

Figure 19-9. MO energy diagram for the d-based orbitals of Coacacen as determined by LFDFT

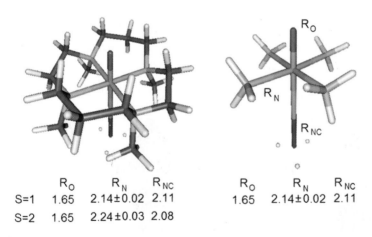

	R_O	R_N	R_{NC}
S=1	1.65	2.14±0.02	2.11
S=2	1.65	2.24±0.03	2.08

	R_O	R_N	R_{NC}
	1.65	2.14±0.02	2.11

Figure 19-10. Fe(IV) = O complexes studied in this work: [FeO(TCM)(CH$_3$CN)]$^{2+}$ (complex I, left) – the geometry of the S = 1 species has been adopted from a BP86 geometry optimization, Ref.[82]; the geometry of the S = 2 species is given for comparison, note that R_O and R_{NC} alter little, but bond lengths to equatorial ligands R_N strongly increase on going from S = 1 to S = 2; [FeO(NH$_3$)$_4$(CH$_3$CN)]$^{2+}$ (complex II, right) – the geometry of the first coordination sphere was taken from complex I (S = 1) but N have been saturated by H – atoms

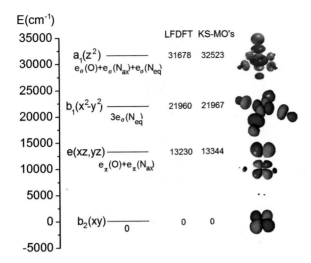

$Figure\ 19-11.$ MO energy diagram for complex I (Figure 19-10) calculated with LFDFT; the energies of LFDFT orbitals are compared with those resulting from the average-of-configuration KS calculation. Orbital contours (pertaining to values of the density of 0.05 a.u.) are plotted, but atoms are omitted for clarity